Smart Cities, Energy and Climate

Smart Cities, Energy and Climate: Governing Cities for a Low-Carbon Future

Edited by

Prof. Oleg Golubchikov
Cardiff University
Cardiff
UK

Dr Komali Yenneti
University of Wolverhampton
Wolverhampton
UK

WILEY

This edition first published 2024

Registered Offices
John Wiley & Sons, Inc., 111 River Street, Hoboken, NJ 07030, USA
John Wiley & Sons Ltd, The Atrium, Southern Gate, Chichester, West Sussex, PO19 8SQ, UK

For details of our global editorial offices, customer services, and more information about Wiley products visit us at www.wiley.com.

Wiley also publishes its books in a variety of electronic formats and by print-on-demand. Some content that appears in standard print versions of this book may not be available in other formats.

Library of Congress Cataloging-in-Publication Data Applied for
Hardback ISBN: 9781118640661
ePDF ISBN: 9781118641118
ePub ISBN: 9781118641170
oBook ISBN: 9781118641156

Cover Design: Wiley
Cover Image: © Pavlo Glazkov/Shutterstock

Set in 9.5/12.5pt STIXTwoText by Straive, Pondicherry, India
Printed and bound by CPI Group (UK) Ltd, Croydon, CR0 4YY

C9781118640661_260624

Dedicated to our parents
Yuri Golubchikov and Nadezhda Dementyeva
Ramarao Yenneti and Jayalakshmi Yenneti

Contents

About the Editors *xiii*
List of Contributors *xv*

1 Introduction: Cities in the Twin Net-Zero and Digital Transition *1*
Oleg Golubchikov and Komali Yenneti
1.1 The Rise of Smart Energy Cities *1*
1.2 Thematical Threads and Issues *3*
1.3 Imagining Smart Urban Energy Systems *6*
1.4 Urban Design, Planning and Policies *7*
1.5 Technologies and Data for Smart and Low-Carbon Urban Futures *9*
1.6 Relevance for Practice and Future Research *11*
References *12*

Part I Imagining Smart Urban Energy Systems *15*

2 Competing Narratives and Interests in Smart Urban Energy Systems *17*
Jess Britton and Emily Judson
2.1 Introduction *17*
2.2 Smart Energy Cities *19*
2.3 Momentum in Energy System Change *20*
2.4 Smart Local Energy Systems in the United Kingdom *21*
2.5 Competing Logics and Interests *25*
2.6 Evolving Smart Energy Governance *26*
2.7 Conclusions *29*
References *30*

3 Where are Smart Sustainable Cities Made? Tracing Wired Socio-Technical Relationships in, Through, Beneath, and Beyond a City *35*
Torik Holmes, Rebecca Windemer, and Carla De Laurentis
3.1 Introduction *35*
3.2 Orientating Ideas, Approaches and Methods *36*

3.3 The City of Manchester *38*
3.4 Connecting 'Smart' Buildings *40*
3.5 'Smart' and Not-So-Smart Large-Scale Network Investments *42*
3.6 Turbulent Urban–Rural Relationships and Contingencies *43*
3.7 Scout Moor Wind Farm: Policy, Social Attitudes and Limiting Growth *44*
3.8 Royd Moor Wind Farm: The Continuation of Ageing Infrastructure *45*
3.9 Addressing 'Hot Areas' *47*
3.10 Conclusions *49*
References *50*

4 Smart Energy Cities: A Perspective from West Africa *53*
Charlotte Ray, Sam Williamson, Zuzana Hrdličková, Derrick Kajjoba, Hillary Kasedde, Lauren Hermanus, Amadu Labor, Joseph Macarthy, and Braima Koroma
4.1 Introduction – Smart Cities: An Urban Panacea? *53*
4.2 Smart Energy City in an African Context *56*
4.3 Current Policy Environment around Smart Cities and SECs in West Africa *58*
4.4 The Need for a More Integrated Approach *63*
4.5 Conclusions *65*
References *66*

5 Beyond Urban Smart Grid Experiments: Replication and Upscaling as Contested Concepts *75*
Harald Rohracher, Gudrun Haindlmaier, Klaus Kubeczko, and Dick Magnusson
5.1 Introduction *75*
5.2 Analysing Context Conditions for the Replicability of Smart Grid Pilot Projects *77*
5.3 Analysis of the Use Cases from the Demo Sites Hartberg and Malmö *81*
5.4 Discussion and Conclusions *88*
Acknowledgements *90*
References *90*

6 The Role of Active Buildings in Smart Energy Imaginaries: Implications of Living Well in Low-Carbon Homes and Neighbourhoods *93*
Kate O'Sullivan, Fiona Shirani, Nick Pidgeon, and Karen Henwood
6.1 Introduction *93*
6.2 Decarbonisation and Smart Energy Systems *95*
6.3 Smart Homes, Smart Occupants? *96*
6.4 Realising Smart Energy Futures: Active Buildings and Homes *99*
6.5 Living Well in Low-Carbon Homes – Initial Insights *100*
6.6 Discussion *105*
References *106*

7 Do Mobility and Sustainability Rhyme in the Autonomous City? *111*
Federico Cugurullo and Alexander Gaio
7.1 Introduction *111*

7.2 From Smart to Autonomous Cities *112*
7.3 Sustainability in the Autonomous City *113*
7.4 Autonomous Vehicle-Induced Urbanism *114*
7.5 Bicycle Urbanism in the Autonomous City *117*
7.6 Conclusions *119*
References *120*

Part II Urban Design, Planning and Policies *123*

8 Re-Defining the Smart City Concept from the Urban Climate Perspectives *125*
Joachim Fallmann, Christopher Holst, Matthias Mauder, and Stefan Emeis
8.1 Introduction *125*
8.2 Existing Urban Studies *127*
8.3 Recent Approaches for Reshaping Building Design *129*
8.4 Suggestions for Urban Planning and Building Design *130*
8.5 Model Approaches *135*
8.6 Conclusion *137*
Funding Sources *138*
Conflict of Interests *138*
References *138*

9 Berlin's Pathway to Climate Neutrality: Scenarios and Measures for a European Metropole *147*
Bernd Hirschl
9.1 Introduction *147*
9.2 The Search for a Climate Policy Target for Berlin *149*
9.3 The Status Quo and Current Trends: Berlin Not Yet on the Path to Climate Neutrality *154*
9.4 Scenarios for 2050, 2030 and 2040 – A Restrictions-Based Approach *155*
9.5 Strategy Recommendations and Measures for a Climate-Neutral Berlin *160*
9.6 Conclusions *162*
Acknowledgement *164*
Literature/References *164*

10 City, Neighbourhood and Citizens: Putting the '20-Minute' Idea to Work in Edinburgh *167*
Alice Creasy, Matthew Lane, and Dan van der Horst
10.1 Introduction *167*
10.2 The 20-Minute Idea *169*
10.3 Case Study: Putting the 20-Minute Concept to Work in Edinburgh *176*
10.4 Discussion *185*
Acknowledgements *188*
References *188*

11 From Smart Urbanism to Sustainable Urban Mobility Plan: A Critical Evaluation of the Case of Cagliari *195*
Chiara Garau, Giulia Desogus, and Vincenza Torrisi
11.1 Introduction: Sustainable Urban Mobility Plan (SUMP) *195*
11.2 Comparison Between the Guidelines for Italian PUMS and the Guidelines of the MCC *197*
11.3 Results and Discussion *208*
11.4 Conclusions *210*
Acknowledgements *210*
References *211*

12 Analysing India's Smart Cities Mission from a Sustainability Perspective *215*
Sarbeswar Praharaj
12.1 Introduction *215*
12.2 Overview of the Smart Cities Mission in India *217*
12.3 Untangling the Indian Smart City Models *219*
12.4 Sustainability Assessment of Indian Smart Cities *221*
12.5 Discussion and Conclusions *231*
References *233*

13 Energy Transitions and Smart Cities in Russia *237*
Irina Ilina and Michinaga Kohno
13.1 Introduction *237*
13.2 National Climate Policy *237*
13.3 Cities in the Climate and Energy Agendas *241*
13.4 The Digitalisation of Energy and Smart Cities *242*
13.5 Conclusions *246*
References *247*

14 Energy Poverty in Cities: A Behaviourally Informed Perspective *249*
Nives Della Valle
14.1 Introduction *249*
14.2 An Additional Lens to Approaching Urban Energy Poverty *251*
14.3 Limitations and Ways Forward *256*
14.4 Conclusions *257*
Disclaimer *257*
References *258*

Part III Technologies and Data for Smart and Low-carbon Urban Futures *263*

15 Smart Energy Future and Smart Cities *265*
Wadim Strielkowski
15.1 Introduction *265*
15.2 The Concept of the Smart City *266*
15.3 Growing Importance of Cities *266*

15.4 Smart Technologies and Energy *268*
15.5 Energy Infrastructure for Smart Cities *269*
15.6 Concluding Remarks *272*
References *273*

16 Governing the Transition Towards Smart Grids Through Organised Industry Events *277*
Suyash Jolly
16.1 Introduction *277*
16.2 Theoretical Background *278*
16.3 Research Method *280*
16.4 India Smart Grid Week 2016 *281*
16.5 Discussion: Role of the Event in Facilitating Future Developments *285*
16.6 Conclusion *287*
References *288*

17 Emission Reduction and Renewables Integration Through Distributed Ledger Technology *293*
Hamid M. Pouran, Komali Yenneti, Mariana Padilha Campos Lopes, Louis Gyoh, and Yong Sheng
17.1 Introduction *293*
17.2 What is the DLT or Blockchain Platform? *293*
17.3 How DLT Could Catalyse Mitigation of GHG Emissions and Integration of Renewables *296*
17.4 Concluding Remarks *300*
References *300*

18 Just Comfort: District Heating and Cooling as a Sustainable Energy Solution *305*
Keith Baker and Ronald Mould
Glossary of Abbreviations *305*
18.1 Introduction – A Brief History of District Heating *305*
18.2 Defining District Heating and Cooling by Generation *307*
18.3 Technologies and Fuel Supplies *308*
18.4 District Cooling *311*
18.5 Lessons from Denmark's District Heating Revolution *313*
18.6 District Heating and Cooling as a Solution to Energy Poverty *316*
18.7 Defining Successful District Heating and Cooling Schemes *318*
18.8 Concluding Remarks *321*
References *322*

19 The Role of Energy-Efficient Buildings in the Post-Carbon Future *327*
Gloria Pignatta and Shayan Naderi
19.1 Introduction *327*
19.2 Building Retrofitting *330*
19.3 Thermal Mass Energy Storage *332*
19.4 Battery Energy Storage *335*

19.5 Solar Energy Utilisation *336*
19.6 Energy Flexibility in Buildings and Districts *339*
19.7 Conclusions *340*
References *341*

20 Using Bottom-Up Digital Technologies in Technical Decision-Making for Designing a Low-Carbon Built Environment *347*
Clarice Bleil de Souza, Camilla Pezzica, and Jakob Hahn
20.1 The Role of Bottom-Up Data in Technical Decision-Making *347*
20.2 Possibilities for Bottom-Up Data Gathering *349*
20.3 Data Gathered by Sensors on Behalf of the People *351*
20.4 Data Gathered by Citizen Scientists Using Digital Technologies *356*
20.5 Challenges in Using Bottom-Up Data in Technical Decision-Making *360*
Acknowledgements *361*
References *361*

21 Street Lighting as a Dimension of Smart Energy Cities *365*
Mary Thornbush and Oleg Golubchikov
21.1 Introduction *365*
21.2 Hardwiring the City *365*
21.3 Street Lighting Efficiency *366*
21.4 Smart Lighting *367*
21.5 System of Street Lights Controller *367*
21.6 Connected Security Lighting System *368*
21.7 The 'Smart Pole' *368*
21.8 Dynamic Lighting Control System *369*
21.9 Conclusions *370*
References *370*

Index *373*

About the Editors

Oleg Golubchikov is Professor of Human Geography at the School of Geography and Planning, Cardiff University. He has over 20 years of experience researching urban and spatial transitions. His publications focus on urban and regional governance and economic development, on green energy economies and low-carbon cities, as well as on smart cities and smart energy systems. One of the strands of his research explores the relationships between decarbonisation and smartization, including how cities, regions and other places respond to the concurrent energy, digital and sustainability transitions. He has a DPhil from the University of Oxford and prior to joining Cardiff University held academic posts at the University of Oxford and the University of Birmingham. He has also undertaken a breadth of policy and strategy assessment, including as a long-term consultant to the United Nations (UN). He has been the lead author of over ten substantive UN publications on topics concerning urban development and housing, climate-neutral cities and people-smart cities.

Dr Komali Yenneti is a Senior Lecturer in Geography, Urban Planning and Environment at the School of Architecture and Built Environment, University of Wolverhampton. She has extensive research, policy and advocacy experience in urban studies, climate change, energy and sustainability. Her current research involves more than 25 academic, government and industry partners around the world. She has published widely and has made policy contributions to the United Nations Office for Disaster Risk Reduction, United Nations Environment Programme and Governments around the world. She has organised and chaired conferences and conferences sessions, delivered keynote addresses and invited lectures in more than 100 events around the world. She holds a PhD in Geography and Environmental Sciences from the University of Birmingham and previously worked at the University of New South Wales (UNSW) and the Nanjing Institute of Geography and Limnology, Chinese Academy of Sciences (NIGLAS). She has received the inaugural International Geographical Union's Early Career Research award and is also the founding chair of the IGU – Young and Early Career Geographers Taskforce. She is an academic secondee at the Department of Energy Security & Net-Zero, Government of the UK.

List of Contributors

Keith Baker
Built Environment Asset Management (BEAM) Centre
Glasgow Caledonian University
Glasgow, Scotland, UK

Jess Britton
School of Social and Political Science
University of Edinburgh
Edinburgh, UK

Alice Creasy
Net Zero Cities consultant at WSP
Edinburgh, UK

Federico Cugurullo
Department of Geography
School of Natural Sciences, Trinity College
Dublin, Ireland

Giulia Desogus
Department of Civil and Environmental Engineering and Architecture (DICAAR)
University of Cagliari
Cagliari, Italy

Stefan Emeis
Karlsruhe Institute of Technology
Garmisch-Partenkirchen, Germany

Joachim Fallmann
Karlsruhe Institute of Technology
Karlsruhe, Germany

Alexander Gaio
Department of Geography
School of Natural Sciences
Trinity College
Dublin, Ireland

Oleg Golubchikov
School of Geography and Planning
Cardiff University
Cardiff, Wales, UK

Louis Gyoh
Faculty of Science and Engineering
University of Wolverhampton
Wolverhampton, UK

Chiara Garau
Department of Civil and Environmental Engineering and Architecture (DICAAR)
University of Cagliari
Cagliari, Italy

Jakob Hahn
Research Institute CENERGIE – Center for Energy Efficient Buildings and Districts
Munich University of Applied Sciences
Munich, Germany

Gudrun Haindlmaier
Austrian Institute of Technology
Vienna, Austria

Karen Henwood
School of Social Science
Cardiff University
Cardiff, Wales, UK

Lauren Hermanus
African Centre for Cities
University of Cape Town
Cape Town, South Africa

Bernd Hirschl
Sustainable Energy and Climate Protection
Institute for Ecological Economy Research
Berlin, Germany
and
Management of Regional Energy Systems
Brandenburg University of Technology
Cottbus-Senftenberg (BTU), Cottbus
Senftenberg, Germany

Torik Holmes
Sustainable Consumption Institute
University of Manchester
Manchester, UK

Christopher Holst
Karlsruhe Institute of Technology
Garmisch-Partenkirchen, Germany

Dan van der Horst
School of Geosciences
University of Edinburgh
Edinburgh, UK

Zuzana Hrdličková
Sierra Leone Urban Research Centre
Freetown, Sierra Leone

Irina Ilina
Faculty of Urban and Regional Development
HSE University
Moscow, Russia

Suyash Jolly
Business Research Group
Nordland Research Institute
Bodø, Norway
and
Division of History of Science, Technology
and Environment
KTH Royal Institute of Technology
Stockholm, Sweden

Emily Judson
Centre for Geography and
Environmental Science
University of Exeter
Exeter, UK
and
The Global Sustainability Institute
Anglia Ruskin University
Cambridge, UK

Derrick Kajjoba
Makerere University
Kampala, Uganda

Hillary Kasedde
Makerere University
Kampala, Uganda

Michinaga Kohno
Michi Creative City Designers Inc.
Hokuto, Japan

Braima Koroma
Sierra Leone Urban Research Centre
Freetown, Sierra Leone

Klaus Kubeczko
Austrian Institute of Technology
Vienna, Austria

Amadu Labor
Sierra Leone Urban Research Centre
Freetown, Sierra Leone

Matthew Lane
School of Geosciences
University of Edinburgh
Edinburgh, UK

Carla De Laurentis
Department of Geography and Environmental Management
University of the West of England
Bristol, UK

Mariana Padilha Campos Lopes
Federal University of Rio de Janeiro
Rio de Janeiro, Brazil

Joseph Macarthy
Sierra Leone Urban Research Centre
Freetown, Sierra Leone

Dick Magnusson
Department of Thematic Studies (TEMA)
Linköping University
Technology and Social Change
Linköping, Sweden

Matthias Mauder
Karlsruhe Institute of Technology
Garmisch-Partenkirchen, Germany

Ronald Mould
Built Environment Asset Management (BEAM) Centre
Glasgow Caledonian University
Glasgow, Scotland, UK

Shayan Naderi
School of Built Environment, Faculty of Art, Design and Architecture
University of New South Wales (UNSW)
Sydney, Australia

Kate O'Sullivan
School of Social Sciences
Cardiff University
Cardiff, Wales, UK

Gloria Pignatta
School of Built Environment, Faculty of Art, Design and Architecture
University of New South Wales (UNSW)
Sydney, Australia
and
City Futures Research Centre (CFRC)
University of New South Wales (UNSW)
Sydney, Australia

Camilla Pezzica
Welsh School of Architecture
Cardiff University
Cardiff, Wales, UK

Nick Pidgeon
School of Psychology
Cardiff University
Cardiff, Wales, UK

Hamid M. Pouran
Faculty of Science and Engineering
University of Wolverhampton
Wolverhampton, UK

Charlotte Ray
School of Electrical, Electronic and Mechanical Engineering
Cabot Institute for the Environment
University of Bristol
Bristol, UK

Harald Rohracher
Department of Thematic Studies (TEMA)
Linköping University
Technology and Social Change
Linköping, Sweden

Yong Sheng
Faculty of Science and Engineering
University of Hull
Hull, UK

Fiona Shirani
School of Social Sciences
Cardiff University
Cardiff, UK

Clarice Bleil de Souza
Welsh School of Architecture
Cardiff University
Cardiff, Wales, UK

Wadim Strielkowski
Department of Agricultural and Resource Economics
University of California
Berkeley, CA, USA
and
Department of Trade and Finance
Faculty of Economics and Management
Czech University of Life Sciences Prague
Prague, Czech Republic

Mary Thornbush
Faculty of Environmental Studies
York University
Toronto, ON, Canada

Vincenza Torrisi
Department of Civil Engineering and Architecture (DICAR)
University of Catania
Catania, Italy

Nives Della Valle
European Commission
Joint Research Centre (JRC)
Ispra, Italy

Sam Williamson
School of Electrical, Electronic and Mechanical Engineering
Cabot Institute for the Environment
University of Bristol
Bristol, UK

Rebecca Windemer
Regen
Exeter, UK

Komali Yenneti
School of Architecture and Built Environment
University of Wolverhampton
Wolverhampton, UK

Sarbeswar Praharaj
Department of Geography and Sustainable Development, University of Miami, FL, USA

1

Introduction: Cities in the Twin Net-Zero and Digital Transition

Oleg Golubchikov[1] *and Komali Yenneti*[2]

[1] *School of Geography and Planning, Cardiff University, Cardiff, Wales, UK*
[2] *School of Architecture and Built Environment, University of Wolverhampton, Wolverhampton, England, UK*

1.1 The Rise of Smart Energy Cities

The future of cities is going to be *digital* and that future is also to be *low-carbon*. To use the language of modellers, there is a high confidence level that the future will be both digital and low-carbon: there are simply too many symptoms pointing us in that direction. But how can we make sense of these two dimensions of the urban future, digital and low-carbon? To what extent are they co-constitutive of one another? For example, how will the cities of the future harness digital innovations to optimise energy consumption, reduce greenhouse gases (GHGs) and achieve overall sustainability? Are those futures already shaping the cities of today? If so, what do these experiences of today tell us about the future, as much as they do about the current trends? These are some of the concerns as well as inspirations underlying this volume.

The global consensus about the need to limit global warming to 1.5 °C and global pressures for climate mitigation have far-reaching implications for cities. Accounting for GHG emissions inside and outside urban areas, the urban share of combined global CO_2 and CH_4 emissions was estimated 67–72% in 2020 (IPCC 2022). These emissions are certainly unevenly distributed across the world, with cities in high- and upper-middle-income countries contributing as much as an estimated 86% of global urban CO_2 emissions in 2015 (Mukim and Roberts 2023). However, as urbanisation is more rapid in the rest of the world, the share of cities in lower-income countries also increases. This means that cities around the world, as a 'global collective', and irrespective of their location, are seen as a key target for climate mitigation and 'deep decarbonisation' to achieve zero or near net-zero emissions (Seto et al. 2021).

Since the 'climate change problem is principally an energy problem' (MacKay 2009, p. 16) – the acknowledgement that it is fossil fuels that make the greatest contribution to GHG emission – actions addressing *energy* (reducing energy demand, switching energy provision to low-carbon supply and improving energy efficiency) are what climate governance at the

Smart Cities, Energy and Climate: Governing Cities for a Low-Carbon Future, First Edition.
Edited by Oleg Golubchikov and Komali Yenneti.

urban level seeks to prioritise. Indeed, many socio-technical systems that contribute to GHG emissions, such as electricity systems, heating and transport, get 'densified' at the scale of the city, overlaying and interplaying with one another. Consequently, the whole variety of urban sectors and urban practices are implicated in the push for low-carbon transition: energy infrastructure, the built environment, the construction sector, industries, business and public services, transport and mobility, natural environments, waste and water management, food and other consumption practices, people's lifestyle and behavioural patterns, urban design and urban planning (Bulkeley et al. 2010; Golubchikov 2011; IPCC 2022; Seto et al. 2021).

The increasing digitalisation of society and the advent of 'smart city' add a new dimension to the urban climate agenda – the promise of leveraging digital innovation for accelerated energy transitions and decarbonisation. While there are no universal definitions for 'smart cities', the common denominator is the use of Information and Communications Technology (ICT) and automation for the improvements of urban processes, services and infrastructure (UNECE and ITU 2016). Wireless connectivity, the Internet of Things (IoT), big data and robotics have all been part of this trend, with artificial intelligence (AI) also becoming increasingly prominent at the urban scale (Cugurullo et al. 2023; Golubchikov and Thornbush 2020, 2022). Smart city has consequently gained wide currency in international practice with the promise of integration, efficiency, sustainability and people-centricity (Thornbush and Golubchikov 2020).

Smart buildings, smart water management, intelligent transport systems and smart waste management systems are a few developments that tap into digitally-enabled technologies in order to contribute to the promise of urban-scale decarbonisation. In parallel developments, energy systems are also becoming 'smart' and transitioning to net-zero and distributed/decentralised/micro-generation systems. Smart urban energy systems become an integral part of urban climate governance, offering solutions that not only optimise energy usage but also promote resilience and adaptability in the face of changing environmental and social dynamics. Smart energy systems can operate on a cross-system level, including the management of electricity via smart grids across different urban systems, or be integrated into particular urban systems, for example, buildings, with the deployment of smart energy meters, smart thermostats, battery storage, smart hot water tanks and electric vehicle charges, to name a few technologies. The relationships between providers of energy and users of energy also change; for example, enabled by smart energy systems the built environment becomes the infrastructure for energy generation and storage.

What essentially combines these various trends and developments is the notion of *smart energy cities* (Thornbush and Golubchikov 2021). Smart energy cities can be seen as part of the evolution of the city-energy-sustainability nexus, also tapping into the opportunities that digital transition brings for managing energy demand, energy supply and energy flows within urban domains, where digitally-enhanced smart cities and energy transitions represent mutually supporting processes. Combining the developments in ICT-led smart cities and sustainable energy, the notion of the smart energy city has come close to represent a digitally-mediated variant of low-carbon cities. Smart energy city, in its different manifestations, is consequently one of the central elements emerging from the discussions in the present volume.

Many cities have already incorporated smart, climate and energy targets into their strategies and plans, with a growing 'club' of cities that declare their intent to become carbon-neutral, climate-neutral, climate-smart, or net-zero in the near future. The European Union's 100 climate-neutral and smart cities (European Union 2022) or India's climate-smart cities initiative (NIUA-C3 2020) are examples of even larger-scale commitments to net-zero cities and sustainable urban development. Other cities and regions around the world are also making strides towards achieving similar goals (Bulkeley and Stripple 2021; Mendes 2022).

The technology-centred visions for net-zero and digital transitions are not without their problems, however. The smart city concept in particular has been widely critiqued for its tendency to glorify technology, where citizens become subordinate to, rather than placed at the heart of, a sustainable city project and where the city's actual needs are often circumvented by the availability of technology (Luque-Ayala and Marvin 2020). As some lines of critique suggested (Yigitcanlar et al. 2019), policymakers often opt for 'Black Box' technology solutions promoted by technology companies, rather than producing long-term visions for the 'good city'. Relatedly, the issues around social justice – in particular, how costs and benefits are distributed and who is included in smart and low-carbon urban transitions – have become an important part of academic and policy conversations (Golubchikov 2020).

In response, the concept of 'smart city' has also morphed over time: from what some now call 'Smart City 1.0' (characterised by a top-down framework, with a focus on ICT infrastructure and deploying solutions promoted by technological companies) to 'Smart City 2.0' (a people-focused, users-friendly framework) and even to 'Smart City 3.0' (a framework for inclusive and participatory urban governance, even if still technology-enhanced) (Golubchikov 2020). The active involvement of stakeholders – including citizens – in the co-creation and implementation of smart city solutions is now claimed, in theory at least, to be the key to improving transparency and incentivising society towards more sustainable practices and behaviours. With regard to technologies, instead of their uncritical deployment, there is now a search for solutions that are tailored to specific needs of individual cities and communities.

This volume further explores how cities – in particular, their energy systems – are being re-shaped by smartisation and decarbonisation (as encapsulated in the notions of the twin net-zero and digital transition and smart energy cities) and with what consequences for wider society. The volume draws together insights and case studies from across a variety of disciplines – from smart energy grids to active and energy-efficient buildings, urban design, sustainable mobility and climate policies. Another objective is to foster an understanding of what these new trends mean for urban imaginaries, urban governance and, ultimately, for urban futures.

1.2 Thematical Threads and Issues

While the book comprehensively covers the role of the 'urban' in the twin net-zero and digital transition, it also generates a number of important common threads, which can be broadly grouped into three dimensions:

- Context-sensitive and contingent urban transitions;
- Complexities and uncertainties coming from the disruptive nature of transitions; and
- The imperative of people-centrality and justice.

These threads are reviewed in this section before the structure and individual chapters are introduced in the subsequent sections.

The very magnitude of the task of transition(s) requires long-term and persistent political, economic and institutional commitments, as well as innovative, creative and often 'alternative' ways of carrying out businesses, producing and consuming goods and services. However, what complicates the landscape is that underlying factors are not merely complex, but are variegated across different localities. There are, for example, wide variations between cities in terms of population densities, climate, inherited economic structures, existing building stock, infrastructure development and capacities of municipal authorities. Besides, cities are shaped by the different national systems of regulation. This means that the transition to a low- or post-carbon city is not linear and each city has its own pathway in this transition (Castán Broto 2019; Haupt et al. 2023; Rutherford and Coutard 2014). The tensions between geographically variegated contextual complexities, technical systems and existing regulations come to the attention of many chapters in this volume (e.g. Chapters 2, 3, 5, 11 and 14).

For technological providers and policymakers, it is the scalability and replicability of different solutions (institutional or technical) that matters (e.g. Chapters 5 and 13). However, translating 'solutions' from one context into another means their de-territorialisation from the whole system of institutional, social and cultural relationships into which these solutions were embedded and re-territorialisation them into the new contexts (this is alternatively conceptualised as disembedding and re-embedding in Chapter 5). This is often problematic and limits the replicability of 'good practices'. Furthermore, technological providers develop their solutions for a universal deployment to ensure that they are deployed 'at scale' and maximise returns on investments. Local context is important, but either for experimentation and testing solutions (upstream) or as one of the many local markets (downstream), with the aim to upscale and replicate solutions across the largest possible geographical area. However, even if built on generalised and replicable solutions, the deployment of smart urbanism involves re-negotiation of these solutions with social and organisational structures and local needs and interests, which are a product of localised processes (Chapter 18). If so, the local context becomes much more than a testbed or a market – but as the place-based arena of negotiation (and often contestations) between what is seen technical and what is seen social. This can (and should) be seen as a process of transforming 'non-political technologies' into being a political, socio-technical actor, where local contexts play the role of the mediator (Chapter 2).

Related to this thread are complexities that accomplish the twin net-zero and digital transition in cities. While the transition brings disruptive changes to many existing institutions and practices, this is happening against the multi-layered palimpsest of different systems that make existing cities, including infrastructural and institutional asynchronies (e.g. different urban-scale systems having their own legacies and operational rhythms) and asymmetries (e.g. tensions between different institutional logics and interests) (e.g. Chapter 3). This makes urban systems open to alternative pathways of development; but this also means uncertainties around urban futures, including room for unintended consequences of different actions.

Many previous notions also acquire new meanings. Even the seemingly very 'mundane' urban services such as street lighting experience deep transformations with the conversion into 'smart' street lighting (Chapter 21). Energy systems in smart cities also experience new types of vulnerabilities than in the 'analogous cities'; this requires new understanding as well as approaches to reduce the vulnerabilities (Chapter 15). Blockchain and distributed ledger technology can be used to improve the reliability of smart grids, as well as facilitate new energy governance regimes such as P2P energy transactions (Chapter 17). The complexity of managing urban systems also increases with the push towards coordination of energy performance at the scale of urban districts rather than solely individual buildings (Chapter 19). Cities also need more nuanced tools for decision-making for sustainable planning and development; for example, even more traditional tools such as scenario-building and modelling need to take into account the increased complexity of the relationships between different urban systems (Chapters 8 and 9). Big data is increasingly available for modelling, but there are new types of complexities such as translating users' data into decision-making (Chapter 20). Exchange of experiences between cities and other actors is important in navigating these complexities, although such exchange may also lead to the propagation of particular narratives (Chapter 16).

Many contributors to this volume are still concerned that the consensus about reducing GHG mission is not sufficiently coordinated with the principles of distributive justice and participatory democracy, while low carbon transition is not immune of its own socio-political struggles. The imposition of particular development strategies for cities often confronts equity, affordability and civic participation. There is a clear understanding that at the core of transitions must be people, citizens and communities and their actual needs (e.g. Chapter 4). This requires no less than 'humanising' relationships in smart cities (Chapter 6), foregrounding deliberative participation and human-centred principles (Chapter 7), as well as addressing placed-based and context-specific energy needs and vulnerabilities (e.g. Chapter 14). Evidence is abundant that without redistributive policies and also without engagement with procedural/participatory justice, low carbon transitions have uneven impacts on different communities and places, with vulnerable social groups and deprived areas being left behind the benefits of transition (Chapters 10 and 12; also see Yenneti et al. 2016; Golubchikov and O'Sullivan 2020; O'Sullivan et al. 2020).

Ideas around 'energy justice' and 'just transition' offer important critique for unpacking the underlying circumstances, focusing the attention on the social costs of climate transition and the imperative of ensuring that the transition towards a climate-neutral economy happens in a fair way (Bouzarovski et al. 2023; Sovacool et al. 2013). Partly for the same reason of ensuring just transition many cities, for example, insist on retaining municipal ownership of key utilities and infrastructure, including those which are energy related. As, for example, Chapters 5 and 18 demonstrate, publicly owned vertically integrated energy providers are not only more socially accountable, but also potentially more conducive of more seamless whole-system transformations. In any case, policies need to be people-smart, prioritising, for example, low-income groups or social housing (Golubchikov 2020; Golubchikov and Deda 2012).

These threads outlined above are common but by far not exhaustive of the whole range of issues raised in this volume. The book consists of 21 chapters, including 20 chapters that follow this introductory chapter. They are divided into three parts:

Part I: Imagining smart urban energy systems;
Part II: Urban design, planning and policies; and
Part III: Technologies and data for smart and low-carbon urban futures.

Below the contribution of each chapter in these three Parts is considered in more detail.

1.3 Imagining Smart Urban Energy Systems

Part I of the book examines the different ways that smart cities, in particular smart urban energy systems, are imagined (or socially constructed) and operationalised. This Part introduces the varieties of socio-technical visions for the sustainable and smart urban futures.

Against the preponderance for one-size-fits-all approaches to smart energy transitions, with cities often framed merely as spaces for experimentation, Britton and Judson (Chapter 2) stress the importance of the geographical and historical contexts on which smart energy developments are being based. The UK government's 'smart local energy systems' (SLES) initiative, for example, emphasises the decarbonisation potential of more distributed energy systems and scalable local energy business models. Local and specific materiality of place are important for SLES in terms of developing functionality, but the overall logic is to homogenise this materiality into scalable business models. Nevertheless, the development of local public-private partnerships and initiatives around SLES already demonstrates a great deal of local divergence in terms of how different cities foreground priorities and 'internalise' SLES differently. The authors argue that national narratives prioritising market-based solutions and end-consumer benefit may conflict with these local narratives and priorities.

Holmes et al. (Chapter 3) discuss the relationships between smart city governance and inherited electricity infrastructures. Taking inspiration from science and technology studies (STS), the authors consider the electricity system as a multitude of interrelated socio-technical components, each with its specific affordances and path dependency. They unpack electricity infrastructure decision-making processes in Manchester. Their study highlights the reliance of Manchester's smart sustainable city aspirations on the material conditions of electricity connections, grid extensions and generation investment. These elements, in turn, depend on the actions of a range of different interest groups, many of whom are outside of the control of the city and even geographically located outside of the city. Local policies are circumscribed by the existing ordering and management of these electricity systems.

Ray et al. (Chapter 4) delve into the topic from the perspectives of West Africa. Physical access to basic services, including electricity, remains among key priorities for the urban poor. The use of 'smart' technologies – particularly mobile-based services or off-grid electricity – allows leapfrogging some stages in traditional infrastructure development. Leveraging technologies such as smart grids, meters and mobile phones to access and pay for energy services is one of the ways to offer services to even the poorest communities. But communities on the fringes are often left out of the debates and access to such technologies and energy services. The discourse on a smart city has rather been dominated by the urban middle class and development agencies who foster their own vision and imaginaries, often prioritising high-tech solutions rather than smartness for the 'masses'.

Rohracher et al. (Chapter 5) take up the issue of upscaling of smart grid experiments. Smart grids are expected to radically change the electricity systems, especially at the local grid level. However, how such grids will be organised in practice (including their institutional configurations) still remains to be decided, with a number of possible alternatives currently being tried in pilot or demonstration projects. Similarly to Chapter 2, the authors argue that upscaling and replicating tend to privilege standardisation and interoperability. However, implementing the same type of projects somewhere else involves cross-contextual translation or disembedding projects from their place-specific conditions and re-embedding emerging new systems into local contexts. This is nothing less than innovation in itself, which also requires mobilisation of stakeholders, as well as negotiation, contesting and reshaping new systems. The cases of two demonstration sites in Sweden (Hyllie in Malmo) and Austria (Okopark Hartberg) show that there are numerous factors which make experimentation projects highly unique and highly embedded in local contexts.

O'Sullivan et al. (Chapter 6) bring the concept of active homes into the conversation with smart city and smart energy. Enabled through ICT-based smart energy grids and in-house smart technologies, active homes have the capacity to produce and store energy, use energy more efficiently and communicate with other infrastructure (e.g. transport and industry). Active homes are also imagined as sites of improved liveability, due to the streamlining of occupant activities and daily needs by smart technologies. However, the authors argue that such assumptions are based on 'generic' occupant lifestyles and behaviours, as well as assumptions about household capability to both understand and digitally manage their energy systems. The authors identify a number of real-life challenges surrounding these imaginaries. For example, smart technologies do not necessarily work as expected; assumptions may not be completely accurate; household needs change over time; households also circumvent their smart energy system to achieve required comfort needs. This highlights the importance of appreciating the role of both occupants and other agents and intermediaries who support them.

Cugurullo and Gaio (Chapter 7) introduce another set of imaginaries – autonomous cities, where AI plays a key role. AI is enabled by traditional smart tech, but it also transcends smart cities in its capacities to learn, think and make decisions, including in normative and ethical domains. However, the authors remind us that there is no one universal AI, there are myriads of different AIs. This is important as it is erroneous to assume the same quality to all AIs, for example, whether they are sustainable or not. By the same token, the same AI can manifest itself differently in different places. The chapter specifically discusses urban mobility in the autonomous city. Autonomous cars can enable a new kind of urbanism, often of a contradictory nature. On the one hand, they may trigger even further suburbanisation; on the other, they may also help vacate parking space for other needs. One possible unintended outcome is marginalisation of cycling. The authors come to the conclusion that AI-enabled city should be rethought in a way that would open rather than close space for deliberative participation and human-centred principles.

1.4 Urban Design, Planning and Policies

Part II of the book includes chapters that are focused on interplay of climate governance, smart systems and energy transitions with urban planning, urban modelling and urban policies.

Fallmann et al. (Chapter 8) put the future city in the context of the interface between climate assessment, climate modelling and urban planning. Sustainable urban planning needs to be 'fit-for-purpose', i.e. recognise both local climate conditions and the historical development of a particular city or region. The chapter reiterates a number of practical solutions for building design, city planning and national infrastructure. It also explores the interrelationships of different urban systems. One example is that a reduced heat island effect in cities (due to adaptation to climate change) will reduce thermally driven vertical ventilation in cities; this will lead to a higher pollutant concentration. It is important that future cities at the same time work on reducing car exhausts to prevent harmful effects on health. Modelling approaches have seen a steady trend towards smaller scales and increased complexity, but the question remains how to facilitate those tools for sustainable planning and development.

Hirschl (Chapter 9) further demonstrates challenges in executing climate governance at the city level. There are no universal standards that can be applied to municipalities to establish their carbon budget. In their absence, arbitrary target years are set for achieving carbon neutrality. Contrary to many cities targeting earlier achievement of carbon neutrality (e.g. 2025–2030), Berlin lacks renewable energy and would struggle with implementing earlier energy transition goals. The chapter shows considerable hurdles for the city economy to restructure itself to accommodate climate neutrality targets. The scenarios show that achievement of climate neutrality is scarcely feasible for the 2030s, but could be done in the 2040s if considerable effort is made, including a much more ambitious rollout at the national level. There is no single sector that can undertake significant reductions on behalf of the others, but each and every sector will have to contribute to a similar extent.

Creasy et al. (Chapter 10) explore the 20-minute city ideas in the context of Edinburgh. The idea is also often connected to smart city, focusing on the decentralisation of cities by bringing essential services into neighbourhoods. However, the infrastructure development is unevenly distributed in the city, so that in combination with a socially fragmented community, the very idea confronts historic spatial and systematic environmental, economic and social justice issues. This chapter also brings examples such as Portland in the United States or Melbourne in Australia to unpack those issues and point out the importance of citizens' involvement and co-design. The chapter then uses the case of Edinburgh to show how a more inclusive 20-minute neighbourhood can be achieved. This essentially requires the development of multi-service 'hubs' located with the historic levels of service development in mind. The use of digital techniques in city analysis and planning may facilitate an understanding of where pedestrian infrastructure upgrades would have the most impact.

Sustainable and smart urban mobility should be seen as a crucial part of urban climate governance and energy transitions, which can also improve social aspects of sustainability. Garau et al. (Chapter 11) review the city-level guidance for the Sustainable Urban Mobility Plan of the Metropolitan City of Cagliari. The chapter identifies a number of gaps in the analysed guidelines, including the discrepancies between the contextual conditions of the city (such as being a maritime location) and the guidelines' actions. The authors also note the absence of info-mobility and innovative intelligent transport solutions. These gaps call for a more targeted approach to urban mobility planning that takes into account the unique characteristics and needs of a city.

India's Smart City Mission (SCM) has been the most ambitious national-level plan to address the problems of urbanisation by making 100+ smart cities. Praharaj (Chapter 12) reviews the policy. The SCM outlines area-based smart cities development components, including, principality: city improvements (e.g. intelligent traffic management, CCTVs), city renewal (e.g. urban density, mixed-use development, cycle infrastructure development) and city extension (e.g. development of urban peripheries and growth attraction). Integrated command and control centres seek to build synergies between various projects and monitor the performance of urban services and infrastructure. The author analyses existing smart city policy documents to assess the four pillars of sustainability – social, environmental, economic and institutional. The author reveals that cities have ventured into the development of a few handpicked technological fixes and the majority of investments are skewed towards exclusive enclaves. This helps little to negate regional disparities across the country and the varying levels of capacities and needs of the different scales of cities.

Ilina and Kohno (Chapter 13) take the case of Russia and investigate the relationships between national climate governance and the development of smart city solutions to tackle climate and energy issues. Cities like Moscow pioneer many technologically sophisticated solutions, with a focus on citizens and their comfort, while the federal government is central for replicating and upscaling smart city solutions from pilot cities to places across the country. The authors demonstrate that heating is one of particularly challenging areas for Russian cities. ICT solutions are also combined with more traditional urban planning solutions for climate mitigation and adaptation measures.

Della Valle (Chapter 14) discusses measures to address energy poverty, basing the analysis on behavioural science. The author argues that policy interventions cannot afford to be designed without detecting the context-specific energy needs and practices. Context-specific conditions may make people act or abstain from acting, which may be further sustained by other people's approval and disapproval. Thus, depending on the prevalent behaviour in the context, social norms may prescribe behaviours that exacerbate suboptimal conditions. Smart solutions that delegate tasks to home automation technologies and make decisions 'easy' can help make optimal choices as a 'default'; but they also reduce people's autonomy and agency. As the condition of deprivation already negatively affects how people perceive the way they control their lives, it is important that behaviour change strategies empower one's agency. For example, people can be trained to design daily environments that induce them to make optimal decisions.

1.5 Technologies and Data for Smart and Low-Carbon Urban Futures

Part III of the book addresses the disruptive potential, but also complexities and challenges, of smart technologies and other innovations, particularly with respect to facilitating low-carbon urban futures.

Strielkowski (Chapter 15) reminds the readers about the role of cities as catalysts for innovations and highlights some of the new opportunities and tensions emerging with smart energy systems within cities. As the IoT is combined with the Internet of Energy (IoE), this brings new considerations for the planning and operation of energy infrastructure. Smart cities need to maintain secure and reliable energy infrastructures that are resilient against a broad spectrum of adversities, both natural and anthropogenic. For example, with increasing digitalisation, the concept of energy security is acquiring a new meaning.

Jolly (Chapter 16) explores the role of the so-called 'field-configuring events' (industry-organised events such as special exhibitions and trade shows) in shaping common understanding and potential of smart grids. Such events not only demonstrate the latest technological trends and stimulate business networks but also propagate different narratives to disrupt the institutional status quo and to shape new agendas.

Pouran et al. (Chapter 17) discuss how the blockchain platform and distributed ledger technology can be used to ensure more transparency in carbon emission data reporting by creating a digital flow of information alongside the physical flow of fossil fuels from the source to end users. The technology can also improve the reliability of smart grid and facilitate the 'prosumer' concept in energy governance and peer-to-peer (P2P) energy transactions. The chapter concludes that in order to make this technology play a significant role in decarbonisation, the challenges related to socio-economic issues, commercial viability, legal complexities and technological shortcomings must be addressed.

Baker and Mould (Chapter 18) note that district heating was already used in the Middle Ages, but remains an attractive option for more energy-efficient and sustainable futures. The penetration of urban district heating and cooling systems is considerable in 'state-managed' societies but represents a remarkable failure in more 'liberal' economies. The impact of policy silos and the lack of a strategic approach to developing new fuel chains still remain a major hindrance for district heating and cooling networks. However, as post-carbon transition reintroduces government interventions in the energy demand-supply relationships (e.g. in the form of subsidies and regulation), a more coordinated approach may be 'retrofitted' into urban economies. This is not straightforward, however. Using the example of Denmark, the chapter stresses the legislative and ownership context for the development of the technical systems, which limits the direct replicability of 'best practices'. Furthermore, ensuring that the system is just (e.g. in terms of mitigating energy poverty) represents an additional challenge.

Reducing energy demand in buildings and improving energy efficiency measures are essential in climate mitigation. Furthermore, with global warming, there is a growing need for cooling in buildings and the use of air conditioning. Pignatta and Naderi (Chapter 19) discuss different technical and building design solutions to shift energy demand from peak to off-peak periods, to reduce energy demand, as well as to generate energy. The chapter then highlights that attention is also now being directed towards the energy performance of building complexes and districts, rather than individual buildings. This shift allows for the economies of scale to be leveraged while highlighting the role of the urban scale as an 'energy collective'. One of the key takeaways is the importance of collaboration, education, awareness and incentives in the transition to a post-carbon future through net-zero and positive energy buildings and communities.

De Souza et al. (Chapter 20) consider users' data collection and how it can inform technical decision-making. The chapter focuses on two approaches to bottom-up data collection: when data are collected by digital sensors 'on behalf' of people and when data are collected by citizen scientists using digital technologies. Through two distinct case studies from Germany and Italy, the chapter discusses some key techniques for data collection as well as the problems that technicians face when translating this data into 'useful' decision-making, which vary from data quality and inferred meaning to making a single holistic view out of disjoint data. Both case studies illustrate how bottom-up and citizen-centric data collection can help decision-making processes and reduce costs involved in data collection. The chapter ends by arguing that a set of 'standardised' guidelines on how data are to be gathered by citizen scientists is required to represent community aspirations and needs.

Finally, Thornbush and Golubchikov (Chapter 21) analyse the role of smart street lighting technologies in making energy efficiency gains. Switching to more efficient lights and making street lighting 'smart' by integrating it with wireless infrastructure of sensors and controllers demonstrate considerable potential for reductions in energy consumption and night-time lighting pollution. More advanced systems include dynamic lighting controls that adjust lighting to ambient conditions; connected lighting systems that can work 'on demand' and/or in a coordinated way; as well as smart poles that become multifunctional observatories able to collect and exchange different types of information. The increasing presence of these street lighting technologies demonstrates how the digitalisation of energy transforms the very nature of seemingly 'mundane' urban services.

1.6 Relevance for Practice and Future Research

To summarise, the chapters in this volume demonstrate cautious optimism around the potential of digitalised technologies, intelligent urban management systems, smart energy systems and other technological options and governance innovation in helping cities and society to decarbonise and in making contribution to a more sustainable and efficient future. However, while smart energy cities have the potential to transform our urban landscapes, without careful attention to justice and equity, these initiatives bear the risk of being turned into dystopian visions.

This volume overall will be of significant use to a range of audiences interested in navigating the complex intersection of energy transition, digitalisation and the city. This includes not only relevance to academic research but also insights and practical guidance to policymakers, business, civil society and international development.

Policymakers play a key role in shaping the future of smart energy cities. In this regard, this volume offers a comprehensive overview of the current state of smart energy systems and cities and can serve as a valuable resource for the development of key policies associated with the topic. The in-depth critical analysis of the case studies, policy innovations and best practices can help policymakers to evaluate existing strategies and frameworks.

The case studies discussed in the volume also offer practical guidance on navigating the complexities of smart energy cities for various industries and business sectors. The volume provides an in-depth discussion on the latest advancements, emerging trends, evolving policy landscapes and citizen perspectives in clean and smart energy technologies and data analytics and automation, providing guidance on developing 'future-proof' sustainable business models in a rapidly changing landscape.

By drawing on chapters from both the Global South and the Global North, this volume also emphasises the importance of international cooperation, multistakeholder engagement and collective action in driving the agendas of digitalisation in energy systems. The volume can support civil society and community organisations with evidence-based knowledge and understanding of intelligent energy technologies for informed advocacy and actions.

As highlighted in this volume, future research needs to continue exploring the avenues of balancing technology and equity in governing cities for a low-carbon future. Research could investigate how energy and digital technologies are distributed across different socio-economic groups and how any disparities impact access to clean and affordable energy and the opportunities that arise from the development of these initiatives (Knox et al. 2022;

Mittal and Sethi 2018). It is critical to address technological and digital divides that may exacerbate pre-existing urban inequalities (Caragliu and Del Bo 2023; Shin et al. 2021; Sovacool and Del Rio 2020). Additionally, privacy concerns and data security must be addressed to avoid further marginalisation of vulnerable groups.

Citizens and local communities should have opportunities to participate in the development of strategies and technologies that impact their lives and places (Corsini et al. 2019). Subsequent studies could explore how institutional structures and power dynamics influence the development and implementation of these technologies. This includes the role of different stakeholders as well as ensuring that justice concerns are adequately addressed. Understanding the mechanisms for involving marginalised communities, ensuring their concerns and needs are taken into account and empowering them to actively participate in the energy transition would be key for just outcomes (Kempin Reuter 2019). Moreover, ownership and co-creation of technologies can foster a sense of empowerment among citizens and communities (Leclercq and Rijshouwer 2022; Ryszawska et al. 2021). This can lead to greater social acceptance, support and long-term success of transition.

Overall, justice and equity are key considerations in ensuring that smart energy systems and technologies can serve as catalysts for positive social change, empowering communities and creating a more inclusive and sustainable urban future. While a one-size-fits-all approach is counterproductive, a critical gaze on the theory and practice can help establish more nuanced perspectives that take into account the context, the needs of people and places and the imperative of justice in the twin net-zero and digital transition (as also defined under the notion of just transition).

References

Bouzarovski, S., Fuller, S., and Reames, T.G. (ed.) (2023). *Handbook on Energy Justice*. Cheltenham: Elgar.

Bulkeley, H. and Stripple, J. (2021). Climate smart city: new cultural political economies in the making in Malmö, Sweden. *New Political Economy* 26 (6): 937–950.

Bulkeley, H., Castan Broto, V., Hodson, M., and Marvin, S. (ed.) (2010). *Cities and Low Carbon Transitions*. London: Routledge.

Caragliu, A. and Del Bo, C.F. (2023). Smart cities and the urban digital divide. *npj Urban Sustainability* 3 (1): 43.

Castán Broto, V. (2019). *Urban Energy Landscapes*. Cambridge: Cambridge University Press.

Corsini, F., Certomà, C., Dyer, M., and Frey, M. (2019). Participatory energy: research, imaginaries and practices on people' contribute to energy systems in the smart city. *Technological Forecasting and Social Change* 142: 322–332.

Cugurullo, F., Caprotti, F., Cook, M. et al. (ed.) (2023). *Artificial Intelligence and the City: Urbanistic Perspectives on AI*. London: Routledge.

European Union (2022). List of applicant cities. https://research-and-innovation.ec.europa.eu/document/download/ed6373ab-d190-4b3c-8615-498fffea7388_en?filename=ec_rtd_he-mission-public-eligible-applicant-cities.pdf (accessed 14 March 2024).

Golubchikov, O. (2011). *Climate Neutral Cities: How to Make Cities Less Energy and Carbon Intensive and More Resilient to Climatic Challenges*. Geneva: United Nations.

Golubchikov, O. (2020). *People-Smart Sustainable Cities*. Geneva: United Nations https://doi.org/10.18356/9789210052658.

Golubchikov, O. and Deda, P. (2012). Governance, technology and equity: an integrated policy framework for energy efficient housing. *Energy Policy* 41: 733–741. https://doi.org/10.1016/j.enpol.2011.11.039.

Golubchikov, O. and O'Sullivan, K. (2020). Energy peripheries: uneven development and the precarious geographies of low carbon transition. *Energy and Buildings* 211: https://doi.org/10.1016/j.enbuild.2020.109818.

Golubchikov, O. and Thornbush, M. (2020). Artificial intelligence and robotics in smart city strategies and planned smart development. *Smart Cities* 3 (4): 1133–1144. https://doi.org/10.3390/smartcities3040056.

Golubchikov, O. and Thornbush, M. (2022). Smart cities as hybrid spaces of governance: beyond the hard/soft dichotomy in cyber-urbanization. *Sustainability* 14 (16): 10080. https://doi.org/10.3390/su141610080.

Haupt, W., Eckersley, P., Irmisch, J., and Kern, K. (2023). How do local factors shape transformation pathways towards climate-neutral and resilient cities? *European Planning Studies* 31 (9): 1903–1925. https://doi.org/10.1080/09654313.2022.2147394.

IPCC (2022). Urban systems and other settlements, Chapter 8. In: *Climate Change 2022: Mitigation of Climate Change. Contribution of Working Group III to the Sixth Assessment Report of the Intergovernmental Panel on Climate Change* (ed. P.R. Shukla, J. Skea, R. Slade, et al.). Cambridge, UK and New York, NY, USA: Cambridge University Press https://doi.org/10.1017/9781009157926.003.

Kempin Reuter, T. (2019). Human rights and the city: Including marginalized communities in urban development and smart cities. *Journal of Human Rights* 18 (4): 382–402.

Knox, S., Hannon, M., Stewart, F., and Ford, R. (2022). The (in)justices of smart local energy systems: a systematic review, integrated framework, and future research agenda. *Energy Research and Social Science* 83: 102333.

Leclercq, E.M. and Rijshouwer, E.A. (2022). Enabling citizens' Right to the Smart City through the co-creation of digital platforms. *Urban Transformations* 4 (1): 2.

Luque-Ayala, A. and Marvin, S. (2020). *Urban Operating Systems: Producing the Computational City*. London: MIT Press.

MacKay, D.J. (2009). *Sustainable Energy – Without the Hot Air*. Cambridge, UK: UIT.

Mendes, V. (2022). Climate smart cities? Technologies of climate governance in Brazil. *Urban Governance* 2 (2): 270–281.

Mittal, S. and Sethi, M. (2018). Smart and livable cities: Opportunities to enhance quality of life and realize multiple co-benefits. In: *Mainstreaming Climate Co-benefits in Indian Cities: Post-Habitat III Innovations and Reforms* (ed. M. Sethi and J.A. Puppim de Oliveira), 245–263. Singapore: Springer.

Mukim, M. and Roberts, M. (ed.) (2023). *Thriving: Making Cities Green, Resilient, and Inclusive in a Changing Climate*. Washington, DC: World Bank http://hdl.handle.net/10986/38295.

NIUA-C3 (2020). Climate-smart cities – Assessment framework. https://smartcities.gov.in/sites/default/files/2023-07/MoHUA%20SCM%20CSCAF-2.0-Brochure_24042020.pdf (accessed 14 March 2024).

O'Sullivan, K., Golubchikov, O., and Mehmood, A. (2020). Uneven energy transitions: understanding continued energy peripheralization in rural communities. *Energy Policy* 138: https://doi.org/10.1016/j.enpol.2020.111288.

Rutherford, J. and Coutard, O. (2014). Urban energy transitions: places, processes and politics of socio-technical change. *Urban Studies* 51 (7): 1353–1377. https://doi.org/10.1177/0042098013500090.

Ryszawska, B., Rozwadowska, M., Ulatowska, R. et al. (2021). The power of co-creation in the energy transition – DART model in citizen energy communities projects. *Energies* 14 (17): 5266.

Seto, K.C., Churkina, G., Hsu, A. et al. (2021). From low- to net-zero carbon cities: the next global agenda. *Annual Review of Environment and Resources* 46 (1): 377–415. https://doi.org/10.1146/annurev-environ-050120-113117.

Shin, S.Y., Kim, D., and Chun, S.A. (2021). Digital divide in advanced smart city innovations. *Sustainability* 13 (7): 4076.

Sovacool, B.K. and Del Rio, D.D.F. (2020). Smart home technologies in Europe: a critical review of concepts, benefits, risks and policies. *Renewable and Sustainable Energy Reviews* 120: 109663.

Sovacool, B.K., Sidortsov, R.V., and Jones, B.R. (2013). *Energy Security, Equality and Justice*. London: Routledge.

Thornbush, M.J. and Golubchikov, O. (2020). *Sustainable Urbanism in Digital Transitions: From Low Carbon to Smart Sustainable Cities*. Cham: Springer https://doi.org/10.1007/978-3-030-25947-1.

Thornbush, M. and Golubchikov, O. (2021). Smart energy cities: the evolution of the energy-city-sustainability nexus. *Environmental Development* 39: 100626. https://doi.org/10.1016/j.envdev.2021.100626.

UNECE and ITU (2016). United 4 smart sustainable cities. https://www.itu.int/en/ITU-T/ssc/united/Pages/default.aspx (accessed 14 March 2024).

Yenneti, K., Day, R., and Golubchikov, O. (2016). Spatial justice and the land politics of renewables: dispossessing vulnerable communities through solar energy mega-projects. *Geoforum* 76: 90–99. https://doi.org/10.1016/j.geoforum.2016.09.004.

Yigitcanlar, T., Kamruzzaman, M., Foth, M. et al. (2019). Can cities become smart without being sustainable? A systematic review of the literature. *Sustainable Cities and Society* 45: 348–365.

Part I

Imagining Smart Urban Energy Systems

2

Competing Narratives and Interests in Smart Urban Energy Systems

Jess Britton[1] *and Emily Judson*[2,3]

[1] *School of Social and Political Science, University of Edinburgh, UK*
[2] *Centre for Geography and Environmental Science, University of Exeter, Exeter, UK*
[3] *The Global Sustainability Institute, Anglia Ruskin University, Cambridge, UK*

2.1 Introduction

Urban climate governance literature increasingly highlights the context-sensitive nature of sustainability transformations, as well as advocating for a multilevel governance perspective that engages with the multiple spaces and spheres through which urban sustainability is constructed and contested (Betsill and Bulkeley 2006). Despite this, there is a preponderance of one-size-fits-all approaches to smart energy transitions with cities often framed merely as spaces for experimentation. In this chapter, we identify a gap in the smart and sustainable cities literature regarding the diversity of city types, geographies and histories on which smart energy developments are being based. To this end, we use the experience of the United Kingdom to explore how narratives of 'smart local energy systems' are being developed at both the national level and across a range of cities.

In tackling climate change, sub-national actors, particularly municipalities, are increasingly highlighted as 'motors of change' with potential to contribute to accelerating decarbonisation and the transformation of social and economic systems (Amundsen et al. 2018; Moloney and Horne 2015; Wolfram 2016). Action by sub-national actors is framed as potentially playing an important role in narrowing the emissions gap between current national determined contributions under the Paris Agreement and the greenhouse gas reductions required to limit global warming to 1.5 °C (Erickson and Tempest 2014). There is a long history of cities, municipalities and local authorities playing a central role in sustainability and climate governance. However, this renewed emphasis on the role of cities in international climate governance puts increased emphasis on the role of smart technologies in addressing urbanisation, sustainability and poverty in integrated ways. Specifically, the United Nations Environment Programme has highlighted the role of sub-national governments in accelerating climate action (Amundsen et al. 2018; Smeds and

Smart Cities, Energy and Climate: Governing Cities for a Low-Carbon Future, First Edition.
Edited by Oleg Golubchikov and Komali Yenneti.

Acuto 2018; UNEP 2018) and both the 2030 Agenda for Sustainable Development and the New Urban Agenda emphasise the role of 'smart' in facilitating urban sustainability transitions (Mora et al. 2020).

This focus on cities as key sites of climate action is based on the argument that many of the socio-technical systems that contribute to greenhouse gas emissions, such as electricity systems, heating and transport, are often organised on a city scale. Additionally, city-scale actors are considered to be better connected (than other scales of governance) to multiple stakeholders in their area, as well as in a position to understand the dynamics of local politics, communities and economies. Cities are therefore often framed as hubs of experimentation, allowing new configurations of actors and technologies to be 'tested' in specific settings. There is also an increasing body of research on urban sustainability transitions that recognises the critical role that spatial and geographical factors play in the shaping of transition dynamics for sustainability (Binz et al. 2020; Coenen et al. 2012).

In addition, while there is a long history of urban municipalities developing sustainability visions, targets, plans and experiments, often in partnership with other local government authorities, corporations or community organisations (Bulkeley et al. 2016), arguably municipal climate action is entering a new phase. Since 2018 many municipal governments have signed Climate Emergency Declarations and committed to ambitious net-zero targets, often earlier than national government targets (see, for example, Committee on Climate Change 2020). This renewed focus on the urban role in sustainability, climate and energy is partly driven by increasing focus on the role of non-state and sub-national actors in climate action (Hale 2018; Sapiains et al. 2020), but there is also recognition that rapid decarbonisation will require data-driven and responsive routes to transforming energy systems at the city scale. This is refocussing debate not just on city-scale action but also on the role of smart technologies, potentially reinvigorating smart city debates which some argue have shown slow progress since their inception over 20 years ago (Kitchin 2015; Mathis and Kanik 2021).

The following sections explore the development of smart energy narratives and practices at the national level and in a range of cities across the United Kingdom. We describe how evolving smart energy governance processes are exhibiting conflicting logics which simultaneously elevate and homogenise socio-spatial relations in cities through a focus on using urban smart energy experiments to develop scalable solutions for widespread adoption. Despite this, we argue that there is great potential for energy transitions to re-energise smart city discourses and practices by making tangible the *potential* individual and collective benefits of smart technologies for sustainability and well-being. We highlight a promising role for municipalities in facilitating smart energy development that is more sensitive to socio-geographic context and can hold space for 'public interest' and/or citizen-led initiatives. This can be valuable in balancing strong commercial interests and 'technological solutionism' that have arguably dominated smart city developments to date. These roles could be vital for building legitimacy in smart energy transitions as well as earning or (re)building public trust. However, we identify several barriers to this role being fulfilled at present, including barriers to data access, resource limitations, and negative consequences of vested interests in smart energy governance. The chapter concludes with a discussion of some of the policy and governance tools that could potentially be employed to address the barriers above.

2.2 Smart Energy Cities

Numerous critical perspectives on smart urbanism have recognised that smart city narratives are often ambiguous, with different definitions and approaches configured to serve a wide range of interests (Glasmeier and Christopherson 2015; Hollands 2008; Yigitcanlar et al. 2019). This has led to considerable debate as to the extent to which smart city technologies are able to contribute to sustainability (Albino et al. 2015; Parks and Rohracher 2019). Concerns have also been raised regarding the technocratic nature of many programmes, which do not adequately consider specific socio-environmental contexts (Söderström et al. 2014).

Techno-centric approaches often highlight the role of local experiments as a starting point, with the aim of upscaling and replicating successful innovations across different contexts. However, smart urbanism cannot be defined as simply an instantaneous technological upgrade (Mora et al. 2020) but also involves shifts in social and organisational structures (Goodspeed 2015). Technical experiments can erase the contested nature of smart cities and minimise questions of how smart processes may be changing the nature of cities, i.e. what 'interests, values, competencies and evidence are being privileged and neglected by smart sustainability?' (Evans et al. 2019, p. 559). Placing greater focus on such questions, however, enables exploration of how different institutional structures, actor networks and capacities are likely to influence the realisation of smart city narratives in different locations (Carvalho 2015).

In line with critiques of the contested nature of the 'smart city', the term 'smart energy city' can also be assigned multiple meanings, spanning scales and vectors (Britton 2017; Luque et al. 2014). However, drawing on the work of Sadowski and Levenda (2020) and Kitchin (2015) we suggest that urban smart energy systems involve integrating and embedding 'data-driven, network-connected, and (semi-) automated' technologies into the fabric of cities to stimulate urban economic development and enable the management of other priorities such as increasing well-being or reducing greenhouse gas emission. In practice, an urban smart energy system can incorporate a wide range of system-level, building-level and policy innovations. For example, system-level innovations could include smart control of electricity transmission and distribution grids to minimise faults or enable more dynamic control of local grid stability. Building-level innovations could incorporate smart energy meters, as well as devices and controls such as smart thermostats, smart hot water tanks and smart electric vehicle charges, which enable consumers to access new energy services and business models. Finally, policy innovations could encompass the integration of granular energy data, analytics and automation into the planning, tailoring and delivery of public policy in energy and related areas such as health. As such, although smart energy systems involve individual interventions, they are generally understood to go beyond these to take a more holistic, systems-based approach (Lund et al. 2017).

Despite the multi-faceted nature of urban smart energy systems, policy and academic narratives are prone to framing 'the urban' in monolithic terms. This privileges a 'one-size-fits-all' approach where cities are assumed to have the same opportunities and challenges in adopting smart technologies to further sustainability priorities (Kitchin 2015). Yet, this universal approach to understanding the smart city has come under critique, with proponents arguing that more nuanced accounts are required that examine how, why, and by and for whom smart is being enacted and resisted (Mora et al. 2020). In reality, there are significant differences among urban, suburban and rural environments, as well as between cities. For smart energy systems, one of the clearest differences is variability of energy needs and resource availability in cities compared to rural or suburban environments. Most cities are hubs of energy demand and infrastructure provision but are limited in their potential to generate energy within the boundary of the city (Balta-Ozkan et al. 2015; Naumann and Rudolph 2020). Similarly, there are wide variations between cities based on variables such as population densities, climate, existing building stock, new build programmes, levels of energy access across the city population, other legacy infrastructure, as well as differences in actor assemblages and the convening capacity of municipal authorities.

These differences can be illustrated through two contrasting examples. First, in the Democratic Republic of Congo (DRC), 43% of the population is currently urbanised – with a quarter of urban dwellers concentrated in the six cities, viz. Kinshasa, Lubumbashi, Mbuji-Mayi, Goma, Kananga and Kisangani – and approximately one million additional people are moving to cities each year (The World Bank 2020). While urbanisation could present opportunities to increase electricity access, this also presents a challenge in terms of balancing supply and demand in rapidly growing urban centres particularly in the absence of a robust national grid. By contrast, recent estimates of UK population distribution have indicated slight shrinkage in large cities as a result of the COVID-19 pandemic, most notably in London where in 2020 a 3.8% (229,000) reduction in workforce jobs outstripped the UK average of 1.8% (O'Connor and Portes 2021). While it is, as yet, unclear how permanent these changes may be, they have potentially significant impacts on infrastructure, including energy, housing and transport.

A further critique of dominant smart (energy) city narratives questions the focus, to date, on dynamics in a small number of global cities, or 'exemplar' locations (Hoppe et al. 2016; Kitchin 2015). These often-studied smart cities include locations such as Masdar, Songho, Toronto and Barcelona. By contrast, relatively limited attention tends to be paid to less high-profile regional cities. Such cities often do not get the funding levels or policy focus of 'global cities' and national capitals but are nonetheless important in sustainability transitions. This is perhaps particularly evident in lower-income countries where, as illustrated in the DRC example above, these cities are often growing extremely fast (Castán Broto 2019).

Notwithstanding these limitations in understanding smart energy transitions in different city contexts, there is an emerging refocus on the detailed dynamics of urban smart energy. This includes a particular emphasis on the roles of policy, business and public–private partnerships in accelerating change. The next section explores drivers for these developments.

2.3 Momentum in Energy System Change

Energy systems around the world are going through a period of fundamental change due to socio-technical transitions (Geels and Schot 2007) surrounding renewable generation technologies and, increasingly, the integration of data and digital technologies. This is producing co-evolutionary patterns (Foxon and Steinberger 2013) whereby the development and adoption of new technologies are entwined with shifts in energy economics, business models, policy and social preferences. In countries where traditional models of power generation have been centralised, there have been notable shifts towards decentralisation – with increasing proportions of generation and flexibility provided locally by renewable sources, electric vehicles (EVs), heat pumps and battery storage (Mitchell 2016; Morstyn et al. 2020). In both the United Kingdom and the European Union, renewable electricity generation outstripped fossil generation for the first time in 2020 with renewables accounting for 43.9% and 38% of total generation, respectively (Agora Energiewende 2021; Department for Business Energy and Industrial Strategy 2021). An increasing proportion of this generation is connected at the distribution grid level leading to challenges in terms of integrating power and grid stability.

The resulting responses in energy policy and practitioner communities have seen a broad shift towards energy system visions and goals focussed on smart, flexible energy systems which can integrate high levels of variable renewable power and manage greater system complexity through widespread electrification of heat and transport systems (Ford and Hardy 2020). But what do smart energy systems actually involve? The concept brings together a range of strategies for localised management of distributed energy resources (DERs), including active (two-way operation of) distribution networks, peer-to-peer trading platforms, distribution- and demand-side flexibility markets, virtual power plants and community energy. Technologies such as smart meters, distribution network automation and smart home appliances have potential to support the flattening of demand peaks, enabling demand to be better matched to the output of variable renewable energy generation. This could enable lower system costs, higher penetration of renewable generation and enable consumers to access new products and services (Joseph and Balachandra 2020; Morstyn et al. 2020).

In addition to benefitting the energy system itself, there is potential for smart energy approaches to contribute to a range of public policy, planning and community-driven uses. These include but are not limited to: data-informed infrastructure planning, the integration of energy decarbonisation scenarios with land use planning policy, improved targeting of measures such as energy efficiency or electric vehicle infrastructure, modelling supply/demand balance in order to explore the viability of different decarbonisation pathways, integrating energy policy with other policy areas such as health, and creating new opportunities for citizen engagement with planning and policy development processes. Cities arguably have a particularly valuable strategic view of the interface between energy system benefits and the wider social and ecological co-benefits of smart energy systems. There has also been an active refocussing on the role of the city scale in debates regarding energy system change. In the United Kingdom, this includes a narrative shift towards the concept 'smart local energy systems' which is explored in the following case study.

2.4 Smart Local Energy Systems in the United Kingdom

While smart city narratives are well established in the United Kingdom since the publication of the 2017 Industrial Strategy, a new narrative has emerged in UK energy policy. This narrative, focussed on the development of 'smart local energy systems' (SLES), emphasises the decarbonisation potential of technology-driven, market-based approaches to managing energy generation, distribution, storage and consumption (HM Government 2017b,c). It also explicitly recognises the momentum towards more complex, decentralised energy systems and targets policy and innovation resources towards the development of scalable local energy business models. This differs significantly from earlier 'local energy' policy approaches in the United Kingdom, which tended to emphasise the role of – often voluntary or non-profit – community energy schemes in energy system change (Devine-Wright 2019).

As part of this new policy focus, the UK government invested £102 million into the Prospering from the Energy Revolution (PFER) innovation challenge, supporting the design, development and demonstration of SLES. The 25 projects funded through the PFER programme between 2018 and 2023 incorporate 11 concept-level designs, 10 detailed designs and 4 active demonstrators. The four demonstrator projects span across different UK locations and operate through multi-actor partnerships between the energy industry, digital innovators, local governments and universities (Energy Systems Catapult 2020). The projects funded via this programme operated with the objective of proving that local approaches to energy system transformation can solve whole system challenges in relation to greenhouse gas emissions, local economic benefit, welfare and wider system costs. The programme specifically aimed to produce investment-ready business models that combine technologies such as renewables, energy storage, electric vehicles, energy efficiency, low carbon heat and digital infrastructure (HM Government 2017a; Morstyn et al. 2020). This activity was further supported by national funding for the development of common smart energy tools. For example, the UK innovation agency, the Energy Systems Catapult (ESC), developed modelling tools that aim to support the integration of smart energy approaches into long-term decarbonisation planning at the city scale (Energy Systems Catapult 2021a). This included the development of the EnergyPath® Networks and Local Energy Asset Representation (LEAR) tools as well as the Energy Revolution Integration Service (ERIS) (Energy Systems Catapult 2021b).

However, despite the well-documented potential for smart energy systems to contribute to UK energy system decarbonisation, the narratives currently surrounding their development at the city scale present a number of issues. First, there is an established literature exploring the challenges in adopting urban sustainability experiments as a means to scale up or scale out innovations (Hölscher et al. 2019; Pickerill 2021). Embedding innovations require new actor relations to be established, the contestation of existing logics, and the parallel construction of new rationalities. Experiments are not value free and, as Evans and Karvonen (2015) emphasise, smart city experiments can play important roles, beyond technical learning, in shaping knowledge production. These critiques are particularly relevant to the development of smart energy cities, which can import assumptions from the digital economy regarding scalability as desirable by default as technological businesses expand through a model of 'solutions looking for a problem' (Morozov 2013). In a highly privatised sphere, the imperative to scale may even be seen as an essential part of digital business models due to the economic and technical advantages conferred by expanding access to data and market phenomena such as 'network effects' (HM Government 2019). Overall, this can lead to stronger drivers to downplay nuances between the social and institutional contexts within and between cities in favour of quicker roll-out of 'smart' solutions. This approach to smart energy city development potentially risks backlash, development of inequalities, and reduced effectiveness of sustainability programmes.

Second, certain disjoints between national and local framings of urban smart energy have emerged in the United Kingdom. National governance has shown strong focus on market-based solutions and consumer outcomes, including scope for smart technology to unlock 'new opportunities to give consumers more control, choice and flexibility over their energy use' (HM Government 2020, p. 21). While city-scale approaches also address consumer outcomes, they are more likely to exercise agency in diversifying the framing of SLES development in a manner tailored to their locality. They also differ in the degree to which they emphasise the public, private, community and environmental outcomes of SLES development. The three short case studies below illustrate these areas of divergence between national and local smart energy city development in more detail across the UK context. Case studies of London and Bristol explore energy-related activity in cities that have approached smart city development as a holistic cross-city topic, while the case of Oxford explores more energy-specific activity explicitly framed by the concept and funding associated with the PFER approach to UK SLES development. Although not addressed within this chapter, city case studies from UK devolved nations (Wales, Scotland and Northern Ireland) could present fruitful topics for further comparative research across country-wide, devolved and city-level framings of smart local energy systems.

2.4.1 London

The Greater London Authority (GLA) uses energy data to inform evidence-based policy, design delivery programmes and projects, evaluate performance and undertake net-zero modelling. This includes the targeting of local energy interventions such as buildings retrofit or installation of renewable generation, monitoring and reducing the performance gap of new builds and fuel poverty interventions. A large amount of work is delivered through the £34 million 'Energy for Londoners' programme which wraps data generated through local SLES developments, as well as wider sources such as the London Data Hub, into further programmes benefitting the city. While these programmes still deliver benefits to end consumers, this framing emphasises collective benefit at the city scale and closer integration with non-energy sectors. For example, the London Fuel Poverty Partnership incorporates 'stakeholders from sectors including local government, social housing, landlords, tenants, health, social care, academic, charities, energy suppliers and the energy efficiency industry' (London City Hall 2021b). The GLA also has a dedicated Deputy Mayor for Environment and Energy, who serves as a place-specific leadership and advocacy figure for these policy areas. Finally, although the GLA's smart energy schemes involve a range of external partners, including commercial organisations such as UK Power Networks which are the distribution network operator (DNO) for London, it is notable that the roles and interests of London residents, the Mayor, GLA, and community groups with strong borough identities are highly visible in GLA communications. For example, the Energy for Londoners website highlights that 'The Mayor wants to provide much-needed support to get community energy projects up and running and deployed faster to reboot the economy and benefit hard-hit communities' (London City Hall 2021a).

2.4.2 Bristol

Bristol is a smaller regional city that has also demonstrated strengths in smart city development – both within and outside the energy sector. Bristol has a strong city culture associated with environmental consciousness and activism, becoming the United Kingdom's first European Green Capital in 2015 and the first Council to declare a Climate Emergency in 2018 (Bristol City Council 2019). Furthermore, the response to climate change is emphasised as a shared, whole community responsibility rather than one solely belonging to government (Bristol City Council 2019). Possibly reflecting this emphasis on engagement, Bristol's approach to smart city development lends particular weight to openness and public scrutiny. For example, Bristol City Council has established an open data portal that currently houses 227 open datasets on subjects including energy, planning, population, democracy, health and social care, and education, among others (Open Data Bristol 2021b). It also includes several 'live' open data dashboards visualised through mapping tools, including an air quality dashboard linked to the city's Clean Air Plan (Open Data Bristol 2021a). Furthermore, the City Council's Energy Service uses data to support projects including renewable installations, local area energy planning, smart energy development, and installation of low-carbon heat networks for social housing (Bristol City Council 2021). The Energy Service also includes significant support for community energy projects which the Council suggests they have a 'moral duty to support' (Bristol Energy Service 2021). Data and learning from some of these projects, such as the SONNET project supporting Bristol City Lab, are in turn being used to feed back into the Council's investment decisions (SONNET project 2021).

2.4.3 Oxford

Energy Superhub Oxford (ESO) is one of the PFER demonstrators in the United Kingdom, showcasing an SLES model of using complex public–private partnerships to enable new smart energy business models. The project incorporates automatable EV charging, heat pumps and battery storage distributed throughout the city. The 'optimisation and trading engine' is a core part of the project designed to enable battery storage services to bid into the UK wholesale electricity market. Four out of the six project partners are commercial businesses, and the project has a clear aim to provide the proof of concept necessary to scale up the approach commercially. ESO does not necessarily exclude community benefit from its aims; indeed, the City Council is a central member of the consortium and was the first local authority in the United Kingdom to hold a citizen's assembly on climate. However, the project is broadly commercially focused, with more limited emphasis on how smart energy data and software can feed back into public policy. In the longer term, questions also remain regarding how sharing and use of data generated through the trial will be governed.

Both Oxford City Council and Oxfordshire County Council are also involved in a second PFER project, Project Local Energy Oxfordshire (LEO), which aims to develop a local marketplace for flexibility as well as local peer-to-peer energy trading. While there is again a focus on developing scalable business models, this project is more explicit in its focus on people, including local social enterprise, Low Carbon Hub, as a project partner and stating that 'Communities and people need to be at the heart of our new energy system' (Low Carbon Hub 2021). While both the ESO and LEO projects are ongoing, they both indicate not only the centrality of commercialisation of PFER projects but also some scope for local actors to emphasise broader community values.

2.5 Competing Logics and Interests

The above case studies demonstrate three distinct approaches to smart local energy systems (SLES) development. However, the degree to which specific dynamics of individual cities shape SLES is yet to be fully determined – both in ongoing trials and as developments expand across UK cities. However, there is emerging evidence of trends that could reduce place-based diversity of SLES, potentially also changing the opportunities and agency of public and community actors in shaping local decarbonisation pathways. It is noteworthy that the leadership of a significant proportion of UK SLES activity is currently undertaken by industry, with smaller roles for public and third-sector actors. From 2009 to 2018, for example, 30% of SLES projects were run by DNOs or energy suppliers, with a further 27% of projects led by other private sector actors (Gupta and Zahiri 2020b). There is also some evidence that smart local energy projects in the United Kingdom favour technological innovation over direct consumer or public participation (Flett et al. 2018; Gupta and Zahiri 2020a).

Beyond considering specific project outcomes, this leads us to examine two competing logics evident in overarching conceptualisations of UK SLES. The first recognises the specific materiality of place through an emphasis on developing a granular local understanding of the built environment, energy demand profiles and the feasibility of various technologies. The second homogenises this materiality in seeking to develop scalable solutions that can be applied anywhere. As Mora et al. (2020) identify, homogenisation de-anchors change processes from their spatial setting and existing socio-technical arrangements. This is not to say that city-based smart energy experiments do not have the potential to build knowledge for both (i) scalable business models and (ii) local learning and deliberation regarding decarbonisation pathways and impacts. However, this 'dual spatial focus, both inwards to address local needs and outwards to provide replicable models to be deployed elsewhere' (Devine-Wright 2019) is often unproblematically taken to be neutral or politics free. Explicitly recognising and exploring the links and contradictions between these two logics could hold the potential to contribute to a more detailed understanding of exactly how place matters for smart energy systems and equitable transitions.

Additionally, as discussed in Section 2.1, there has been a phase shift in local climate politics since 2018, with large numbers of cities committing to local net-zero targets. Despite proactive action from city authorities, and the potential for local empowerment through SLES, data gaps and data access difficulties continue to affect the ability of municipalities to action 'public interest' in use of smart energy data for decarbonisation. This includes data gaps relating to building-level consumption, demand patterns (seasonal and daily) and peaks (Britton 2017), as well as the lack of a central smart meter data access mechanism (Frerk et al. 2019). Additionally, limited smart energy skills and resources within municipalities can constrain activity or lead to expensive contracting (Kuzemko and Britton 2020). Many city authorities have become adept at developing data workarounds to inform policy and planning. However, improvements to data quality, granularity, timeliness and affordability would support city authorities to meet net-zero targets and empower them to act meaningfully in a field increasingly populated by commercial actors.

The opportunity for cities to hold space in smart energy development is perhaps particularly poignant given the power of municipalities to instigate deliberative processes with local communities to negotiate ambitious greenhouse gas reduction pathways. For example, Oxford City Council, Camden Council and Devon County Council have all formed citizen's assemblies to inform local approaches to decarbonisation. This approach places politics, negotiation and the envisioning of our future places at the centre of processes of change and is somewhat at odds with the technocratic and business-scaling focus of many United Kingdom's smart local energy system discourses. Despite these emerging participatory approaches to exploring decarbonisation pathways, the potential for smart energy systems to contribute to local debates on energy transitions is still largely untapped. Increased data openness and critical analysis of the values and objectives of smart energy systems could help to ensure that this agenda is mobilised to support the inclusion of diverse voices and perspectives in processes of change, rather than further embed a metrics and efficiency-based narrative. Indeed, Cowley and Caprotti (2019), in their exploration of smart cities initiatives in Bristol, Manchester and Milton Keynes, suggest that the disruptive business model innovation central to many smart city technologies can challenge existing technocratic processes for achieving environmental and social goals and have the potential to contest established power relations in urban (and beyond) contexts.

However, despite potential positive consequences of local empowerment, neither national consumer-focussed nor local planning-focussed approaches are necessarily more adept at handling the influence of vested interests in smart energy developments. The national-level emphasis on technology-enabled new business models offering new value to consumers may be vulnerable to being shaped by business elites and interests of proprietary platforms. Meanwhile, the planning focus of much of city activity risks domination by local elites, urban managers, and corporate interests with local lobbying power. There is also currently insufficient evidence to determine if either the national or local narratives result in more inclusive smart local energy systems. However, there are persistent asymmetries in data access and engagement in governance reforms between national and commercial actors, on one hand, and local and community actors, on the other. The following section further explores governance issues in the context of ongoing smart energy reforms, which have the potential to reconfigure actor networks and power relations in energy systems across the United Kingdom.

2.6 Evolving Smart Energy Governance

UK SLES developments integrate with several national programmes designed to improve the enabling infrastructure for smart energy. Two repeated themes within national policy development in this area are data access and interoperability.

Data access policy in the United Kingdom has two main work streams, both stemming from recommendations made by the influential Energy Data Taskforce (EDTF) in 2019. The EDTF highlighted both visibility and accessibility of data as significant barriers to energy system digitalisation at all scales and recommended the adoption of a new 'Presumed Open' principle (Energy Data Taskforce 2019). The modernising energy data (MED) coalition among the Department for Business, Energy and Industrial Strategy (BEIS); economic regulator, The Office for Gas and Electricity Markets (Ofgem); and government innovation agency, 'Innovate UK', has since developed the Principle of Presumed Open and associated system of 'triage' assessment used by organisations to determine the level of openness under which datasets should be shared. This information, incorporated in the MED Data Best Practice Guidance, went to public consultation in May 2021 (Ofgem 2021a). Once finalised, compliance with the guidance will be included as a new license condition for energy networks. This accompanies two other digital regulatory changes: the first requiring electricity distribution networks to submit Digitalisation Strategies and Actions Plans for Ofgem review and publication (Ofgem 2021b), and the second requiring networks to submit Privacy Plans for Ofgem approval prior to accessing smart meter data (Ofgem 2021c).

Data access improvement has also received £2 million of government investment into the Modernising Energy Data Access innovation competition (HM Government 2021). Competition winner, Open Energy (run by Icebreaker One), is now building two new capabilities for the UK energy ecosystem: a public search function for datasets and a membership-based governance service designed to reduce friction in sharing/accessing more sensitive datasets under a system of pre-emptive licensing (Open Energy 2021a). Although Open Energy is a non-profit entity, it is strongly influenced by the market-based ethos common to national smart energy approaches. The project narrative emphasises its role in facilitating a competitive market for energy data sharing in the United Kingdom. Open Energy must also become commercially competitive (financially self-sufficient) by the end of its funded period.

Interoperability policy development is less defined by particular work streams, but increasingly appears as a common 'horizontal' theme across policy work. This form of policy development is most clearly exemplified by the UK Government's Smart Data initiative (SDI). SDI is a cross-sector initiative designed to integrate learning about data sharing and digitalisation across key industries and challenges outlined by the United Kingdom's 2017 Industrial Strategy. Work accelerated in 2018 with the establishment of the Open Banking programme, which translated the EU's Second Payment Services Directive (PSD2) into the UK regulatory landscape (BEIS 2020d). SDI has subsequently included initiatives such as Midata in the energy sector, the Pensions Dashboard, and Open Finance (an extension of Open Banking) and has commissioned research reports on consent, liability and authentication issues applicable across sectors (BEIS 2020a,b,c). In 2020, the government formed a cross-sector working group designed to promote further cross-sector learning, 're-use of assets' and integration (BEIS 2020e). Membership of the group incorporates government departments and regulators spanning banking, pensions, communications, utilities and transport sectors.

These areas of national policymaking present opportunities for city-based smart energy development. First, as outlined in Section 2.5, data access is an ongoing issue for city authorities. As such, improvements hold clear potential for SLES development, particularly regarding inclusivity of public, community and third-sector actors most affected by current data access barriers. Increased availability of energy network data could also be of particular value in supporting a wider range of organisations to actively collaborate in the technical side of SLES planning. Financial accessibility of the Open Energy membership model, which proposes to be progressively tiered with 'members charged according to their size and ability to contribute' (Open Energy 2021b), could further help these organisations discover and access a wider range of datasets.

Supportive frameworks promoting interoperability are also particularly important for SLES as they can help prevent the formation of siloes in which data, software, systems and skills develop in ways that limit their usability and learning potential. This can both help ensure smooth transfer of information between actors working at a local scale and support scale up of tools and learning where appropriate. Work such as SDI also holds potential to support holistic approaches to smart city development whereby standards and common processes could enable city authorities to combine data and draw insights in innovative ways. For example, data interoperability could help cities integrate SLES with policy areas such as health, finance or water. As the adoption of digital twin modelling progresses, interoperability will also be key to ensuring compatibility of twins working within the same geographic areas (e.g. integrated utility modelling) or across different geographic areas (e.g. integrating different SLES systems) (Bolton 2018).

However, it is important to acknowledge that the smart energy policy landscape remains nascent and subject to gaps that could affect the character of, and approaches to, building smart energy cities. First, smart energy governance is more advanced for the electricity sector than for heat and transport. This presents a risk of fragmentation that is particularly pertinent to city authorities seeking to take a coherent systems view of decarbonisation in their area. Second, as outlined above, national approaches to improving energy data access remain philosophically dominated by market-based solutions and narratives regarding end-consumer benefit. This can lead to challenges for actors lacking capital or market power. For example, while Open Energy can make the design choice to offer financially inclusive services, the project is not permitted to determine the pricing of data products themselves (i.e. the data flowing through the sharing ecosystem). As such, it may remain possible for actors such as city authorities to be ‘priced out’ of data access in favour of commercial actors who are able to pay more to exploit valuable data. While this issue may in some cases be addressed through public–private partnerships that include data sharing, these arrangements are not guaranteed and still risk leading city authorities into partnerships of unequal weighting.

Finally, fairness issues associated with market-based approaches risk being further exacerbated if governance of market function is subject to vested interests, particularly through participant self-governance. For example, although data triage and the Presumed Open Principle are being incorporated into network regulation, triage decisions will be self-governed and lack a neutral dispute resolution mechanism. As such, there is a risk that networks may gravitate towards data protectionism in the absence of a more prescriptive regulatory mandate. This may ultimately serve to entrench their power as emerging energy data, as well as physical asset, monopolies.

These three issues are by no means the only governance challenges that urban smart energy projects face, both in the United Kingdom and beyond. However, they serve to illustrate the influence of national governance on the distribution of power between different partners in developments, as well as the guiding philosophies or development pathways with which local smart energy narratives are incentivised to align. These factors hold significant potential to influence the extent to which place-sensitive or homogenising approaches dominate urban smart energy development, altering the degree to which urban context is considered either a central factor or mere background to the transition processes. It may also determine the ways in which locally held imaginaries and discourses about energy system change are either integrated into, or are led to compete with, national visions and governance of the future energy system (Levenda et al. 2019).

2.7 Conclusions

This chapter has discussed the importance of municipalities in smart energy development, including notable strengths in their ability to take a more holistic view of the particularities of their socio-geographic areas. Taking the United Kingdom as a case study, the chapter explored the extent to which place-based context is incorporated into new approaches to 'smart local energy system' development. Case studies demonstrate that tailoring smart energy approaches to variation in 'technical' factors – such as energy resources, demand or building stock – is common across local approaches, including influential new narratives advanced through the PFER programme. However, responsiveness to local social context, including city identity and politics, is less universally recognised, despite potentially affecting their legitimacy. New narratives accompanying UK smart energy development are noted to potentially risk backgrounding municipalities, along with associated local socio-political knowledge and avenues for citizen representation, in favour of development pathways emphasising scalable technologies and business models.

The chapter further explored how national narratives in the United Kingdom, prioritising market-based solutions and end-consumer benefit, can conflict with local narratives foregrounding different priorities such as community identity or cross-sector service cohesion within a city. It is suggested that these conflicts are significant, and should be noted as a productive part of sensitive smart energy development. In a sphere that is rapidly becoming dominated by commercial entities and interests prone to promoting homogenised tech-solutionism, city authorities can play a valuable role in carving out space both for the representation of the local and stronger prioritisation of public interest and/or citizen-led aspects of energy transitions.

In order to play this role effectively, city authorities require empowerment. This includes better resources with which to navigate current barriers to data access and use. These ambitions are being supported to some extent by national developments in smart energy governance aiming to broaden visibility, accessibility and interoperability of data. However, smart energy governance reforms are unlikely to fully rebalance the domination of the agenda by commercial interests. The consequences of this governance gap could include development of new data monopolies, disempowerment of small or non-commercial entities, and the distortion or gaming of markets. Within city authorities themselves, a lack of checks and balances regarding vested interests also risks the capture of municipalities by urban elites.

To conclude, it is important to recognise that the urban contexts are not merely background to the energy transition and development of smart technologies, but central to its politics. Diverse interests and priorities, operating across local, national and international scales, are at play in urban smart energy transitions and local experimentation can play an important role in shaping knowledge production and politics across scales (Evans and Karvonen 2015). This emphasises both the need for further research into the processes by which smart energy transitions are constituted and contested in cities and the need for improved urban smart energy policy frameworks to ensure that a wide range of interests and voices are integrated into the urban smart energy development processes.

References

Agora Energiewende (2021). *The European Power Sector in 2020*. Berlin, Germany: Agora Energiewende.

Albino, V., Berardi, U., and Dangelico, R.M. (2015). Smart cities: definitions, dimensions, performance, and initiatives. *Journal of Urban Technology* 22 (1): 3–21. https://doi.org/10.1080/10630732.2014.942092.

Amundsen, H., Hovelsrud, G.K., Aall, C. et al. (2018). Local governments as drivers for societal transformation: towards the 1.5°C ambition. *Current Opinion in Environmental Sustainability* 31: 23–29.

Balta-Ozkan, N., Watson, T., and Mocca, E. (2015). Spatially uneven development and low carbon transitions: insights from urban and regional planning. *Energy Policy* 85: 1–11. https://doi.org/10.1016/j.enpol.2015.05.013.

BEIS (2020a). *BEIS: Smart Data Research. Report: Consent*. London: BEIS.

BEIS (2020b). *BEIS: Smart Data Research – Liability*. London: BEIS.

BEIS (2020c). *BEIS: Smart Data Research Report: Authentication and Trust*. London: BEIS.

BEIS (2020d). *Next Steps for Smart Data Putting Consumers and SMEs in Control of Their Data and Enabling Innovation 2*. London: BEIS.

BEIS (2020e). *Smart Data Working Group: Terms of Reference*. London: BEIS.

Betsill, M.M. and Bulkeley, H. (2006). Cities and the multilevel governance of global climate change. *Global Governance* 12 (2): 141–159.

Binz, C., Coenen, L., Murphy, J.T., and Truffer, B. (2020). Geographies of transition – from topical concerns to theoretical engagement: a commentary on the transitions research agenda. *Environmental Innovation and Societal Transitions* 34: 1–3. https://doi.org/10.1016/j.eist.2019.11.002.

Bolton, A. (2018). *The Gemini Principles*. Cambridge, UK: Centre for Digital Built Britain.

Bristol City Council (2019). *Mayor's Climate Emergency Action Plan*. Bristol: Bristol City Council.

Bristol City Council (2021). Strategic energy unit – Bristol.Gov.Uk. https://www.bristol.gov.uk/policies-plans-strategies/the-energy-service (accessed 9 June 2021).

Bristol Energy Service (2021). Community – energy service Bristol – renewable energy and sustainability. https://www.energyservicebristol.co.uk/community/ (accessed 9 June 2021).

Britton, J. (2017). Smart meter data and equitable energy transitions – can cities play a role? *Local Environment* 1–15. https://doi.org/10.1080/13549839.2017.1383372: 595–609.

Bulkeley, H., McGuirk, P.M., and Dowling, R. (2016). Making a smart city for the smart grid? The urban material politics of actualising smart electricity networks. *Environment and Planning A: Economy and Space* 48 (9): 1709–1726. https://doi.org/10.1177/0308518X16648152.

Carvalho, L. (2015). Smart cities from scratch? A socio-technical perspective. *Cambridge Journal of Regions, Economy and Society* 8 (1): 43–60. https://doi.org/10.1093/cjres/rsu010.

Castán Broto, V. (2019). Climate change politics and the urban contexts of messy governmentalities. *Territory, Politics, Governance* 8 (2): 241–258. https://doi.org/10.1080/21622671.2019.1632220.

Coenen, L., Benneworth, P., and Truffer, B. (2012). Toward a spatial perspective on sustainability transitions. *Research Policy* 41 (6): 968–979. https://doi.org/10.1016/j.respol.2012.02.014.

Committee on Climate Change (2020). *Local Authorities and the Sixth Carbon Budget*. London: Committee on Climate Change.

Cowley, R. and Caprotti, F. (2019). Smart city as anti-planning in the UK. *Environment and Planning D: Society and Space* 37 (3): 428–448. https://doi.org/10.1177/0263775818787506.

Department for Business Energy and Industical Strategy (2021). *Energy Trends. UK 2020*. London: Department for Business Energy and Industical Strategy.

Devine-Wright, P. (2019). Community versus local energy in a context of climate emergency. *Nature Energy* 4 (11): 894–896. https://doi.org/10.1038/s41560-019-0459-2.

Energy Data Taskforce (2019). *A Strategy for a Modern Digitalised Energy System*. London: Energy Data Taskforce.

Energy Systems Catapult (2020). *Smart Local Energy System Composition. A Portfolio Review of PFER Concept Projects*. Birmingham, UK: Energy Systems Catapult.

Energy Systems Catapult (2021a). Energy revolution integration service (ERIS). https://es.catapult.org.uk/service-platforms/energy-revolution-integration-service/ (accessed 9 June 2021).

Energy Systems Catapult (2021b). Local energy system modelling. https://es.catapult.org.uk/capabilities/modelling/local-energy-system-modelling/ (accessed 9 June 2021)

Erickson, P. and Tempest, K. (2014). *Advancing Climate Ambition: How city-scale actions can contribute to global climate goals. Stockholm Environment Institute*, Working Paper. 2014-06.

Evans, J. and Karvonen, A. (2015). 'Give me a laboratory and I will lower your carbon footprint!' – urban laboratories and the governance of low-carbon futures. *International Journal of Urban and Regional Research* 38 (2): 413–430. https://doi.org/10.1111/1468-2427.12077.

Evans, J., Karvonen, A., Luque-Ayala, A. et al. (2019). Smart and sustainable cities? pipedreams, practicalities and possibilities. *Local Environment* 24 (7): 557–564. https://doi.org/10.1080/13549839.2019.1624701.

Flett, G., Kelly, N., and McGhee, R. (2018). *Review of UK Energy System Demonstrators*. London: UK Energy Research Centre.

Ford, R. and Hardy, J. (2020). Are we seeing clearly? The need for aligned vision and supporting strategies to deliver net-zero electricity systems. *Energy Policy* 147: 111902. https://doi.org/10.1016/j.enpol.2020.111902.

Foxon, T.J. and Steinberger, J.K. (2013). *Energy, Efficiency and Economic Growth: A Coevolutionary Perspective and Implications for a Low Carbon Transition*. Leeds, UK: University of Leeds.

Frerk, M., Ward, J., Roberts, S., and Hodges, N. (2019). Smart meter energy data: public interest advisory group. *Final Report – Phase 1*. London.

Geels, F.W. and Schot, J. (2007). Typology of sociotechnical transition pathways. *Research Policy* 36 (3): 399–417. https://doi.org/10.1016/j.respol.2007.01.003.

Glasmeier, A. and Christopherson, S. (2015). Thinking about smart cities. *Cambridge Journal of Regions, Economy and Society* 8 (1): 3–12. https://doi.org/10.1093/cjres/rsu034.

Goodspeed, R. (2015). Smart cities: moving beyond urban cybernetics to tackle wicked problems. *Cambridge Journal of Regions, Economy and Society* 8 (1): 79–92. https://doi.org/10.1093/cjres/rsu013.

Gupta, R. and Zahiri, S. (2020a). Evaluation of user engagement in smart local energy system projects in the UK. *Energy Evaluation Europe 2020 Conference*. 29 June–1 July 2020,

London, UK. https://energy-evaluation.org/wp-content/uploads/2020/07/eee2020-paper-rajat-gupta-abstract77.pdf.

Gupta, R. and Zahiri, S. (2020b). Meta-study of smart and local energy system demonstrators in the UK: technologies, leadership and user engagement. *IOP Conference Series: Earth and Environmental Science*, 588. Beyond2020 – World Sustainable Built Environment conference, 2–4 November 2020, Gothenburg, Sweden. https://doi.org/10.1088/1755-1315/588/2/022049.

Hale, T. (2018). *The Role of Sub-State and Non- State Actors in International Climate Processes.* The Royal Institute of International Affairs.

HM Goverment (2017a). Industrial strategy challenge fund: for research and innovation. https://www.gov.uk/government/collections/industrial-strategy-challenge-fund-joint-research-and-innovation#prospering-from-the-energy-revolution (accessed 14 October 2019).

HM Government (2017b). *Industrial Strategy. Building a Britain Fit for the Future*. London: HM Government.

HM Government (2017c). *The Clean Growth Strategy: Leading the Way to a Low Carbon Future.* London: Crown Copyright.

HM Government (2019). *Unlocking Digital Competition. Report E of the Digital Competition Expert Panel*. London, UK: HM Government.

HM Government (2020). *Energy White Paper: Powering Our Net Zero Future*. London: HM Government.

HM Government (2021). Modernising energy data. https://www.gov.uk/government/groups/modernising-energy-data (accessed 9 June 2021).

Hollands, R.G. (2008). Will the real smart city please stand up? *City* 12 (3): 303–320. https://doi.org/10.1080/13604810802479126.

Hölscher, K., Frantzeskaki, N., McPhearson, T., and Loorbach, D. (2019). Tales of transforming cities: transformative climate governance capacities in New York City, U.S. and Rotterdam, Netherlands. *Journal of Environmental Management* 231 (October 2018): 843, 857. https://doi.org/10.1016/j.jenvman.2018.10.043.

Hoppe, T., van der Vegt, A., and Stegmaier, P. (2016). Presenting a framework to analyze local climate policy and action in small and medium-sized cities. *Sustainability (Switzerland)* 8 (9): 847. https://doi.org/10.3390/su8090847.

Joseph, A. and Balachandra, P. (2020). Energy internet, the future electricity system: overview, concept, model structure, and mechanism. *Energies* 13 (4242): 1–26. https://doi.org/10.3390/en13164242.

Kitchin, R. (2015). Making sense of smart cities: addressing present shortcomings. *Cambridge Journal of Regions, Economy and Society* 8: 131–136. https://doi.org/10.1093/cjres/rsu027.

Kuzemko, Caroline, and Jess Britton. 2020. "Policy, politics and materiality across scales: a framework for understanding local government sustainable energy capacity applied in England." *Energy Research and Social Science* 62, 101367. https://doi.org/10.1016/j.erss.2019.101367.

Levenda, A.M., Richter, J., Miller, T., and Fisher, E. (2019). Regional sociotechnical imaginaries and the governance of energy innovations. *Futures* 109: 181–191. https://doi.org/10.1016/j.futures.2018.03.001.

London City Hall (2021a). London community energy fund. https://www.london.gov.uk/what-we-do/environment/energy/london-community-energy-fund#acc-i-62962 (accessed 14 June 2021).

London City Hall (2021b). London fuel poverty partnership. https://www.london.gov.uk/what-we-do/environment/energy/london-fuel-poverty-partnership (accessed 9 June 2021).

Low Carbon Hub (2021). Project LEO. https://www.lowcarbonhub.org/p/programmes/project-leo/ (accessed 15 June 2021).

Lund, H., Østergaard, P.A., Connolly, D., and Mathiesen, B.V. (2017). Smart energy and smart energy systems. *Energy* 137: 556–565.

Luque, A., McFarlane, C., and Marvin, S. (2014). Smart urbanism: cities, grids and alternatives? In: *After Sustainable Cities?* (ed. M. Hodson and S. Marvin), 74–90. London: Routledge.

Mathis, S. and Kanik, A. (2021). Why you'll be hearing a lot less about 'smart cities'. *City Monitor*. https://citymonitor.ai/government/why-youll-be-hearing-a-lot-less-about-smart-cities (accessed 23 May 2021).

Mitchell, C. (2016). Momentum is increasing towards a flexible electricity system based on renewables. *Nature Energy* 1: 15030. https://doi.org/10.1038/NENERGY.2015.30.

Moloney, S. and Horne, R. (2015). Low carbon urban transitioning: from local experimentation to urban transformation? *Sustainability* 7: 2437–2453. https://doi.org/10.3390/su7032437.

Mora, L., Deakin, M., Zhang, X. et al. (2020). Assembling sustainable smart city transitions: an interdisciplinary theoretical perspective. *Journal of Urban Technology* 1–27. https://doi.org/10.1080/10630732.2020.1834831.

Morozov, E. (2013). *To Save Everything, Click Here: The Folly of Technological Solutionism*. New York, USA: PublicAffairs.

Morstyn, T., Collett, K.A., Vijay, A. et al. (2020). OPEN: an open-source platform for developing smart local energy system applications. *Applied Energy* 275: 115397. https://doi.org/10.1016/j.apenergy.2020.115397.

Naumann, M. and Rudolph, D. (2020). Conceptualizing rural energy transitions: energizing rural studies, ruralizing energy research. *Journal of Rural Studies* 73: 97–104. https://doi.org/10.1016/j.jrurstud.2019.12.011.

O'Connor, M. and Portes, J. (2021). Estimating the UK population during the pandemic – ESCoE: ESCoE. *Economic Statistics Centre of Excellence*. https://www.escoe.ac.uk/estimating-the-uk-population-during-the-pandemic/ (accessed 8 June 2021).

Ofgem (2021a). *Data Best Practice Guidance*. London: Ofgem.

Ofgem (2021b). Digitalisation strategies for modernising energy data. https://www.ofgem.gov.uk/publications-and-updates/digitalisation-strategies-modernising-energy-data (accessed 9 June 2021).

Ofgem (2021c). *Open Letter on DNOs' Privacy Plans for Access to Smart Meter Data*. London: Ofgem.

Open Data Bristol (2021a). Air quality dashboard – open data Bristol. https://opendata.bristol.gov.uk/pages/air-quality-dashboard-new/map#air-quality-now (accessed 9 June 2021).

Open Data Bristol. (2021b). Explore – open data Bristol. https://opendata.bristol.gov.uk/explore/?sort=modified (accessed 9 June 2021).

Open Energy (2021a). Building the web of energy data. https://energydata.org.uk/ (accessed 9 June 2021).

Open Energy (2021b). *Webinar: Unlocking Energy Data – an Early Look at the Open Energy Service Architecture*. Open Energy.

Parks, D. and Rohracher, H. (2019). From sustainable to smart: re-branding or re-assembling urban energy infrastructure? *Geoforum* 100 (February): 51–59. https://doi.org/10.1016/j.geoforum.2019.02.012.

Pickerill, J. (2021). Hopefulness for transformative grassroots change. *Environmental Policy and Governance* 1–3: https://doi.org/10.1002/eet.1933.

Sadowski, J. and Levenda, A.M. (2020). The anti-politics of smart energy regimes. *Political Geography* 81: 102202. https://doi.org/10.1016/j.polgeo.2020.102202.

Sapiains, R., Ibarra, C., Jiménez, G. et al. (2020). Exploring the contours of climate governance: an interdisciplinary systematic literature review from a southern perspective. *Environmental Policy and Governance* 31 (1): eet.1912. https://doi.org/10.1002/eet.1912.

Smeds, E. and Acuto, M. (2018). Networking cities after Paris: weighing the ambition of urban climate change experimentation. *Global Policy* 9 (4): 549–559. https://doi.org/10.1111/1758-5899.12587.

Söderström, O., Paasche, T., and Klauser, F. (2014). Smart cities as corporate storytelling. *City* 18 (3): 307–320. https://doi.org/10.1080/13604813.2014.906716.

SONNET Project (2021). Bristol – SONNET energy. https://sonnet-energy.eu/portfolio-item/bristol/ (accessed 9 June 2021).

The World Bank (2020). *Increasing Access to Electricity in the Democratic Republic of Congo: Opportunities and Challenges*. Washington, DC: The World Bank.

UNEP (2018). *Emissions Gap Report 2018*. Nairobi, Kenya: UNEP.

Wolfram, M. (2016). Conceptualizing urban transformative capacity: a framework for research and policy. *Cities* 51: 121–130. https://doi.org/10.1016/j.cities.2015.11.011.

Yigitcanlar, T., Hoon Han, M., Kamruzzaman, G.I., and Sabatini-Marques, J. (2019). The making of smart cities: are Songdo, Masdar, Amsterdam, San Francisco and Brisbane the best we could build? *Land Use Policy* 88: 104187. https://doi.org/10.1016/j.landusepol.2019.104187.

3

Where are Smart Sustainable Cities Made? Tracing Wired Socio-Technical Relationships in, Through, Beneath, and Beyond a City

Torik Holmes[1], Rebecca Windemer[2], and Carla De Laurentis[3]

[1] *Sustainable Consumption Institute, University of Manchester, Manchester, UK*
[2] *Regen, Exeter, UK*
[3] *Department of Geography and Environmental Management, University of the West of England, Bristol, UK*

3.1 Introduction

Sustainability is now a central component of smart city agendas, and smart cities have been identified as playing a central role in enabling urban areas to achieve net-zero (Angelo and Vormann 2018; Uspenskaia et al. 2021). As Trindade et al. (2017, p. 11) write, 'today, the smart city concept is viewed as a vision, manifesto or promise aiming to constitute the twenty-first century's sustainable and ideal city form'. Recognising the significance of smart sustainable cities for attempts to achieve climate change goals, a plethora of research has focused on cities and city regions to explore programmes of change. Key topics include smart city ideologies (Hollands 2008; Söderström et al. 2014), local policy initiatives (Angelidou 2014; Caragliu and Del Bo 2019), and the deployment of 'smart' technological innovations in context (Beretta 2018; Kandt and Batty 2021). However, the wider ordering and management of key service infrastructures and the provision of renewable sources of energy have received less attention. This is the case even though these are critical for smart and sustainable city ambitions. In quite rudimentary terms, without an ability to connect to electricity networks and without the provision of clean energy, smart and sustainable city buildings, technologies, policies and initiatives would not (and will not) materialise. The ordering and management of electricity supply and generation are thus critical topics.

It makes sense, in turn, to take service networks and in particular the electricity system as a starting point of empirical analysis in order to explore where, when and how smart sustainable cities are made. This is a complicated route – electricity networks are characteristically defined by 'reach' and 'scope' (Star 1999, p. 381). Indeed, such systems are scripted into the fabric of technologies and buildings and stretch underneath and beyond cities into rural settings. Electricity systems thus cut across various socio-spatial scales and are caught up in the production of such scales, while also being ensnarled in and integral to the practices of multiple actors and organisations. Together, these actors, organisations

Smart Cities, Energy and Climate: Governing Cities for a Low-Carbon Future, First Edition.
Edited by Oleg Golubchikov and Komali Yenneti.

and technologies help form a 'seamless web' of socio-technical relations (Hughes 1986, p. 281). Examining where, when and how smart sustainable cities are made necessarily involves exploring such seamless webs. However, given the complexity noted, where does it make sense to start?

We take as our starting point several key ideas generated in science and technology studies (STS). These ideas lead us to identify several socio-technical '*hot areas*, where the constitution of the world is still being debated', and in the case of this chapter, where smart sustainable city ambitions are contested, configured and calibrated (Kjellberg 2008, p. 157). These areas are the product of and linked through socio-technical relationships and thus by various aims, ambitions, initiatives and policies, and by technologies, wires and cables.

We do so by exploring the case of Manchester. This is a city with a relatively long-held smart sustainable and ambitious 2038 net-zero agenda. We outline a set of challenges that the city has struggled to overcome as part of attempts to accelerate smart sustainable transitions. This is our first 'hot area'. The second hot area we visit is an emblematically smart and sustainable building in the centre of the city. We consider what it takes to get such buildings connected to the electricity network and what the connection process means for future smart sustainable transitions. From this building, we follow wires out onto the electricity network, arriving at our third hot area – a key piece of investment in infrastructure. We reveal that such instances of investment, which are critical for smart sustainable ambitions, do not always clearly serve or support such aims and objectives. Indeed, we argue that core components of investment decisions do not always support smart sustainable shifts but instead perpetuate unsustainable patterns of everyday life.

Continuing to follow the wires, we arrive at our fourth hot area: two wind farms. We reveal, in the first instance, how policy changes and wider societal values combine to restrict the growth of existing sites and thus increased generating capacity. In the second instance, favourable subsidies, coupled with a fraught policy and public context, are shown to help hold in place relatively inefficient infrastructure, which could be replaced with more efficient assets that produce greater amounts of clean energy. We argue that these socio-technical dynamics, which fetter the deployment of larger wind farms, act to slow the fruition of Manchester as a smart, sustainable and net-zero city. Having revealed several points of tension, which mediate and slow broader smart and sustainable transitions, we etch out a set of speculative responses in the penultimate section of this chapter. In the concluding section, we consider what this chapter calls for in future smart and sustainable city-oriented research.

3.2 Orientating Ideas, Approaches and Methods

As briefly noted, the fruition of smart sustainable cities clearly depends on the deployment and development of various technologies and buildings and the related organisation of key service infrastructures. Electricity is particularly crucial because it enables the digitalisation and smartification of networked technologies, buildings and services. Electricity is also a critical component of sustainable ambitions, with clean energy at the core of attempts to make more efficient and environmentally friendly cities and social futures. Thanks to this context, the electricity system forms an 'obligatory passage point' (Callon 1984, p. 196);

it is both an essential medium and mediator of smart and sustainable ambitions. In short, smart sustainable ambitions are entangled in and beholden to the socio-technical ordering of electricity networks.

Conceptualised as a socio-technical system, electricity networks are specifically made up of 'physical artefacts, such as the turbogenerators, transformers, and transmission lines in electric light and power systems [. . . and] organisations, such as manufacturing firms, utility companies, and investment banks' (Hughes 1989, p. 51). Electricity provision thus involves multiple social and material components. These components are the product of various histories and have specific capacities and affordances, and have been shown to entail path dependencies and shape future possibilities (Foxon et al. 2010; Meadowcroft 2016; Verbong and Geels 2010).

The bulky materiality of networked hardware means, for example, that once a 'high-voltage electricity distribution system is in place, it is hardly conceivable to deconstruct it and shift to a decentralized system of windmill power generation' (Hommels 2005, p. 329). Such inabilities are also to do with the ordering of the 'invisible grid' (Summerton and Coutard 1994, p. 55). This is at least as 'strong' as the physical grid (Summerton and Coutard 1994, p. 55). It 'consists of equipment suppliers, politicians on national and local levels, a multitude of regulators, interest groups, and [. . .] heterogeneous customers or subscribers' (Summerton and Coutard 1994, p. 55). Such actors and groups have expectations and vested interests that prove obdurate to and shape change. As Summerton and Coutard (1994, p. 55) note, thanks to political, social and economic interests, related to, for example, high sunk costs: 'the invisible grid often turns out to be less "malleable" than the physical one: "hard" pipelines have to bend to follow large customers, and "hard" heat or power plants perhaps have to be relocalized in response to immovable regulators' (Summerton and Coutard 1994, p. 55). Thanks to the strong links described above, smart sustainable city-based ambitions necessarily take shape in reference to and are caught up in the 'invisible' and 'physical' socio-technical dynamics of electricity supply and use.

The geographic dimensions of such dynamics match the characteristic reach and scope of electricity systems. Because electricity networks simultaneously reach into and stretch out across multiple physical and institutional spaces, 'local' challenges tend to hinge on things going on elsewhere (Dowling et al. 2018). Indeed, local challenges are rooted in multiple histories and caught up in various socio-technical scales, timeframes and related 'hot areas' (Kjellberg 2008, p. 157). In the case of this chapter, these hot areas are places where smart sustainable city-based futures are contested, configured and calibrated. Thanks to the reach and scope of electricity provision, these are distributed across space and sited inside and outside cities.

Acknowledging the reach and scope of electricity provision and the distributed contingencies of smart sustainable city-based ambitions, we embrace an oligoptic approach. As Latour (2005) explains, panoptic approaches assume a central vantage point and frame of reference. In smart city research, the 'city' tends to play this role. By contrast, an oligoptic approach is premised on the relational nature of all phenomena and the idea that there are always multiple and connected hot areas where challenges are contested, configured and calibrated in reference to each other.

The hot areas we identify have roots in the key orientating ideas introduced and our interest in exploring where, when and how smart cities are made. The first hot area we visit is Manchester, a city with quite clear and relatively long-held smart and sustainable aspirations. We describe the history of these ambitions and key challenges. Recognising the fruition of Manchester's smart sustainable ambitions as a socio-technical challenge, embroiled in electricity provision, we zoom in on a building project that is indicative of current smart and sustainable ideals. In this instance, we describe aspects of the electricity connection process, which, we argue, can act to slow the broader deployment of smart and sustainable buildings across the city. Continuing to follow an oligoptic approach, we trace wires out from the building to a primary substation. Accepting that smart sustainable aims hinge on the increasing use of electricity and thus the provision of additional network capacity, we explore why such substations are built. We show that logics of investment do not necessarily map neatly with sustainability ambitions. Continuing to look elsewhere and for hot areas of contestation and contingency, we visit two onshore wind farms in order to explore the social, political and technical dimensions of renewable provision. As noted, smart and sustainable ambitions hinge on the provision of clean electricity, which wind farms provide. We show, in this instance, that the scale of provision is a complicated socio-technical affair, involving multiple contingent factors. We argue that these, in combination, not only shape the scale of provision but, in turn, the speed of broader city-based smart sustainable transitions.

A range of secondary sources, including city strategies, electricity network plans, schematic diagrams and land-use planning applications, were drawn on as part of our study of each hot area. This chapter is also informed by research conducted as part of two projects concerning the generation and distribution of electricity. These involved over 40 expert interviews, which although not at the core of this chapter, provided foundational insight regarding the socio-technical dynamics of electricity supply in and beyond Manchester.

The hot areas we visit are not the only sites where smart sustainable futures are configured and contested; other hot areas exist along the electricity network (and other networks too). This multiplicity lies at the core of the socio-technical and oligoptic approach we embrace. We return to this point in the concluding section of this chapter and consider its implications for future research.

3.3 The City of Manchester

Manchester has clear ambitions to become an 'inclusive, sustainable and resilient smart city' with the aim of becoming 'a world leading Digital City within the next decade' (Manchester City Council 2021). The city of Manchester can be considered as a 'hot area' or site of contestation as it attempts to achieve its smart sustainable city ambitions. Indeed, Manchester is illustrative of many ambitious smart cities, relying upon a wide range of investments and decision-makers within and beyond the city boundary to provide the investments and policy change needed to ensure that smart sustainable city ambitions are achieved.

Manchester has experienced significant regeneration over the past 35 years with an increase in city centre businesses and residents. A key focus of this growth has centred on digital technology and service provision (Carter 2013). Over the past decade, the city has scaled up its digitalisation and smart city ambitions, building upon digital strategies in 2008 and 2012 and a 2016 smarter city programme (Cowley et al. 2018). Manchester is now the United Kingdom's biggest digital sector outside of London, with 100,000 people employed in the digital tech sector by 2019 (Prolific North 2020). While the council owns approximately 30–40 smart city projects (Caird and Hallett 2019), the wider activities of the smart city rely on a wider network of local organisations, including digital and high-tech organisations, universities, transport operators and engineering firms (Cowley and Caprotti 2019). Such organisations, their decisions and investments exert a significant influence on how quickly the city can achieve its smart and sustainable aims.

In recent years, Manchester's smart city and climate change aims have aligned, creating not only opportunities but also areas of challenge. The *Manchester Digital Strategy* (2021) sets out its smart city ambitions under four key themes: smart people, digital places, future prosperity and sustainable resilience. It recognises the city's commitment to become a zero-carbon city through innovative ways to generate and use renewable energy, as set out in The Manchester Climate Change Framework 2020–2025 (Manchester Climate Change Agency 2020). In doing so, it also identifies a number of significant challenges to achieving this. These include increasing the deployment of renewable sources of generation (i.e. solar photovoltaic [PV] and wind), electricity storage capacity (i.e. batteries), and the use of low-carbon technologies (electric vehicles and heat pumps) (Manchester Climate Change Agency 2020). Challenges also include facilitating peer-to-peer energy trading (to support the deployment of renewable technologies) and the need to develop strong relationships with cooperatives and green energy investment companies. All these challenges further hang on the constitution of a smarter, decentralised and more intuitive electricity network. The network is not, however, under the control of the local authority, Nor is the local authority solely in control of the design, development and deployment of smart sustainable technologies.

The scale and contingencies of the net-zero challenges faced by Manchester are particularly acute in regard to the pressing need for clean energy generation. While the city has long acknowledged the need for increased renewable production, 'in 2019, only 1% of Manchester's electricity demand was met by local renewable generation' (Manchester Climate Change Agency 2020, p. 25). The need for a greatly increased capacity for renewable generation inside the city thus remains a critical challenge, which history suggests is clearly not easy for the local authority to solve. The local authority acknowledges this, identifying a pressing need for the UK government to rapidly drive decarbonisation across the National Grid (Manchester Climate Change Agency 2020). It is clear, in this regard and in respect of key local policy documents and energy strategies, that smart sustainable city ambitions are contingent on relationships and forms of action and change that are not of the city's making or only sited therein.

While there are a number of studies that have explored Manchester's smart city ambitions (e.g. Caird and Hallett [2019], Cowley et al. [2018], Carter [2013]), these have typically focused on actions occurring within the city boundary. In doing so, such studies lack consideration of the multiple points of tension that are occurring beyond the city but impacting the city's smart sustainable future. Our approach thus differs from these existing studies by tracing the wires out from the city to explore the wider factors and 'hot areas' impacting the feasibility of Manchester's smart sustainable city ambitions.

3.4 Connecting 'Smart' Buildings

The development of smarter sustainable buildings in Manchester is a key component of broader smart sustainable shifts across the city. Plot 9 First St, Manchester is such a development (Figure 3.1) (Ask Real Estate 2021). Located at the heart of a burgeoning mixed-use neighbourhood in Manchester City Centre, this £150 million 'building will deliver 2,60,766 ft^2 of offices and 21,111 ft^2 of retail across nine storeys' (Europe Real Estate 2020). The building will have an integrated and bespoke 'smart' operating system designed by *Smart Spaces* – a leading master systems integrating technology firm (Smart Spaces 2021). The operating system will offer 'a 3D interactive replica of the building [. . . to] provide contactless access and 360 control for building users', supporting real-time HVAC and lighting energy metering and management (Europe Real Estate 2020). Smart building services will include touch-free access, lift, and climate and lighting controls; smart desk and meeting room booking; a seamless smart visitor management system; real-time occupancy with journey mapping; predictive fault management; and real-time energy footprints (Europe Real Estate 2020). As a consequence of providing these services, John Hughes (Managing Director of Ask Real Estate) believes the development 'will be one of the smartest and most sustainable buildings outside London [. . ., ensuring] efficiency and sustainability' (Europe Real Estate 2020). Plot 9 First St is indicative of the type of developments that are seen as crucial to the fruition of the aims outlined above and thus Manchester's smarter and more sustainable future.

Plot 9 First St and the smart sustainable objectives embedded in the building depend on the provision of electricity. The provision of electricity to a new building does not simply happen. In the first instance, a connection application for supply must be made to Electricity North West (ENW), the distribution network operator (DNO). Regulation means that

Figure 3.1 Plot 9 first St, Manchester city centre (Europe Real Estate 2020).

connection costs are footed by connecting customers. As outlined in *The Electricity (Connection Charges) Regulations* (Department of Trade and Industry 2002):

> Regulations [. . .] provide for an electricity distributor to recover from subsequent users of electric lines and electrical plant first provided for the purpose of giving a connection to the premises or distribution system of another person an amount in respect of the expenses incurred in first providing the line or plant.

These regulations mean that there is no single connection charge. Instead, connection costs are unevenly distributed across the network in reference to the sited capacities of existing wires, cables and substations. On 'hot spots' – segments of network where peak demand is already approaching asset capacities – new connections can carry heavy costs. These costs differ from those on 'colder' segments of the network, where asset ratings and headroom are strong. Connecting smart sustainable buildings on 'hot spots' is, in turn, typically more costly than the price of plugging in on a 'colder spot' of the network.

Given the strong relationship between costs of investment and the viability of built developments, the realisation of smart sustainable buildings in Manchester, such as Plot 9 First St, is thus partly contingent on the established capacities of electricity infrastructure. Thanks to connection regulations and the adjudication of costs, the material capacities of infrastructure make particular smart building projects more or less viable in certain areas and in reference to specific segments of network. In the case of Plot 9 First St, connection costs have not proved inhibitive to development. A reading of ENW's heat map tool suggests that this is because there is existing transmission headroom across primary substations in the city centre (Electricity North West 2021b). This is a positive for smart sustainable transitions taking shape through the deployment of more efficient buildings. However, this is not to say that challenges do not lie on the horizon, with the arrival of smart sustainable buildings, like Plot 9 First St, taking up network capacity and therefore increasing the need for expensive network investment that could prove inhibitive to future developments.

This is all the more likely, thanks to the boom in development continuing to unfold across Manchester (Deloitte 2021). Indeed, this boom and the related spread of smart sustainable buildings contribute to, in the terms described and thanks to connection regulations, the constitution of future hotspots. And, these hotspots may very well act to slow the deployment of smart sustainable buildings across the city more broadly, playing the role of a restrictor placed on engines to limit acceleration. This quite ironic point of contingency is a product of socio-technical relationships held among smart ambitions, buildings, asset capacities and regulations concerning network connection charges.

Other points of contingency are spread out across the electricity network. The wires in the emblematically smart Plot 9 First St project will lead underground and onto a low-voltage network. This network is fed by a pre-existing and largely subterranean tributary of cables that lead onto a high-voltage circuitry of primary substations, bulk and grid supply points. These substations play the crucial role of stepping down electricity voltages, ready for commercial and non-commercial purposes. These assets also link up with points of generation that feed electricity onto the system. Shorter- and longer-term programmes of large-scale investment across this extended system are critical for cities with smart sustainable city ambitions. This is largely due to core changes, such as the aggregate digitalisation of services and the growing electrification of transport and heating as part of decarbonisation, specifically hinging on the development of a suitably sized and intuitive network (Manchester Climate Change Agency 2020). It is accepted, for example, that Manchester's net-zero future depends on the development of an electricity system that is appropriately sized to accommodate new forms of distributed generation, battery storage, the uptake of electric vehicles and heating and cooling technologies (ENW 2018, Manchester Climate Change Agency 2020). Instances of large-scale investment to increase network capacity in (and indeed beyond) Manchester, therefore, represent other 'hot areas' where smart city transitions and ambitions are made possible. Such investments are typically rare and expensive. And, as we demonstrate in the next section, though crucial for smart sustainable city-oriented aspirations, bulky investments do not transpire solely in support of, nor do they necessarily represent, particularly smart sustainable futures.

3.5 'Smart' and Not-So-Smart Large-Scale Network Investments

Bulky investment in electricity infrastructure is sporadic. There has, for example, only been three 33 kV primary substations built in Manchester city centre since 1972. The Central Manchester 33 kV Primary Substation is one of these. The Primary was commissioned in 2007 and completed in 2009 at an estimated cost of £1.8 million (IQS 2017).

Bulky investments in primary substations are, in part, to do with load growth linked with planned development. In the short term, applications for connection trigger network tests to see if the load growth associated with the proposed project can be facilitated by pre-existing hardware. In the case of the Central Manchester Primary, the 2005 Piccadilly Place regeneration project proved informative (Bennetts Associates 2009). At the time of its inception, Piccadilly Place represented 'one of the largest urban regeneration projects to be built outside London in the last 20 years' (Austin-Smith: Lord, 2019).

Proposed developments may or may not be particularly smart or sustainable, and it is not a DNO's responsibility to pass judgement in this respect. DNOs are simply obliged to respond to connection applications and provide relative quotes in a timely fashion. In effect, this means that buildings that are not particularly smart, efficient or sustainable are still given an opportunity to materialise.

Investments such as the Central Manchester Primary are also the product of more speculative calculations to do with longer-term load growth. The inclusion of 'future' circuit breakers, connection opportunities and spare capacity built into the Central Manchester Primary is the outcome of such calculations.

The DNO uses econometric modelling to forecast trends in electricity demand and spot where future challenges are likely to transpire (ENW 2018a). In 2018, it was forecasted that 'the Manchester region will experience the greatest year on year growth in demand' across ENW's network, with growth of '40% by 2023' (ENW 2018a, p. 15). Such forecasts are premised on modelling centred around national and local authority population and economic gross value added (GVA) calculations and the associated diffusion of technologies including, in particular, EVs, heat pumps and air conditioning units (ENW 2018). The extent to which these informative factors support smart sustainable shifts is debatable.

On the one hand, network investments to increase capacity that supports the spread of smart sustainable buildings, the uptake of EVs and heat pumps align with the current sustainability paradigm. On the other, and going against the grain of dominant ideas of sustainability, adding network capacity to support population and economic growth, alongside the diffusion of carbon-embedded technologies, does not necessarily support particularly sustainable shifts and is not necessarily that 'smart'. Indeed, adding network capacity to support population and economic growth, regardless of the sustainable credentials of such change, could very well act to reproduce and hold in place arguably unsustainable patterns and practices of everyday life. This danger is made all the more likely by building in capacity to support the forecasted diffusion of energy-hungry technologies, including air conditioning units. As discussed elsewhere, the global diffusion of air conditioning technologies, which ratchets up energy demand and the production of carbon-embedded materials, is a problematic and arguably very unsustainable trend (Rutherford and Coutard 2014, Walker et al. 2014). Similar worries have been raised in reference to the diffusion of electric vehicles (EVs), with production and use remaining carbon dependent (Thomas 2012). In this regard, by planning and producing a network to facilitate the estimated uptake of 'over 3 million EVs in the North West' by 2050 a *less* carbon-based system of automobility is certainly supported, but other even less energy-intensive futures are overlooked (ENW 2021a, p.7). These include, for example, urban futures premised more on the uptake of bicycles and cycling (Behrendt 2019). More broadly and following the concept of 'induced demand' from transport studies (Lee et al. 1999), whereby network extension is understood to generate additional consumption, adding capacity for sustainably questionable trends in consumption makes problematic load growth possible.

The problematic sustainable value of network investments that come in response to and support broader growth agendas and the diffusion of 'smart' (but not necessarily that sustainable) technologies is particularly acute thanks to a reliance on fossil fuels. Continuing to follow wires and cables out of the city, in the next section, we discuss this reliance in reference to the politics of onshore wind generation. As we show, this is another 'hot area' where smarter and more sustainable futures are contested, configured and calibrated.

3.6 Turbulent Urban–Rural Relationships and Contingencies

Manchester's smart and sustainable city aspirations hinge on the growing use of renewable sources of energy (Manchester City Council 2021). Cities typically accommodate small-scale renewables such as rooftop solar PV. However, due to the high competition for land use, large-scale renewable energy generation in cities is rare. As noted, in 2019, 1% of Manchester's electricity demand was met by local renewables (Manchester Climate Change Agency 2020). There are aspirations to increase the amount of renewable energy generated across the city region. It is accepted, however, that hopes of more sustainable futures are contingent on the wider deployment of renewable sources of generation, including wind.

The deployment of large-scale wind farms is a complex and contingent process involving various social, political, economic and material dimensions (Van der Horst and Toke 2010). Manchester's smart and sustainable city aspirations are embroiled in and hang on the complicated socio-material dynamics of wind farm deployment. Cables leading out from the Central Manchester Primary connect up to two wind farms sited on the moorland that surrounds the city region. These sites form two other 'hot areas' where smart and sustainable city aspirations are implicitly contested and configured.

3.7 Scout Moor Wind Farm: Policy, Social Attitudes and Limiting Growth

Scout Moor wind farm (Figure 3.2), near Rochdale, is approximately 25 miles from Manchester. Operational since 2008, it is one of England's largest wind farms. It comprises 26 turbines and generates 65 MW. This generating capacity is estimated to be enough to power 40,000 homes.

While the wind farm was developed at a time when there was policy support for onshore wind, political attitudes have since changed, limiting opportunities to increase generation from this (and other) sites. In 2015, permission was granted to increase the wind farm's size and output. These plans fell victim to a national policy *U-turn* that effectively created a barrier to new onshore wind developments. UK policy changes in 2015 specifically imposed two strict requirements. First, sites have to have first been identified as suitable for wind farms in a local or neighbourhood plan. Updating such plans is not something that can be achieved quickly or easily, particularly for already overstretched local (and often rural) planning authorities. Second, all planning impacts identified by local people must be

Figure 3.2 Scout Moor wind farm. *Source:* Paul Anderson/Wikimedia Commons/CC BY-SA 2.0.

addressed and there must be community backing. This is not easy to achieve or evidence. And, the impact of this policy change is clear. Just 16 new turbines were granted planning permission between 2016 and 2020, in 7 separate locations across England. Comparatively, prior to this policy change, between 2011 and 2015, 435 turbines were permitted to be built on 108 sites in England. This reflects a 96% decrease since the policy change.

At Scout Moor, more specifically, plans to add another 16 turbines (with the capacity to power an additional 22,500 homes) resulted in the deployment of only 2 (equivalent to powering an additional 2,813 homes and thus only 13% of the output linked with the original plan). The 14 that failed to materialise were refused due to their impact on landscape and views as well as the policy change which at that time was in the form of a Written Ministerial Statement (WMS). The decision letter providing the Secretary of State's refusal identified that:

> The overall benefits of the scheme would not be sufficient to also outweigh the harm identified to the character and appearance of the area, which attracts significant weight against the proposal [. . .] the proposal does not accord with the WMS and this attracts significant weight. (DCLG 2017)

This example shows clearly how national policy changes and aesthetic values concerning rural landscapes contrive to limit the size of onshore wind farms and the delivery of renewables that are critical for Manchester's sustainable and smart city ambitions. Indeed, while onshore renewables have a huge potential to enable cities to meet their smart and sustainable aims, thanks to policy changes and political opinions, obtaining consent is increasingly challenging. In effect, the speed of smart and sustainable city transitions is restricted by a convolution of national energy policy and social values concerning nature and landscapes. These are not, however, the only factors that limit the size and scale of wind farms and thus the broader scope of renewable generation to power the city, as sites are also impacted by time-limited planning consents and economic challenges.

3.8 Royd Moor Wind Farm: The Continuation of Ageing Infrastructure

Royd Moor wind farm (Figure 3.3), Barnsley (approximately 27 miles outside of Manchester City Centre), is one of England's oldest wind farms. Operational since 1993, it comprises 13 turbines, with a total output of 6 MW (enough to power 3,300 homes). Royd Moor demonstrates the complexities of making decisions on ageing infrastructure in a policy context that is not favourable to onshore wind farms, adding additional temporal complexity to considerations of how we power smart cities. As with most UK wind farms, the planning consent for Royd Moor was granted for 25 years with a requirement to decommission after this period, raising the question of how decisions should be made about the future and how a long-term supply of renewables for the smart city can be guaranteed. Decisions regarding the future of existing onshore renewable sites are not unproblematic and matter greatly for longer-term ambitions concerning decarbonisation and how quickly this will be achieved.

Figure 3.3 Royd Moor wind farm. *Source:* © Christine Johnstone.

Wind farm owners and operators typically face two options to continue energy generation from an existing site: to repower or life extension. Repowering involves decommissioning the existing infrastructure and replacing it with new technology. For onshore wind, this usually involves a smaller number of larger, more efficient turbines. At the time of writing in 2021, repowering the oldest UK wind farms has on average increased the maximum capacity of a site by 162%, while reducing the number of turbines by 41% and increasing the average height of the turbines by 98%. Crucially, this provides a key opportunity to significantly increase the energy output of an existing site. However, repowering is not easy in all locations. There are multiple contingencies, e.g. grid constraints, geographic challenges preventing larger turbines, changes in land designations and/or local opposition. Moreover, policy support for repowering in England was only introduced in 2018 and is creating challenges for decision-makers through a lack of detail (Windemer 2019).

Asset life extension is the alternative and lower-risk option for owners and operators. It involves increasing the duration of the original planning consent in order to keep the existing infrastructure in place. However, sites cannot keep extending life as the infrastructure will wear out and replacement parts will often no longer be available due to the significant changes in technology that have occurred in the past 20 years. Significantly, if existing sites do not repower, there is a risk that the supply of onshore wind energy will fall over time.

This lower-risk but problematic option was taken at Royd Moor. In 2017, permission was granted to extend the planning permission from 25 to 30 years (with decommissioning in

2023). In 2021, a further application was submitted to extend the duration of the existing wind farm again, for another 4 years, to end in 2027. Accordingly, the developer is likely to continue benefitting from a subsidy, which for many UK wind farms is due to end in 2027, meanwhile, the energy provided from the site is likely to keep decreasing due to the decreasing efficiency of the ageing infrastructure. Thus, over time, this site will be doing increasingly less to contribute to Manchester's renewable energy goals.

A review of wind farm applications in the surrounding area reveals why developers pursue such a lower-risk strategy, despite decreasing levels of energy output. Obtaining planning permission for onshore wind farms is challenging, particularly when taller turbines are involved, due to visual impacts on valued landscapes. Evidencing this, close to Royd Moor, a 2008 application for a new 15 MW wind farm – 'Sheephouse Heights' – with capacity to power almost 10,500 homes was refused. The reasons for refusal were fourfold. They included visual impact on the green belt, the impact on the character and appearance of the nearby Peak District National Park, the proximity to residential properties and harm caused to the characteristics of a local conservation area. Common throughout these reasons was the height of the turbines, something that is often non-negotiable if developers are going to ensure that a project is viable without subsidy. Notably, this refusal occurred before the 2015 policy U-turn on onshore wind, revealing the significance of visual impact even in a more favourable policy context. A repowering application for the nearby Royd Moor wind farm may attract similar objections as it is situated within a very similar landscape, particularly as significantly larger turbines would be required in order to make it viable. Life extension for Royd Moor avoids such opposition, while existing subsidies make this option financially lucrative.

The decision to extend the life of Royd Moor and thus hold in place ageing assets, with outdated generating capacity, is indicative of the types of choices that limit the scale and scope of renewable generation. Given the strong links between smart and sustainable city ambitions and the growth of renewable energy provision, such decisions are critical, if somewhat veiled, components of decarbonisation transitions in and beyond cities. As it stands, life extending arguably acts to slow sustainable transitions, with repowering promising to quicken such shifts.

The provision of renewable energy to a smart city is thus not as simple as generating the maximum energy from an area of land as political, economic and public attitudes all create complex challenges. The ability of the smart city to have a long-term supply of renewable energy is caught in these different temporal challenges of ageing infrastructure, changing policy contexts, changing site contexts and potentially changing preferences of local communities.

3.9 Addressing 'Hot Areas'

Thus far we have identified four different 'hot areas' where Manchester's smart sustainable futures are contested, configured and calibrated. As shown, the speed of broader smart sustainable urban transitions is informed by the types of socio-technical relationships discerned and discussed. In this section, we offer a set of short and speculative provocations in response to each hot area discussed. These aim at supporting and speeding up smart and sustainable shifts that align with the pressing need to achieve net zero in and beyond Manchester.

3.9.1 Speeding Up the Pace of Smart and Sustainable Ambitions

While Manchester's digitalisation and net-zero ambitions have become aligned, achieving these ambitions depends largely on the actions of a range of different interest groups (Cowley and Caprotti 2019), including (but not limited to) engineering and technology businesses, transport operators, property developers, renewable energy developers and investment firms. Success is also dependent on the actions of investors beyond the city region and on the policy decisions made at a national level.

In order to speed up the pace of change, there is a need for the city to develop a specific planning policy that prioritises developments that align with these goals. Due to the pace and scale of change required, the city may also benefit from developing partnerships with the private sector that facilitate investment in large-scale smart sustainable city projects such as local renewable energy development and storage and smart grid development. On a wider scale, there is also a pressing need to increase the intensity of lobbying aimed at the UK government to speed up the decarbonation of the national grid.

3.9.2 Re-Regulate Connection Charges

Changes to connection regulations, which offset costs relative to the sustainable value of a project, in essence representing a sustainability concession on buildings, could act to support rather than inhibit development. Sustainable developments are favoured through planning processes that call for and reward alignment with sustainable building standards (see Caird and Hallett 2019). Connection regulations could be organised to explicitly reward such good practices, making sustainable development and innovation attractive to developers. This would of course depend on a new rubric for calculating connection costs.

3.9.3 Only Provide for Clearly Sustainable Futures

The danger of predicting and providing for arguably unsustainable futures, through large-scale network investments, could be challenged by questioning and discarding calculative factors that do not clearly support sustainable transitions. The network would not be extended, in this case, in response to forecasted growth that does not clearly represent and support sustainable shifts. In the case of Manchester, economic and population growth, and the uptake of unsustainable technologies, would not be simply accepted and afforded by network extension, which in effect perpetuates unsustainable ways of life long into the future. This shift could involve a move away from current predict and provide models and the privileging of certain views of 'the' future. Instead, other, even more sustainable futures could also be considered, including those not dependent on cars. Such futures arguably deserve more attention and could be offered up to local stakeholders and publics for discussion and consideration.

3.9.4 Supporting a Reliable Long-Term Supply of Renewable Energy

Tracing the wires out from Central Manchester to nearby wind farms has shown that achieving an adequate, reliable supply of renewable energy to supply a smart city is not straightforward. Obtaining consent for onshore renewables involves a careful management of relationships and systems far beyond the city boundary. Here, speculative suggestions involve changes in regulation, policy and practice, which clearly have social and material roots that extend beyond any city but could very well speed up positive moves therein. A longer-term approach is needed in order to make the most out of existing renewable energy sites and technology, e.g. through increasing generation capacity or through battery storage. In order to achieve this, there must be a supportive policy in place both for new and repowered onshore renewable sites, and a quicker, adequately resourced planning system to prevent delays.

The long-term supply of renewables also requires diversification through planning both within and outside of the city. Part of this approach should involve ensuring that opportunities to locate renewables within the city, e.g. through rooftop solar, are maximised. More broadly, there are other barriers to ensuring that there is an adequate supply of renewable energy that also need addressing such as grid capacity, financial support and incentives and gaining community support.

3.10 Conclusions

Smart sustainable urban futures are contested, configured and calibrated in various hot areas simultaneously. Local political policies and initiatives are important. So too are those to do with the ordering and management of electricity generation and supply, with the electricity system forming an obligatory point of passage for smart sustainable urban futures. Recognising this to be the case, we adopted an oligoptic approach that involved tracing out wired connections and identifying different, yet connected, hot areas, in particular to do with electricity, where Manchester's smart sustainable ambitions are contested, configured and calibrated. Having identified such areas, we articulated a set of speculative provocations aimed at speeding up smart and sustainable transitions across the city.

Other hot areas, not discussed in this chapter, lie along the electricity network. Defining such points of contingency, alongside producing more and less speculative responses, represents a means of further accelerating smart sustainable transitions. Other hot areas undoubtedly also lie across other socio-technical networks and obligatory points of passage. These include, inter alia, those to do with the provision of gas, telecommunications, water and mobility infrastructures.

The hotspots that lie across these socio-technical networks could be identified using the approach taken in this chapter. This would involve following other wires, pipes, sewers and roads in and out of cities, as well as visiting junction boxes, data centres, transport intersections and reservoirs. It would also involve asking what is happening at particular sites along the way and specifically what the implications are for attempts to bring about more sustainable futures in and indeed beyond cities. Different hot areas not only necessitate different responses but they also become sites for political contests, creating political effects that might influence change and smart city transitions. The chapter shows that tracing out wired socio-technical relationships and connections can benefit future research into understanding smart cities infrastructure decision-making processes, including the power and influence of different actors.

References

Angelidou, M. (2014). Smart city policies: a spatial approach. *Cities* 41: 3–11.

Angelo, H. and Vormann, B. (2018). Long waves of urban reform: putting the smart city in its place. *City* 22: 782–800.

Ask Real Estate (2021). Plot 9 first street: Manchester city centre. https://www.askrealestate.co.uk/projects/plot-9-first-street/ (accessed 1 March 2021).

Behrendt, F. (2019). Cycling the smart and sustainable city: analyzing EC policy documents on Internet of Things, mobility and transport, and smart cities. *Sustainability* 11 (3): 763. https://doi.org/10.3390/su11030763.

Bennetts Associates (2009). Mayfield strategic regeneration framework draft revision A. https://www.bennettsassociates.com/assets/lib/2017/04/21/1211_Mayfield-SRF.pdf (accessed 5 October 2017).

Beretta, I. (2018). The social effects of eco-innovations in Italian smart cities. *Cities* 72: 115–121. https://doi.org/10.1016/j.cities.2017.07.010.

Caird, S.P. and Hallett, S.H. (2019). Towards evaluation design for smart city development. *Journal of Urban Design* 24 (2): 188–209.

Callon, M. (1984). Some elements of a sociology of translation: domestication of the scallops and the fishermen of St Brieuc Bay. *The Sociological Review* 32 (1): 196–233. https://doi.org/10.1111/j.1467-954X.1984.tb00113.x.

Caragliu, A. and Del Bo, C.F. (2019). Smart innovative cities: the impact of smart city policies on urban innovation. *Technological Forecasting & Social Change* 142: 373–383.

Carter, D. (2013). Urban regeneration, digital development strategies and the knowledge economy: Manchester case study. *Journal of the Knowledge Economy* 4 (2): 169–189.

Cowley, R. and Caprotti, F. (2019). Smart city as anti-planning in the UK. *Environment and Planning D: Society and Space* 37 (3): 428–448.

Cowley, R., Joss, S., and Dayot, Y. (2018). The smart city and its publics: insights from across six UK cities. *Urban Research & Practice* 11 (1): 53–77.

Deloitte (2021). Maintaining momentum: Manchester crane survey 2021. https://www2.deloitte.com/content/dam/Deloitte/uk/Documents/realestate/deloitte-uk-manchester-crane-survey-2021.pdf (accessed 1 February 2021).

Department for Communities and Local Government (DCLG) (2017). Final decision letter for Scout Moor wind farm. London: UK Government.

Department of Trade and Industry (2002). The electricity (connection charges) regulations 2002 (ECCR). http://www.legislation.gov.uk/uksi/2002/93/pdfs/uksi_20020093_en.pdf (accessed 30 July 2017).

Dowling, R., McGuirk, P., and Maalsen, S. (2018). Multiscalar governance of urban energy transitions in Australia: the cases of Sydney and Melbourne. *Energy Research & Social Science* 44: 260–267.

ENW (2018). Demand scenarios and innovation projects at electricity north west. https://www.enwl.co.uk/globalassets/get-connected/network-information/dfes/distribution-future-electricity-scenarios-and-regional-insights-november-2018.pdf (accessed 16 August 2018).

ENW (2021a). Electric vehicle strategy: April 2021. https://www.enwl.co.uk/globalassets/go-net-zero/net-zero/ev-strategy/electric-vehicle-strategy-2021.pdf (accessed 9 March 2021).

ENW (2021b). Our HeEat map tool. https://www.enwl.co.uk/get-connected/network-information/heatmap-tool/ (accessed 5 March 2021).

Europe Real Estate (2020). Smart spaces to deliver Manchester's first smart building (GB). https://europe-re.com/smart-spaces-to-deliver-manchester-s-first-smart-building-gb/67983 (accessed 9 March 2021).

Foxon, T.J., Hammond, G.P., and Pearson, P.J.G. (2010). Developing transition pathways for a low carbon electricity system in the UK. *Technological Forecasting & Social Change* 77 (8): 1203–1213.

Hollands, R.G. (2008). Will the real smart city please stand up? Intelligent, progressive or entrepreneurial? *City* 12 (3): 303–320.

Hommels, A. (2005). *Unbuilding Cities. Obduracy in Urban Sociotechnical Change.* London and Cambridge, MA: The MIT Press.

Hughes, T. (1983). *Networks of Power: Electrification in Western Society 1880–1930.* Baltimore, MD: Johns Hopkins University Press.

Hughes, T. (1986). The seamless web: technology, science, etcetera, etcetera. *Social Studies of Science* 16 (2): 281–292.

Hughes, T.P. (1989). The evolution of large technological systems. In: Hughes T.P., Bijker W.E., Pinch T. (Eds.), The Social Construction of Technological Systems: New Directions in the Sociology and History of Technology, The MIT Press, pp. 51–82. http://www.jstor.org/stable/j.ctt5vjrsq.9

IQS (2017). Central Manchester primary (travis street) substation. http://iqs.co.uk/32/ (accessed 5 October 2017).

Kandt, J. and Batty, M. (2021). Smart cities, big data and urban policy: towards urban analytics for the long run. *Cities* 109: 102992. https://doi-org.ezproxy.lancs.ac.uk/10.1016/j.cities.2020.102992.

Kjellberg, H. (2008). Market practices and over-consumption. *Consumption Markets & Culture* 11 (2): 151–167.

Latour, B. (2005). *Reassembling the Social: An Introduction to Actor-Network-Theory.* Oxford: Oxford University Press.

Lee, D.B., Klein, L.A., and Camus, G. (1999). Induced traffic and induced demand. *Transportation Research Record* 1659 (1): 68–75. https://doi.org/10.3141/1659-09.

Manchester City Council (2021). Manchester's digital strategy: creating an inclusive, sustainable and resilient smart city (draft). https://www.manchester.gov.uk/downloads/download/7344/manchesters_digital_strategy (accessed 1 March 2022).

Manchester Climate Change Agency (2020). Manchester Climate Change Framework 2020–25 Manchester Climate Change Agency. https://www.manchesterclimate.com/sites/default/files/Manchester%20Climate%20Change%20Framework%202020-25.pdf (accessed 14 July 2023).

Meadowcroft, J. (2016). Let's get this transition moving! *Canadian Public Policy* 42 (S1): S10–S17.

Prolific North (2020). Manchester now second only to London when it comes to tech investment. https://www.prolificnorth.co.uk/news/tech-news/2020/11/manchester-now-second-only-london-when-it-comes-tech-investment (accessed 15 February 2022).

Rutherford, J. and Coutard, O. (2014). Urban energy transitions: places, processes and politics of socio-technical change. *Urban Studies* 51 (7): 1353–1377. https://doi.org/10.1177/0042098013500090.

Smart Space (2021). Smart Spaces to Deliver Manchester's First Smart Building at Ask Real Estate's First Street. https://www.smartspaces.app/ (accessed 9 March 2021).

Söderström, O., Paasche, T., and Klauser, F. (2014). Smart cities as corporate storytelling. *City* 18 (3): 307–320.

Star, S.L. (1999). The ethnography of infrastructure. *American Behavioral Scientist* 43 (3): 377–391. https://doi.org/10.1177/00027649921955326.

Summerton, J. and Coutard, O. (1994). Social shaping in large technical systems: an interview with Jane Summerton. *Flux* 10 (17): 54–56.

Thomas, C.E.S. (2012). How green are electric vehicles? *International Journal of Hydrogen Energy* 37 (7): 6053–6062. https://doi.org/10.1016/j.ijhydene.2011.12.118.

Trindade, E.P., Hinnig, M.P.F., da Costa, E.M. et al. (2017). Sustainable development of smart cities: a systematic review of the literature. *Journal of Open Innovation* 3: 11. https://doi.org/10.1186/s40852-017-0063-2.

Uspenskaia, D., Specht, K., Kondziella, H., and Bruckner, T. (2021). Challenges and barriers for net-zero/positive energy buildings and districts—empirical evidence from the smart city project SPARCS. *Buildings* 11 (2): 1–24. https://doi.org/10.3390/buildings11020078.

Van der Horst, D. and Toke, D. (2010). Exploring the landscape of wind farm developments; local area characteristics and planning process outcomes in rural England. *Land Use Policy* 27 (2): 214–221.

Verbong, G.P.J. and Geels, F.W. (2010). Exploring sustainability transitions in the electricity sector with socio-technical pathways. *Technological Forecasting & Social Change* 77 (8): 1214–1221.

Walker, G., Shove, E., and Brown, S. (2014). How does air conditioning become 'needed'? A case study of routes, rationales and dynamics. *Energy Research & Social Science* 4: 1–9.

Windemer, R. (2019). Considering time in land use planning: an assessment of end-of-life decision making for commercially managed onshore wind schemes. *Land Use Policy* 87.

4

Smart Energy Cities: A Perspective from West Africa

Charlotte Ray[1], Sam Williamson[1], Zuzana Hrdličková[4], Derrick Kajjoba[2], Hillary Kasedde[2], Lauren Hermanus[3], Amadu Labor[4], Joseph Macarthy[4], and Braima Koroma[4]

[1] *School of Electrical, Electronic and Mechanical Engineering, Cabot Institute for the Environment, University of Bristol, Bristol, UK*
[2] *Makerere University, Kampala, Uganda*
[3] *African Centre for Cities, University of Cape Town, Cape Town, South Africa*
[4] *Sierra Leone Urban Research Centre, Freetown, Sierra Leone*

4.1 Introduction – Smart Cities: An Urban Panacea?

The smart city concept has come under significant criticism, particularly with respect to its relevance to developing country contexts, and African urban centres, specifically. This technology-driven approach to solving complex societal challenges has, at times, been hailed as a catch-all panacea to urban issues. What has been identified as a kind of techno-determinism assumes that the correct application of the right technologies will necessarily lead to 'better' development outcomes, with clear pathways from implementation to outcomes (Odendaal 2021). In its extreme expression, the combined use of data and technology to improve services, transport, employment, and planning is often presented as a requirement for development and modernisation (Watson 2014).

There is no single universally accepted definition for a smart city (Cavada et al. 2014; Hammons and Myers 2019), allowing scholars, politicians, and businesses to define their own interpretation of the term, often to pursue their own agenda. A typical definition incorporates the ideas of data collection and processing to improve the efficiency of urban processes, together with the concepts of sustainability and good quality of life (Calvillo et al. 2016), such as the following from the International Telecommunication Union, 'an innovative city that uses Information and Communication Technologies (ICTs) and other means to improve the quality of life, efficiency of urban operation and services and competitiveness. . .' (ITU 2014). It is also recognised that to achieve a truly smart city, the city must upgrade and update all systems simultaneously in order to cope with the incoming challenges (Morvaj et al. 2011), placing a large infrastructure and financial burden on the city governance for the transition. Data are the backbone of the smart city concept,

Smart Cities, Energy and Climate: Governing Cities for a Low-Carbon Future, First Edition.
Edited by Oleg Golubchikov and Komali Yenneti.

providing evidence for decision-making and different temporal models of urban systems (Batty 2013). However, these definitions often do not highlight who the beneficiaries of the smart city paradigm are, often leaning on technological innovations rather than taking a more people-centric approach. Literature has increasingly recognised a societal dimension where there is a strong sustainability component (Toli and Murtagh 2020), with a grass-roots, bottom-up drive required for successful roll-out (Schaffers et al. 2012).

Smart cities are presented by policymakers as the solution to rapid urbanisation growth (Fernandez-Anez et al. 2018). There are several benefits identified by smart city scholars, including planning and managing increasingly complex and interdependent urban networks, rapid change mapping utilising ground-up sensing technology, understanding travel behaviour, improving the structure and usage of urban land for individuals and organisations, developing flexible commercial marketplaces, and developing social networks to build communities (Batty et al. 2012). It must be noted that often evidence for these benefits is not clear within the literature, however, real care needs to be taken to ensure that all stakeholders are brought along on the smart journey, leaving no group behind.

In sub-Saharan Africa, 40% of the population lives below the poverty line (Schoch and Lakner 2020), over half of the urban population resides in informal settlements (World Bank n.d.) and less than half of the population completes their secondary education (UNESCO 2019). Access to electricity and ability to pay for electricity in informal settlements is often lower than the urban average (Kovacic et al. 2016; Lemanski 2009). Priorities are therefore often different from the majority of the populations in the Global North. Access to basic services – food security, shelter, energy, water, and sanitation – as well as education, healthcare, and regular employment, are often the key priorities for the urban poor (Mitlin and Satterthwaite 2012). As with many regions in the Global South, access to financial services is difficult for the urban poor, causing informal and exploitative practices to occur. However, the recent rise of mobile financial transactions has enabled financial inclusion for these populations, opening the door to more regulated financial services (Ky et al. 2021).

Urban planning requirements in sub-Saharan African cities often follow similar priorities. As an example, Freetown City Council, the municipal government of Sierra Leone's capital city and largest urban area, has identified four main priorities: resilience, human development, healthy city, and urban mobility (FCC 2019). The plan is to make the city more resilient through better environmental management, revenue mobilisation, and urban planning and housing. Increased revenue started to be collected and over 90% of new buildings need to abide by appropriate regulations. Human development is to be supported through education, skills development, job creation, and support for people with disabilities. The city also plans to decrease maternal mortality, improve access to water supply and waste disposal, and promote recycling of plastic waste. The city is also taking steps to decrease traffic congestion at five key locations in the city. These requirements do not immediately appear to fit with the initial ethos of a smart city. However, the backbone to achieving these priorities is through data- and evidence-based decision-making, which arguably fits directly in the remit of a smart city.

Scholars such as Watson (2014, 2015), Datta (2015, 2018, 2019), and Odendaal (2021) have previously argued that the smart city narrative, especially in a Global South context, is largely based on the premise of urban regeneration and prosperity with the use of ICT to address urban problems including governance, communication, and infrastructure.

Indeed, Datta (2015) has interestingly highlighted the changing semantics of the urban space using terms such as 'knowledge city', 'intelligent city', 'global city', 'eco city', and more recently, 'smart city' which in turn has translated into urban spaces that operate as standalone entities. As such, there are already a plethora of examples of 'smart city' initiatives on the African continent, and calls for Africans to 'construct a "new middle ground" of liveable urban spaces' (Slavova and Okwechime 2016, p. 15) ring true throughout the continent. Purported to be 'the key to Africa's third revolution' (Bearing Point Institute 2016), the smart city concept has captured the hearts and imagination of both public policymakers and the growing urban middle class (Watson 2014).

These initiatives come in a variety of forms starting from enthusiastic political promises from elected officials to entire city regeneration plans such as the 'Smart City Masterplan' in Kigali, Rwanda. Indeed, many African countries have already developed, deployed, and embraced 'smart' technologies into their existing infrastructure. M-Pesa in Kenya is the most notable example of how mobile money and payments have been transformed across the continent from a simple communication tool to a financial necessity (Guma 2019; Poggiali 2017). Other examples include the creation of technology parks or centres such as the Cabo Verde Technology Park, or fully fledged standalone satellite cities, such as Eko Atlantic City in Nigeria, Konzo City in Kenya, and Waterfall City in South Africa, as a metropolis of technocratic innovation to encourage economic growth and development (Solomon and van Klyton 2020).

The idea of smart city in Africa has also attracted interest among development agencies. For example, in 2019, the French Development Agency funded the African Smart Town Networks project supporting the digital transition of African local authorities, such as e-administration and digitisation of urban services such as mobility and waste control (Haushofer 2019). Multi-nationals are also providing the drive to more smart infrastructure in Africa. Huawei is focussing on setting up Internet of Things, communication, and intelligent video surveillance for African smart cities (Takouleu 2018).

The reality, however, of the smart city paradigm has not always lived up to its promising expectations. Many technology parks and satellite city plans rely on the need for foreign investments which does little to ensure sustainability (Tavares 2009) and plans often fail to materialise, such as Hope City, Ghana, due to reasons such as finance, infrastructure, contractual problems, and low energy supply (TechNova 2020; The Tech Observer 2016). Too often in these debates around smart cities and sustainability, the concept often ignores the human and social dimension of 'smart', excluding the role of human agency in the creation of urban spaces (Aurigi and Odendaal 2021; Watson 2014) when in fact they need to be integrated. Indeed, it is 'the role of social capital and networks of trust and reciprocity that are prerequisites for innovation' (ibid, p. 225). On a simplistic level, it is impossible to not consider the human dimension in the smart city rhetoric, especially as Africa has a high youth population who have accepted, invested in, and even designed elements of smart technology (Echendu and Okafor 2021). At the same time, it is impossible to ignore the notion that the smarter a system becomes, the less it is able to interact with the very people it is designed for, not only highlighting the need for flexibility and adaptation (Deville et al. 2014) but also recognising the potential to widen the inequality gaps in some of these spaces.

Energy is a key service within a city and has been shown to be a fundamental component in delivery of the Sustainable Development Goals (McCollum et al. 2018). Energy is also key to enabling the smart city concept, powering the sensing, ICT, and interaction required from stakeholders. In sub-Saharan Africa, with 24% of the urban population unable to access electricity (IEA 2020b) and only 17% of the total population accessing clean cooking facilities (IEA 2019), energy is a key service to support improving quality of life. This article, therefore, discusses the smart city concept through the energy lens to understand how the smart city concept interfaces with the sustainable energy transition within cities in sub-Saharan Africa and how policymakers and urban populations, especially those living on the fringes such as those living in informal settlements/slums, engage and interact with the smart city narrative (if they do at all). We do not seek to position ourselves as 'for' or 'against' the smart city paradigm, rather ensuring that key human and socio-economic dimensions are recognised to ensure that the smart city paradigm does not widen the inequality gap in these cities.

4.2 Smart Energy City in an African Context

With the drawbacks of a whole smart city in mind, Odendaal (2021) and Aurigi and Odendaal (2021) highlight that the 'smart city in a box' concept is a simplified, one-size-fits-all narrative of what a city should/could look like. As they suggest, perhaps the phrase 'smart city out of the box' should be used to acknowledge and incorporate digital technologies into complex 'context-specific city making'. As such, the role of energy is subject not only to its own sustainable development goal (SDG 7) but is also seen as critical to the success of the smart city (Haarstad and Wathne 2019), especially as energy and the environment are key drivers for a smart city (Lai et al. 2020). As such the concept of 'Smart Energy Cities' (SEC) can be broadly defined as a 'component of smart city development aiming at a site-specific continuous transition towards sustainability, self-sufficiency, and resilience of energy systems while ensuring accessibility, affordability and adequacy of energy services. . .' (Mosannenzadeh et al. 2017, p. 57). To achieve the transition to smart energy cities, especially through a net-zero emission lens, it requires all actors (governments, energy companies, investors, and citizens) to have a clear vision of how the transition might be achieved. Technologies need to be researched and developed, policies need to be adjusted, business operations need to evolve, and exponential rates of change must be realised (Teske et al. 2020). Therefore, in a smart city, it is very important to increase energy efficiency and decrease total energy consumption while simultaneously maintaining an open and fair internal energy market. Furthermore, it should be seemingly capable of solving complex social issues through algorithms and technical innovation (Odendaal 2021).

The development of a smart city energy system requires the interaction of intervention areas such as generation, storage, infrastructure, facilities, and transport/mobility. These factors relate to a smart energy city development where generation provides energy, storage secures the energy availability, infrastructure provides the distribution of energy and the required information and communication technologies, while the facilities and transport are the major energy consumers of energy since they need it to operate (Calvillo Munoz 2017). The concept of SECs relies majorly on renewable energy sources (RES). Energy sources such as wind power and photovoltaic (PV) are expected to be important components in this case. The potential for biomass for bioenergy is important but limited considering its other potential uses such as food and materials, as well as the use of land for agriculture, etc. (Bentsen and Felby 2012; Hamelin et al. 2019). A smart energy system is defined as an approach in which energy grids (electricity, gas, and heat) are combined with storage technologies and coordinated to identify synergies between them in order to achieve an optimal solution for each individual sector as well as for the overall energy system (Connolly et al. 2013). In addition, all the above-mentioned fuel-consuming sectors must interact together and communicate with each other in order to better balance demand and supply and contribute to the system's stability in a smart energy system approach (Lund and Mathiesen 2009). Indeed, the concept of an SEC historically represents a dependence on technology (Thornbush and Golubchikov 2021) and does not necessarily account for the very citizens it is intended to serve. As Golubchikov (2020) has suggested, 'people-smart' cities allow for full participation in urban life.

There are already examples of adoption of smart technologies to address current energy challenges in African cities. For example, Guma (2019) has documented examples of prepaid electricity meters for Kenyan consumers to manage their credit and consumption, as well as the interface between energy and telecommunication whereby mobile-based services have emerged for Kenyan consumers to pay for and settle electricity bills. Other smart meter examples include Ghana and Nigeria (Kambule and Nwulu 2021), South Africa (Nwaiwu 2021), and Kenya (Muigua 2021) to name a few. At the same time, the introduction and adaptation of smart grids are under development and interrogation in countries such as Ghana (Appiah Addae et al. 2019; Nartey 2016), Morocco (Meliani et al. 2021), and South Africa (Masembe 2015). Other examples include smart sensors for energy efficiency in South Africa and Kenya (Outside Insight 2020) and solar mapping in Durban, South Africa, to gather information about power generation and electricity savings (Sustainable Energy Africa 2017).

The move towards the realisation of smart energy cities in an African context is faced with five key barriers: financial, market, policy, technology, and social (Appiah Addae et al. 2019). Moreover, besides the identification of such barriers, there is need to organise these barriers and prioritise them to enhance strategic decision-making. The technology barriers to be overcome mainly dwell on aspects of the nature of new technology, cost of technology, lack of infrastructure, stakeholder involvement, legislative framework of the technology, access to the technology, and presence of skilled personnel for deployment of the technology (ibid).

Energy is therefore central not only to the creation and operation of smart cities but also meeting the everyday energy needs of people. What is largely missing from the current literature, and indeed warrants greater research, is the energy (generation, storage, and demand) needed for smart cities to succeed. What are the power requirements of creating and operating technoparks and satellite cities and how does that impact energy access in cities, especially those communities on the fringes of these spaces? At the same time, there is still a technocentric approach to an SEC and a distinct lack of interaction with more user-based priorities, demands, and aspirations (McIlvennie et al. 2020), which highlights uncertainties on whether these technologies will actually make a positive impact in the Global South (WEF 2019). This also aligns with broader energy access literature, especially the energy for cooking literature, that suggests that smarter, efficient, and cleaner technologies are often out of reach to the beneficiaries they are intended to serve (Fuso Nerini et al. 2017; Sharma 2020) and are more exclusionary rather than inclusionary (Cloke et al. 2017). However, concepts such as a commons-based energy system[1] may enable a more inclusive smart energy system (Giotitsas et al. 2020).

With this in mind, what does the enabling environment look if African nations wanted to adopt the smart energy city narrative? What policies, technologies, and governance mechanisms are in place to drive the SEC concept from idea to reality, and how ultimately might it impact the very communities it intends to serve?

4.3 Current Policy Environment around Smart Cities and SECs in West Africa

Notwithstanding that the definition of smart cities and smart energy cities vary depending on the context, stakeholder involvement, existing infrastructure, policy, governance mechanisms, and technology, they all require an enabling policy, regulatory, and investment environment to ensure their successful deployment. This environment will then impact how African cities are able to utilise the smart city narrative and adopt smart energy solutions to meet the needs of urban populations. Importantly, it will then help to develop an understanding of how these smart technologies interface with the very people they are intended to serve and how does that compare with a mainly Global North narrative on what a smart city should be?

Given the current drive towards net-zero greenhouse gas emissions, there is limited research linking smart city policies with the net-zero agenda beyond the examination of zero-carbon urban zones (e.g. Uspenskaia et al. 2021) and the recognition that smart planning and systems can support a reduction in emissions (e.g. Calvillo et al. 2016). In sub-Saharan Africa, the net-zero agenda is often seen as a much lower priority, behind improving human development, resilience, and health, especially as a region it contributes only 4% of global emissions, half of which comes from one country – South Africa (Thomson Reuters 2020). Therefore, energy policy in the region is directly often targeting access to energy to improve quality of life than directly considering environmental issues. (As an example, following the publication of the IEA's Roadmap to Net-Zero [IEA 2021], the Africa Energy Chamber – a regional energy industry body – pushed back on the suggestion of no new fossil fuel investments after 2021, and that only through fossil fuel-based solutions would allow African nations to reach their SDG targets.)

Referring to literature in the West African context, with particular attention paid to Sierra Leone, we analyse the current energy policy and investment environment within this region and start to unpack what the implications of these will be to support smart regional and national energy portfolios.

1 This system would involve a more bottom-up approach to energy generation and system management, operating for the service of people rather than profit maximisation, therefore removing the complex market structures currently built around energy systems. It would require smart energy management systems to maintain the balance of supply and demand, but as it would be operating as social service, it would enable better integration with consumers.

4.3.1 Regional Efforts in West Africa

While North African countries have nearly achieved universal access to clean cooking and electricity, most of the rest of the continent is unlikely to achieve SDG 7 by 2030 (United Nations 2019). Between 2014 and 2019, electricity access in sub-Saharan Africa[2] had doubled to almost 20 million and had started to outpace population growth (IEA 2020a)[3]. However, population growth continues to outpace the annual increase of people with access to clean cooking (SEforAll 2020), and although electricity access is increasing in rural areas, population growth in urban cities is outpacing the rate of growth of energy access (IEA 2020b). According to UNECA, to achieve improved energy access, an investment of $35 billion a year is required along with creation of an enabling environment for investment from the private sector promoting the increased penetration of variable renewable power. Countries should encourage innovation and improve their capacity in areas of energy research, energy planning, and management. Their infrastructure planning should integrate climate resilience strategies. And they should synergise, share, and coordinate good practices concerning on- and off-grid energy programmes. Another key priority is improvement of energy efficiency overall and of the electrical distribution grid (UNECA 2018).

Taking a closer look at West Africa, the Economic Community of Western African States (ECOWAS) region has been facing interrelated challenges of climate change, energy access, and energy security. The reality on the ground is that for most citizens of West Africa, clean energy is inaccessible, unavailable, and unaffordable. In 2019, only 56% people of the region had access to electricity, with a rate of 28% in rural areas (IEA 2020b), and only 12% have access to clean cooking solutions (IEA 2019). The infrastructure to meet demand is lacking. The region is heavily dependent on traditional biomass, mainly wood fuel, leading to deforestation and threatening livelihoods, as well as sustainable environment of the region (Olorunfemi et al. 2021). In 2010, the ECOWAS Centre for Renewable Energy and Energy Efficiency (ECREEE) was established, and it has formulated two policies in the area of energy access: the ECOWAS Renewable Energy Plan (ECOWAS 2013b) and the ECOWAS Energy Efficiency Policy (ECOWAS 2013a). These policies feed down into the member nations' renewable energy and energy efficiency plans, aiming to increase the renewable energy share, develop appropriate regulatory frameworks for renewable technologies, reduce electrical distribution losses, and phase out inefficient incandescent bulbs. The policies have the ambition to achieve universal access to electricity and clean cooking. ECOWAS also aims to address the gendered differences in energy access in their Policy for Gender Mainstreaming in Energy Access (ECOWAS 2017).[4] However, their monitoring

---

2 It is important to note that these figures relate to a relatively small subset of African countries such as Kenya, Rwanda, Ethiopia, Ghana, and Senegal.

3 The same IEA reports estimate that as a result of the global COVID-19 pandemic, there will be an increase, for the first time since 2013, in the number of people in the global population without access to electricity.

4 The aim of this policy is to enhance equal access to modern energy services by all, effectively seeing it as a right, regardless of socio-economic status, age, or gender. This is achieved by accelerating adoption of new sustainable energy practices, introducing legislation ensuring gender equality in energy access, and increasing equal participation of men and women in energy value chains, providing them equal opportunities to contribute to all economic levels (ECOWAS 2017, pp. 15–16).

reports (ECREEE 2018, 2019) reveal that the progress is nowhere near rapid enough to achieve the stated goals, and progress monitoring in itself is difficult due to inadequate data collection practices. Despite these challenges, ECOWAS countries take part in multinational energy access projects, such as the regional electricity access and battery energy storage systems project (The World Bank 2019). To understand how these regional policies and plans are followed through, we consider the case of Sierra Leone.

4.3.2 Case Study Nation: Sierra Leone

Sierra Leone is among the least developed countries in the world, with high poverty levels and significant inequalities (UNDP 2012). Country statistics on urban electricity access vary greatly ranging from 10% to 53% and it is often hard to determine which figures are the most accurate. However, what is clear is that the current situation of electricity supply is unsustainable (Ministry of Energy 2017). Recently, there has been a political push for access to electricity in health centres, especially in rural areas, in light of the Ebola outbreak (Liu and Bah 2021), and the government of Sierra Leone has noted the huge potential for generation of renewable energy which has so far been untapped due to a number of reasons including lack of government direction, financial resources, and poor R&D of renewable energy technologies. The country is aiming to make progress, with their National Renewable Energy Policy (Ministry of Energy 2016) vision being: *'An energy economy in which renewable energy increases its share in the energy mix, is reliable and affordable throughout Sierra Leone, thus contributing to sustainable development and environmental conservation'.*

As one Ministry of Energy official put it, '*the government cannot do it alone*'. The 2016 policy provides a blueprint for further private investment in research and development driving renewable energy solutions including hydropower[5], bioenergy, and solar and wind energy, on- and off-grid. Two notable successes from this policy are the energy revolution taskforce, enabling private sector investment in the decentralised renewable energy market, providing policy support and regulation alignment (Power For All 2018), and the recently completed 6 MW Newton Solar Park near Freetown (IRENA and ADFD 2020). Although these outcomes are very positive, Sierra Leone still has a long way to go in order to provide the remaining three-quarters of its population with electricity access. *Sustainable energy for all* has made a number of key recommendations around policy, tariffs, and subsidies, along with the productive end use of electricity and site selection (SEforALL 2021), to make further progress on the increase in electricity access, especially in rural areas. Recent analysis has also indicated that the state-owned model of energy utilities presents bias towards politically motivated decision-making and the risk of key stakeholders (such as private sector actors and community groups) being excluded from decision-making processes (Hirmer et al. 2021). It is also important to note here that, as is often the case with energy access, the elephant in the room is that of clean cooking, with over 90% of the population cooking using biomass-based solutions (Muchiri 2019).

5 It has previously been estimated that the potential for hydropower in West Africa is mainly concentrated in five countries, with Sierra Leone being one of them (IAEA 2016).

Recognising that digitalisation can support Sierra Leone's targets of improving development indices and their achievement of the SDG targets, in 2018 the government of Sierra Leone set up the Directorate of Science, Technology and Innovation (DSTI). The directorate published the National Innovation and Digital Strategy in 2019, in line with policies from the African Union, ECOWAS, and similar countries, with a strapline of 'Digitalisation for All' (DSTI 2019). The policy includes all members of society through three principles: utilising mobile technology to enable the benefits of digitalisation, developing the country as an artificial intelligent laboratory with local academics and private sector organisations, and building optimised hybrid systems (e.g. on- and off-grid electrification solutions) to achieve universal service access. This policy has made some strong initial progress, with an Educational Data Hub (DSTI 2019) to support education decision-making, a blockchain-based database to enable access to finance (Reuters 2019), and the prospective use of drones to strengthen medicine delivery to remote inaccessible areas (DSTI 2020). The Sierra Leonean state is also making strides towards becoming 'smart' in the conventional meaning by setting up the National Civil Registration Authority, which has conducted repetitive blanket registration exercises for all its citizens[6] consolidating citizen data in one digital system (National Civil Registration Authority 2020; Sierra Leone Telegraph 2021; Thomas 2021).

The literature on mobile phone penetration in sub-Saharan Africa is rapidly growing (Mignamissi and Djijo T 2021; Porter 2015; Porter et al. 2012) and there are numerous debates around the impact of mobile phone use, especially with regard to income and inequality (Asongu 2013; Asongu and Odhiambo 2019). Importantly, the use of mobile phones has allowed a nexus between top-down and bottom-up development that has provided populations with additional opportunities to communicate, streamline supply chains, bridge financial gaps, and interestingly, potentially increase access to energy services. As such, we are already aware of the importance of the mobile phone. The Sierra Leonean government is too aware of this and has highlighted 'The singular technology that connects all Sierra Leoneans is mobile technology. As such, at the heart of this strategy is the development and availability of services that are optimized for mobile devices both in the public and private sector' (DSTI 2019). There is a case for energy pay-as-you-go solutions offering financing for energy products and paying for energy utilities (GSMA 2014), but a previous report by Lighting Global has analysed that Sierra Leone has one of the weakest Pay Go markets (IFC 2019; WEF 2019), attributing poor access to finance for commercial entities which could suggest that the use of smart applications (such as mobile money) to pay for or finance energy services has significant way to go.

4.3.3 The Role of External Investors and African Diaspora in Building Future Cities in Africa

Within the smart city narrative, it is clear that large investments in infrastructure are needed to respond to rapid urbanisation and the demands for urban accessibility (Mboup 2019). There are clear policy drives, especially in the energy sector and for

6 And foreign residents in 2017 as well as in 2021.

promoting renewables, but often the high targets set are not always matched with the investment required to drive these policies forward (Pueyo 2018). Indeed, the informal economy which serves the majority of urban populations in many African cities is often supplied with services that have minimal investment and minimal public support, e.g. transportation and energy (Mboup 2019), and so the role of national actors is siloed into 'translator roles' tasked with implementing and evaluating smart city initiatives which are often led by external and/or private sector investors (Söderström et al. 2021). Khan et al. (2021) have documented that foreign direct investment in the energy sector has led to greater private sector investment. The long-term strategy therefore lies within the public–private partnerships that are key within a competitive global market (Bélissent 2010; Dupar et al. 2021). More effective governance and policy implementation may indeed lower the investment requirements needed from external actors (Dagnachew 2021), and arguments have been made for the possibility of complementarity whereby external stakeholders invest to further their environmental goals as well as address high priorities for policy stakeholders (e.g. increasing electricity access) (Baker et al. 2021). However, in practice, local assets have often been crowded out by foreign investment (Rahman 2021), with trade-offs made among priorities of external investors, state actors, and the urban populations whom these technologies and initiatives are intended to serve. These trade-offs are entrenched in neo-liberal, colonial, and patriarchal systems (Dunlap and Arce 2021; Khodadadzadeh 2017; Ojong 2021).

At the same time, transnational theory recognises that migrants maintain important links to their countries of origin (Horevitz 2009), which often include acts of philanthropy, such as contributions towards various development and building projects, as well as donating their own time, knowledge, and skills (Newland et al. 2010). The role of the diaspora is considered important for social, political, and economic development (Magocha 2020) and often key figures in the diaspora serve as leading members of the community at home. As an example, Sierra Leonean diaspora Yvonne Aki-Sawyer has become Freetown City Mayor, whose family is otherwise based in the United Kingdom. At the same time, the African Diaspora Development Institute (ADDI) has leveraged the popularity of the Black Panther film and the vision of Wakanda, the futuristic African city within the film, to propose a number of Wakanda-inspired city destinations to encourage diaspora to return to Africa and contribute to its growth (The Washington Informer 2018; African Vibes 2020) evoking notions of pilgrimage, identity, and heritage (Travel Noire 2020).

In terms of energy, there are two major examples of celebrity diaspora individuals supporting future smart city projects. The Senegalese-American R&B singer Akon is backing the construction of the green and futuristic Akon City in western Senegal. Akon City will have certified high environmental quality buildings, operating exclusively with the Akoin crypto-currency, and it has been designated a 'special tax zone' by the Senegalese government. It will also be integral part of tourism vision of the country. On a 2000-hectare site, it will have parks, roads, hospitals, university, schools, shopping malls, residences, hotels, a police station, a landfill, and an industrial zone, including a photovoltaic solar power plant to supply the whole area (Takouleu 2020). This leads on from his 2014 initiative entitled 'Akon Lighting Africa' whereby the strategy to harness solar energy to provide electricity, utilising African expertise (especially youth) and recognising that awareness, acceptability, affordability, and accessibility were key. This has become a Senegalese energy success story (Ahmed et al. 2019). Subsequent investment and political will have led to the creation of Akon City and indeed the plans for a second satellite city in Uganda (The Washington Post 2021).

In Sierra Leone, plans are underway for the Sherbro Island City project, which is promoted as a public–private partnership between the Sierra Leonean government and a company set up by the British actor Idris Elba and entrepreneur Siaka Stevens (State House Media and Communications Unit 2019). The city plans to have its own governance system, intelligently applying smart technology to improve infrastructure and transparency of decision-making. The government of Sierra Leone has partnered on the project to highlight Sierra Leone as a credible investment opportunity (SmartCitiesWorld 2019).

4.4 The Need for a More Integrated Approach

There seems to be a general consensus within the wider literature that smart cities are beneficial (Mboup and Oyelaran-Oyeyinka 2019). The long-term opportunities and benefits seem to far outweigh the current challenges and as such smart cities have been embraced globally (Lin et al. 2019). This generalisation should not deter from the clear challenges that are being faced in order to fully realise the smart city. In an African context, challenges around infrastructure, investment, hardware requirements, governance, etc. are commonly noted and seem to be reasons why African cities 'lag behind' their Global North counterparts, but the literature is still favourable towards the smart city concept if these barriers were to be overcome. The message is seemingly clear, smart cities 'are better-governed cities' (Chirisa and Matamanda 2019) and are more accessible (Balkaran 2019).

We understand that there are policies in the West African region and in countries such as Sierra Leone that are creating an enabling environment for smart city realisation. Regulation of the ICT sector and state policies in West Africa have provided legal frameworks to support the operation of smart cities (Franck and Beda 2019), and collective ECOWAS policies have encouraged harmonisation and collective thinking (ibid). Other regional policies, such as ECOWAS Renewable Energy Policy (EREP) and ECOWAS Energy Efficiency Policy (EEEP), support the push for renewable generation within its members, however, to deploy at a large scale requires enormous investment, often from foreign external investors or members of the diaspora.

To that end, what indication do we have that African states and regional collectives are driving the smart city agenda in sub-Saharan Africa rather than potentially mimicking strategies that have been deployed in the Global North (Njoh 2021)? In some ways, we can engage with literature around post-colonialism and decolonisation that potentially suggest that concepts such as 'smart cities' and 'smart energy cities' are Global North terms imposed on African governments to highlight 'best practice' and how development in these spaces should happen (Radcliffe 2017), and rather the critiques around decolonising urban governance remain 'presentist' without really examining how 'imaginations of the future drive contemporary urbanism' (Datta 2019, p. 7).

At the same time, the literature provides a rather lack lustre account of how people, both as contributors and consumers of the smart city, interact and engage with the narrative and how it features as part of the city that they imagine for themselves. This also begs questions about who these smart cities are designed for and for what purpose. This has led some to argue that it serves the purposes of the business community rather than the urban population the city is supposed to serve (Grossi and Pianezzi 2017). Indeed, scholars such as Musakwa and Mokoena (2017) argue that top-down strategies do not take into consideration the needs of people, and in the quest for smartness, there are widening inequality gaps where ultimately those on the fringes continue to be excluded from planning processes and service provision. There are also a number of examples whereby residents occupying the land acquired for smart city development have been unlawfully evicted (Tan and Taeihagh 2020). At the same time, there are examples of smart city initiatives such as transportation and commercial and residential buildings that are simply unaffordable for the majority of urban citizens and often appeal to the middle-class community.

Despite having the fastest growing middle-class population with approximately 34% being categorised as middle class in sub-Saharan Africa (SSA) (AfDB 2011), this still only represents a small portion of society and indeed does not account for the great variances in the definition of the African middle class (Campbell 2020), many of whom still earn between $2 and $4 a day, and are likely to fall back into poverty as a result of varying shocks and trends (e.g. COVID-19 – AfDB 2021). Specifically, when thinking about energy, not only do we need to consider the ways in which energy access and service provision (e.g. smart meters/grids, etc.) are being utilised by consumers as part of the smart city narrative but also the ways in which energy systems planning continues to be a challenge in smart city development in regards to both generation and consumption (Clever et al. 2018; Mudau and Mhangara 2019; Yamamura et al. 2017). It is important to note that previous studies have noted that access to traditional fossil fuels has positively impacted on consumption of renewable energy services (Khan et al. 2021) and it is crucial to recognise that this process may lead to more people without basic access to energy. What is largely missing from the literature is the interaction between the smart city narratives and energy for cooking. Recent research has advocated for the use of pay-as-you-go LPG smart meters (Shupler et al. 2021) which can be an effective consumer finance mechanism, especially in resource-poor urban households. Planners, investors, and researchers of smart cities will need to engage with the body of literature around energy for cooking and the implications it has on smart cities when (i) the majority of cooking fuel in African urban spaces is derived from fossil fuels (often including access to grid electricity) and (ii) the use of cleaner alternatives such as LPG is not sustainable due to long-term environmental implications (Mansoori 2021) and challenges around subsidies.

It is also important to remind ourselves of the ethical implications of collecting data from citizens, many of whom live below the poverty line and often sit outside of formal networks and boundaries. In line with the above, there are ethical considerations to be made considering that there is evidence to support that smart city implementation can reinforce class boundaries and exclude poorer households from engaging with, connecting to, and benefiting from services (Clever et al. 2018). At the same time, smart monitoring of energy use can lead to privacy violations including not only what sort of devices households and individuals use but also other habits such as time spent away from the home (ibid). Other ethical considerations include social equity (Chang 2021), the inference of particular social, political, or religious affiliation based on data collected from individuals can lead to inaccurate profiling which may lead to stigma and/or harm, the capture of data without consent and often the benefits of 'shared' data are often reaped by those collecting it (Kitchin 2016).

Advocating for a more human-centric element in the smart city/smart energy city narrative often results in unintended consequences, with the divide between technical and human persisting. Rather, greater thought is needed on how these approaches can be better integrated into urban planning where the smart city agenda will most likely continue. It is acknowledging that the smart city itself is not a 'one-size-fits-all' product but rather small incremental steps to enhancing a city's ability to thrive and for its citizens to benefit. To do that, there is a need to continue to recognise the socio-technical connections that underpin cities and that there is a risk of deepening inequality if these complex societal networks are not considered. It is also most importantly, understanding how people themselves engage with smart cities and their initiatives as not only a contributor to the paradigm itself but also how they interact with different initiatives as consumers and whether it ultimately benefits them.

4.5 Conclusions

The smart city agenda has attracted global appraise and appears to be a key policy approach to address the challenges around rapid urbanisation and the transformation of urban spaces. Driven by the Global North, it is unsurprising that the smart city agenda has gained significant traction in the Global South with national governments and key investors adding smart cities to national agendas. This in itself is not a new phenomenon; the development agendas of the Global North have often become bargaining chips in exchange for trade, aid, markets, investment, technology, healthcare, etc. and while we can agree that the 'North-South' paradigm is binary and outdated, agenda's such as smart cities are imposed in the Global South with results not always living up to expectations. There is a strong need for evidence-based decision-making in urban planning, but many of the smart concepts that are investigated in the Global North are not necessarily relevant in an African context. This in turn can create exclusionary processes which focus on the gains of businesses and investments of foreign actors rather than the growth of healthy, resilient societies. Embedded within this debate is the role of energy, as it is increasingly recognised that 'smartness' does not only incorporate ICT alone, energy is fundamental to this concept. Energy underpins the sustainability and success of a smart city (especially in terms of net-zero, low-carbon, and energy-efficient cities) but is essential to the everyday lives of people and so being able to manage the demand for energy services and ensure (at least basic) access is essential.

In this chapter, we have interrogated the smart energy city in a West African context, unpacking how regional and national (Sierra Leone) energy policies contribute to the smart city rhetoric by adopting smart technologies (such as prepaid meters) and fostering their own vision of smart cities through ambitious energy targets and the help of private–public partnerships, including engaging African diaspora. We observe that leveraging technologies such as smart grids, meters, and mobile phones to access and pay for energy services is one of the few ways that consumers can feel in control of the narrative and, in principle, offer services to even the poorest communities. The reality, however, is that regional and national stakeholders are not only tasked with meeting international targets but also with addressing the challenges of rapid urbanisation and achieving universal energy access. Ultimately, it is those communities on the fringes (in places such as informal settlements) already marginalised and barely benefitting from city life that pays the price as often they are left out of the debates and access to technologies, energy services, and supporting finance mechanisms remain out of reach. Clean cooking in the context of smart cities is just one example of energy that is often left out of the smart city debate and warrants urgent attention. Only time will tell whether ongoing efforts are in the realm of unrealistic fantasies or genuinely sustainable solutions.

More than anything, this chapter calls for greater research and practice, whereby citizens are at the centre of the smart energy city agenda, and utilising human and social capital to drive a bottom-up approach to the smart city enables communities, grassroots activities, and individual creativity at the very heart of the agenda with governance acting as mediators and ICT and technology as the hardware (Schaffers et al. 2012). It is imperative when leveraging technologies such as mobile phones and more complex infrastructure that inequality gaps are not widened and ensure people are not excluded from engaging with the smart city.

References

Addae, B.A., Zhang, L., Zhou, P., and Wang, F. (2019). Analyzing barriers of smart energy city in Accra with two-step fuzzy DEMATEL. *Cities* 89: 218–227. https://doi.org/10.1016/j.cities.2019.01.043.

AfDB (2011). *Middle of the Pyramid: Dynamics of the Middle Class in Africa*. African Development Bank.

AfDB (2021). *African Economic Outlook 2021. From Debt Resolution to Growth: The Road Ahead for Africa*. African Development Bank.

African Vibes (2020). *Wakanda One Village Project Gets a Boost*. African Vibes.

Ahmed, J.U., Talukder, N., Ahmed, A., and Hoque, M.T. (2019). Sustainable energy solutions: Akon Lighting Africa. *Decision* 46: 253–266. https://doi.org/10.1007/s40622-019-00220-x.

Asongu, S.A. (2013). The impact of mobile phone penetration on African inequality (AGDI Working Paper, No. No. WP/13/021). African Governance and Development Institute (AGDI), Yaoundé.

Asongu, S.A. and Odhiambo, N.M. (2019). How enhancing information and communication technology has affected inequality in Africa for sustainable development: an empirical investigation. *Sustainable Development* 27: 647–656. https://doi.org/10.1002/sd.1929.

Aurigi, A. and Odendaal, N. (2021). From "smart in the box" to "smart in the city": rethinking the socially sustainable smart city in context. *Journal of Urban Technology* 28: 55–70. https://doi.org/10.1080/10630732.2019.1704203.

Baker, E., Nock, D., Levin, T. et al. (2021). Who is marginalized in energy justice? Amplifying community leader perspectives of energy transitions in Ghana. *Energy Research & Social Science* 73: 101933. https://doi.org/10.1016/j.erss.2021.101933.

Balkaran, S. (2019). Smart cities as misplaced priorities in South Africa: a complex balance of conflicting societal needs. *Journal of Management & Administration* 2019 (2): 1–30.

Batty, M. (2013). Big data, smart cities and city planning. *Dialogues in Human Geography* 3: 274–279. https://doi.org/10.1177/2043820613513390.

Batty, M., Axhausen, K.W., Giannotti, F. et al. (2012). Smart cities of the future. *European Physical Journal: Special Topics* 214: 481–518. https://doi.org/10.1140/epjst/e2012-01703-3.

Bearing Point Institute (2016). *Smart Cities: The Key to Africa's Third Revolution*. Bearing Point Institute.

Beda, K.F.G. (2019). Information and communication technologies in Africa: levels, trends and perspectives. In: *Smart Economy in Smart African Cities, Advances in 21st Century Human Settlements* (ed. G. Mboup and B. Oyelaran-Oyeyinka), 447–479. Singapore: Springer Singapore https://doi.org/10.1007/978-981-13-3471-9_15.

Bélissent, J. (2010, 2010). Getting clever about smart cities: new opportunities require new business models. *Forrester Research* 3: 1–31.

Bentsen, N. and Felby, C. (2012). Biomass for energy in the European Union - a review of bioenergy resource assessments. *Biotechnology for Biofuels* 5: 25. https://doi.org/10.1186/1754-6834-5-25.

Calvillo Munoz, C.F. (2017). *Energy Management in Smart Cities*. KTH Royal Institute of Technology.

Calvillo, C.F., Sánchez-Miralles, A., and Villar, J. (2016). Energy management and planning in smart cities. *Renewable and Sustainable Energy Reviews* 55: 273–287. https://doi.org/10.1016/j.rser.2015.10.133.

Campbell, J. (2020). How to think about Africa's "rising middle class" amid COVID-19. https://www.cfr.org/blog/how-think-about-africas-rising-middle-class-amid-covid-19 (accessed 19 April 2021).

Cavada, M., Rogers, C., and Hunt, D. (2014). Smart cities: contradicting definitions and unclear measures. *Proceedings of the 4th World Sustainability Forum*. Presented at the 4th World Sustainability Forum, MDPI, Sciforum.net, p. f004. https://doi.org/10.3390/wsf-4-f004.

Chang, V. (2021). An ethical framework for big data and smart cities. *Technological Forecasting and Social Change* 165: 120559. https://doi.org/10.1016/j.techfore.2020.120559.

Chirisa, I. and Matamanda, A. (2019). Smart cities in Sub Saharan Africa: opportunities and challenges. In: *Industrial and Urban Growth Policies at the Sub-National, National, and Global Levels*. Advances in Public Policy and Administration. IGI Global https://doi.org/10.4018/978-1-5225-7625-9.

Clever, S., Crago, T., Polka, A. et al. (2018). Ethical analyses of smart city applications. *Urban Science* 2: 96. https://doi.org/10.3390/urbansci2040096.

Cloke, J., Mohr, A., and Brown, E. (2017). Imagining renewable energy: towards a social energy systems approach to community renewable energy projects in the global south. *Energy Research & Social Science* 31: 263–272. https://doi.org/10.1016/j.erss.2017.06.023.

Connolly, D., Lund, H., Vad Mathiesen, B. et al. (2013). Smart energy systems: holistic and integrated energy systems for the era of 100% renewable energy. Smart Energy Systems, Aalborg University.

Dagnachew, A.G. (2021). *The Power of Energy: Synergies and Trade-Offs of Achieving SDG7 Targets in Sub-Saharan Africa – A Model-Based Analysis*. Utrecht University.

Datta, A. (2015). New urban utopias of postcolonial India: 'Entrepreneurial urbanization' in Dholera smart city, Gujarat. *Dialogues in Human Geography* 5: 3–22. https://doi.org/10.1177/2043820614565748.

Datta, A. (2018). The digital turn in postcolonial urbanism: smart citizenship in the making of India's 100 smart cities. *Transactions of the Institute of British Geographers* 43: 405–419. https://doi.org/10.1111/tran.12225.

Datta, A. (2019). Postcolonial urban futures: imagining and governing India's smart urban age. *Environment and Planning D: Society and Space* 37: 393–410. https://doi.org/10.1177/0263775818800721.

Deville, J., Guggenheim, M., Hrdličková, Z., 2014. Concrete governmentality: shelters and the transformations of preparedness, in: Tironi, M., Rodriguez-Giralt, I., Guggenheim, M. (Eds.), Disasters and Politics. Materials, Preparedness and Governance, Sociological Review Monograph. Wiley, London, pp. 183–210.

DSTI (2019). *Sierra Leone National Innovation & Digital Strategy (2019 – 2029)*. DSTI.

DSTI (2020). The Bill and Melinda Gates Foundation supports Drones in Sierra Leone for medical supply chain innovation. https://www.dsti.gov.sl/the-bill-and-melinda-gates-foundation-supports-drones-in-sierra-leone-for-medical-supply-chain-innovation/ (accessed 19 April 2021).

Dunlap, A. and Arce, M.C. (2021). 'Murderous energy' in Oaxaca, Mexico: wind factories, territorial struggle and social warfare. *The Journal of Peasant Studies* 49 (2): 1–26. https://doi.org/10.1080/03066150.2020.1862090.

Dupar, M., Weingartner, L., and Opitz-Stapleton, S. (2021). *Investing for Sustainable Climate Services: Insights from African Experience*. London, UK: Overseas Development Institute.

Echendu, A.J. and Okafor, P.C.C. (2021). Smart city technology: a potential solution to Africa's growing population and rapid urbanization? *Development Studies Research* 8: 82–93. https://doi.org/10.1080/21665095.2021.1894963.

ECOWAS (2013a). *ECOWAs Energy Efficiency Policy*. ECOWAS.

ECOWAS (2013b). *ECOWAs Renewable Energy Policy*. ECOWAS.

ECOWAS (2017). *ECOWAS Policy for Gender Mainstreaming in Energy Access*. ECOWAS.

ECREEE (2018). *Regional Progress Report on Renewable Energy, Energy Efficiency and Energy Access in ECOWAS region Monitoring year: 2016*. Cabo Verde: ECOWAS Centre for Renewable Energy and Energy Efficiency.

ECREEE (2019). *Regional Progress Report on Renewable Energy, Energy Efficiency and Energy Access in ECOWAS region Monitoring year: 2017*. Cabo Verde: ECOWAS Centre for Renewable Energy and Energy Efficiency.

Fernandez-Anez, V., Fernández-Güell, J.M., and Giffinger, R. (2018). Smart city implementation and discourses: an integrated conceptual model. The case of Vienna. *Cities* 78: 4–16. https://doi.org/10.1016/j.cities.2017.12.004.

Freetown City Council (2019). *Transform Freetown: An Overview 2019-2022*. Freetown City Council.

Fuso Nerini, F., Ray, C., and Boulkaid, Y. (2017). The cost of cooking a meal. The case of Nyeri County, Kenya. *Environmental Research Letters* 12: 065007:https://doi.org/10.1088/1748-9326/aa6fd0.

Giotitsas, C., Nardelli, P.H.J., Kostakis, V., and Narayanan, A. (2020). From private to public governance: the case for reconfiguring energy systems as a commons. *Energy Research & Social Science* 70: 101737. https://doi.org/10.1016/j.erss.2020.101737.

Golubchikov, O. (2020). *People-Smart Sustainable Cities*. Geneva: United Nations Economic Commission for Europe.

Grossi, G. and Pianezzi, D. (2017). Smart cities: Utopia or neoliberal ideology? *Cities* 69: 79–85. https://doi.org/10.1016/j.cities.2017.07.012.

GSMA (2014). *The Synergies Between Mobile, Energy and Water Access: Africa*. GSMA.

Guma, P.K. (2019). Smart urbanism? ICTs for water and electricity supply in Nairobi. *Urban Studies* 56: 2333–2352. https://doi.org/10.1177/0042098018813041.

Haarstad, H. and Wathne, M.W. (2019). Are smart city projects catalyzing urban energy sustainability? *Energy Policy* 129: 918–925. https://doi.org/10.1016/j.enpol.2019.03.001.

Hamelin, L., Borzęcka, M., Kozak, M., and Pudełko, R. (2019). A spatial approach to bioeconomy: quantifying the residual biomass potential in the EU-27. *Renewable and Sustainable Energy Reviews* 100: 127–142. https://doi.org/10.1016/j.rser.2018.10.017.

Hammons, R. and Myers, J. (2019). Smart cities. *IEEE Internet of Things Journal* 2: 8–9: https://doi.org/10.1109/MIOT.2019.8892761.

Haushofer, C. (2019). *AFRICA: AFD and ANRU Create "Smart Cities" Network for 12 African Cities*. Afrik 21.

Hirmer, S.A., George-Williams, H., Rhys, J. et al. (2021). Stakeholder decision-making: understanding Sierra Leone's energy sector. *Renewable and Sustainable Energy Reviews* 145: 111093. https://doi.org/10.1016/j.rser.2021.111093.

Horevitz, E. (2009). Understanding the anthropology of immigration and migration. *Journal of Human Behavior in the Social Environment* 19: 745–758. https://doi.org/10.1080/10911350902910914.

IAEA (2016). *Sustainable Electricity Supply Scenarios for West Africa (No. IAEA-TECDOC-1793), IAEA TECDOC Series*. Vienna: International Atomic Energy Ag.

IEA (2019). *World Energy Outlook 2019*. International Energy Agency.

IEA (2020a). *SDG7: Data and Projections: Access to Affordable, Reliable, Sustainable and Modern Energy for All*. France: International Energy Agency.

IEA (2020b). *World Energy Outlook 2020*. International Energy Agency.

IEA (2021). *Net Zero by 2050: A Roadmap for the Global Energy Sector*. France: International Energy Agency.

IFC (2019). *Pay-As-You-Go Market Attractiveness Index 2019*. International Finance Corporation.

IRENA, ADFD (2020). *Advancing Renewables in Developing Countries: Progress of Projects Supported Through the IRENA/ADFD Project Facility*. Abu Dhabi: International Renewable Energy Agency; Abu Dhabi Fund for Development.

ITU (2014). *Smart Sustainable Cities: An Analysis of Definitions*. International Telecommunication Union.

Kambule, N. and Nwulu, N. (2021). Western Africa Region. In: *The Deployment of Prepaid Electricity Meters in Sub-Saharan Africa*, Lecture Notes in Electrical Engineering, 79–102. Cham: Springer International Publishing https://doi.org/10.1007/978-3-030-71217-4_5.

Khan, A., Chenggang, Y., Hussain, J., and Kui, Z. (2021). Impact of technological innovation, financial development and foreign direct investment on renewable energy, non-renewable energy and the environment in belt & road initiative countries. *Renewable Energy* 171: 479–491. https://doi.org/10.1016/j.renene.2021.02.075.

Khodadadzadeh, O. (2017). *China in Africa: A Modern Story of Colonization?: A Case Study of China's Engagement in Angola*. Uppsala Universitet.

Kitchin, R. (2016). The ethics of smart cities and urban science. *Philosophical Transactions of the Royal Society A* 374: 20160115. https://doi.org/10.1098/rsta.2016.0115.

Kovacic, Z., Smit, S., Musango, J.K. et al. (2016). Probing uncertainty levels of electrification in informal urban settlements: a case from South Africa. *Habitat International* 56: 212–221. https://doi.org/10.1016/j.habitatint.2016.06.002.

Ky, S.S., Rugemintwari, C., and Sauviat, A. (2021). Friends or foes? Mobile money interaction with formal and informal finance. *Telecommunications Policy* 45: 102057. https://doi.org/10.1016/j.telpol.2020.102057.

Lai, C.S., Jia, Y., Dong, Z. et al. (2020). A review of technical standards for smart cities. *Clean Technologies* 2: 290–310. https://doi.org/10.3390/cleantechnol2030019.

Lemanski, C. (2009). Augmented informality: South Africa's backyard dwellings as a by-product of formal housing policies. *Habitat International* 33: 472–484. https://doi.org/10.1016/j.habitatint.2009.03.002.

Lin, C., Zhao, G., Yu, C., and Wu, Y. (2019). Smart city development and residents' well-being. *Sustainability* 11: 676. https://doi.org/10.3390/su11030676.

Liu, Y. and Bah, Z. (2021). Enabling development impact of solar mini-grids through the community engagement: evidence from rural Sierra Leone. *Energy Policy* 154: 112294. https://doi.org/10.1016/j.enpol.2021.112294.

Lund, H. and Mathiesen, B.V. (2009). Energy system analysis of 100% renewable energy systems—The case of Denmark in years 2030 and 2050. *Energy* 34: 524–531. https://doi.org/10.1016/j.energy.2008.04.003.

Magocha, M. (2020). The conceptualised role of African diaspora in the renaissance of the African continent. *The Journal for Transdisciplinary Research in Southern Africa* 16: 1–13. https://doi.org/10.4102/td.v16i1.733.

Mansoori, S.A. (2021). Challenges and new research directions to the development of smart cities: systems-of-systems perspective. *Journal of Physics: Conference Series* 1828: 012136. https://doi.org/10.1088/1742-6596/1828/1/012136.

Masembe, A. (2015). Reliability benefit of smart grid technologies: a case for South Africa. *Journal of Energy in Southern Africa* 26: 2–9. https://doi.org/10.17159/2413-3051/2015/v26i3a2124.

Mboup, G. (2019). Africa's smart city foundation: urbanization, urban form and structure, land tenure and basic infrastructures. In: *Smart Economy in Smart African Cities: Sustainable, Inclusive, Resilient and Prosperous, Advances in 21st Century Human Settlements* (ed. B. Oyelaran-Oyeyinka). Singapore: Springer https://doi.org/10.1007/978-981-13-3471-9.

Mboup, G. and Oyelaran-Oyeyinka, B. (2019). Relevance of smart economy in smart cities in Africa. In: *Smart Economy in Smart African Cities: Sustainable, Inclusive, Resilient and Prosperous, Advances in 21st Century Human Settlements* (ed. G. Mboup and B. Oyelaran-Oyeyinka). Singapore: Springer https://doi.org/10.1007/978-981-13-3471-9.

McCollum, D.L., Echeverri, L.G., Busch, S. et al. (2018). Connecting the sustainable development goals by their energy inter-linkages. *Environmental Research Letters* 13: 033006. https://doi.org/10.1088/1748-9326/aaafe3.

McIlvennie, C., Sanguinetti, A., and Pritoni, M. (2020). Of impacts, agents, and functions: an interdisciplinary meta-review of smart home energy management systems research. *Energy Research & Social Science* 68: 101555. https://doi.org/10.1016/j.erss.2020.101555.

Meliani, M., El Barkany, A., El Abbassi, I. et al. (2021). Smart grid implementation in Morocco: case study. *Materials Today: Proceedings* 45: 7675–7679. https://doi.org/10.1016/j.matpr.2021.03.176.

Mignamissi, D. and Djijo T, A.J. (2021). Digital divide and financial development in Africa. *Telecommunications Policy* 45: 102199. https://doi.org/10.1016/j.telpol.2021.102199.

Ministry of Energy (2016). *Renewable Energy Policy of Sierra Leone*. Ministry of Energy.

Mitlin, D. and Satterthwaite, D. (2012). *Urban Poverty in the Global South.* Routledge https://doi.org/10.4324/9780203104316.

Morvaj, B., Lugaric, L., and Krajcar, S. (2011). Proceedings of the 2011 3rd International Youth Conference on Energetics: (IYCE 2011), Leiria, Portugal (7–9 July 2011). Presented at the International Youth Conference on Energetics, IEEE, Piscataway, NJ.

Mosannenzadeh, F., Bisello, A., Vaccaro, R. et al. (2017). Smart energy city development: a story told by urban planners. *Cities* 64: 54–65. https://doi.org/10.1016/j.cities.2017.02.001.

Muchiri, J. (2019). Clean cooking: the forgotten link in universal energy access. https://intpolicydigest.org/clean-cooking-the-forgotten-link-in-universal-energy-access/ (accessed 19 April 2021).

Mudau, N. and Mhangara, P. (2019). Smart energy for smart infrastructure development. In: *Smart Economy in Smart African Cities, Advances in 21st Century Human Settlements* (ed. G. Mboup and B. Oyelaran-Oyeyinka), 219–249. Singapore: Springer Singapore https://doi.org/10.1007/978-981-13-3471-9_7.

Muigua, K. (2021). Delivering Clean and Affordable Energy for All. http://kmco.co.ke/wp-content/uploads/2021/05/Delivering-Clean-and-Affordable-Energy-for-All-Kariuki-Muigua-Ph.D-24th-April-2021-1.pdf (accessed 19 April 2021).

Musakwa, W. and Mokoena, B.T. (2017). Smart cities in South Africa! A case of misplaced priorities? *Presented at the Computers in Urban Planning and Urban Management*, Adelaide, Australia, July 2017.

Nartey, E. (2016). Energy management system to prevent blackout in smart grid network - a case study of Kotoka International Airport. *JEEE* 4: 83. https://doi.org/10.11648/j.jeee.20160404.11.

National Civil Registration Authority, 2020. National Civil Registration Authority (Sierra Leone).

Newland, K., Terrazas, A., and Munster, R. (2010). *Diaspora Philanthropy: Private Giving and Public Policy*. Washington, DC: Diasporas & Development Policy Project. Migration Policy Institute.

Njoh, A.J. (2021). Renewable energy as a determinant of inter-country differentials in CO2 emissions in Africa. *Renewable Energy* 172: 1225–1232. https://doi.org/10.1016/j.renene.2021.03.096.

Nwaiwu, F. (2021). Digitalisation of energy systems within the context of existing policy frameworks: an assessment of sustainable energy transitions in Africa (preprint). In review. https://www.researchsquare.com/article/rs-148680/v1. https://doi.org/10.21203/rs.3.rs-148680/v1

Odendaal, N. (2021). Everyday urbanisms and the importance of place: exploring the elements of the emancipatory smart city. *Urban Studies* 58: 639–654. https://doi.org/10.1177/0042098020970970.

Ojong, N. (2021). The rise of solar home systems in sub-Saharan Africa: examining gender, class, and sustainability. *Energy Research & Social Science* 75: 102011. https://doi.org/10.1016/j.erss.2021.102011.

Olorunfemi, I.E., Olufayo, A.A., Fasinmirin, J.T., and Komolafe, A.A. (2021). Dynamics of land use land cover and its impact on carbon stocks in Sub-Saharan Africa: an overview. *Environment, Development and Sustainability* 24 (1): 40–76. https://doi.org/10.1007/s10668-021-01484-z.

Outside Insight (2020). Smart cities Africa: Cape Town and Nairobi take the lead. https://outsideinsight.com/insights/smart-cities-africa-cape-town-and-nairobi-take-the-lead/.

Poggiali, L. (2017). Digital futures and analogue pasts? Citizenship and ethnicity in techno-utopian Kenya. *Africa* 87: 253–277. https://doi.org/10.1017/S0001972016000942.

Porter, G. (2015). Mobile phones, mobility practices, and transport organization in Sub-Saharan Africa. *Mobility in History* 6: 81–88. https://doi.org/10.3167/mih.2015.060109.

Porter, G., Hampshire, K., Abane, A. et al. (2012). Youth, mobility and mobile phones in Africa: findings from a three-country study. *Information Technology for Development* 18: 145–162. https://doi.org/10.1080/02681102.2011.643210.

Power For All (2018). *Case Study: Energy Access Market Activation Sierra Leone*. PowerForAll.

Pueyo, A. (2018). What constrains renewable energy investment in Sub-Saharan Africa? A comparison of Kenya and Ghana. *World Development* 109: 85–100. https://doi.org/10.1016/j.worlddev.2018.04.008.

Radcliffe, S.A. (2017). Decolonising geographical knowledges. *Transactions of the Institute of British Geographers* 42: 329–333. https://doi.org/10.1111/tran.12195.

Rahman, M.M. (2021). The dynamic nexus of energy consumption, international trade and economic growth in BRICS and ASEAN countries: a panel causality test. *Energy* 229: 120679. https://doi.org/10.1016/j.energy.2021.120679.

REUTERS (2019). *San Francisco Crowdfunder Kiva Sets Up Sierra Leone Credit Database*. REUTERS.

Schaffers, H., Ratti, C., and Komninos, N. (2012). Special issue on smart applications for smart cities-new approaches to innovation: guest editors' introduction. *Journal of Theoretical and Applied Electronic Commerce Research* 7: 9–10. https://doi.org/10.4067/S0718-18762012000300005.

Schoch, M. and Lakner, C. (2020). The number of poor people continues to rise in Sub-Saharan Africa, despite a slow decline in the poverty rate. https://blogs.worldbank.org/opendata/number-poor-people-continues-rise-sub-saharan-africa-despite-slow-decline-poverty-rate (accessed 6 January 2021).

SEforALL (2020). SEforALL Analysis of SDG7 Progress - 2020. Sustainable Energy for All, Vienna

SEforALL (2021). *Increasing Energy Access in Sierra Leone: Mini-Grid Survey Analysis on Tariffs, Subsidies and Productive Use*. Vienna: Sustainable Energy for All.

Sharma, A., 2020. 'We do not want fake energy': the social shaping of a solar micro-grid in rural India. *Science, Technology and Society* 25, 308–324. https://doi.org/10.1177/0971721820903006

Shupler, M., O'Keefe, M., Puzzolo, E. et al. (2021). Pay-as-you-go liquefied petroleum gas supports sustainable clean cooking in Kenyan informal urban settlement during COVID-19 lockdown. *Applied Energy* 292: 116769. https://doi.org/10.1016/j.apenergy.2021.116769.

Sierra Leone Telegraph (2021). Confirmation and registration of citizens and non-citizens in Sierra Leone to commence soon. https://www.thesierraleonetelegraph.com/confirmation-and-registration-of-citizens-and-non-citizens-in-sierra-leone-to-commence-soon/ (accessed 19 April 2021).

Slavova, M. and Okwechime, E. (2016). African smart cities strategies for Agenda 2063. *Africa Journal of Management* 2: 210–229. https://doi.org/10.1080/23322373.2016.1175266.

SmartCitiesWorld (2019). *Sierra Leone Signs Charter City Development Deal*. Smart Cities World.

Söderström, O., Blake, E., and Odendaal, N. (2021). More-than-local, more-than-mobile: the smart city effect in South Africa. *Geoforum* 122: 103–117. https://doi.org/10.1016/j.geoforum.2021.03.017.

Solomon, E.M. and van Klyton, A. (2020). The impact of digital technology usage on economic growth in Africa. *Utilities Policy* 67: 101104. https://doi.org/10.1016/j.jup.2020.101104.

State House Media and Communications Unit (2019). The Government of Sierra Leone and Sherbro Alliance Partners forge alliance to deliver economic prosperity to Sherbro Island. https://statehouse.gov.sl/the-government-of-sierra-leone-and-sherbro-alliance-partners-forge-alliance-to-deliver-economic-prosperity-to-sherbro-island/ (accessed 19 April 2021).

Sustainable Energy Africa (2017). *Sustainable Energy Solutions for South African Local Government: A Practical Guide*. Cape Town: Sustainable Energy Africa.

Takouleu, J.M. (2018). *AFRICA: Huawei Sets Up a $1.5 Billion Fund to Boost African Smart Cities*. Afrik 21.

Takouleu, J.M. (2020). *SENEGAL: Akon City, the Green City Project is Launched at a Cost of $6 Billion*. Afrik 21.

Tan, S. and Taeihagh, A. (2020). Smart city governance in developing countries: a systematic literature review. *Sustainability* 12: 899. https://doi.org/10.3390/su12030899.

Tavares, R. (2009). Science and technology parks: an overview of the ongoing initiatives in Africa. *African Journal of Political Science and International Relations* 3: 208–223. https://doi.org/10.5897/AJPSIR.9000095.

TechNova (2020). The rise and fall of RLG. https://technovagh.com/the-rise-and-fall-of-rlg/ (accessed 19 April 2021).

Teske, S., Niklas, S., Atherton, A. et al. (2020). *Sectoral Pathways to Net Zero Emissions*. University of Technology Sydney, Institute for Sustainable Futures.

The Tech Observer (2016). What happened to hope city? https://revolver360.wordpress.com/2016/06/21/what-happened-to-hope-city/ (accessed 19 April 2021).

The Washington Informer (2018). *African Union to Launch First Real Wakanda*. The Washington Informer.

The Washington Post (2021). *Akon Hasn't Even Built His First 'Futuristic' City Yet, But Uganda is Giving Him Land for a Second*. The Washington Post.

The World Bank (2019). *ECOWAS-Battery Energy Storage Systems and Synchronization (BE3S) Project (P167569): Project Information Document*. Washington, DC: World Bank.

Thomas, A.R. (2021). Confirmation and registration of citizens and non-citizens in Sierra Leone to commence soon. Sierra Leone Telegraph. https://www.thesierraleonetelegraph.com/confirmation-and-registration-of-citizens-and-non-citizens-in-sierra-leone-to-commence-soon/ (accessed 6 August 2021).

Thomson Reuters (2020). *Africa Shrugs Off Net-Zero Emissions Push Without Finance to Follow*. Thpmson Reuters Foundation News.

Thornbush, M. and Golubchikov, O. (2021). Smart energy cities: the evolution of the city-energy-sustainability nexus. *Environment and Development* 100626: https://doi.org/10.1016/j.envdev.2021.100626.

Toli, A.M. and Murtagh, N. (2020). The concept of sustainability in smart city definitions. *Frontiers in Built Environment* 6: 77. https://doi.org/10.3389/fbuil.2020.00077.

Travel Noire (2020). *Ghana to Build 'Wakanda City' to Serve as a Pilgrimage for People of African Descent*. Travel Noire.

UNDP (2012). *Sierra Leone Sustainable Energy for All: Rapid Assessment Gap Analysis*. Sierra Leone.

UNECA (2018). Policy Brief #18: achieving SDG 7 in Africa. *Accelerating SDG7 Achievement: Policy Briefs in Support of the First Sdg7 Review at the UN High-Level Political Forum 2018*. United Nations, pp. 141–148.

UNESCO (2019). New methodology shows that 258 million children, adolescents and youth are out of School (No. UIS/2019/ED/FS/56).

United Nations (2019). Accelerating SDG7 Achievement Policy Briefs In Support of the First SDG7 Review at the UN High-Level Political Forum 2018.

Uspenskaia, D., Specht, K., Kondziella, H., and Bruckner, T. (2021). Challenges and barriers for net-zero/positive energy buildings and districts—empirical evidence from the smart city project SPARCS. *Buildings* 11: 78. https://doi.org/10.3390/buildings11020078.

Watson, V. (2014). African urban fantasies: dreams or nightmares? *Environment and Urbanization* 26: 215–231. https://doi.org/10.1177/0956247813513705.

Watson, V. (2015). The allure of 'smart city' rhetoric: India and Africa. *Dialogues in Human Geography* 5: 36–39. https://doi.org/10.1177/2043820614565868.

WEF (2019). Africa is facing an electricity crisis – a pay-as-you-go model could solve the problem. https://www.weforum.org/agenda/2019/07/pay-as-you-go-africas-solar-energy (accessed 19 April 2021).

World Bank (n.d.). Population living in slums (% of urban population) - Sub-Saharan Africa. https://data.worldbank.org/indicator/EN.POP.SLUM.UR.ZS?locations=ZG&name_desc=true (accessed 19 April 2021).

Yamamura, S., Fan, L., and Suzuki, Y. (2017). Assessment of urban energy performance through integration of BIM and GIS for smart city planning. *Procedia Engineering* 180: 1462–1472. https://doi.org/10.1016/j.proeng.2017.04.309.

5

Beyond Urban Smart Grid Experiments: Replication and Upscaling as Contested Concepts

Harald Rohracher[1], Gudrun Haindlmaier[2], Klaus Kubeczko[2], and Dick Magnusson[1]

[1] *Department of Thematic Studies (TEMA), Linköping University, Technology and Social Change, Linköping, Sweden*
[2] *Austrian Institute of Technology, Vienna, Austria*

5.1 Introduction

The concept of smart cities has historically been closely connected to digitalisation in general, but more specifically to the digitalisation of the energy system. It has been introduced in the European research and innovation context through the European Strategic Energy Technologies (SET) plan, which was organised around different industrial initiatives and one of these dedicated to Smart Cities. Smart grids are perceived as an essential element of future sustainable energy systems, and many smart city initiatives have the digitalisation of the electricity grid in combination with smart homes at their centre. Such electricity networks are characterised by a pervasive deployment of intelligent (i.e. information technology based) communication, monitoring and management systems which enable a two-way flow of both information and power exchange between electricity suppliers and customers (Sigrist et al. 2016). These grid technologies are the basis for a range of applications and services. In particular, the flexible shifting of electricity loads of households or industries helps to match electricity demand with a volatile electricity supply due to a large share of intermittent renewable energy sources. These new developments are expected to radically change the electricity system, especially at local grid level (Gangale et al. 2017).

However, how such a future grid will be organised in practice – which technological configurations it builds upon, which services and incentive structures will be offered to its users, which new actors will become part of the electricity system (e.g. storage providers), and how the role and relations of old and new actors will be changed – is still in flux, and a broad range of possible configurations of smart grids at urban or at household level are currently under investigation in various pilot or demonstration projects (WEF 2010). One of the surveys of smart grid pilot projects in the European Union counts 950 different cases (Gangale et al. 2017). Such pilot projects and urban smart grid experiments have become a key element of governing the transition towards smart grids and smart cities (see, for

Smart Cities, Energy and Climate: Governing Cities for a Low-Carbon Future, First Edition.
Edited by Oleg Golubchikov and Komali Yenneti.

example, Ansell and Bartenberger 2016; Evans et al. 2016), often in the form of urban living labs (Bulkeley et al. 2019; Voytenko et al. 2016) or as an object of learning between cities with different levels of experience (Kern 2019).

A key question in this context of pilot projects and urban experiments is thus how to move beyond these projects and contribute to a systemic transformation to a more active, flexible and sustainable grid system. This issue of expanding and transferring urban smart grid projects has been mostly framed in terms of upscaling and replication, particularly in engineering-driven research at a European level (as exemplified in the review of Sigrist and Rouco 2012). Such discourses tend to understand smart grids and smart cities in a highly decontextualised way that mainly focuses on technological configurations and 'solutions' which need to be rolled out in further cities and regions.

However, rolling out urban smart energy systems is not only a question of 'getting the technologies right'. Energy grids are deeply intertwined with regulations and other institutional structures (e.g. ownership structures and tariff systems), the types and networks of actors who operate the grid or supply electricity, the social practices and cultures of energy use, or the models and cognitive frames of actors and stakeholders, including the business models and interests of companies participating in the energy system. Studies of the scalability and replicability of smart grid technologies, i.e. the potential of implementing technologies from pilot projects for the whole electricity system or parts of it, even if carefully done and insightful, thus necessarily fall short of capturing the dynamics of electricity system transformation and the requirements of moving from limited smart grid pilot projects to a full-scale smart electricity grid.

Moving beyond this still dominant discourse of almost 'mechanical' replicability or transferability shifts attention to the embedding of smart energy systems in specific urban contexts, including their geographical, cultural and socio-political dimensions. This also opens up for a broader involvement of different groups of actors in the design and implementation of urban smart grid projects and potentially turns institutional features such as ownership and operation of these systems or the relationship between utilities and energy users into contested political questions. As Peng et al. (2019, p. 302) put it, such a transfer to new contexts 'requires mobilization and realignments of actors, resources, and institutional arrangements within the process of learning'. Because such a re-embedding is much more than a duplication, the emerging new arrangements can often be seen as an innovation in themselves. A more nuanced understanding of concepts of replication, upscaling, or transfer can thus have profound implications for the governance of urban change and development.

The question of how to scale up and replicate pilot projects and urban experiments has more recently been taken up in several research projects from a socio-technical perspective (e.g. Naber et al. 2017; van Doren et al. 2018; van Winden and van den Buuse 2017). These contributions mostly investigate different patterns of upscaling, factors driving the upscaling process, and the contribution of these upscaling processes to a long-term smart grid transition. In this chapter, we will build on these analyses. However, instead of focussing on patterns and long-term transition dynamics, we will zoom in on the next phase of pilot projects and discuss preconditions of replication and growth of these experiments. Given the complexity and local embedding of such smart grid projects, replication in a strict sense obviously is not possible. Implementing the same type of project somewhere else will necessarily involve elements of translation, or in other words, disembedding projects from place-specific conditions and re-embedding them in a new context. Such a translation cannot be done in isolation but requires engaged stakeholders and publics negotiating, contesting and shaping this process.

In this chapter, we aim for a better understanding of such processes and will work out a number of context conditions that are relevant to understand the local embedding of smart grid pilot projects. Based on two empirical case studies we will then try to identify critical contextual elements of smart grid projects which are of crucial importance for transferring such configurations to other urban contexts. Certain smart grid pilot projects may, for example, critically depend on spatial-geographical preconditions such as highly densified urban contexts, or on certain institutional preconditions such as time-variable tariff schemes or organisational structures like the unbundling of grid operators and energy providers, as is the case for most electricity grids in EU countries. Without these preconditions, the same type of project will have very little chance of implementation in other places.

The typology and analysis we suggest allow for a better assessment of the transferability potential of smart grid experiments and should be considered in the dissemination of results gained from such pilot projects. Furthermore, it may help to compare the portfolio of existing pilot projects and identify contexts of smart grid applications, which have not been sufficiently studied yet and would require further testing and development. Not least, a better understanding of the replicability of smart grid solutions developed in pilot projects and the institutional or regulatory conditions, they require, may inform energy and innovation policies, which aim at more supportive context conditions of smart grid applications.

5.2 Analysing Context Conditions for the Replicability of Smart Grid Pilot Projects

Much of the ongoing discussion on the replication and upscaling of smart grids has a predominantly technical focus (e.g. Rae et al. 2020) – although usually acknowledging the importance of regulation or market design (for a good example, see Rodriguez-Calvo et al. 2018). As Sigrist et al. (2016) point out in a comparative analysis of smart grid demonstration projects in the European Union (GRID+ project, Sigrist and Rouco (2012)), scalability and replicability are the two qualities of smart grid pilot projects which reduce barriers for the growth and reuse of the solutions tested. They conclude that scalability as the 'ability of the system to maintain its performance and function (...) when its scale is increased' (p. 2) requires a modular design because centrally organised systems cannot be easily increased in size. Replicability in turn, which 'denotes the property of a system that allows to be duplicated at another location or time' (p. 2), largely depends on standardisation and interoperability. Only if a smart grid solution is interoperable with existing grid infrastructures at other places, and solutions as well as interfaces are sufficiently standardised, does it have a chance to be replicated at a variety of other places from an engineering perspective. In both described circumstances of replication and upscaling, they identify economic factors (e.g. the economic viability of scaling up small-scale solutions and some market-related institutional factors such as similarities of market designs along with regulations that define the role of different actors, tariff structures, etc.) as factors for the viability of business models and replicability of solutions (see also May et al. 2015). As a final factor, stakeholder acceptance is pointed out as a precondition to make projects larger or replicate them.

5.2.1 Socio-Technical Concepts of Upscaling

However, from a socio-technical perspective, social, economic, or institutional factors are not discussed in sufficient depth and differentiation in these technology-oriented analyses. Techno-economic performance of a smart grid solution is an important precondition for economic viability, but not sufficient for successful diffusion or upscaling. From a socio-technical perspective, which takes the entanglement of new technological configurations with a range of socioeconomic and cultural dimensions seriously, the question of replication and upscaling becomes more complex. Moreover, the replication and upscaling of solutions from pilot projects need to be integrated with a broader perspective of systemic change to live up to the longer-term aim of transforming the existing electricity grid, also beyond the replication and economic success of specific smart grid solutions (Naber et al. 2017). Table 5.1 gives an overview of the different processes of upscaling they define in a transition context.

These patterns closely follow concepts of strategic niche management (Schot and Geels 2008; Smith and Raven 2012), where radically new socio-technical configurations initially take form and gain strength in protected niches such as publicly subsidised pilot or demonstration projects. Niches contribute to processes of social learning, the formation of social networks, and the alignment of expectations of different actors. Through processes of interlocal learning (Geels and Raven 2006), the variety of local experiments contributes to the development of a global niche eventually challenging the established regime. In analogy to these niche development processes, new smart urban energy arrangements might find their way into a global niche or become part of a modified or transformed regime. Pilot projects may grow in size (e.g. include additional actors or households) and similar arrangements may be implemented in other places. Eventually, these projects may be aggregated, e.g. through the work of intermediary organisations, which facilitate the knowledge flow and learning across smart grid pilot projects and may finally transform

Table 5.1 Different patterns of upscaling smart grid experiments.

Pattern of upscaling	Description
1. Growing	The experiment continues and more actors participate, or the scale at which technologies are used increases
2. Replication	The main concept of the experiment is replicated in other locations or contexts
3. Accumulation	Experiments are linked to other initiatives
4. Transformation	The experiment shapes wider institutional change in the regime selection environment

Source: Naber et al. (2017)/with permission of Elsevier.

current structures of the electricity grid. Analysis of a variety of case studies shows that it is extremely rare to repeat an entire experiment, but often parts of a project (e.g. technologies, routines and institutional arrangements) are circulated between multiple contexts.

Also, other researchers have suggested socio-technical concepts to better understand and categorise scaling up processes. Van Winden and van den Buuse (2017) identify three types of upscaling in their review of literature: (i) roll-out mainly refers to manufactured smart city products and service innovations and their scaling up on the market (market roll-out) or in organisations (organisational roll-out); (ii) expansion refers to increasing the size of existing smart city projects (such as mobility platforms); and (iii) replication refers to the implementation of solutions developed in pilot projects in other contexts (other organisations, other parts of the city and other cities). By taking the social, cultural and institutional local structures and relations (despite a highly complex, interwoven and networked society) into account, also regional theories and theories on urban competitiveness have increasingly paid attention to the 'place-basedness' of local experiments, policies and implementation projects in the last decade (see, for example, Camagni 2019; Friedmann 2002).

A further distinction of upscaling patterns is made by van Doren et al. (2018) regarding cases of urban low-carbon initiatives. They distinguish a horizontal pathway of scaling up, which refers to the spatial growth of initiatives and includes replication, diffusion, or other ways of scalar expansion, from vertical pathways of scaling up, which rather focus on 'structural learning' and institutional change and includes related terms such as translation, mainstreaming, or institutionalisation (p. 178). Vertical scaling up can also mean a successive embedding of smart grid projects in wider institutional fields (regional, national and international) which in this process are adapted to meet the requirements of smart grids (see also van Doren et al. 2016). In relation to Naber et al. (2017), horizontal expansion means remaining in the same niche, while horizontal upscaling means going beyond the niche and transforming regime structures.

Summing up, current literature, on the replication/scaling up of urban energy pilot projects or sustainability initiatives, makes a main distinction between different forms of scaling up within given institutional and socio-cultural contexts on the one hand, and more far-reaching upscaling strategies which involve different forms of system transformation and institutional change on the other. The first category includes different scaling strategies such as the expansion of pilot projects, the replication of smart grid solutions in new contexts, or the roll-out and diffusion of new products and services. A key tenet of these studies is the importance of socio-cultural, institutional, political and economic contexts as a precondition, incentive and shaping factor of upscaling processes.

5.2.2 Contextual Conditions for the Transfer of Smart Grid Solutions

In our following framework and analysis of empirical cases, we are thus mainly interested in how solutions developed in smart grid pilot projects can be taken up as a source of learning in different places. In relation to smart grid pilot projects, it is crucial to ask: Which elements and solutions of the pilot project can be taken up somewhere else? Which context dimensions are critical for the transferability of these solutions? To do such an analysis, we have specified a number of dimensions that make the framework sensitive to specific institutional, economic, technological, geographical and stakeholder contexts of pilot projects, both in the place of the experiment and in the new context where similar arrangements are implemented.

For the (partial) transfer of smart grid experiments, we regard the following dimensions as significant:

- *Spatial-structural dimension*: In many cases, geographical characteristics such as climate zone and landscape are crucial for the implementation of certain smart grid solutions. Also, the spatial scale of the pilot project may play a significant role in its replicability, as well as whether it is an urban or rural area that is being constructed or is already built. In certain scale-specific cases, the possibilities for upscaling are limited.
- *Technological dimension*: The functionalities of the technological components and the whole system relevant to the pilot project case (e.g. grid infrastructure, energy sources, storage equipment and loads).
- *Mission and economic dimension*: Key (long-term) missions, vision and non-commercial strategies and policies of public and private actors implementing the specific smart grid solution (e.g. commitment to climate reduction goals and sustainable development goals); macro-economic effects (benefits and costs) of a solution for third parties; relevant market and contractual relations (e.g. between energy supplier and customer, between grid operator and flat owner, between grid owner and grid operator, etc.); and the value added for the economic actors driving the use case.
- *Actor constellations*: Actors involved in the pilot project and kind of involvement; concrete ownership structure; relevant stakeholders and their positions and controversies they are involved in; and potential exclusion of certain actor groups.
- *Institutional dimension*: For the institutional dimension, we particularly refer to the 'field' concept of Beckert (2010), which allows us to analyse the dynamics among three social forces (a) institutions (mostly formal), (b) social networks and (c) cognitive frames:
  a) Formal institutions include legislative regulations, ownership and possession rights, market rules of the involved markets, organisational structures, technical standards, as well as formally agreed strategies;
  b) Social networks of incumbent and new actors involved in the experiment and stakeholders affected by it or influencing its context. This is going beyond the role of individual actors, which is dealt with separately (see above);
  c) Cognitive frames, as the 'culturally shaped meaning', collectively shape the way formal institutions, habits and practices are built on (e.g. mental models of how energy systems and markets shall look like, locked-in social practices influencing energy consumption, or acceptance criteria for new market rules or privacy risks of end-user groups).

A crucial question for the transfer of pilot projects is the 'criticality' of these dimensions or elements of it for a transfer of the use case. Often solutions only function under particular regulatory preconditions (e.g. the possibility of time-variable tariffs), ownership structures (e.g. ownership by the municipality), social network characteristics (e.g. hierarchical vs. distributed power), or spatial characteristics (e.g. dense settlements). To further flesh out the context specificity of smart energy projects, we now turn to two empirical cases from Austria and Sweden.

5.3 Analysis of the Use Cases from the Demo Sites Hartberg and Malmö

The smart grid case studies analysed in this chapter are both related to the flexible energy management of renewable energy-based energy systems in the context of increased volatility of energy generation. Thus, the solutions developed in the pilot projects are not intended to test single components independently of the systemic nature of the solution. In the case of Hartberg, the demo site is Ökopark Hartberg, a business park in the industrial zone of a small town, in Austria, aiming at energy system integration with smart grids. In the case of Malmö, Sweden, the demo site is the new smart city district of Hyllie in Malmö, where a platform for the smart integration of district heating and electricity system infrastructures has been developed.

For both case studies, a combination of methods was used for the analysis along the above-mentioned contextual dimensions. Data on the specific socio-technical configurations have been collected in semi-structured interviews with project managers of the local pilot projects, site visits, workshops with key stakeholders and document studies of municipal plans and project descriptions. The mixed-method approach has been considered most suitable to analyse the embeddedness and place dependence of these experiments as well as the limitations and challenges this creates for replication and transfer in a holistic manner.

5.3.1 The Case of Ökopark Hartberg

5.3.1.1 Spatial-Structural Dimension

Hartberg is a typical small town in Austria with around 11,000 inhabitants, a historic town centre, and commercial areas on the outskirts. It is situated in Styria, a region with significant forest resources. The demo site 'Ökopark Hartberg' is a business park in the south of the city centre with office buildings, a cinema, and museum (Figure 5.1). On site is also a small biomass-fuelled CHP plant, which also serves the district heating grid of the town.

The urban development measures and projects of Hartberg in recent years have been strongly focussed on linking the issues of energy and environment to life quality for the citizens. This is a common trend to be observed in Austrian cities of similar size (see ÖROK 2018; Österreichischer Städtebund 2017), along with efforts regarding mobility, economy and city centre revitalisation as well as (affordable) housing as the main topics of urban development strategies.

Regarding the energy sector, the utility 'Stadtwerke Hartberg' and its affiliated organisations are 100% owned by the City of Hartberg. As municipal utility, they operate the grid and are also the local provider of electric energy. One of the particularities of the demo site is that the business park is also run by the Stadtwerke, thus they have a close relationship with end users on site: firms renting the offices, aquarium in the entertainment complex (cinema/museum) and owners of electric cars charging them in the car park.

In this spatial-structural perspective, Hartberg is rather typical example of a small-sized city in Austria, whereas the specific ownership and stakeholder constellation of the business park is rather unusual in Austria. Other business parks in Austria are either focussed on high-tech companies (sometimes combined with a research centre) or are intended as

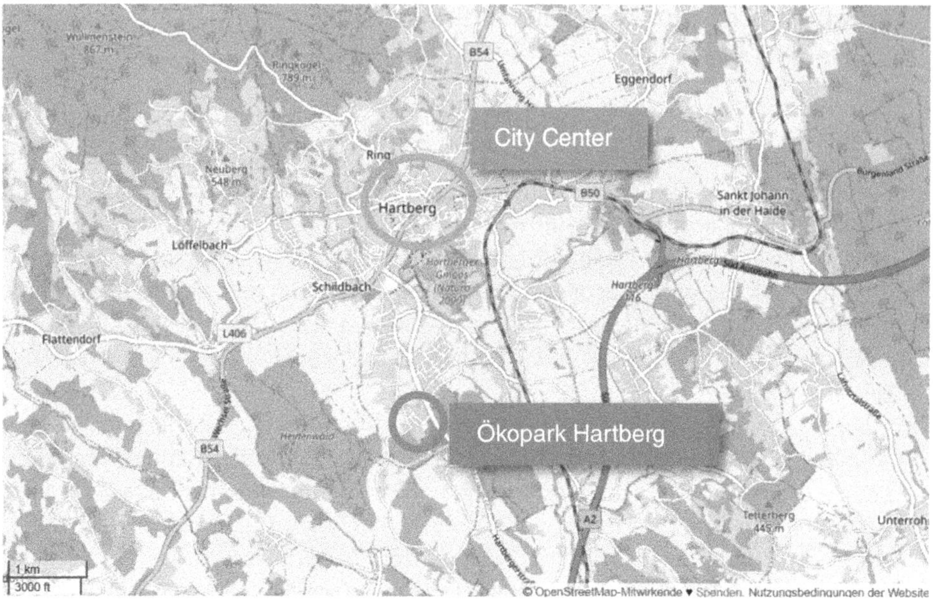

Figure 5.1 Location of Ökopark Hartberg (open street map).

driving forces for regional development ('clusters'). Some of them also aim for innovative energy business model solutions via crowdfunding or 100% ownership by citizens. Even though their spatial-structural components may be similar, their aims, business models and stakeholder configurations deviate significantly. This indicates the embeddedness into local context and specific configuration of the demonstration project, which of course makes it difficult to copy–paste solutions and provide easy guidelines for replication.

5.3.2 Technological and Functional Characteristics

Due to its neighbourhood with a biobased combined heat–power plant (CHP) on the campus of the business park, Ökopark Hartberg also serves as a local energy hub. The energy infrastructure includes a heat network, photovoltaics and E-mobility charging stations. All buildings are designed to provide a significantly higher standard of energy efficiency. With several demonstration projects, Stadtwerke aims to develop innovative energy management concepts and flexible solutions for the local energy system and is trying to act as energy hub with storage and energy conversion capacities for the business park and its vicinity. For one of the solutions tested, parallel to the local low-voltage AC electricity grid, a direct power line between the heating system of the aquarium and the PV charging station was established. This allows to make use of excess electricity and test the potential of peer-to-peer exchange of energy.

5.3.3 Mission and Economic Dimensions

The city of Hartberg has a vision and commitment to become CO_2 neutral. To reach this goal, a roadmap and action plan have been developed including measures to achieve zero

CO_2 emissions until 2040 by using effectively local resources within the city and its surrounding communities. For more than 25 years Hartberg has been a member of the Austrian network for climate protection (Netzwerk Klimabündnis). The development of the Ökopark is also linked to this vision. Since its start, the manager of the Stadtwerke was personally taking care of the implementation of the vision of climate neutrality and the related city strategy, which proved to be one of the success factors in promoting the envisaged smart city development in Hartberg. In their innovation approach, Stadtwerke Hartberg uses the business park as a testbed for its sustainable business model – a concept for optimised energy management including PV-supported E-mobility charging stations. At the same time, Ökopark is also intended to be an attractive campus for local firms. Overall, urban and regional policies have been strong drivers and important elements of the mission of the business park as smart energy system. As part of the regional and economic development strategy, the Ökopark also hosts research and firms dealing with environmental technologies and greening the economy.

One of the main success factors of this place for innovation and experimentation is the continuity in the implementation of the long-term vision shared by the City of Hartberg, Stadtwerke and Ökopark. This collective cognitive frame had been in place well before the solutions described in this case study. The orchestration of a shared vision and mission turned out to be crucial for experimenting with innovative collaborative business models. For replicating a sustainable business park model such as in Hartberg, this might be a critical factor as long as replication involves co-creation and experimenting to adapt the model to new context conditions.

5.3.4 Social Networks Involved and Actor Constellation

With respect to social networks involved, the close collaboration between actors and stakeholders is rather exceptional. The municipality as owner of the energy utility and its subsidiaries, as well as the municipality as energy consumer, allows for special conditions for experimenting. Moreover, the long-term commitment of key individuals as caretakers also seems to have played an important role to orchestrate the complex network of stakeholders and actor groups in the city. In Hartberg, the manager of Stadtwerke took on this role over several years, particularly linking municipal policy-making with the local business sector and the various actors within the utility.

Another aspect that makes replication challenging is that the business park is also run by the Stadtwerke. Thus, not only do they have a close relationship as grid operators with end users in an energy supply contract (electricity, heat and electric car charging) but also as landlords with tenants (firms renting the offices, operation of the entertainment complex, and renting space in a car park).

5.3.5 Institutional Setting

With respect to formal institutional settings, ownership of the grid and its operation are two main factors. In this respect, Hartberg's energy infrastructure is owned by the municipality and operated by its own energy utility. 'Stadtwerke' as municipally owned energy utilities are a peculiarity in the European context and are mostly found in Germany and Austria.

For less than 1,00,000 customers, the European legal framework allows for exemptions from the obligatory unbundling among distribution networks, energy generation and supply. In Hartberg, grid operation and electricity supply are vertically integrated into one company.

A further institutional dimension to be highlighted here is that of the energy regulation and the room it provides for contractual arrangements. In general, the energy regulator is aiming to set the rules for both, the monopolistic grid operator's rights and obligations. An example from the Ökopark is the above-mentioned set-up of a direct line for peer-to-peer exchange of electricity. Under Austrian regulation, the monopolistic distribution system operator has the right to forbid two grid users to exchange energy through the electricity grid or to build a separate power line. Thus, many aspects of the pilot project would need exceptional contractual arrangements between involved businesses in an unbundled energy system context. This would either increase transaction costs, making the potential business model of the use case less attractive, or would be impossible at all.

The concurrence of almost all relevant actors within one or two organisations, including the ownership of the business park infrastructure, allows for experimenting in ways that are rather exceptional. To take up the example of peer-to-peer exchange of electricity, testing solutions with direct DC connection between a PV site and a large end user would be hampered by the interest of one of the owners or operators of generation, distribution, or consumption, be it because of conflicting interests (e.g. grid operators not wanting to lose income), too high transaction cost, or too little profit to be shared. In Hartberg, no regulatory hurdles are hindering the installation nor are there significant transaction costs between different units within the organisation. Furthermore, sharing the benefits with a tenant with a sufficiently high load of DC power (e.g. heating an aquarium) will also be much easier.

5.3.6 Replication and Upscaling

If we consider the potential of this smart energy experiment for replication and upscaling, one of the critical conditions seems to be the 'bundled' context as Hartberg's local grid is operated by its own energy utility, Stadtwerke Hartberg, which is also the energy supplier. While the use case could be replicated in other places with vertically integrated municipal utilities, replication would be limited to locations in rural or small urban areas and not suitable for significant upscaling. For other countries outside of the EU with less restrictive legislation on the separation of grid operation and energy supply services, this limitation would not become an obstacle to the replication of the business park use case itself. In an 'unbundled' context with an operator of the electricity grid in charge of connecting a business park, which does not share the same goals as the electricity supplier, the specific solutions chosen would face a more complex situation with a higher risk of failure. Even more challenging is the situation when different suppliers provide electricity and heat to the business park which would require a quite sophisticated collaborative business model.

5.3.7 The Case of Hyllie, Malmö

5.3.7.1 Spatial-Structural Dimension

In the new urban district, Hyllie in the city of Malmö, Sweden, a smart grid platform has been implemented which in the longer term will integrate both the electricity and

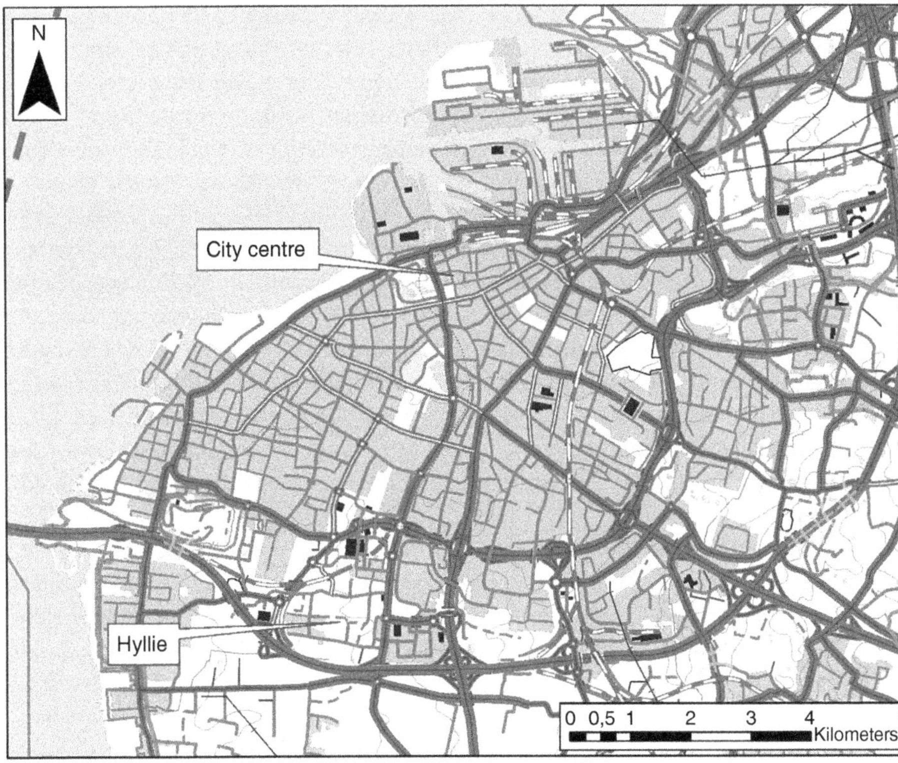

Figure 5.2 Location of Hyllie. © Lantmäteriet.

the district heating grid. Currently, the system is mainly applied to manage heat loads of the district heating system by using building structures for heat storage.

Malmö is located in southern Sweden and with around 3,40,000 inhabitants, it is Sweden´s third-largest city. Hyllie is Malmö's largest development area, located in southern Malmö (Figure 5.2), and in the final phase, will comprise around 12,000 new homes and around 15,000 office spaces. Hyllie is seen to be a 'lighthouse' for Malmö's target to become 100% renewable by the year 2030. Further aims and procedures on the way to becoming Öresund region's most climate-smart city district have been laid out in a climate contract between the city of Malmö and the companies VA Syd and EON as well as in the Hyllie Environmental Programme (HEP). The aim for Hyllie is to provide 100% renewable and 'recovered' energy, sourced by a large proportion from local sources; develop a sustainable local transport system based on electricity and gas; create an integrated energy system of electricity, gas, heating and cooling; and reduce energy consumption by connecting energy-efficient buildings to a smart grid and facilitate a climate-smart lifestyle.

5.3.7.2 Technological Dimension

The developer, German-owned E.ON, presents the Hyllie project as a holistic smart city concept, including smart homes, local production, integrated networks, local storage and sustainable mobility. In terms of smart grid configurations, the main part is based on the district heating system. The system is applied to manage heat loads of the system by using building structures for heat storage. To date, the smart electricity grid features are limited to a small number of smart home-equipped apartments. The most advanced smart grid feature which has been implemented is a 'smart district heating grid platform' which allows to manage district heating loads through demand-side measures. In the longer run, the aim is to develop a smart grid platform operating across the heat and electricity grid.

The main feature of the smart district heating system is a smart residential gateway (control box with mini-computer) connected to the buildings' energy management systems allowing to use the thermal storage capacity of building structure by slightly overheating the building in advance of expected peak demand and thus reducing heat demand during the peak hours. The increase in temperature (usually below 1 °C) is at a level where it goes unnoticed by residents in the building. The technical challenge beyond building a smart control platform has been to make the residential gateways compatible with different types of building energy management systems. In the longer run, the residential gateways are planned to provide other types of smart grid functionalities to the buildings and households.

5.3.7.3 Mission and Economic Dimensions

From the Swedish Energy Agency's perspective, the pilot project is part of a wider strategy to develop smart grids in Sweden as central element of a future sustainable energy system. However, making smart grids into an export industry is an explicitly stated aim as well. From the perspective of Malmö municipality, the main goal is developing the urban district of Hyllie. The main vision is 'With communication at the centre, Hyllie shall be a driving force for growth and sustainable development of the new Malmö – a place that crosser borders and is facing outward with the whole world as the arena'. Sustainable development and the mitigation of climate change are key elements of this vision, and the smart grid is regarded as an important part of this.

The smart district heating system has so far turned out to be the only smart grid application in Malmö with a viable business case. Managing the district heating loads by shaving off peak loads allows for more efficient use of the district heating grid system and reduced need to expand its capacity. A broader installation of load management would have further potential and is currently implemented also in existing residential areas in Malmö and other Swedish cities. This is even more so the case since E.ON also owns energy companies with predominantly district heating in 13 other Swedish cities as the energy infrastructure has been privatised since 1991.

5.3.7.4 Social Networks Involved and Actor Constellation

In terms of contractual relations between energy suppliers and customers, there is a difference between district heating and electricity. District heating customers are 'passive users'. In multi-dwelling buildings in Sweden, the heating cost is included in the rent, and residents cannot choose another heat provider. In electricity, the customers can choose whom they buy their electricity from, but the grids are monopolised. In district heating, grid and production are not separated.

The smart energy model in Hyllie is embedded in a broad partnership of the municipality, E.ON as the owner of the district heating system and heat provider, and municipal or private building developers and owners. A foundation for these collaborative relationships is the Climate Contract and Hyllie Environmental Programme as well as a long tradition of collaborative public–private partnerships for the provision of different types of infrastructure services in Malmö municipality. Most buildings in Hyllie and Malmö are connected to the district heating system. The Swedish Energy Agency is a key partner in terms of funds for the pilot; they supported the project with 47 million SEK (4.4 million Euros) from 2011 onwards.

Key economic actors in this constellation are thus the municipality, building developers (public or private), energy companies (mainly E.ON) and customers. Customer incentives are rather limited as they are passive in terms of technological configuration.

5.3.7.5 Institutional Setting

The electricity market in Sweden is liberalised with competition in production and sales but monopolised for transmission and distribution. The district heating infrastructure is vertically integrated with one supplier who also owns the grid. Although competition may occur if all parties agree, this has not been implemented anywhere yet in Sweden. Regulations are strict and significant standardisation has occurred in the district heating sector throughout the twentieth century, meaning that the differences between different electric systems or DH systems are not that large. From the national perspective, the smart grid development is supported in various ways, through institutional channels as well as informally.

The privatisation of energy companies has led to a changed relationship between municipalities, the main actor in urban planning and often the largest owner of multi-dwelling buildings in the municipalities, and the energy companies. If the energy company is not owned by the municipality, which was mainly the case when building the energy systems, the only way for the municipality to influence the decisions is through partnership and negotiation.

5.3.7.6 Replication and Upscaling

In terms of the potential for replication, the smart energy model in Hyllie is thus a smart grid platform for the load management of district heating with an IT infrastructure for control and prediction of heat loads and special devices interacting with the building energy system to store heat in the building structure in advance of peak load of the district heating system. Despite a number of technical limitations (control infrastructure needs to be built up; interfaces with building energy management systems and their technical specifications need to be developed and are currently not commercially available), only few of the contextual conditions appear to be critical for replication or upscaling of this smart grid solution.

Obviously, such solutions are only relevant for cities with district heating systems (and preferably in situations where a reduction in peak loads is useful, e.g. by reducing the need to increase the capacity of heat pipes despite expansion of the DH system) and only buildings with an energy management system can be integrated. The close strategic and organisational coordination of grid operation and energy supply is a precondition, although this is usually the case with district heating systems, which are – unlike in the case of electricity grids – vertically integrated utilities. A cooperative relationship between utility and building owners is of importance because the latter have to voluntarily approve the installation of such a system in their building without financial benefits (otherwise the business case for this solution in Hyllie might be in jeopardy). It is thus also advantageous to have only a limited number of building owners as contract partners in the supply area. As has already been tested by E.ON, the system is scalable and can be expanded to further parts of the city and it can also be applied to the existing building stock. For E.ON itself there are possibilities to replicate the system in other cities in which they own the energy system.

The aspect of testing the solution in a newly built area is something to take into account, as this provides a 'blank slate' for system development without limitations of existing physical structures. Moreover, in a large project like this, extensive negotiations are required in the planning process, and the commitment to over-arching goals of sustainability seems crucial in order to keep focus on developing innovative smart technical features, otherwise 'business as usual' would be a logical path to follow.

5.4 Discussion and Conclusions

The analysis of our two cases of smart grid demonstration projects makes clear that the replicability of solutions developed in these projects, i.e. their contribution to the development of smart grids in other places, depends on a combination of technical and non-technical (social, institutional and economic) factors. These factors turned out to be highly context specific and depend on the local constellation of actors, the historic development of these places, the role of key individuals, and the shared visions and future plans for reaching a more sustainable energy system. As Ryghaug et al. (2019, p. 11) find in a similar analysis of pilot projects, the specific project contexts often have 'properties that produce new alignments between users and technologies that create new expectations about the future energy system'. Current concepts of replication and transfer of pilot projects often uphold the idea that such smart energy arrangements can be disentangled from these local contexts and reproduced somewhere else. In light of concrete cases of smart cities or smart grids, such a transfer is, however, severely restrained in practice. What would rather be required – and in both cases we only saw minor attempts to do so – are more systematic efforts and strategies to facilitate learning processes between and beyond such pilot projects. An element of such learning would be to better understand the situatedness of smart energy projects and the variety of context conditions that have contributed to the success or failure of existing cases. This is what we tried to demonstrate in a structured way in our case analyses.

In particular, we tried to identify additional dimensions which appear to be essential to better understand and assess the replicability of solutions from pilot projects: a spatial-structural dimension, a mission dimension and the three social dimensions taken up from Beckert's field concept (Beckert 2010), i.e. formal institutions, social networks and cognitive frames. As those social forces are integrated into a dynamic relationship, they take the structures of social relations and relational patterns into account just as they also put emphasis on the constraining rules and norms that influence the structure of social networks and cognitive frames of individuals. Furthermore, they refer to legitimation through collective power and commonly shared meanings and sense-making processes. Those dimensions not only allow for a more differentiated and context-sensitive approach to replication and upscaling, but they also establish a framework for the evaluation and identification of success factors for replicating smart grid experiments. By facilitating learning from experiments, such analysis can be important elements in governance or sustainable urban energy transitions.

The structuring of the analysis of the two smart grid case studies from Malmö/Hyllie in Sweden and Ökopark Hartberg in Austria along different dimensions of their socio-technical configuration demonstrates the applicability of this approach. Making these dimensions explicit in the case analysis provides a basis for further reflection and analysis of the potential to implement these use cases in other contexts. While the possibility of replication was hardly influenced by certain dimensions (e.g. the ownership structure in the case of smart district heating grid load management), other dimensions turned out to be highly critical (e.g. certain regulatory contexts and shared vision between municipality and utility) for the implementation of these solutions.

In the end, the analysis of use cases of smart grids as sketched out in this paper is supposed to support processes of transfer and learning across different sites of smart grid applications. In transition studies, such processes have been conceptualised as inter-local learning (Raven and Geels 2010) and have been identified as key elements of the growth and upscaling of socio-technical niches. Replication and upscaling of smart grid experiments can be seen as targeted, problem-solving attempts of learning from one experiment and transferring the (successful) elements of this demonstration project to another site. As with other examples of technology transfer between different contexts (see, for example, Ulsrud et al. 2018), the contextualised nature of such experiments requires elements of decontextualisation and recontextualisation in the process of replication and upscaling.

Thereby, *contextualisation* refers to the act or process of putting information or specific factors into context (making sense of information from the situation or location in which the information was found), while *decontextualisation* occurs when those factors become separated from their social, cultural and institutional context. However, if we view context as a process rather than a static concept (see Gumperz 1982), this allows the actors who replicate the smart grid experiment to create, maintain, and change which contextual factors are relevant to them, whereas we need to consider that it is never one single context but usually a multiplicity of contexts (such as set of actions, stakeholders and preconditions) that changes due to interaction. In practice, strategies of upscaling and transfer face a series of challenges as the 'work' of embedding needs the close collaboration of local actors and stakeholders and thus a translation between different types of knowledge (academic vs. practical), the alignment of bottom-up initiatives, etc. (Augenstein et al. 2020). Such active processes of transfer as translation thus indeed rather resemble processes of re-innovation than dissemination or replication of a stable configuration (Peng et al. 2019).

In terms of replication and upscaling, both described case studies showed the importance of a caretaker role as well as political/strategic support to orchestrate the complex network of stakeholders and projects in the city, as well as to ensure continuous development of follow-up initiatives and the shaping of the socio-technical framework for upscaling and new experiments. Obviously, such strategies of replication and implementation of solutions in other places are just one step in moving beyond pilot projects, and more fundamental changes of the contextual conditions, which have been treated as stable in our analysis, are needed for a socio-technical transition of the electricity system. Nevertheless, the spreading and replication of new arrangements developed in pilot projects are a crucial step in this way and require the development of more nuanced replication strategies than we can observe today.

Acknowledgements

We are grateful to the ERA-net Smart Grid plus programme for funding the project 'Replicability Concept for Flexible Smart Grids' which provided the case study data for this chapter, as well as the working group 'IEA ISGAN Annex 7: Smart Grids Transitions' for valuable feedback.

References

Ansell, C.K. and Bartenberger, M. (2016). Varieties of experimentalism. *Ecological Economics* 130: 64–73.

Augenstein, K., Bachmann, B., Egermann, M. et al. (2020). From niche to mainstream: the dilemmas of scaling up sustainable alternatives. *GAIA - Ecological Perspectives for Science and Society* 29 (3): 143–147.

Beckert, J. (2010). How do fields change? the interrelations of institutions, networks, and cognition in the dynamics of markets. *Organization Studies* 31 (5): 605–627.

Bulkeley, H., Marvin, S., Palgan, Y.V. et al. (2019). Urban living laboratories: conducting the experimental city? *European Urban and Regional Studies* 26 (4): 317–335.

Camagni, R. (2019). Territorial capital and regional development. In: *Handbook of Regional Growth and Development Theories* (ed. R. Capello and P. Nijkamp), 118–132. Cheltenham: Edward Elgar.

van Doren, D., Giezen, M., Driessen, P.P.J., and Runhaar, H.A.C. (2016). Scaling-up energy conservation initiatives: barriers and local strategies. *Sustainable Cities and Society* 26: 227–239.

van Doren, D., Driessen, P.P., Runhaar, H., and Giezen, M. (2018). Scaling-up low-carbon urban initiatives: towards a better understanding. *Urban Studies* 55 (1): 175–194.

Evans, J., Karvonen, A., and Raven, R. (ed.) (2016). *The Experimental City*. London: Routledge.

Friedmann, J. (2002). *The Prospect of Cities*. Minneapolis, MN: University of Minnesota Press.

Gangale, F., Vasiljevska, J., Covrig, C.F. et al. (2017). *Smart Grid Projects Outlook 2017: Facts, Figures and Trends in Europe*. Luxembourg: Publications Office of the European Union.

Geels, F.W. and Raven, R.P.J.M. (2006). Non-linearity and expectations in niche-development trajectories: ups and downs in Dutch biogas development (1973–2003). *Technology Analysis & Strategic Management* 18 (3/4): 375–392.

Gumperz, J.J. (1982). *Discourse Strategies*. Cambridge: Cambridge University Press.

Kern, K. (2019). Cities as leaders in EU multilevel climate governance: embedded upscaling of local experiments in Europe. *Environmental Politics* 28 (1): 125–145.

May, K., Sigrist, L., Vingerhoets, P. et al. (2015). Improving scalability and replicability of smart grid projects. *23rd International Conference on Electricity Distribution. CIRED Conference Proceedings* (15–18 June 2015), Lyon: CIRED - Congrès International des Réseaux Electriques de Distribution.

Naber, R., Raven, R., Kouw, M., and Dassen, T. (2017). Scaling up sustainable energy innovations. *Energy Policy* 110: 342–354.

ÖROK ed. (2018). *15. Raumordnungsbericht: Analysen und Berichte zur räumlichen Entwicklung Österreichs 2015–2017*. Vienna: Österreichische Raumordnunskonferenz.

Österreichischer Städtebund ed. (2017). *Weißbuch Innenstadt. Chancen und Herausforderungen der Innenstadtbelebung für Klein- und Mittelstädte*. Graz: Österreichischer Städtebund, Landesgruppe Steiermark.

Peng, Y., Wei, Y., and Bai, X. (2019). Scaling urban sustainability experiments: contextualization as an innovation. *Journal of Cleaner Production* 227: 302–312.

Rae, C., Kerr, S., and Maroto-Valer, M.M. (2020). Upscaling smart local energy systems: a review of technical barriers. *Renewable and Sustainable Energy Reviews* 131: 110020.

Raven, R.P.J.M. and Geels, F.W. (2010). Socio-cognitive evolution in niche development: comparative analysis of biogas development in Denmark and the Netherlands (1973–2004). *Technovation* 30 (2): 87–99.

Rodriguez-Calvo, A., Cossent, R., and Frías, P. (2018). Scalability and replicability analysis of large-scale smart grid implementations: approaches and proposals in Europe. *Renewable and Sustainable Energy Reviews* 93: 1–15.

Ryghaug, M., Ornetzeder, M., Skjølsvold, T.M., and Throndsen, W. (2019). The role of experiments and demonstration projects in efforts of upscaling: an analysis of two projects attempting to reconfigure production and consumption in energy and mobility. *Sustainability* 11 (20): 5771.

Schot, J. and Geels, F.W. (2008). Strategic niche management and sustainable innovation journeys: theory, findings, research agenda, and policy. *Technology Analysis & Strategic Management* 20 (5): 537–554.

Sigrist, L. and Rouco, L. (2012). GRID+ (Project no. 282794). Supporting the development of the European Electricity Grids Initiative. *Report D 4.1 Review of the State of the Art*, European Commission.

Sigrist, L., May, K., Morch, A. et al. (2016). On scalability and replicability of smart grid projects – a case study. *Energies* 9 (3): 195.

Smith, A. and Raven, R. (2012). What is protective space? Reconsidering niches in transitions to sustainability. *Research Policy* 41 (6): 1025–1036.

Ulsrud, K., Rohracher, H., and Muchunku, C. (2018). Spatial transfer of innovations: south-south learning on village-scale solar power supply between India and Kenya. *Energy Policy* 114: 89–97.

Voytenko, Y., McCormick, K., Evans, J., and Schliwa, G. (2016). Urban living labs for sustainability and low carbon cities in Europe: towards a research agenda. *Journal of Cleaner Production* 123: 45–54.

WEF (2010). *Accelerating Successful Smart Grid Pilots*. Geneva: World Economic Forum.

van Winden, W. and van den Buuse, D. (2017). Smart city pilot projects: exploring the dimensions and conditions of scaling up. *Journal of Urban Technology* 24 (4): 51–72.

6

The Role of Active Buildings in Smart Energy Imaginaries: Implications of Living Well in Low-Carbon Homes and Neighbourhoods

Kate O'Sullivan[1], Fiona Shirani[1], Nick Pidgeon[2], and Karen Henwood[1]

[1] *School of Social Sciences, Cardiff University, Cardiff, Wales, UK*
[2] *School of Psychology, Cardiff University, Cardiff, Wales, UK*

6.1 Introduction

Decarbonisation to reduce carbon dioxide (CO_2) emissions and mitigate the effects of climate change is a widespread policy goal, reflected in ambitions at international, national, regional and even city scales. In the United Kingdom, early commitments to limit greenhouse gas (GHG) emissions have been superseded by new targets of 100% (or net zero) reduction in GHG emissions by 2050 (Climate Change Act 2008, 2050 Target Amendment). To achieve such ambitious targets necessitates decarbonisation across all sectors of society and the economy. Multi-faceted and multi-scale decarbonisation is viewed as an opportunity to address several cross-sector issues: climate change, energy security, energy vulnerability and the development of a sustainable low-carbon economy (c.f. Welsh Government 2019; HM Government 2020).

Decarbonisation of the energy system has for some time been a policy priority and has seen significant reductions in its GHG emissions; conversely, decarbonisation of buildings has seen unstable and changeable policy commitments (CCC 2020a). Between 2008 and 2018, the buildings sector reduced GHG emissions by 13%, attributed to policy directives. However, the phase out and 'scrapping' of some policy directives, especially for new homes, have slowed the reduction of emissions from the sector, with the 1.8 million new homes built since 2007 requiring low-carbon retrofit (CCC 2020a). While there is no agreed pathway to net zero, future scenarios – such as the UK government's 2020 energy white paper 'Powering our net zero future' and energy regulator Ofgem's (2020) 'Decarbonisation programme action plan' – anticipate significant electrification of not just the energy system (including heating) but also of transport and buildings (c.f. BEIS 2020; CCC 2019; Ofgem 2020). In 2019, buildings accounted for 18% of UK GHG emissions (CCC 2020a) and 40% of carbon emissions, most of which was from residential buildings (CCC 2020b). As such, decarbonisation pathways highlight that buildings, especially homes, must be increasingly energy efficient, powered by low-carbon and 'smart' energy (CCC 2020a;

Smart Cities, Energy and Climate: Governing Cities for a Low-Carbon Future, First Edition.
Edited by Oleg Golubchikov and Komali Yenneti.

National Grid ESO 2020a; Ofgem 2020). Furthermore, these changes indicate that how and when energy is demanded will alter in the future, both caused by and causing consumer 'behaviour change' (National Grid ESO 2020a; Ofgem 2020).

The symbiotic relationship between the energy sector (as a provider of energy) and the building sector (as site of significant energy demand) means that decarbonisation of one affects the other. These sectors are also becoming increasingly entangled as buildings transform into sites of energy storage and production as well as demand. Electrification necessitates development of intelligently digitalised (or smart) energy systems that can manage new distributed and multi-scalar forms of energy production, storage and demand, available variably through time. Thus, pathways for decarbonisation through electrification are now also interlinked with visions of smart energy futures and smart city imaginaries. At the convergence between these increasingly mutually reaffirming visions are buildings as homes and the people living within them. Envisaged as active elements within a national energy system, through intelligent digitalisation and capacity to produce, store and demand energy, Active Homes are an integral component of grid flexibility, increasingly required as electricity demand and renewable energy rise. At the same time, homes and their occupants are an important element of the city.

Apart from distanced and generic imaginings of smart city occupants, there are few detailed imaginings of what life will be like in these future scenarios. The same can be said for smart homes, where life within them is largely imagined as unproblematic, and actually improved, due to the streamlining of occupant activities, daily needs, and broader quality of life by smart technologies. Active Homes are already being designed and developed, incorporating low-carbon energy production, energy storage and smart technologies towards the achievement of decarbonised, smart energy futures. These innovations could signify steps towards both decarbonisation pathways and smart city imaginaries. However, such innovative and novel developments rely on assumed and generic occupant lifestyles and behaviours to inform their combination of materialities, including energy and smart technologies. These assumptions are often informed by limited understandings of how, why, and to what effect people carry out energy-related and other daily behaviours. Reliance on such assumptions may impair the ability of the homes to perform as expected, in terms of energy consumption, CO_2 emissions and grid flexibility, as well as to provide homes that people can, and want to, live well within.

This chapter draws on early insights from our Living Well in Low Carbon Homes research project, in which we are exploring lived experiences within a range of differently located, imagined and materialised Active Homes through time. Drawing on early research insights, we illustrate how intelligent digitalisation of the homes' energy systems is expected to provide comfortable, affordable, low-carbon living while also being integral elements of a flexible electricity grid. Contrary to these assumptions, smart technologies do not always work as planned, or necessarily integrate with new Active Home technologies as seamlessly as imagined. Furthermore, in line with already established critiques of the smart imaginary, we highlight a potential risk of the technologies being exclusionary to some households or certain members of households. First, we outline the intersection between decarbonisation pathways and smart city imaginaries, exploring the role of homes and households in this evolving context. We then move to describe how Active Homes may represent a step towards the realisation of a decarbonised, smart energy future. By drawing on empirical evidence from our research, we then explore how stakeholder expectations of the future energy and building sectors interplay with widespread assumptions around household capacity and capability to both understand and digitally manage their energy systems. We elucidate how these perceptions directly influence the materialisation of the homes and the daily lives of those people living within them. Finally, we consider the implications of current Active Home development for both the daily lives of households and for wider city-scale smart imaginaries.

6.2 Decarbonisation and Smart Energy Systems

Electrification of the energy network is a dominant pathway towards whole-system decarbonisation, yet there are a number of challenges to overcome. Electricity demand is predicted to substantially increase as it is used for heating, replacing some of the existing daily and seasonal peaks in gas demand and increasing the peak electricity load overall (Wilson et al. 2014). In addition, the grid's ability to meet such demand will be more difficult as renewable energy production is at times misaligned with periods of peak energy demand (Wilson et al. 2014). To overcome this, energy demand needs to be modifiable over time, either through demand reduction or demand shifting; in other words, there needs to be increased 'flexibility' in the energy system. One way of creating such flexibility involves developing information communication technologies (ICT) that can support real-time communication between energy production, energy storage and energy demand at multiple scales (Ofgem 2020). Conceptualised as a 'smart grid', such a system can integrate and more efficiently manage many distributed energy producers and storage, while simultaneously achieving multiple energy demands.

The development of a smart grid can enable the digital integration of buildings and their users, as well as other sectors (e.g. transport and industry) into the grid, thus working towards the decarbonisation of multiple sectors. Digitalised connection and communication are not confined to building grid arrangements, it is multi-faceted, allowing communication between buildings and other infrastructure. In addition, it enables energy production, demand and storage aggregation at neighbourhood and even city scales. Cities, with their concentrations of human populations, buildings and infrastructure, are responsible for almost 80% of global energy consumption and over 60% of GHG emissions (Harris et al. 2020). Yet, simultaneously, cities are perceived to be the key sites for enacting cross-sector decarbonisation, envisaged as sustainable, socially fair and economically efficient (Pulselli et al. 2021). It is here where decarbonisation pathways converge with smart city socio-technical imaginaries, which, while constantly evolving (Sadowski and Bendor 2019), are based on the understanding that intelligent digital communication and automation between multiple elements of the city, material and social hold benefits for decarbonisation, the environment, human well-being and economic growth:

> The safe, secure, environmental and efficient urban centre of the future with advanced information and communication technologies to stimulate sustainable development in the economic growth and a high quality of life.
>
> *(Tan et al. 2017, p. 1921)*

As socio-technical imaginaries or visions for the future, smart cities encompass 'assemblages of materiality, meaning, and morality' (Jasanoff 2015, p. 4) and represent a widely accepted, practiced and institutionally informed vision of a desirable future (Jasanoff 2015) attained through the advancement and integration of ICT. Indeed, the wide-ranging remit of smart cities in solving economic, environmental and social problems positions them as a form of technological saviour (Cugurullo 2018; Sadowski and Bendor 2019). As ICT can, in theory, be applied to almost any aspect of our material, built environment, and used by and for social endeavours and fulfilment of social practices, smart cities are flexible in how exactly they are interpreted and developed in practice. In this way, smart cities are transferable and customisable between different geographies, political economies and cultures (Cugurullo 2018), being reflective of, while also transforming, the place they are within.

Increasingly, it is perceived that smart city is inevitable, partly because it is already embedded within long-standing views of cities as future-oriented and progressive places of innovation (Sadowski and Bendor 2019). However, the strength of 'smart city' as an imagined future and pragmatic policy goal means that alternative imaginaries may be locked out (Sadowski and Bendor 2019). Critiques of smart cities point out that they can place strong focus on 'hard infrastructure' such as utilities, urban spaces and buildings, leaving little space for 'soft infrastructures' such as institutional, social and personal processes (Corsinia et al. 2019, p. 323). Frequently excluded from smart city narratives are how to address already existent social problems or even cater to different social groups (Gonella 2019), raising further concerns for social inclusivity and place making. Indeed, as around the world smart cities are being realised, the benefits can be regressive in their distribution, creating or reinforcing already present, socio-economic and spatial advantages and disadvantages (Gonella 2019). Together, these inconsistencies and exclusions raise questions as to whether smart cities can really deliver on all fronts: social, economic and environmental.

6.3 Smart Homes, Smart Occupants?

Smart homes have been described as the basic building block for smart cities (Hui et al. 2017). Like smart cities, smart homes are imagined as enabling communication between the material and social elements within and outside of a home, holding multiple benefits for occupants. Aldrich (2003, p. 17) conceptualises smart homes as places where 'computing and information technology [. . .] anticipates and responds to the needs of the occupants, while working to promote their comfort, convenience, security, entertainment, healthcare, education, and communication through the management of technology within the home and connections to the world beyond'. However, questions have been raised about who smart technologies are designed for, particularly in the use of features such as remote control of heating (Shirani et al. 2020), which has been described by Hui et al. (2017) as 'hype' rather than a necessity, providing little incentive for users to upgrade from ordinary homes to smart homes.

The smart city imaginary stems from the rapid advancements in science and digital technology which saw new visions of urban development emerge (Cugurullo 2018). But, as with past technological utopian visions, and similar to critiques of the smart city, the domestic 'smart utopian' (Strengers 2013, p. 2) or 'domestic-machine utopia' (Darby 2018, p. 141) can be exclusionary to those outside of the dominant vision (Wilson et al. 2014) as we consider further below. Furthermore, smart city imaginaries often present a streamlined, generic and uncomplicated version of the city, where distinctive place-specific characteristics are excluded (Valdez et al. 2020). This simplification extends to how the occupants of smart cities are imagined, as an 'audience' to this new city and 'reduced to manageable flows and units rather than complicated beings' (Valdez et al. 2020, p. 232).

Smart city visions of passive populations managed and streamlined through intelligent digitalisation are shared in visions of smart homes and their occupants (Goulden et al. 2018; Gram-Hanssen and Darby 2018; Hine 2020; Strengers 2013), as well as in domestic energy demand models (McKenna et al. 2017). Smart home occupants are often assumed to be either logical, reasoned and IT literate (Strengers 2013) or 'unable or unwilling to make lifestyle changes' (Gram-Hanssen and Darby 2018, p. 97). Strengers (2013) describes the former archetype as 'resource man', who is imagined to be a responsive and rational economic agent, an ideal as entrenched within the energy industry (Verkade and Höffken 2017), whereas the latter archetype has been termed the 'indifferent consumer' who is disengaged and irrational (Goulden et al. 2018). In combination with 'designer expectations', which are not always empirically informed (McKenna et al. 2017, p. 1584), models are drawn on to inform the types of technological and energy services and management tools or devices that become integrated into building design (Gaetani et al. 2020; McKenna et al. 2017). Social science research suggests these misconceptions of consumers are problematic, misrepresenting the realities of most people's energy consumption and failing to give due attention to social dynamics and processes relevant to everyday energy use (Goulden et al. 2018; Strengers 2013).

Prior smart energy visions at household scale have focused on demand-side responses (DSR) that enable the shift of energy demand (Calver and Simcock 2021), including smart meters that communicate between the home, the energy supplier and the grid (BEIS 2020; Tirado Herrero et al. 2018). In their critical overview of DSR in the context of energy justice, Calver and Simcock (2021, p. 2) point to the way DSR is expected to significantly change household interaction with the energy system; moving from passive consumers to active 'energy engagers'. Through being more 'informed' of their energy demand and the cost of consumption, households are 'empowered' to manage this (Tirado Herrero et al. 2018, p. 67), reducing their demand to reduce their bills. In shifting demand to reduce peaks, lower prices can be offered for energy consumption, otherwise referred to as time of use (ToU) tariffs (Calver and Simcock 2021; Tirado Herrero et al. 2018). As Ozaki (2018) points out, adopting a 'rational choice' view of consumers leads to the assumption that inadequate information is the dominant 'cause' of consumers failing to act as expected. Alternatively, direct load control assumes a passive role for households, where household energy demand is managed by agents outside of the home, such as energy supply companies or energy service companies, network operators, or energy aggregators (Calver and Simcock 2021). These agents take action to increase household energy efficiency and reduce energy costs on behalf of the household. Finally, taking passive consumer and digital management a step further, automated appliances over time learn household energy demand routines, autonomously meeting these while communicating with the wider grid (Gram-Hanssen and Darby 2018; Tirado Herrero et al. 2018).

Intelligent digitalisation of the energy system and the installation of smart meters are increasingly perceived as normative action towards decarbonisation. Widespread installation of smart meters combined with the rapid uptake of new smart speaker assistants (SSAs) such as Amazon Echo and Amazon Alexa (Brause and Blank 2019) may indicate a quickening towards 'smart utopian' visions and the changing role of households in future energy systems. However, less discussed is that smart energy systems are being developed within an uneven socio-economic landscape and that differences between household 'energy needs, vulnerabilities and capacities' means that the impacts on households' daily lives will also be variable (Calver and Simcock 2021, p. 2). Importantly, the assumption that smart systems will universally hold financial and wider social benefits to all households is misplaced, raising questions around energy justice (Calver and Simcock 2021), with the potential for people without access to a 'smart new world of energy services' being excluded, or becoming the new fuel poor (Sovacool et al. 2019, p. 499). As Hine (2020, p. 23) notes, there are 'significant silences within the simplistic story of empowerment and control that tends to permeate public discourse around smart home technologies'. For example, households living with constrained financial budgets are often already acutely aware of their energy demand and have to take actions to limit it as much as possible, meaning that monitoring their energy using smart devices is seen to hold little additional benefit (Shirani et al. 2020). Furthermore, for households with pre-payment energy meters, it may not be possible to use smart monitoring devices due to incompatibility and ToU tariffs may not be available (Sovacool et al. 2019).

Beyond concerns for the distribution of costs and benefits associated with smart homes are concerns for technology reliability, risks to household privacy and security, and changes to household autonomy (Shirani et al. 2020). Social science work has highlighted the intertwining of household energy demand with other 'logics, constellations and processes' (Skjølsvold et al. 2017, p. 3). Mundane energy use and demand reduction can also involve dynamic intra-, inter- and transgenerational transmissions. Embryonic pointers exist to the importance of these for understanding spatio-temporal patterning of psychological investments in consumer-centric, familial and other archetypes (Henwood et al. 2016). Whenever cultural constellations and identificatory processes converge in the deeper psychosocial patterns and logics at work in everyday life and sense making, forms of 'lock in' to energy demand practices can be generated (Henwood 2019) that limit capability to take advantage of ToU tariffs (Calver and Simcock 2021). Relying on simplistic archetypes of consumers and failing to account for the complexities of people's everyday lives create a risk that contemporary and evolving smart energy systems and smart building design will lead to homes that do not perform as expected, and fundamentally that people do not wish to live within. Furthermore, as building use models are often scaled up to inform neighbourhood-, district- and city-scale models (Happle et al. 2018), assumptions of occupancy behaviour at building scale also hold wider implications. Imaginaries of future consumers and expectations of their interactions with smart technology play a crucial role in the way homes are designed (Shirani et al. 2022), which will have an impact on how smart cities may be realised.

6.4 Realising Smart Energy Futures: Active Buildings and Homes

Active Buildings directly seek to meet occupant energy demands in ways that minimise carbon emissions and financial costs, by intelligently digitally managing both the buildings' capacity to produce and store energy, occupants' energy demands and signals from the energy grid, thus linking them directly with concepts of smart buildings and smart grids as discussed in the preceding sections. Elliott et al. (2020) provide the following definition of Active Buildings:

> Active Buildings support the wider energy system by intelligently integrating renewable energy technologies for heat, power and transport. Active Buildings are designed to be energy efficient, with novel ways of creating, controlling and releasing energy. Active Buildings have the potential to be energy self-sufficient. When connected with other Active Buildings in a network, they could have the ability to trade energy.
>
> *(Elliott et al. 2020, cited in O'Sullivan et al. 2020)*

Active Buildings present a logical development of other predecessors sustainable and low-carbon building concepts, including smart buildings (for an in-depth review of this, see O'Sullivan et al. 2020). While Active Buildings are imagined as a step towards decarbonisation, they also encapsulate other ambitions. For example, through their design, construction and occupied 'life-course'; to protect the environment; create communities of place; safeguard occupants and other building users' health and well-being; and make a positive socio-economic contribution to building users and local economies.

Active Homes are essentially Active Buildings that are lived in, but, as homes, they have specific materialities (technology combinations, layouts, size and energy systems) that vary according to how the homes are imagined to be 'used' or lived in by their occupants (Shirani et al. 2022). As different Active Homes are developed, varying expectations for the role of occupants in their energy management are incorporated into their design. Most require that some new occupant behaviours are adopted to achieve their low-/zero-carbon and grid flexibility outputs. Despite possible changes to the daily lives of Active Home occupants, there is also an ambition to maintain or improve their lives. For this ambition to be realised, elucidation of how people value their homes, what people expect from their homes, and how and why people carry out certain daily routines and energy behaviours must be understood and incorporated into their design.

Our Living Well in Low Carbon Homes research aims to understand these issues by exploring the lived experience of Active Homes. We adopt a qualitative methodological approach to generate in-depth understanding of the motivations behind and influences on the development of Active Homes, as well as elucidating the experiences of those who reside in them. The first of the project's research strands involves interviews with expert stakeholders – such as developers, architects and registered social landlords (RSLs) – to consider the rationale behind the Active Home developments, including how they envisage future residents interacting with the homes and technology. The second research strand comprises qualitative longitudinal interviews with residents prior to moving into their new

Active Home and twice within the first year afterwards to explore their experience of living in the home over time. Resident interviews are also interspersed with creative activity packs, providing participants with different mediums of expression. By researching with households over time and gaining insight into their lived experiences, we aim to elucidate how the homes maintain or alter daily life and expectations of home, how households adapt to life within the homes, and how well the homes perform for the household and their (low carbon and other) performance goals.

Our case sites consist of a number of different housing developments across South Wales. These developments have different materialities (size, floorplan, building materials, aesthetics and energy systems) and varying expectations around the role of the household in managing their home energy systems. This variability is reflective of the mix of stakeholders involved in each development and the political, economic and social context at the time of each development's conceptualisation. In addition, each of the case sites holds multiple ambitions, including those aimed directly at improving the daily lives of their occupants, for example, aiming to achieve reduced energy vulnerabilities through providing energy sufficiency or providing a healthy indoor environment for occupants' health and well-being. Other ambitions, while holding benefits for building occupants, are wider in their scope, e.g. fostering closer connections between occupants and nature while minimising environmental harm, disrupting existing socio-technical systems (energy sector, construction sector, or housing market) through system change, or using locally sourced resources to encourage local economic development.

Within this range of ambitions, materialities and stakeholders, there are areas of similarity that position all our case site developments within the active conceptualisation. All are low or zero carbon, all hold the capacity to provide grid flexibility, and all can be replicated or 'scaled up' to neighbourhood or city scales. It is through the integration of smart energy technologies into the building design that these active capabilities can be realised. The integration of smart technologies into these different low-carbon home designs also signals an acceptance and integration of a smart ontology within the active conceptualisation. Furthermore, as the digital integration of homes into a smart system is an integral element of smart city conceptualisations, Active Homes could represent a step towards this vision. In the following section, we present some initial insights stemming from our research at our first case sites. Here, we elucidate the way expectations and experiences of smart control systems have been discussed by both development stakeholders and residents in our research to date.

6.5 Living Well in Low-Carbon Homes – Initial Insights

The inclusion of smart meters in Active Homes is universally perceived by the developers interviewed for our research as an essential technology, as the future energy system, and indeed future homes will require smart energy management:

> [They] can't be on a half hourly tariff without a smart meter in the home. So the half hourly energy tariffs, one of the ways that we manage to give the residents low energy bills. So without a smart meter then we can't be on a half hourly tariff. (Expert 5)

As Expert 5 indicates, smart energy management was seen as an important way of reducing energy costs, which was an ambition held by all our case sites, as well as an expectation of future residents:

> And obviously, you know, utility bills, like electric and things would be cheaper. Well, they're expected to be cheaper, because obviously, they're supposed to be eco houses. (Emma, pre-occupancy)[1]

For developers, there was a view that the energy and building sectors are on the pathway towards decarbonisation, signalled by net-zero policy goals, and that the development of Active Homes was almost inevitable. This contributed to a perception that building low-carbon homes with smart capabilities and a mix of energy production and storage technologies would mean that the homes would be 'future-proof'. Future proofing creates a more attractive personal economic investment for prospective owner occupiers, or in the case of socially rented homes, a more secure economic investment to RSLs, as it negates the need for subsequent retrofitting. These concerns impact multiple design decisions including the energy technologies used and the level of and mode of interaction between the occupants and their energy systems. For example, while acknowledging that most people would not currently have an electric vehicle (EV), homes were equipped with EV charging points, in the expectation that in future they would become more commonplace.

> So every home of ours has a vehicle charger and our user interface has the ability to smart charge in a very smart way. And it'll be bundled in as part of their kind of package. (Expert 4)

> We didn't expect everybody to buy a house and turn up with an electric car, but it's there future proofing for society and the environment as well. (Expert 10)

Beyond smart EV charging, some case sites sought to encourage the use of smart appliances (as an additional purchase) in the hope of creating further energy efficiencies:

> What they can then have is a washer, a dryer, or a washer/dryer, and there are Samsung smart appliances. So basically, they're wi-fi enabled so that they can be controlled, triggered, stopped, paused during a cycle, should it be beneficial to the grid to do so. (Expert 6)

While our initial resident interviews indicate many were enthusiastic about EVs and smart appliances that could run on power-saving settings – seeing them as potentially beneficial both environmentally and financially – the cost was deemed by them to be prohibitively expensive, putting them out of reach for most people in the short term. Smart appliances, therefore, remained a future ideal for many of our participants.

During our expert interviews, households were often (although not exclusively) considered to be generally uninterested in how their energy system works. In limiting the

1 Participants are referred to by pseudonyms.

information given to them about their energy production, storage, and demand, developers perceive that they are making life easier for households. More than a liberation from managing their energy resources, there is also a perception that in many instances, households will be incapable of managing a more complicated energy system. By limiting their ability to alter and change settings, and managing it on their behalf, it is believed that households will achieve greater carbon and financial savings:

> I think that's the aim. Because [the homes] are that different . . . you want to try and make them as simple as possible for the users. They don't need to be worried about how full the batteries are and that kind of thing. (Expert 1)
>
> I guess the key thing here is taking what is more complex as a building but then actually making it even more simple and convenient for the customer, because you know, we've got access to more data because there's more, and we've got access to better controls. So, we can use that information to improve the experience for the people that live there. (Expert 4)

These extracts illustrate the way that information was seen as beneficial for the developers and energy service providers but not necessarily of relevance to the households. For example, one expert argued against the utility of smart in-home displays, suggesting they did not inspire long-term changes to energy use (a conclusion supported by some existing social science research, e.g. Hargreaves et al. 2013):

> So all of the homes do have smart meters. That's necessary for us to be able to run them on the half hourly tariff that we need. . . What we don't particularly worry about putting in is the smart display . . . Which is typically part of a smart meter you get a little display that tells you what energy. All the research on those tends to suggest that you ignore them within three to six months and so they don't have a long-term behaviour change impact . . . You know, we're not hiding any of that data from them, we just recognise that most people don't care about it, if brutally honest. (Expert 5)

In some of our case sites, households were able to sign up for an energy service platform, a form of direct load control. This service assumes a passive role for households, balancing on their behalf their energy demand, production and storage, with the flexibility demands of the national grid, while also accounting for external influences on occupant energy demand, such as weather forecasts. Prior to moving into their property, household members are 'onboarded', whereby the energy system and service are explained to the household, and the household sets preferences over their energy demand (e.g. heating and bathing regimes). Once the homes are occupied, households can alter their energy regime using an app downloaded onto a smart device, or by telephoning an agent at the energy service platform who can make the changes on their behalf. In addition, there are wall-mounted thermostats around the home that facilitate manual changes to the temperature. Where this energy service platform was available, all the participants we spoke to prior to moving in had opted to sign up despite varying levels of understanding about what this entailed. The uncertainty over how the service would work was tempered by the expectation of lower energy bills, and that if they were unsatisfied with the service, they could switch to a standard energy company and personally manage their energy system.

Since moving in, some residents felt that they would benefit from more information about their energy system, in order to alter some energy behaviours to take advantage of low-carbon energy production and storage:

> After speaking to a couple of the other residents we'd like to get to a point where we know when our solar panel heating isn't being used or is running out, when the battery kicks in, when we get to a point where we're ending the battery and then we're clicking into terrestrial, or whatever they want to call it, ordinary electric, you know? [. . .] for example, when I use the washing machine I want to try and work out when my washing machine' on how much does it draw down because I try to use it now in a day, [I've] stop[ped] using it at night. So theoretically I shouldn't really use much more now than what I'm getting from my solar. (Sally, post-occupancy)

> I wonder, like, how much energy's in there [battery], whether we're like, running off that or whether we're running off the grid. So, it does make me think, yeah, and I think I just wish I had some more, like, information about it because we don't really know a lot about it or what it's doing, when or whatever, yeah. (Sophie, post-occupancy)

These extracts suggest that while information about their energy use and the operation of the various energy systems within their homes was not seen as particularly relevant to participants before they moved in, once living in the homes, they sought a more detailed picture of how their home operated. The smart app, designed for residents to monitor and adjust their energy systems, has been described by many as simple and easy to use, with some perceiving that control of their energy system using an app is 'the way forward' (Sophie). However, initial glitches with the app mean that some participants had avoided using it and resorted to making manual adjustments (such as opening windows and doors to cool homes, rather than interacting with the app to reduce temperatures). Such technological glitches can feed into assumptions of indifferent consumers, that some occupants are uninterested in managing their energy, or that the energy system is too complex for them, when this may not be the case.

Other issues with infrastructure were evident in our expert interviews. For example, in one case site, the technical specifications of the homes appeared to be incompatible with smart meters:

> what we've actually discovered is that because there's some residents that are really keen to use the Tesla apps, they have actually gone to the likes of Octopus to get the Green tariffs but there aren't any energy suppliers out there at the moment that can currently fit a three-phase smart meter and because of that, none of our residents can actually get on to any of these smart tariffs. . . so they can't actually download energy overnight, say at a cheap time, and store it in the battery to then run the home off the next day. (Expert 14)

> we've got the home of the future which people actually can't put smart meters into . . . So there's a real issue for the sector, everybody, the government even, to really talk to the grid providers about what's preventing them and how can we overcome this. (Expert 13)

The above experts described their frustration that they had sought to create a future-proof home, but the benefits were not able to be fully realised by current residents due to lack of supportive infrastructure. In these accounts, residents are not indifferent consumers but actively attempting to engage with smart technology to improve energy (and relatedly financial) efficiencies.

In addition to those who are prevented from fully utilising smart systems due to challenges with the technology, some households are not digitally connected, e.g. lacking a device that enables the home control app to be downloaded. While this again feeds into the indifferent consumer assumption, it creates challenges and privacy issues for those households. In one instance, a family member external to the household managed their energy preferences on their behalf. This not only creates inconvenience for the household but presents challenges for those managing the energy service having to adopt different ways of working to ensure the residents remain comfortable. Reliance solely on smart management systems can result in the digital exclusion of such households, potentially creating new energy vulnerabilities. In overcoming the discrepancies between a smart, low-carbon energy system and variable household digital capacities and energy needs, experts in our research emphasised the importance of personal communication to understand the decisions residents make in relation to their energy use and to support them to get the best out of their smart home technologies:

> I think there will always need to be a human element to be honest with you . . . you have to build that relationship before you can answer their real questions. I had a call yesterday, came through to me, 'I want everything down 15°' I said to him, 'right, why you going to 15? it's 17 to 24 at the minute, that's quite a change? Are you sweating?' 'No, it's lovely in here'. 'Why then change it?' 'Because I can't afford it.' 'yeah, you can, that'll be fine'. 'But the heating's on all the time, I couldn't afford that with a gas boiler'. 'It's a different system'. And unless you take the time to talk to people and understand that, really the thing to do would have just been 'okay, I can do that for you, or you can do it from the app by yourself', but actually, because I'm understanding what is causing the query, I'm supporting them to find their best solution because some of the things that they will do instinctively, from using a gas boiler, will actually create reverse results in their home . . . I do think there will always be a role . . . for providing coaching and support, I don't think we will be ever at a point where it is just a billing function and full automation. (Expert 12)

As opposed to relying on the smart system to understand and manage household energy needs, it is the information exchange in conversation that is important in this process. Similar to the research findings of De Wilde (2021, p. 2), during the onboarding process and post-occupancy, a 'symmetrical relationship' is formed between the agent and the household. In our above example, the agent can provide meaningful assistance and reassurance to the household only after learning about the household's experiential knowledge of both their previous and new energy systems. In this instance, without the personal explanation offered, it is likely that the customer would have significantly reduced their households' temperatures making them more vulnerable to a variety of social-, economic- and health-related harms associated with energy deprivation (c.f. Grey et al. 2017).

6.6 Discussion

Active Homes, through their scale-up and aggregation, could represent a crucial component of pathways towards net zero and through their smart capabilities are foundational elements of smart cities. However, their crucial role in mitigating climate change can only be realised if Active Homes can function as homes that occupants can live well within. As we have argued in this chapter, it is imperative that as Active Homes begin to be created and occupied, the lived experience of those occupants is understood and insights are applied to future developments.

At the highest scale, Active Homes is a concept that, if proven, aggregated and scaled up, can support national plans of net zero by 2050 (BEIS 2020) through enabling the balancing of national grid flexibility demands with household energy demands (National Grid ESO 2020a; Ofgem 2020). More than this, there is a common understanding that the smartness of the homes will hold multiple, multi-faceted benefits. This includes the development of a new low-carbon economy (HM Government 2020; Welsh Government 2019), and a future that is more environmentally and socially just, broadly and indirectly benefiting all of society (National Grid ESO 2020b). In line with smart energy city visions, a smart energy system is one of several steps towards a streamlined way of life where human health and well-being are more easily achieved (National Grid ESO 2020b; Pulselli et al. 2021; Tan et al. 2017). However, as we have found, practical implementation of smart technologies is not always straightforward or streamlined for Active Home developers, energy service companies, or residents.

In installing smart technologies into these new homes, practical decisions must be made about what technologies should be used and the level of and mode of interaction between the technologies and the household. In these decisions, assumptions about the wants, needs and capabilities of the households become apparent. Our research illustrates how these assumptions may not be completely accurate or that household needs may all change over time, highlighting a challenge to reliance on consumer archetypes. As our interviews with experts involved in the development and management of Active Homes have illuminated, the installation of smart capacity, whether smart meters, smart monitors and displays, smart appliances, or EV charging points, is considered necessary to 'future proofing' the homes. Here, the perceptions of our experts are shared by many of the households, who while unable to financially afford much of the smart appliances or EVs that their homes can accommodate, plan to do so in the future. However, our research also highlights the challenges that can arise when future-proofed homes do not fulfil functions that residents expect in the present.

Our research has also illuminated the fallibility of smart technology. While our example of a glitch with a smart control app is hoped to be a temporary issue with an imminent resolution, already it has led to a cooling of interest from some residents. Those resigned to not using the app manually control their heating using in-home wall-mounted thermostats, most closely resembling heating controls they have used in the past, or simply opening windows and doors to naturally ventilate and regulate the temperature of their home. These actions parallel other findings concerning how other households living in low-carbon homes have circumvented their energy systems to meet their comfort needs (c.f. Botti 2017; Costanzo et al. 2020). Linked to this is the unintended exclusion of people from interaction with their energy system due to either digital illiteracy or not owning a smart device to download the app. In both instances, the reasons why households do not engage with the smart technology are 'silenced' (Hine 2020) as the assumption that they are passive consumers is reaffirmed.

This widespread misrecognition of households is at risk of becoming solidified and embedded within future home developments (Goulden et al. 2018). With the aggregation of such homes into neighbourhoods and cities, inaccurate assumptions reified at the household scale will affect the realisation of smart city futures, as places that are not liveable for the whole of society (Valdez et al. 2020). This works against the aspirations of a fully digitalised energy system, which for homes is supposed to more easily achieve occupant comfort, and for cities, be fair and inclusive. Thus, important still is the role of people in smart energy systems – not just people within the smart households, but agents and intermediaries who support the communication between households and their home energy system creating a triangular experiential knowledge exchange (De Wilde 2021). As our research has found, without this, some households would remain excluded from their energy systems and vulnerable to energy deprivation (Sovacool et al. 2019).

These findings bring to the fore what should be a central concern for Active Homes – people; residents, developers and intermediaries who are still needed to translate human comfort along with other, related personal priorities and social needs, into smart energy systems and in the latter instances, explain these new systems to households. Indeed, central to energy demand, the development of buildings as homes and the formation of cities are people. The digital intelligent integration of homes into the energy system through the development of Active Homes holds multiple potential benefits for people, both individually and at societal scales. However, unless people can live well within them, there is a very real danger that these opportunities will be lost. It is important that in these early stages of Active Home development, assumptions of household capabilities and preferences become less channelled by narrow archetypes so that unintended negative consequences that curtail their ability to live well and comfortably in their homes are limited. This will be critical not only for the mitigation of energy vulnerabilities but also for the success of Active Buildings and the wider cities and communities that they are located within.

References

Aldrich, F. (2003). *Smart Homes: Past, Present and Future, Inside the Smart Home* (ed. R. Harper), 17–39. London: Springer-Verlag.

Botti, A. (2017). Thermal comfort and overheating investigations on a large-scale Passivhaus affordable housing scheme. In: *Design to Thrive: Proceedings of 33rd PLEA International Conference 3-5 July 2017, Volume III* (ed. L. Brotas, S. Roaf, and F. Nicol), 3722–3729.

NCEUB (Network for Comfort and Energy Use in Buildings). Available at: https://discovery.ucl.ac.uk/id/eprint/10059148/ (accessed 07 July 2023).
Brause, S.R. and Blank, G. (2019). Externalized domestication: smart speaker assistants, networks and domestication theory. *Information, Communication & Society* 23 (5): 751–763.
Calver, P. and Simcock, N. (2021). Demand response and energy justice: a critical overview of ethical risks and opportunities within digital, decentralised, and decarbonised futures. *Energy Policy* 151: 1–12.
Climate Change Act (2008). *The Climate Change Act 2008 (2050 Target Amendment) Order 2019, No. 1056*. United Kingdom. Available at: https://www.legislation.gov.uk/ukdsi/2019/9780111187654 (accessed 07 July 2023).
Committee on Climate Change (CCC) (2019). Net zero – the UK's contribution to stopping global warming. https://www.theccc.org.uk/publication/net-zero-the-uks-contribution-to-stopping-global-warming/ (accessed 15 July 2020).
Committee on Climate Change (CCC) (2020a). Reducing UK emissions, progress report to parliament. https://www.theccc.org.uk/publication/reducing-uk-emissions-2020-progress-report-to-parliament/ (accessed May 2021).
Committee on Climate Change (CCC) (2020b). Buildings. https://www.theccc.org.uk/wp-content/uploads/2020/12/Sector-summary-Buildings.pdf (accessed June 2021).
Corsinia, F., Certomàb, C., Dyerc, M., and Freya, M. (2019). Participatory energy: research, imaginaries and practices on people' contribution to energy systems in the smart city. *Technological Forecasting & Social Change* 142: 322–332.
Costanzo, V., Carrillo Gomez, J.E., Evola, G., and Marletta, L. (2020). Suitability of Passivhaus design for housing projects in Colombia. In: *Sustainability in Energy and Buildings: Proceedings of SEB, 4–5 July, 2019*, 97–107. Singapore: Springer.
Cugurullo, F. (2018). The origin of the smart city imaginary: from the dawn of modernity to the eclipse of reason. In: *The Routledge Companion to Urban Imaginaries* (ed. C. Lindner and M. Meissner), 117–129. London: Routledge.
Darby, S.J. (2018). Smart technology in the home: time for more clarity. *Building Research & Information* 46 (1): 140–147.
De Wilde, M. (2021). "A heat pump needs a bit of care": on maintainability and repairing gender–technology relations. *Science, Technology, & Human Value* 46 (6): 1261–1285. https://doi.org/10.1177/0162243920978301.
Department for Business, Energy and Industrial Strategy (BEIS) (2020). Energy white paper: powering our net zero future. https://www.gov.uk/government/publications/energy-white-paper-powering-our-net-zero-future/energy-white-paper-powering-our-net-zero-future-accessible-html-version (accessed May 2021).
Elliott, T., Geske, J., and Green, R. (2020). Can you make money from active buildings? Challenges facing business models. *Active Building Centre Research Programme (Energy Systems Catapult and Imperial College Business School)*. Available at: https://abc-rp.com/impact/white-papers/ (accessed 7 July 2023).
Gaetani, I., Hoes, P., and Hensen, J.L.M. (2020). A stepwise approach for assessing the appropriate occupant behaviour modelling in building performance simulation. *Journal of Building Performance Simulation* 13 (3): 362–377. https://doi.org/10.1080/19401493.2020.1734660.
Gonella, F. (2019). The smart narrative of a smart city. *Frontiers in Sustainable Cities* 1 (9): 1–7. https://doi.org/10.3389/frsc.2019.00009.

Goulden, M., Spence, A., Wardman, J., and Leygue, C. (2018). Differentiating 'the user' in DSR: developing demand side response in advanced economies. *Energy Policy* 122: 176–185.

Gram-Hanssen, K. and Darby, S.J. (2018). "Home is where the smart is?" Evaluating smart home research and approaches against the concept of home. *Energy Research & Social Science* 37: 94–101.

Grey, C.N.B., Schmeider-Gaite, T.J., Nascimento, C., and Poortinga, W. (2017). Cold homes, fuel poverty and energy efficiency improvements: a longitudinal focus group approach. *Indoor and Built Environment* 26 (7): 902–913.

Happle, G., Fonseca, J.A., and Schlueter, A. (2018). A review on occupant behavior in urban building energy models. *Energy and Buildings* 174: 276–292. https://doi.org/10.1016/j.enbuild.2018.06.030.

Hargreaves, T., Nye, M., and Burgess, J. (2013). Keeping energy visible? Exploring how householders interact with feedback from smart energy monitors in the longer term. *Energy Policy* 52: 126–134.

Harris, S., Weinzettel, A., Bigano, A., and Källmén, A. (2020). Low carbon cities in 2050? GHG emissions of European cities using production-based and consumption-based emission accounting methods. *Journal of Cleaner Production* 248: 119206. https://doi.org/10.1016/j.jclepro.2019.119206.

Henwood, K. (2019). Investigating risk – methodological insights from interpretive social science and sustainable energy transitions research. In: *Researching Risk and Uncertainty: Methodologies, Methods and Research Strategies* (ed. A. Olofsson and J. Zinn), 129–152. Basingstoke: Palgrave MacMillan.

Henwood, K., Groves, C., and Shirani, F. (2016). Relationality, entangled practices, and psychosocial exploration of intergenerational dynamics in sustainable energy studies families. *Relationships and Societies* 5 (3): 393–410. https://doi.org/10.1332/204674316X147584383416945.

Hine, C. (2020). Strategies for reflexive ethnography in the smart home: auto ethnography of silence and emotion. *Sociology* 54 (1): 22–36.

HM Government. (2020). The ten point plan for a green industrial revolution. https://www.gov.uk/government/publications/the-ten-point-plan-for-a-green-industrial-revolution (accessed June 2021).

Hui, T.K., Sherratt, R.S., and Sánchez, D.D. (2017). Major requirements for building smart homes in smart cities based on internet of things technologies. *Future Generation Computer Systems* 76: 358–369.

Jasanoff, S. (2015). Future imperfect: science, technology, and the imaginations of modernity. In: *Dreamscapes of Modernity: Sociotechnical Imaginaries and the Fabrication of Power* (ed. S. Jasanoff and S.-H. Kim), 1–33. Chicago: The University of Chicago Press.

McKenna, E., Higginson, S., Grunewald, P., and Darby, S.J. (2017). Simulating residential demand response: improving socio-technical assumptions in activity-based models of energy demand. *Energy Efficiency* 11: 1583–1597. https://doi.org/10.1007/s12053-017-9525-4.

National Grid ESO (2020a). Future energy scenarios. https://www.nationalgrideso.com/document/173821/download (accessed 12 August 2020).

National Grid ESO (2020b). Digitalisation strategy. https://www.nationalgrideso.com/document/186426/download (accessed 12 June 2021).

O'Sullivan, K., Henwood, K., and Pidgeon, N. (2020). *Active Buildings in the Changing Policy Landscape: Conceptual Challenges and Social Scientific Perspectives*. Active Building Centre Research Programme, Cardiff University.

O'Sullivan, K., Shirani, F., Pidgeon, N., and Henwood, K. (2022). Why active buildings? Realising the potentials of energy networked homes – a social scientific perspective. In: *Active Building Energy Systems: Operation and Control* (ed. V. Vahidinasab and B. Mohammadi-ivatloo). Cham: Springer https://doi.org/10.1007/978-3-030-79742-3_2.

Ofgem (2020). Ofgem decarbonisation programme action plan. https://www.ofgem.gov.uk/publications-and-updates/ofgem-s-decarbonisation-action-plan (accessed 15 July 2020).

Ozaki, R. (2018). Follow the price signal: people's willingness to shift household practices in a dynamic time-of-use tariff trial in the United Kingdom. *Energy Research & Social Science* 46: 10–18.

Pulselli, R.M., Broersma, S., Martin, C.L. et al. (2021). Future city visions. The energy transition towards carbon-neutrality: lessons learned from the case of Roeselare, Belgium. *Renewable and Sustainable Energy Reviews* 137: 110612. https://doi.org/10.1016/j.rser.2020.110612.

Sadowski, J. and Bendor, R. (2019). Selling smartness: corporate narratives and the smart city as a sociotechnical imaginary. *Science, Technology & Human Values* 44 (3): 540–563. https://doi.org/10.1177/0162243918806061.

Shirani, F., Groves, C., Henwood, K. et al. (2020). 'I'm the smart meter': perceptions and experiences of smart technology amongst vulnerable consumers. *Energy Policy* 144: 111637.

Shirani, F., O'Sullivan, K., Hale, R. et al. (2022). Transformational innovation in home energy: How developers imagine and engage with future residents of low carbon homes in the United Kingdom. *Energy Research & Social Science* https://doi.org/10.1016/j.erss.2022.102743.

Skjølsvold, T.M., Jørgensen, S., and Ryghaug, M. (2017). Users, design and the role of feedback technologies in the Norwegian energy transition: an empirical study and some radical challenges. *Energy Research & Social Science* 25: 1–8.

Sovacool, B.K., Lipson, M.M., and Chard, R. (2019). Temporality, vulnerability, and energy justice in household low carbon innovations. *Energy Policy* 128: 495–504. https://doi.org/10.1016/j.enpol.2019.01.010.

Strengers, Y. (2013). *Smart Energy Technologies in Everyday Life, Smart Utopia?* Hampshire: Palgrave Macmillan.

Tan, S., Yang, J., Yan, J. et al. (2017). A holistic low carbon city indicator framework for sustainable development. *Applied Energy* 185: 1919–1930.

Tirado Herrero, S., Nicholls, L., and Strengers, Y. (2018). Smart home technologies in everyday life: do they address key energy challenges in households? *Current Opinion in Environmental Sustainability* 31: 65–70.

Valdez, A., Wigley, E., Zanetti, O., and Rose, G. (2020). Learning lessons for avoiding the inadvertent exclusion of communities from smart city projects. In: *Shaping Smart for Better Cities* (ed. A. Aurigi and N. Odendaal), 221–237. Academic Press https://doi.org/10.1016/B978-0-12-818636-7.00003-2.

Verkade, N. and Höffken, J. (2017). Is the resource man coming home? Engaging with an energy monitoring platform to foster flexible energy consumption in the Netherlands. *Energy Research and Social Science* 27: 36–44.

Welsh Government (2019). Prosperity for all: a low carbon wales. https://gov.wales/sites/default/files/publications/2019-06/low-carbon-delivery-plan_1.pdf (accessed 12 August 2020).

Wilson, C., Hargreaves, T., and Hauxwell-Baldwin. (2014). Smart homes and their users: a systematic analysis and key challenges. *Personal and Ubiquitous Computing* 19: 463–476.

7

Do Mobility and Sustainability Rhyme in the Autonomous City?

Federico Cugurullo and Alexander Gaio

Department of Geography, School of Natural Sciences, Trinity College, Dublin, Ireland

7.1 Introduction

Artificial intelligence (AI) has rapidly become a potent driver of urban change. Incubated in numerous smart city initiatives, AI has now gained considerable prominence in the world, leading to the formation of a new type of urbanism called *autonomous urbanism* (Cugurullo 2021). This urbanism is characterised by the presence of autonomous and intelligent technologies which are capable of functioning without human supervision (Barns 2021; Cugurullo 2020; Yigitcanlar et al. 2020). In addition, these autonomous AI technologies are managing an increasing number of urban spaces, sectors and services. This is happening in an autonomous manner, in a way that human agency occupies only a marginal role. In these terms, what we are witnessing is the emergence of *autonomous cities* intended as spatial formations operated largely by AI, where the role, vision and intelligence of human actors are being overshadowed by non-biological intelligences (Cugurullo 2021).

Since AI has become an increasingly important dimension in governing cities towards smartness, sustainability and low-carbon future, this chapter discusses the implications for future urbanism and mobility in the autonomous city as well as their implications for sustainability. We discuss these implications in four areas. First, we outline the shifting dynamics from the smart city to the autonomous city and how urban artificial intelligence is manifesting itself in cities. Second, we discuss the main alterations that AI can inflict on the sustainability of the city and specific implications for AI-enabled mobility. In the third subsection, we delve deeper into how the autonomous vehicle will change three fundamental areas of urban life around the tempo of the city, the form of the city and the right to the city. Fourth, we draw attention to the power dynamic between autonomous vehicles and cyclists and how this will implicate future mobility in the autonomous city. We ultimately conclude with the potential for AI-enabled mobility to impact sustainability, highlighting the need for an approach to autonomous cities that incorporates participatory planning and human-centred principles.

Smart Cities, Energy and Climate: Governing Cities for a Low-Carbon Future, First Edition.
Edited by Oleg Golubchikov and Komali Yenneti.

7.2 From Smart to Autonomous Cities

In recent years, the diffusion of AI in cities has substantially changed the dynamics of smart urbanism (Cugurullo 2021). The nature of these changes and the reason behind their emergence can be appreciated by looking at the nature of AI and how this radical technology has penetrated the built environment and influenced numerous urban services. Although a universal definition of AI does not exist, artificially intelligent entities share five defining characteristics. First, they are not the product of a natural process. They are either human-made or machine-made and reside in non-organic artifacts. Second, they are capable of learning by acquiring information from the environment that is around them (Russell and Norvig 2016). Third, AIs can make sense of the information that they gain, by extracting a wide range of concepts from it, such as *danger*, *pedestrian*, *intruder* and *sky* (Bostrom 2017). Fourth, information and concepts are used by AIs to act rationally and handle uncertain situations in real-life environments (Cugurullo 2020; Pearl 2014). Fifth and most importantly, an artificial intelligence is in theory capable of performing its skills and acting in an autonomous manner, without being directed or supervised by a human being (Levesque 2017).

AI is manifesting itself in cities through what has been defined as *urban artificial intelligence*: 'artifacts operating in cities, which are capable of acquiring and making sense of information on the surrounding urban environment, eventually using the acquired knowledge to act rationally, in complex urban situations when some information might be missing or incomplete' (Cugurullo 2020, p. 3). In addition, urban AIs act autonomously: 'they make decisions in an unsupervised manner, thus displaying a rudimentary form of thinking, and take actions which can potentially trigger radical changes in the city' (Cugurullo 2020). In practice, there are three main types of urban AI, namely autonomous cars, robots and city brains (Cugurullo 2021). Autonomous cars are vehicles animated and controlled by an artificial intelligence, and their employment is already tangible in a number of cities (Acheampong et al. 2021; Dowling and McGuirk 2020). They can sense the surrounding urban environment via cameras, radar and lidar systems and, at the highest level of autonomy (level 5), they should be capable of navigating complex urban spaces without human inputs. Robots are a multiform category that includes drones, humanoid machines, nanobots and androids. The urban spaces that they perceive and operate tend to be characterised by a strong social dimension, as robots are frequently employed in the frontline and communicate directly with humans in the fields of retail, customer service, hospitality and education (Del Casino Jr et al. 2020; Tiddi et al. 2020). Finally, city brains are large-scale AIs located in a digital platform through which they control entire portions of the city and its infrastructure (Caprotti and Liu 2020; Cugurullo 2021). They sense the city by means of hundreds of smart cameras and, while their actions are currently limited to the control of traffic lights, city brain developers are keen (particularly in China) to employ this technology to render urban governance as a whole (transportation, health, safety, security, environmental preservation, urban planning, finance and more) fully autonomous.

On the one hand, there are strong connections between smart urbanism and the emerging autonomous urbanism that AI is pushing forward. Urban AIs require traditional smart tech in order to function. They resort to large datasets, i.e. *big data* to increment their knowledge, they need *sensors* to perceive the surrounding environment and they rely on the Internet of Things (IoT) to share data. As a number of studies have noted, this *big data-sensors-IoT trinity* has been the core of smart city initiatives for decades (Karvonen et al. 2018; Kitchin 2014). On the other hand, AI tech contains elements of innovation that go beyond the realm of smart urbanism. Urban AIs possess a rudimentary capacity of thinking that is unprecedented. Albeit in a problematic manner (i.e. far from being functional and safe), these are technologies that are also capable of understanding what is right or wrong, thus entering the field of ethics. Most importantly, although these AIs still rely on human agents to become operative, learn and deal with complex moral dilemmas, their autonomy place them in situations that are very different from what smart cities have experienced in the past. Urban artificial intelligences act in the city and impact urban spaces and their inhabitants, without being manipulated by humans. They are not puppets. They are autonomous agents whose sustainability implications will be discussed in the next section.

7.3 Sustainability in the Autonomous City

AI is a radical technology (Greenfield 2018). It is radical in the sense that it can significantly alter the environment and domain in which it operates. It has an impact whose scale ranges from the life of individuals to that of cities and states. It is because of its power to change environments and domains that it is important to look at AI from a sustainability perspective. AI can influence physical spaces, natural habitats, social dynamics, politics, financial systems and, more broadly, those key environmental, social, political, cultural and economic dimensions that constitute the notion and practice of sustainability. Nowadays, AI can be anywhere and so can its influence.

Particularly in relation to cities and their sustainability, there are three important aspects to consider. First, the general concept of *artificial intelligence* is misleading because it conjures up the image of a universal artificially intelligent species. Such image is false. As in science, we do not generalise biological intelligence, we should apply the same conceptual and analytical approach to AI. As there are many varieties of biological intelligence on the planet (human intelligence constituting only a variegated portion of Earth's intelligence), there are numerous varieties of AI. As we mentioned above, there are three main categories of urban artificial intelligence, but these are *macro* categories that include a lot of diversity. For example, there are numerous models of autonomous cars, ranging from pods to saloon vehicles and possessing distinct levels of autonomy. Robots exist in many different shapes and sizes, comprising minuscule nanobots and human-like androids. City brains are far from manifesting a universal architecture, especially when we compare how this technology is being developed in Eastern and Western countries. In essence, when we look at cities and urban services, there is no such thing as an AI, intended as a singular and universal typology of intelligence. This observation matters because, from a sustainability perspective, it would be erroneous to place all the myriad of existing AIs into a single one-size-fits-all category. By the same token, it would be erroneous to say that AI is either sustainable or unsustainable. When it comes to AI, the question of sustainability should be addressed not in generic terms, but rather in relation to a specific example of AI (whose sustainability or unsustainability would not be representative of the whole AI world).

Second, AI, its multiple incarnations and its related impact on the sustainability of cities, vary geographically. AI is spreading unevenly across the world, and we cannot expect the same type of AI to always manifest itself in the same identical way and to have the same exact impact on different spaces. The same type of urban AI can be developed and implemented differently according to the specific space where AI tech is becoming operative. Similarly, the same urban AI can trigger heterogeneous environmental, social, cultural, political and economic changes that depend on the particular city in which that AI is acting. For example, the global diffusion of autonomous cars is highly uneven. Not every city in the world is adopting this new technology and, where a transition to autonomous transport is actually taking place, the patterns of adoption tend to be locally specific. In some countries, autonomous cars are tested in confined experimental spaces acting as testbed facilities, while, in others, the same technology is being gradually released in the old heart of the city, next to ordinary urban spaces and activities (Cugurullo and Acheampong 2020; Dowling and McGuirk 2020). Moreover, the actual transport modes through which autonomous vehicles are being made available, i.e. private ownership, car sharing and public transport, are largely determined by the cultural heritage of the place where the new AI technology is landing. It is expected, for instance, that cities with a tradition of sharing services will adopt autonomous cars via car-sharing services, contrary to places in which the culture of car ownership is deeply rooted (Acheampong et al. 2021; Acheampong and Cugurullo 2019; Narayanan et al. 2020).

Third, when it comes to urban sustainability, AI does not have to be particularly intelligent to make a change. Several scholars look at the future of AI and its impact on human societies, contemplating the birth of artificial intelligences that meet or even surpass human intelligence (Bostrom 2017; Lovelock 2019; Tegmark 2017). There are those who imagine the possibility of artificial general intelligence (AGI) as a synthetic intellect virtually indistinguishable from a human one, gifted with consciousness and capable of doing everything that humans do (Naudé and Dimitri 2020). Others consider the advent of a *superintelligence* intended as an AI whose cognitive power radically exceeds human limits (Bostrom 2017). Both scenarios, although possible, are very problematic from an urbanistic point of view. They focus our attention on types of AI that do not exist yet and, more problematically, they cultivate (in academia and among the general public) the assumption that only very advanced and smart AIs are what we need to be careful with. In urbanistic terms, this is not the case at all. Even a basic AI bereft of consciousness and able to autonomously perform simple computations can trigger enormous socio-environmental, political and economic changes in cities (Cugurullo 2021). In the next section, we explore this issue more in-depth by focussing on autonomous cars and the urban transformations that their diffusion is triggering.

7.4 Autonomous Vehicle-Induced Urbanism

The diffusion of autonomous cars seemingly raises more questions than it has the potential to solve. With any novel technology, particularly as it pertains to transport and mobility, it is important to take a holistic approach when assessing the impact that it will have on the urban environment – where more than half of the global population live. With an anticipated growth to 68% of the share of global urban population by 2050 (Ritchie and Roser 2018), the autonomous city will need to be human centred and, naturally, sustainable. Any ulterior motive would be misguided and lead to negative outcomes. There are three general areas that we argue will be impacted by the advent of autonomous car ubiquity. First, the *tempo of the city* stands to be recalibrated, in concert with other smart city technologies such as faster wireless connectivity and rise in prevalence of flexible working arrangements. Second, there will be new opportunities to change the *form of the city* from what was once an automobile-dominated urbanism. Third, new conversations around the *right to the city* will emerge as autonomous cars in the form of pods or saloons begin to be defined as public, private or shared spaces. This will trigger discussions of how the existing traditional public spaces and rights of ways are opened, or replaced by autonomous cars, and how egalitarianism is applied. Autonomous cars can enable a new kind of urbanism that will open large swaths of land for repurposing. There exists a sustainability 'triple bottom line' in the changing dynamics of tempo, form and rights in the city. The new dynamics of the changing *tempo* of the city will impact economic productivity, the built environment will be reshaped by the new dynamics of a shifting urban *form* and the *right* to the city will change dynamics of social equity.

The tempo and pace of the city hold significant stake in the proliferation of the autonomous city. With a particular focus on autonomous cars, the changes in labour will be a key point of contention, especially as humans stand to be relieved from most tasks associated with operating a vehicle. Under autonomous driving levels 4 and 5, the concern of lost productivity from aggravated traffic congestion will be nullified if the division of work and personal lives are blurred and the space of an autonomous car is an extension of home and/or work. France, for instance, has endeavoured to set clear separation parameters for work under laws that consider the 'right to disconnect' (Hesselberth 2018). These laws were created with the intention to enable employees to disconnect outside working hours. However, their implementation has created a scenario in which this right is not mandatory or legally binding. Ironically, this law could jeopardise free time and, when considered with social norms of hyperconnectivity, lead to all available time being turned into unwaged labour (ibid). There is much at stake around how individuals will use their time in an age of autonomous cars. The division of work and free time will form a large part of the discourse which has implications for the tempo and pace of the urban economy.

Some of the largest swaths of public land in cities are allocated to the movement and storage of cars, and it has been provided at great expense (Scharnhorst 2018). Widespread automobility has also contributed to low-density urban development and more resource-intensive individual transport hampering the overall sustainability of the transport system (Kenworthy and Laube 1996). Cars remain parked 95% of the time (Shoup 2021), and because conventional cars are largely privately owned and require humans to operate them, they are parked in proximal locations in cities at the convenience of their owners. This requirement to provide space for private cars has defined much of the modern city and has the potential to dramatically shift when cars can operate themselves. Autonomous cars open the possibility that a significant portion of parking in urban settings will be made redundant (Zhang et al. 2015; Zhang and Guhathakurta 2017). This is particularly salient should autonomous cars be implemented as *shared* autonomous cars that are not owned by an individual. Shared autonomous cars enable daisy-chaining trips so that when a trip is completed, the car does not sit unused 95% of the time (like conventional cars) and can be put to productive use straight away in a subsequent trip.

The autonomous city will create new opportunities for time and space, both contributing to a new wave of urban form. In freeing individuals from time-consuming tasks of driving, future commutes will enable more peripheral dwelling arrangements that will result in more dispersed cities that encroach onto greenfield sites. Adequate environmental protection and urban containment will be necessary to curtail urban expansion that is enabled by the autonomous car so that the autonomous city does not have the unintended consequence of demanding more energy for more transport. Outside the urban core of cities, and in considering the rising cost of housing, blending of work and free time and elimination of the act of driving, there is sufficient probability that the combination of these factors will enable longer commutes or 'extreme commuting' (Bai et al. 2020). Autonomous cars stand to usher in a new era of suburbanisation and low-density cities that continue to exist counter to societal goals around advancing sustainable land use, energy consumption and economic infrastructure development. The same way that the conventional car opened up 'unfathomable and unfeasible options' (Swan 2019, p. 66) for commuting in low-density suburbs and exurbs, the autonomous car will also play a role in future suburbanisation. The other opportunity that the autonomous car presents that implicates the form and structure of cities is new free space that is presently dedicated to both on-street and off-street parking. The opportunity that additional space presents would allow for addressing urban issues like a conversion of surface parking lots to address a shortage in housing supply, or on-street parking repurposed to address the lack of provision for low-stress, safe, separated cycling infrastructure thereby helping to create a more just and equitable city. There are many conversations that will need to be had in ensuring the autonomous city does not leave marginalised groups behind (Sadoway and Shekhar 2014). In doing so, there will need to be a larger conversation around the right to the city (Lefebvre 1967).

This will undoubtedly apply similar arguments from Lefebvre's original argument where the capitalist commodification of space is at odds with co-created space and citizen-led urbanism (Lefebvre 1967). The new spatial inequality is between motorised and non-motorised transport and how streetscape should be allocated. It involves shifting the assumption that the car is the de facto transport mode for all, and instead enables the individual to choose what transport option best suits their needs. This will only be possible with the provision of safe multi-modal transportation infrastructure that is suitable for all ages and abilities. Continuing on the assumption that every individual will drive and that the car is the most appropriate transport option is misguided because it often is not (Maibach et al. 2009). Many cities, by their design, nudge (Thaler and Sunstein 2021) their residents to drive due to the amount of road space dedicated to cars making up the majority of public space.

Given the opportunity that this imminent availability of space in cities presents, the right to the city will need to have a more prominent place in discourse surrounding the advancement of smart and autonomous city technology. The concept of shared autonomous cars gives rise to ideas that may see these new types of pseudo-public space come to fruition. Saloon-type autonomous cars present a new kind of shared space that would weaken arguments against re-allocating road space to more equitable public uses. Public rights-of-way (streets, roads and highways) that are currently dedicated to private conventional cars, if populated by shared autonomous cars, stand to create a new kind of 'third place' (Oldenburg and Brissett 1982) in a future city of blended work and leisure where commutes would likely grow longer. Aside from the traditional first place of 'home', and the second place of 'work', the shared autonomous vehicle could augment or replace the role of a 'third place' of a library, pub or community centre. Even if autonomous vehicles stand to make the street a pseudo-public space or new kind of 'third place', the argument still exists that a disproportionate share of land in cities has been allocated to cars and all other modes of transportation have been neglected (Creutzig et al. 2020; Lefebvre-Ropars et al. 2021). In a future where autonomous vehicles are the dominant form of transport, the arguments of neglected non-auto modes present a real threat where previous generation transport methods such as cycling are further marginalised in favour of novel (autonomous) transport (Cugurullo et al. 2020). This marginalisation of cycling in an autonomous city is discussed further in the next section.

7.5 Bicycle Urbanism in the Autonomous City

The autonomous city has much to gain from bicycle urbanism, particularly with the population health outcomes that bicycles afford (Bourne et al. 2018; Garrard et al. 2012; Götschi et al. 2016; Oja et al. 2011), as well as their contributions to space-efficient and sustainable transport system (Cox and Koglin 2020). However, the modern and contemporary cities have not fully capitalised on these benefits, instead preferring to prioritise car infrastructure which creates a myriad of negative externalities in urban settings. In prioritising car infrastructure, it is not unreasonable to anticipate that the natural trajectory for urban mobility will prioritise autonomous vehicles, complete with all of their negative externalities for urban design, safety and inequity (Acheampong et al. 2021). Progress has been made in recent years by cities that have begun to capitalise on the benefits of bicycle urbanism, however, there remains a significant gap in complete, connected and safe cycling infrastructure when compared to the availability of infrastructure extended to cars in cities. Cyclists have been historically, and continue to be, a minority group that is marginalised and considered secondary to automobile transport (Prati et al. 2017) which has the inertia of existing planning processes and policies. Automobile defaults are also deeply engrained into the psyche of elected officials and members of the public. Significant efforts will need to be made to change the course of urban infrastructure planning to create environments that are more inclusive and less marginalising for a growing number of individuals who choose a bicycle as their primary mode of transportation in the autonomous city.

The transition to autonomous vehicles presents tremendous opportunity to shift the urban mobility paradigm. A major focus of autonomous transport has been on the future potential users of the technology as a natural progression from conventional automobility. What the prevailing discourse does not place significant focus on is the perception of other vulnerable road users (VRUs) who would otherwise interact with autonomous mobility as external actors (Hagenzieker et al. 2020). Vulnerable road users are grouped into a single group which does not consider the unique characteristics and impacts that will be felt unevenly among pedestrians, cyclists and other VRUs who walk or roll. In particular, cyclists will experience a notably different environment when autonomous vehicles form the dominant user on public roadways. There are varying degrees to this change as autonomous vehicles are in the stages of early adoption, transition and ubiquity. There will be times when a cyclist will not know intuitively how to act in each of these stages. In these cases, a bicyclist may not be able to differentiate an autonomous car from a conventional car. In the former case, there will also be a variation in how a cyclist will interact with regard to their level of defensiveness, for example. The implications of this will extend to the attitude of cycling, the travel behaviours and habits of cyclists and preferences for cycling infrastructure in an autonomous urban future.

Planners and researchers will need to make an earnest effort to better understand implications for cyclists and other vulnerable road users and how they will interact with or be impacted by autonomous cars. There are still uncertainties on how cyclist behaviour will be impacted in an autonomous urban future, and if it will further impact people bicycling in cities. Cyclist attitudes are particularly salient to understand how active transport travel behaviour will be impacted. Behaviour demonstrated under conventional automobile scenarios will likely be distinct from the behaviour demonstrated in a world of autonomous mobility (Hagenzieker et al. 2020). One of the most frequently cited challenges for cyclists in an AI-mediated transport landscape is the communication gap that will exist. Of note, while explicit communication from the autonomous vehicle would be helpful, a study found a clear weakness in the technology's ability to detect small signals, hints or no signals from cyclists when operating in real-life traffic scenarios (Botello et al. 2019). These small signals such as making eye contact, nodding or small hand waves are crucial for cyclists to establish trust and right-of-way to avoid conflict between road users. Without a way of establishing trust, there is likely to be diminishing confidence among cyclists who may have implications that further suppress bicycling and other non-motorised forms of transport. In addition, the technology's inability to recognise current communication strategies from cyclists stands to create an environment with more cyclist–AV conflict.

Cyclist infrastructure preferences around comfort and confidence are closely interconnected and will also need to form part of the discourse to strengthen areas of autonomous vehicle–cyclist interaction research as identified by several scholars. On the one hand, research suggests that there is a need for continuing to build separate cycling infrastructure, and other scholars have suggested that VRUs may consider autonomous cars more trustworthy than human drivers (Pammer et al. 2021), which may suggest less of a need for mode separation. On the other hand, this trust may not be sufficient grounds to undo the progress made on safe separated cycling infrastructure because there will likely continue to be the need for it in the future (Botello et al. 2019), especially depending on the variables used as predictors of cyclist behaviour.

Should adequate holistic foresight not be afforded to *comprehensive* urban mobility, there exists a tangible risk that the future of autonomous cars in the autonomous city will have a negligible impact on the overall challenges of urban transportation that they are touted to solve. We argue that greater attention and prioritisation of bicycle urbanism is a key component to ensure that better transportation in the autonomous city will materialise. Taking a laissez faire approach in letting the inertia of automobility run its course will extend negative transportation externalities into the autonomous city. Critically, and in considering transportation planning models, there is reason to pay closer attention to planned behaviour theory (Ajzen 1991) variables that contribute to transportation forecasting models that contemplate trip generation, distribution, mode choice and route assignment. Cyclist attitudes, preferences (informed by subjective norms) and infrastructure preferences (that dictate comfort and difficulty) for riding in urban environments will ultimately determine the extent they will exist in the autonomous city. This approach can also be applied to other vulnerable road users but, given the recent emphasis placed on bicycle urbanism, cycling is where we argue immediate emphasis should be given.

7.6 Conclusions

This chapter has shed light on the complexity of the many challenges that the advent of the autonomous city is posing from a sustainability perspective. By focussing on a relatively small part of autonomous urbanism, i.e. autonomous urban transport, we have shown how a single AI-operated urban technology can trigger enormous urban changes. An autonomous car is not a particularly large technology. It does not look ominous either. It might appear like a neutral artefact that simply traverses our cities and leaves no trace. This perception would be false. In our discussion, we have argued that autonomous cars are altering significant aspects of the city. The *tempo of the city*, the *form of the city* and the *right to the city* are prone to be largely influenced by the diffusion of AI in urban transport, to the point of being redefined. The effects of such changes might become obvious in a couple of decades, but the process that is driving them is already taking place and the time to be proactive is now.

Virtually no space or human being will be immune to the impact of AI. Aspects of our life and our city that we take for granted, including the most banal and ordinary ones, are about to be hit by a wave of change. Our mobility in everyday urban spaces is a good example of the risks that we are facing. We have discussed cycling, in particular, to highlight the challenges that cyclists will face in the autonomous city, as well as the challenges that AI will face when interacting with vulnerable road users. The interface between AI-enabled transport and non-motorised transport will define the extent of sustainability outcomes that can be achieved. In essence, mobility does not automatically rhyme with sustainability in the autonomous city, and autonomous urbanism is not necessarily equal to sustainable urbanism. On these terms, a focus on cycling is again helpful to understand how the advent of AI can replicate socio-environmental problems that have affected cities since the early automobiles began to dominate and reshape urban spaces.

From a sustainability point of view, the crux of the matter is that our needs as citizens and human beings, as well as the needs of the natural environment, might not be in synch with autonomous cities that are operated and managed by non-biological intelligences. The compatibility between human intelligence and AI is not something that we can take for granted. This will not be the default scenario unless we strive for it. When it comes to urban development, the only way forward that we see to align the development of autonomous cities with sustainable development is *participatory planning*. This is a matter of politicising and democratising the use of AI in cities, making sure that human needs and human visions of the future are taken into account, together with ecological insights (Yigitcanlar and Cugurullo 2020). Of course, this will not be an easy path to follow. Studies in explainable artificial intelligence (XAI) stress the scarce intelligibility of this innovative technology (Arrieta et al. 2020). AI is an arcane technology that is difficult to understand and explain, and developing a conversation about AI with the general public and, above all, creating a dialogue between human and artificial intelligences are extremely difficult tasks. However, postponing these tasks is only going to make them harder. We still live in an era of *cold machines* lacking empathy and consciousness and showing a rudimentary capacity for thinking, but as innovation in AI moves forward, so does the level of intelligence of autonomous technologies. We should lead these vital conversations now that we still have empathic and cognitive advantages.

References

Acheampong, R.A. and Cugurullo, F. (2019). Capturing the behavioural determinants behind the adoption of autonomous vehicles: conceptual frameworks and measurement models to predict public transport, sharing and ownership trends of self-driving cars. *Transportation Research Part F: Traffic Psychology and Behaviour* 62: 349–375.

Acheampong, R.A., Cugurullo, F., Gueriau, M., and Dusparic, I. (2021). Can autonomous vehicles enable sustainable mobility in future cities? Insights and policy challenges from user preferences over different urban transport options. *Cities* 112: 103134.

Ajzen, I. (1991). The theory of planned behavior. *Organizational Behavior and Human Decision Processes 50* (2): 179–211.

Arrieta, A.B., Díaz-Rodríguez, N., Del Ser, J. et al. (2020). Explainable Artificial Intelligence (XAI): concepts, taxonomies, opportunities and challenges toward responsible AI. *Information Fusion* 58: 82–115.

Bai, X., Zhai, W., Steiner, R.L., and He, Z. (2020). Exploring extreme commuting and its relationship to land use and socioeconomics in the central Puget Sound. *Transportation Research Part D: Transport and Environment 88*: 102574.

Barns, S. (2021). Out of the loop? On the radical and the routine in urban big data. *Urban Studies* 58 (15): 3203–3210.

Bostrom, N. (2017). *Superintelligence*. Oxford: Oxford University Press.

Botello, B., Buehler, R., Hankey, S. et al. (2019). Planning for walking and cycling in an autonomous-vehicle future. *Transportation Research Interdisciplinary Perspectives* 1: 100012.

Bourne, J.E., Sauchelli, S., Perry, R. et al. (2018). Health benefits of electrically-assisted cycling: a systematic review. *International Journal of Behavioral Nutrition and Physical Activity* 15 (1): 1–15.

Caprotti, F. and Liu, D. (2020). Platform urbanism and the Chinese smart city: the co-production and territorialisation of Hangzhou City Brain. *GeoJournal* 87 (3): 1559–1573.

Cox, P. and Koglin, T. (ed.) (2020). *The Politics of Cycling Infrastructure: Spaces and (in) equality*. Policy Press.

Creutzig, F., Javaid, A., Soomauroo, Z. et al. (2020). Fair street space allocation: ethical principles and empirical insights. *Transport Reviews* 40 (6): 711–733.

Cugurullo, F. (2020). Urban artificial intelligence: from automation to autonomy in the smart city. *Frontiers in Sustainable Cities* 2: 38.

Cugurullo, F. (2021). *Frankenstein Urbanism: Eco, Smart and Autonomous Cities, Artificial Intelligence and the End of the City*. London: Routledge.
Cugurullo, F. and Acheampong, R. (2020). Smart cities. In: *Handbook of Urban Mobilities* (ed. O. Jensen, C. Lassen, V. Kausfmann, et al.), 389–297. London: Routledge.
Cugurullo, F., Acheampong, R.A., Gueriau, M., and Dusparic, I. (2020). The transition to autonomous cars, the redesign of cities and the future of urban sustainability. *Urban Geography* 42 (6): 833–859.
Del Casino Jr, V.J., House-Peters, L., Crampton, J.W., and Gerhardt, H. (2020). The social life of robots: the politics of algorithms, governance, and sovereignty. *Antipode* 52 (3): 605–618.
Dowling, R. and McGuirk, P. (2020). Autonomous vehicle experiments and the city. *Urban Geography* 43 (3): 409–426.
Garrard, J., Rissel, C., and Bauman, A. (2012). Health benefits of cycling. *City Cycling* 31: 31–56.
Götschi, T., Garrard, J., and Giles-Corti, B. (2016). Cycling as a part of daily life: a review of health perspectives. *Transport Reviews* 36 (1): 45–71.
Greenfield, A. (2018). *Radical Technologies: The Design of Everyday Life*. London: Verso Books.
Hagenzieker, M.P., van der Kint, S., Vissers, L. et al. (2020). Interactions between cyclists and automated vehicles: results of a photo experiment. *Journal of Transportation Safety & Security* 12 (1): 94–115.
Hesselberth, P. (2018). Discourses on disconnectivity and the right to disconnect. *New Media & Society* 20 (5): 1994–2010.
Karvonen, A., Cugurullo, F., and Caprotti, F. (ed.) (2018). *Inside Smart Cities: Place, Politics and Urban Innovation*. London: Routledge.
Kenworthy, J.R. and Laube, F.B. (1996). Automobile dependence in cities: an international comparison of urban transport and land use patterns with implications for sustainability. *Environmental Impact Assessment Review* 16 (4-6): 279–308.
Kitchin, R. (2014). The real-time city?. Big data and smart urbanism. *GeoJournal* 79 (1): 1–14.
Lefebvre, H. (1967). Le droit à la ville. *L'Homme et la société* 6 (1): 29–35.
Lefebvre-Ropars, G., Morency, C., and Negron-Poblete, P. (2021). A needs-gap analysis of street space allocation. *Journal of Transport and Land Use* 14 (1): 151–170.
Levesque, H.J. (2017). *Common Sense, the Turing Test, and the Quest for Real AI*. Cambridge: MIT Press.
Lovelock, J. (2019). *Novacene: The Coming Age of Hyperintelligence*. London: Allen Lane.
Maibach, E., Steg, L., and Anable, J. (2009). Promoting physical activity and reducing climate change: opportunities to replace short car trips with active transportation. *Preventive Medicine* 49 (4): 326–327.
Narayanan, S., Chaniotakis, E., and Antoniou, C. (2020). Shared autonomous vehicle services: a comprehensive review. *Transportation Research Part C: Emerging Technologies* 111: 255–293.
Naudé, W. and Dimitri, N. (2020). The race for an artificial general intelligence: implications for public policy. *AI & SOCIETY* 35 (2): 367–379.
Oja, P., Titze, S., Bauman, A. et al. (2011). Health benefits of cycling: a systematic review. *Scandinavian Journal of Medicine & Science in Sports* 21 (4): 496–509.
Oldenburg, R. and Brissett, D. (1982). The third place. *Qualitative Sociology* 5 (4): 265–284.

Pammer, K., Gauld, C., McKerral, A., and Reeves, C. (2021). "They have to be better than human drivers!" motorcyclists' and cyclists' perceptions of autonomous vehicles. *Transportation Research Part F: Traffic Psychology and Behaviour* 78: 246–258.

Pearl, J. (2014). *Probabilistic Reasoning in Intelligent Systems: Networks of Plausible Inference*. Amsterdam: Elsevier.

Prati, G., Puchades, V.M., and Pietrantoni, L. (2017). Cyclists as a minority group? *Transportation Research Part F: Traffic Psychology and Behaviour* 47: 34–41.

Ritchie, H. and Roser, M. (2018). Urbanization. OurWorldInData.org. https://ourworldindata.org/urbanization (accessed 9 July 2021).

Russell, S.J. and Norvig, P. (2016). *Artificial Intelligence: A Modern Approach*. Harlow: Pearson Education Limited.

Sadoway, D. and Shekhar, S. (2014). (Re) prioritizing citizens in 'smart cities' governance: examples of smart citizenship form India. *The Journal of Community Informatics* 10 (3): 1–29.

Scharnhorst, E. (2018). *Quantified Parking: Comprehensive Parking Inventories for Five US Cities*. Research Institute for Housing America.

Shoup, D.C. (2021). *The High Cost of Free Parking*. New York: Routledge.

Swan, B. (2019). Connected and automated vehicles: urbanization versus suburbanization. *New Visions for Public Affairs* 11 (1): 63–73.

Tegmark, M. (2017). *Life 3.0: Being Human in the Age of Artificial Intelligence*. London: Penguin.

Thaler, R.H. and Sunstein, C.R. (2021). *Nudge*. New York: Penguin.

Tiddi, I., Bastianelli, E., Daga, E. et al. (2020). Robot–city interaction: mapping the research landscape—a survey of the interactions between robots and modern cities. *International Journal of Social Robotics* 12 (2): 299–324.

Yigitcanlar, T. and Cugurullo, F. (2020). The sustainability of artificial intelligence: an urbanistic viewpoint from the lens of smart and sustainable cities. *Sustainability* 12 (20): 8548.

Yigitcanlar, T., Kankanamge, N., Regona, M. et al. (2020). Artificial intelligence technologies and related urban planning and development concepts: how are they perceived and utilized in Australia? *Journal of Open Innovation: Technology, Market, and Complexity* 6 (4): 187.

Zhang, W. and Guhathakurta, S. (2017). Parking spaces in the age of shared autonomous vehicles: how much parking will we need and where? *Transportation Research Record* 2651 (1): 80–91.

Zhang, W., Guhathakurta, S., Fang, J., and Zhang, G. (2015). Exploring the impact of shared autonomous vehicles on urban parking demand: an agent-based simulation approach. *Sustainable Cities and Society* 19: 34–45.

Part II

Urban Design, Planning and Policies

8

Re-Defining the Smart City Concept from the Urban Climate Perspectives

Joachim Fallmann[1], Christopher Holst[2], Matthias Mauder[2], and Stefan Emeis[2]

[1] *Karlsruhe Institute of Technology, Karlsruhe, Germany*
[2] *Karlsruhe Institute of Technology, Garmisch-Partenkirchen, Germany*

8.1 Introduction

Cities have become the main living space for humans on Earth and will be even more so in the future (Kii 2021). Understanding how these urban spaces can be governed needs holistic approaches which mirror and combine a broad range of fields that include societal, economic and ecological challenges. Changing features of urban and global climate are a major part of these challenges. The interface and interaction between climate assessment, on the one hand, and urban planning together with building design, on the other, require both research and action in order to shape sustainable cities for tomorrow. Climate resilience of an urban entity has to be assessed on the basis of scenarios of future development of these fields which are developed by dedicated forecasting systems linked to local experiences. That is per se a difficult task, as urban agglomerations are very complex entities which are difficult to govern even apart from any climate issues.

A broad range of studies exist that focus on sustainable urban planning strategies (Bibri and Krogstie 2017), and the way cities have to be designed in order to best cope with global change aspects. This Chapter looks at the situation from an atmospheric scientist's point of view. It intends to provide suggestions for sustainable urban planning under present and future climate conditions, by providing a review of existing approaches and best practice examples and also addresses requirements for holistic model systems directed to local decision support. Rather than defining specifically dedicated and detailed measures or technical realisations, this Chapter aims to display and offer ideas summarising results from five decades of scientific research in the field of urban planning with regard to local air quality and local and global climate.

Traditional city design has been the result of a long historic evolution (Bandarin and van Oers 2012). As local climate had been one of the major external factors for city design since

Smart Cities, Energy and Climate: Governing Cities for a Low-Carbon Future, First Edition.
Edited by Oleg Golubchikov and Komali Yenneti.

ancient times, climate-adapted city designs evolved in many parts of the world. Just to name a few examples of cities in very warm and hot areas, one could look at white Greek cities or the densely built desert cities in the Sahara and the Middle East. More examples can be found in Linhares de Siqueira (2015). Also, early techniques of air conditioning such as towers serving as wind catchers were used in some hot places to keep city buildings cool and inhabitable (Azami 2005). Sustainable urban planning as such has to be treated with respect to the historical stage of development of the city or region in focus, thus enabling a 'fitness-for-purpose'.

Many terms have been coined to describe cities that are preparing for the above-mentioned challenges. Among these terms are words such as 'smart', 'sustainable', 'climate-friendly', 'de-fossilised' and many more (Bunning 2014). The term 'de-fossilised' is probably more appropriate than the well-established term 'decarbonised' because organic carbon may play a role in future cities. 'Sustainable' seems to be the most general one of these terms. The meaning of this term, introduced into the modern discussion by Brundtland (1987), is strongly dependent on the context in which it is applied and on whether its use is based on a social, economic or ecological perspective (Brown et al. 1987). As shaping of modern cities is different from those of the past, this calls for a re-evaluation in order to shape sustainable cities to be simultaneously economically viable, socially just, politically well managed and ecologically sustainable to maximise human comfort (Panagopoulos et al. 2016). Figure 8.1 summarises the most important topics for shaping cities under a changing climate seen in a holistic approach.

The frequent use of the term 'smart city' in recent years can be partly seen as a linguistic development as well. In the urban context, the meaning of 'smart' is often narrowed down

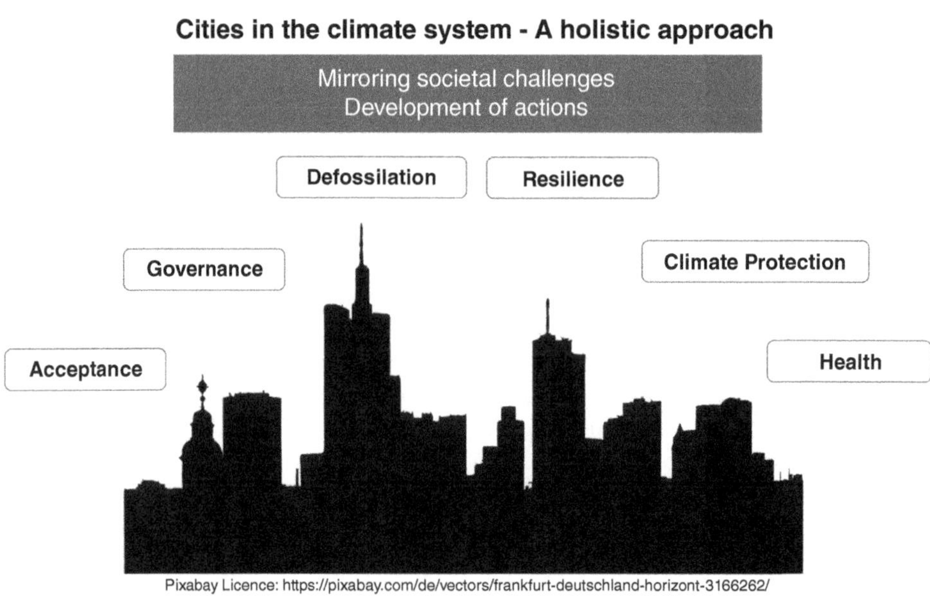

Figure 8.1 Six main topics of shaping sustainable urban agglomerations. *Source:* GDJ/15184 images/Pixabay.

to better information and communication infrastructure and enhanced IT applications which should lead to better governance (Silva et al. 2018). But in this narrow sense, 'smart' does not necessarily mean 'sustainable'. Bibri and Krogstie (2017) give a literature review on the designations 'smart' and 'sustainable' (and all possible combinations thereof) for cities. A recent review (Angelidou et al. 2018) analyses that the links between smart applications and sustainable cities are still fragmented and that there are at least six fields of action which need further research. These are as follows: (i) green mobility, (ii) waste management, (iii) air pollution, (iv) energy consumption, (v) urban biodiversity and (vi) water management.

The look into the future is always difficult. Today, modelling approaches to derive future scenarios have become a frequently used tool (Walloth et al. 2016). The vast complexity of urban systems does not yet allow for models depicting the full complexity. Therefore, often single processes and effects are selected and modelled with otherwise given and fixed boundary conditions. While this may be helpful on a local scale for short-term projections, this does not work properly for larger scales and longer time periods due to the strong non-linear interactions between different compartments of larger urban systems. In the final part of this Chapter, we will thus address some of the presently available modelling tools for the atmospheric and climatic parts of urban systems.

This Chapter is structured as follows: A selection of existing urban studies is presented in Section 8.2. The missing link between climatic studies and city planning is addressed. Section 8.3 features a few recent approaches for reshaping our cities. The main section is Section 8.4 which presents a long list of suggestions that could be helpful to include the findings of atmospheric and climatic research into future city planning. Section 8.5 sketches recent development in urban modelling with explicit consideration of the urban atmosphere. A few conclusions are formulated in Section 8.6.

8.2 Existing Urban Studies

8.2.1 Studies from Atmospheric Sciences

Atmospheric sciences have dealt with the special features of urban climate for about 200 years starting with the seminal book by Howard (1818–20) in 1818. More modern studies start with the book of Kratzer in 1937 (Kratzer 1937), followed by extended bibliographies compiled, e.g. by Chandler in 1970 (Chandler 1970) and Oke in 1974 (Oke 1974), leading to the famous compendium by Landsberg in 1981 (Landsberg 1981). The most important features of urban climate include higher air and surface temperatures, changes in radiation balances, lower humidity and restricted atmospheric exchange, which cause accumulations of pollutants from local sources. The four main causes of a special urban climate are as follows: (i) replacement of natural soil by sealed surfaces, mostly artificial and having a strong 3D structure; (ii) reduction in the surface area covered by vegetation; (iii) reduction in long-wave emission of the surface by street canyons; and (iv) release of gaseous, solid and liquid atmospheric pollutants and waste heat. These factors all have severe impacts on the urban energy budget, water storage and atmospheric exchange in near-surface layers (Marzluff et al. 2008).

Urban climate studies in the last decades have usually focussed on specific features such as the formation of urban heat islands (UHI; Hidalgo et al. 2008; Levermore et al. 2018; Li et al. 2019), the circulations driven by these heat islands (Fan et al. 2017), the enhancement of local rainfall by the heat island and increased surface roughness (Holst et al. 2016), degraded air quality (emission, transport, chemical transformation and deposition of reactive gases and particles [Parrish et al. 2011; Parrish et al. 2016; Andrade et al. 2017]) and pedestrian comfort in street canyons and close to high-rise buildings (e.g. local wind conditions [Stathopoulos et al. 2004] or human thermal comfort [Chatzidimitriou and Yannas 2016]) often in connection with public health issues (Salmond et al. 2018; Steeneveld et al. 2018; Tan et al. 2010). Some of these studies discussed the interrelation between regional and global climate and climate change (Akbari et al. 2016; Cleugh and Grimmond 2012; Founda and Santamouris 2017; Ward et al. 2016). Several studies also considered in-door climate (see, e.g. Gross 2021 and references therein).

8.2.2 Studies on Sustainable Cities

Sustainability studies referring to cities have mainly concentrated on energy consumption and greenhouse gas emissions. About 80% of all greenhouse gas emissions is generally attributed to cities (Martos et al. 2016), although, according to Satterthwaite (2008), most likely less than half of anthropogenic greenhouse gas emissions are actually generated within the city limits. However, cities account for this significantly higher percentage of these emissions, if emissions from power stations or industries are assigned to the location of the actually consuming persons or institutions. Some studies also addressed matter fluxes and cycles (water, energy and commodities) but rarely recyclable buildings and infrastructure (see, e.g. Petit-Boix et al. 2017).

8.2.3 Missing Link Between Climate and Sustainability Studies and City Planning

There seems to be a consensus that nearly none of the abovementioned studies, neither climate studies nor sustainability studies, had considerable influence on actual city planning so far (Eliasson 2000; Mills et al. 2010; Parsaee et al. 2019). The low impact is said to be a result of several constraints which could be related to five explanatory variables, i.e. conceptual and knowledge based, technical, policy, organisational and the market (Eliasson 2000). UHI mitigation strategies have had negligible contributions to urban development policies and action plans (Parsaee et al. 2019) as they necessarily have to involve the dynamic nature of the urban landscape with regard to social and economic aspects. Bringing together concepts from various disciplines, sustainability of cities – leading to human health and well-being – should be manifested in economic viability, social justice, ecological sustainability and transparency of political management (Amado et al. 2016; Panagopoulos et al. 2016).

Amado et al. (2016) even state that current urban planning is not a process that ensures energy-efficient cities, and they demand that both reducing energy consumption and supporting the integration of solar energy systems and smart grid technologies in urban context should be part of urban planning. Martos et al. (2016) analysed more specifically that the impact on the environment and the elevated energy consumption generated by the dominant use of private motor vehicles in the cities is one of the most pressing demands for making urban policies more sustainable. Notwithstanding the high interdependencies between urban issues and sustainability strategies, cities have not been analysed comprehensively from a life cycle perspective so far due to their complexity (Petit-Boix et al. 2017).

8.3 Recent Approaches for Reshaping Building Design

Despite the non-action discussed in Eliasson (2000); Mills et al. (2010); and Parsaee et al. (2019), some progressive ideas in building design have found widespread interest across disciplinary borders. For example, 'biophilic design' (Kellert et al. 2011) is said to enhance human well-being by fostering connections between people and nature in the built environment. Studies report that people living in proximity to natural spaces report fewer health and social problems, independent of rural and urban residence, level of education and income (Kellert et al. 2011). The 'Bosco Verticale' (vertical forest), two buildings in Milano, Italy, and several buildings in Singapore are listed as examples by Hayles and Aranda-Mena (n.d.). Bosco Verticale consists of two separate towers reaching heights of 111 and 78 metres. The towers are composed of 26 and 18 floors, respectively, and the facades are covered with plants, shrubs and trees. The buildings were designed by the Milanese architect Stefano Boeri in cooperation with the architects Giovanni La Varra and Gianandrea Barreca (Visser 2019). Biophilic design can be viewed as the missing link in prevailing approaches to sustainable design when working in complementary relation with low-environmental-impact design, thus achieving true and lasting sustainability. With major objectives of low-environmental-impact design focussing on goals such as energy and resource efficiency, sustainable products and materials, safe waste generation and disposal, pollution abatement, biodiversity protection and indoor environmental quality, a detailed specification of design strategies to achieve these goals, for instance, has been incorporated into certification systems such as the US Green Building Council's 'Leadership in Energy and Environmental Design' (LEED) rating approach (Kellert et al. 2011). Biophilic design can thus be an overarching strategy, incorporating various needs for sustainable building design thus providing holistic solutions for sustainable planning needs.

Although not explicitly noted by Hayles and Aranda-Mena (n.d.), biophilic design helps to meet some of the requirements due to atmospheric and climate impacts, e.g. reducing urban heat islands, enhancing outdoor air quality and, via shading, also providing a better indoor climate. In reducing the need for air conditioning, biophilic design also helps to create more sustainable cities by reducing energy consumption and by reducing greenhouse gas and gaseous reactive nitrogen compounds emissions.

Another trend is using wood as building material. Tupėnaitė et al. (2020) give an overview of existing buildings made from timber. Their ranking of the buildings also includes the saved carbon dioxide in comparison to concrete buildings. The highest timber building is presently the Mjøsa Tower in Norway, an 18-floor building which is 85 m high. Recently, the Sumitomo Forestry Co. in Tokyo, Japan, announced plans to build a 70-storey, 350-m-tall skyscraper made of 90% wood and 10% steel (Buffi and Angelini 2019; Naser 2019), although the project is planned to be completed only in 2041. Again, it seems that nearly no climate or sustainability arguments are driving the architects. The undoubtedly good rationale given instead tells that avant-garde designers are reshaping the historical relationship between nature and architecture fostering a natural approach to design that results in structural lightness, rational use of energy and elegance. Wood seems to be a perfect material to engage with this new period of design research, and while timber towers are getting higher, there is a strong interest in understanding to which extent wood can represent a valuable alternative to those materials that have characterised the recent architectural debate (Buffi and Angelini 2019).

Climate or sustainability arguments for skyscrapers made of timber instead of metal and concrete involve less carbon dioxide emissions during the production of these materials and the construction phase of the building. And in addition, timber buildings store biologically extracted carbon from the atmosphere for a long time (Churkina et al. 2020). CO_2 emissions from global cement production sum up to about 4% of those from fossil fuel combustion and contributed 1.45 Gt CO_2 in 2016 [42]. Buffi and Angelini (2019) briefly mention this and give a share of 8% of the global CO_2 emissions for the combination of global cement and steel production. Anyhow, using wood as construction material is a much better strategy to use biomass than to use it as a substitute for fuel oil or natural gas in residential heating (Searchinger et al. 2018).

8.4 Suggestions for Urban Planning and Building Design

Changes in building design in order to adapt to a changing climate are often narrowed down to issues caused by extreme winds, lightning and extreme precipitation (see, e.g. the guidelines given in Wallbaum et al. 2011). In this context, 'extreme precipitation' comprises phenomena such as local flooding due to heavy downpours, hail and heavy snow loads. This definitely helps to erect resilient buildings which can withstand extreme weather events and reduces the costs for owners and insurance companies.

Resilience to extreme weather events is only one small facet of a sustainable city which is to become resilient in a changing local and global climate and with respect to changes in local and global energy and matter fluxes. Stress to human health by high temperatures, bad air quality and noise has to be reduced as well in order to shape resilient cities. Unwanted interactions between strategies and actions to achieve each of these goals have to be paid attention to (e.g. air quality can become worse as a consequence of heat island mitigation measures [Fallmann et al. 2016]) and require holistic approaches which assess the overall impact of several (or ideally all) strategies and actions together. Such a consideration of the overall complexity of urban systems is addressed in Petit-Boix et al. (2017) as well.

Climate resilient cities should come into existence through co-ordinated efforts and not just as beneficial side effects of other sustainability efforts. The following part of this Chapter lists ideas and options to directly address air quality and local and global climate issues. The list complements the more general list in Petit-Boix et al. (2017) directed towards sustainable cities. The below-listed measures are sorted into four fields from the building scale to the nationwide scale: measures to modify building design, measures to enhance urban green and blue, measures to replan cities and finally, measures to secure the overall resilience of urban areas. Within each of the action fields, the measures are sorted by descending importance from an atmospheric scientist's point of view. A more detailed form of this list can be found in Fallmann and Emeis (2020).

8.4.1 Action Field 'Building Design'

1) *Urban surfaces*: White and reflecting roofs and walls (Akbari et al. 1997, 2009) and road pavements (Rossi et al. 2016; Santamouris 2013), or green and transpiring roofs and walls (Georgescu et al. 2014) prevent urban structures from heating up as well as rooftop PV and solar thermal devices. In addition, green façades and roofs can help to remove particulates from the urban air by dry deposition, but need sufficient water supply (Cascone et al. 2019; William et al. 2016) and more maintenance (Zhang et al. 2012).
2) *Passive solar architecture – cooling*: Protruding roofs reduce sunshine into the windows in summer but enable it in winter (Goetzberger and Schmid 2010). This reduces the energy demand for air conditioning and heating. Well-insulated lightweight constructions reduce heat gain and loss, making energy use more efficient. Heavy mass buildings have damp temperature amplitudes inside the building.
3) *Passive solar architecture – heating*: Darker colours for those parts of the facade only hit by the sun in winter supply radiative heat gains. An enhancement of this idea is 'Trombe' walls coated with a dark surface shortly behind glazing (Dimassi and Dehmani 2012). Heavy mass buildings can release heat at night to warm interior spaces (Haggard et al. 2016). Clear orientation of buildings with respect to the sun's daily culmination point and enabling in-door fresh air ventilation (Gross 2021) is one of the prerequisites for some of the measures listed here. But buildings should not be arranged in such a way that they produce excessive wind channelling and nozzle effects which decrease pedestrian comfort (Blocken et al. 2016).
4) *Passive solar architecture – phase change materials (PCM)*: Phase change materials such as salts or paraffin may be integrated into building walls. The melting temperature can be adjusted as necessary. If PCM is made to store heat at a useful temperature, it will maintain that temperature level until the whole phase change process has taken place, thereby reducing overheating or cooling of the internal air (Gillott et al. 2010).
5) *Building materials – greenhouse gases*: Reduce the usage of steel and concrete wherever possible, because steel and concrete production are responsible for a considerable part of global greenhouse gas emissions. Buildings at least partly made of wood (Tupėnaitė et al. 2019) help to store carbon for a long time and they are a means of carbon capture helping to reduce atmospheric carbon dioxide concentrations (Naser 2019).

6) *Building materials – recycling*: The selection of building materials and building construction methods should already have in mind the recycling of these materials later. This will help to reduce the demand for commodities. Even sand and gravel are no longer unlimited resources (Torres et al. 2017).
7) *Passive Aeolian architecture*: Wind catchers (Azami 2005) should be reconsidered as a meaningful technology for cities in hotter areas with regular winds, e.g. near sea shores or closer to larger mountain ranges. Night-time evaporation from wet surfaces would be the only passive cooling mechanism in hot humid climates, achieved by allowing sufficient airflow through the building (Haggard et al. 2016).
8) *Roof shape*: In canyon, flow conditions can be strongly dependent on geometric features of buildings and the shape of the roofs (Llaguno-Munitxa et al. 2017). Turbulent structures in the urban canopy can affect pollution levels inside the urban canyon. Pollution levels were found to decrease with increasing roof slopes (Takano and Moonen 2013).

8.4.2 Action Field 'Urban Green and Blue'

1) *Green close to buildings*: Urban green moderates the local climate. But trees should not be much higher than the buildings themselves in order to avoid damage from falling trees due to extreme wind gusts, and not to shadow reflecting or green roofs for cooling and/or renewable energy supply. Urban green on the side lighted by the midday sun should consist of deciduous species in order to allow for shade in summer and incoming light in winter. Coniferous trees provide shelter against cold winds on the opposite side. Such vegetation meaningfully complements passive solar architecture. See Wang et al. (2014) for a review.
2) *Urban green*: Small parks in cities reduce urban heat islands (Fallmann et al. 2013) and offer recreation spots. Chosen shrub and tree species should be resistant against heat, temporal water shortage and air pollutants (Quigley 2004). Dense tree canopies in narrow streets can lead to a relative warming of the surface underneath (Meili et al. 2020). In canyons with much traffic exhausts, they could impede the lateral and vertical air exchange too much (Vranckx et al. 2015). Tree species known to emit hydrocarbons such as isoprene (e.g. plane trees/sycamores) should not be planted in urban environments because sunshine into air masses polluted with nitric oxides and isoprene leads to rapid formation of ozone (Simon et al. 2019; Zhang et al. 2016).
3) *Urban blue*: Water in cities moderates temperature extremes due to the large heat capacity of water. But it is not easy to maintain the water quality of lakes with small water exchange. Additionally, water bodies can be breeding areas for insects which could serve as vectors for diseases. Therefore, urban blue may be limited to areas with sufficient fresh water supply. Roof ponds could act as passive heating and cooling systems as well (Sharifi and Yamagata 2015).
4) *Urban gardening*: Local supply from urban farming may help to provide fresh food (Feola et al. 2020; Martin and Molin 2019; Valley and Wittman 2019) in addition to greening the city. Urban gardening can contribute to perceived attractiveness of urban areas (Lindemann-Matthies and Brieger 2016) and within the frame of the compact city, aim at an efficient and resource-saving (re)organisation of urban space (Tappert et al. 2018).

8.4.3 Action Field 'City Planning'

1) *Energy supply*: Cities with reduced urban heat island have to be cities with strongly reduced exhausts from combustion processes because cooler cities do not foster so much thermally driven vertical ventilation (Fan et al. 2017) which helps diluting high pollutant concentrations. Electricity can be provided by photovoltaic devices. Especially, transparent solar photovoltaic technologies could use the part of the vast window surfaces in megacities (Husain et al. 2018). An option for residential heating could be electrical heat pumps. A presently widely unused heat reservoir is the urban heat island in the groundwater underneath cities (Grant et al. 2008; Zhu et al. 2011). Grant et al. (2008) discusses the installation of roof-mounted ducted wind turbines, which use pressure differentials created by wind flow around a building.
2) *Fresh air corridors*: Extension of existing city quarters and planning of new quarters should not hamper the free flow of air from regional environments into cities. Nocturnal venting of fresh and cooler air from the outside into cities, e.g. gives relief from nocturnal heat stress and enhances human living conditions and air quality (Barlag and Kuttler 1990; Schau-Noppel et al. 2020).
3) *Reduce urban sprawl*: Cities should be built in denser ways in order to reduce traffic needs and to save the natural environment of cities (Nechyba and Walsh 2004). Part of such strategies has to be to keep shopping malls within residential areas and not to erect them off city limits (Ye et al. 2018).
4) *Overcome strict functional separation*: Today's industries do not emit so much detrimental substances and noise as they did in former times. Thus, residential areas and industrial areas could be closer together reducing commuting traffic. City planners should overcome the paradigms of the charter of Athens 1933 (Gold 1998). Architects had developed the idea of a functional city which strictly separated residential, industrial, recreational and transport areas in the early 1930s. The Leipzig Charter of 2007 (Eltges 2010) is a first and considerable step away from the Athens Charter and towards more sustainable cities. In a further step, a New Leipzig Charter (Geppert 2020) was adopted at the end of November 2020.
5) *Privileged public transport*: Public surface transport has to be privileged in larger cities in order to reduce pollutant emission from individual traffic (Nieuwenhuijsen 2020). The organisation of public transport will become easier if the previous point is put into reality.

8.4.4 Action Field 'Nationwide Planning of Infrastructure'

1) *Regional traffic*: Larger cities should be interconnected by efficient high-speed railway lines, separated for passenger and goods transport (Baker et al. 2010). Air traffic should be reserved for intercontinental passenger transport.
2) *City planning – sea level*: New cities and industrial plants must not be planned directly at low sea shores as they are usually designed to exist for centuries. But on this time scale, sea level rise will become a major problem from global warming. The present rate (2011–2020) of sea level rise is close to 4 mm per year and is expected to increase further (Ablain et al. 2017; Siegert et al. 2020). Existing cities and industrial hubs have to be prepared and hardened to rising sea levels (McEvoy et al. 2021).

This list is not complete and future research will identify further possible actions. Although the sequence of the measures within the action fields above has been ranked according to their importance for air quality and climate, the actual choice of appropriate measures from the above list and their importance for a given city may depend on the specific characteristics of the city in question.

Figure 8.2 summarises links between different action fields, inspired by Figure 2 in Petit-Boix et al. (2017), but introducing the outer bubbles 'air', 'local climate' and 'global climate' which were completely missing in the respective Figure in Petit-Boix et al. (2017). Full arrows point to intersections of the outer bubbles with the central bubble 'Urban Planning', and dashed arrows point to intersections between neighbouring outer bubbles. The Figure should not be misinterpreted. The world is not two dimensional. This means, there are many intersections between bubbles which are not direct neighbours in this Figure. Just to give one example for this multi-dimensionality: 'Mobility' has a strong influence on 'Air'. The way we organise traffic in a city has strong repercussions on urban air quality. As recent example, Amouei Torkmahalleh et al. (2021), for instance, found a global increase in near-surface ozone concentrations due to the limitations of NO_2 exhaust from traffic during Covid-19 lockdown. Although these intersections cannot be plotted here, they have

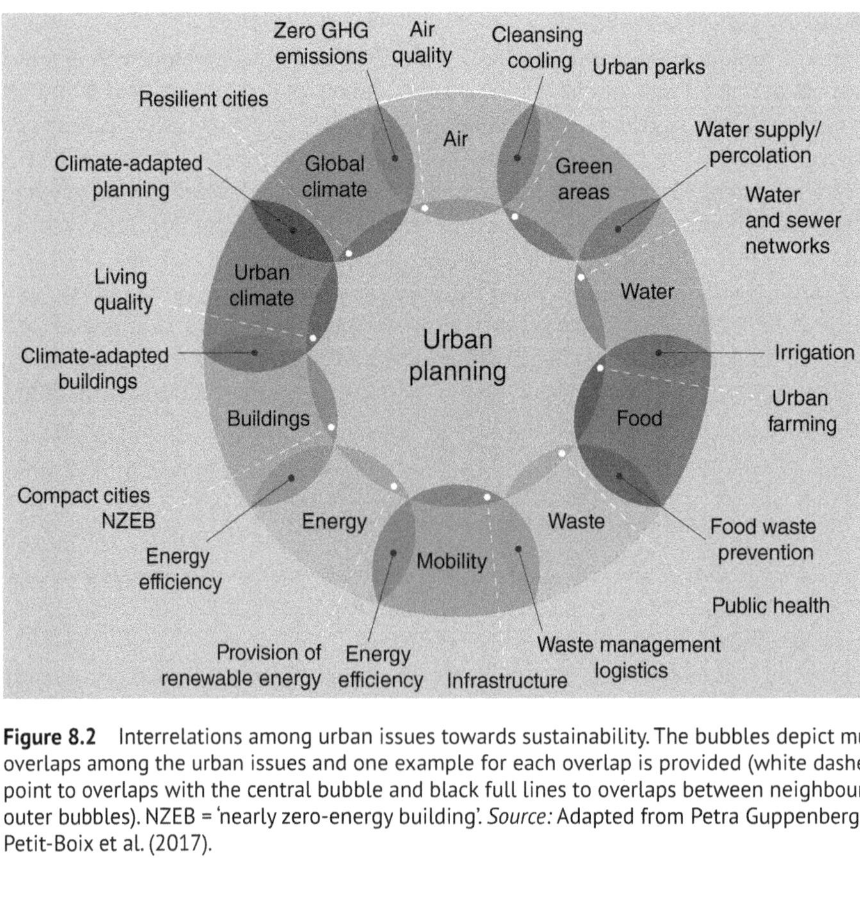

Figure 8.2 Interrelations among urban issues towards sustainability. The bubbles depict mutual overlaps among the urban issues and one example for each overlap is provided (white dashed lines point to overlaps with the central bubble and black full lines to overlaps between neighbouring outer bubbles). NZEB = 'nearly zero-energy building'. *Source:* Adapted from Petra Guppenberger; Petit-Boix et al. (2017).

to be taken into account in the abovementioned holistic approaches exploring and paving the way towards more sustainable and climate-friendly cities.

8.5 Model Approaches

Climate sciences have started to pave the way for holistic numerical model approaches by designing Earth system models (Bonan and Doney 2018; Claussen et al. 2002) in order to bring together the atmosphere, the oceans, the ice masses, the vegetation and other forms of land use on a global scale. Urban areas are one special form of land use. Fully coupled (ocean, wave, atmosphere and land) earth-system-type models have lately also been developed for regional-scale applications in the United Kingdom (Lewis et al. 2018, 2019), and their potential to improve the forecast of small-scale atmospheric features in the order of 1–10 km has been assessed (Fallmann et al. 2017).

Convective-scale (i.e. resolving spatial scales of a few hundred metres to a few kilometres and temporal scales of minutes; in atmospheric sciences, the term convection denotes thermally driven upwards vertical motions such as in showers and thunderstorms) atmospheric and chemical transport models have recently been used in atmospheric sciences to analyse urban features mentioned in Section 8.2.1, covering a range of modelling systems and research questions. Although not being able to provide ready-to-use applications for urban planning purposes, they, however, can offer sophisticated information on urban–rural interactions and on the feedbacks among regional climate, meteorological processes and small-scale phenomena such as sea breeze, urban heat island and mountain–valley circulation. Thanks to efforts in the development of more detailed urban canopy parametrisations for these kind of models (e.g. WRF-BEP [Chen et al. 2011], TERRA_URB [Wouters et al. 2015], WRF-TEB [Meyer et al. 2020], BEP-Tree [Krayenhoff et al. 2020] and COSMO-BEP-Tree [Mussetti et al. 2020b], or more recently, CCLM-DCEP [Mussetti et al. 2020a]) and various evaluation studies (Jandaghian and Berardi 2020; Trusilova et al. 2016), the urban energy balance and the diurnal features of the urban boundary layer are widely understood at the convective scale. With more and more studies critically assessing the sources and quality of input data and model systems needed (Masson et al. 2020) in order to also provide sophisticated information for societal needs (Hidalgo et al. 2019), the link between the regional and the local scale (meso- and microscale) and the transfer to actual planning processes remains an ongoing challenge. Current studies try to overcome the scale dependency of processes and link atmospheric and planning perspectives by coupling meso-scale models with CFD-based approaches, e.g. analysing micro-scale flow patterns during heat-wave episodes (Piroozmand et al. 2020), investigating the evolution of extreme wind induced by building structures (Javanroodi and Nik 2020) or with Lagrangian modelling systems, e.g. evaluating urban NOx pollution (Veratti et al. 2020).

Recent efforts in downscaling climate simulations to a sub-kilometre resolution have been made by several communities and weather services (Schau-Noppel et al. 2020) in order to generate climate impact studies for a selected urban region with results underpinning decisions at a municipal level. Although providing more insight into features relevant to regional-scale urban planning such as the definition of cold air catchments or areas particularly prone to climate extremes (by linking model results to population statistics one can analyse the vulnerability of certain city quarters to urban heat stress), these kinds of models fail to allow for a street-scale analysis.

Fully integrated CFD-based 3D-microclimate models have recently been developed, e.g. for analysing the benefit of reflective pavements at building scale, for isolated street canyons under dry and wet conditions (Ferrari et al. 2020), investigating street ventilations (Shirzadi et al. 2020) or drag induced by street trees (Zeng et al. 2020). First steps towards a holistic street-scale and building-resolving urban climate model covering spatial scales from metres to many kilometres and temporal scales from seconds to years have recently been made in Germany (Maronga et al. 2019; Scherer et al. 2019). This microscale model can be coupled to mesoscale models, such as COSMO or WRF, and hence can be used for simulating the impact of realistic weather events on the city at the street scale (Kadasch et al. 2020; Lin et al. 2020; Resler et al. 2020). Also including air chemistry, such a model can simulate the interactions between buildings and the atmospheric flow, the urban radiation and heat budget and the urban air quality (Khan et al. 2021). The implementation of a multi-agent system and a biometeorology module allows for more specific analysis of impacts on humans in these models. With this tool, many of the suggestions listed in Section 8.4 above can be tested for their suitability and effectiveness in given urban situations. Coupling such urban climate models with regional and global Earth system models will offer the opportunity to consider climate change issues in urban planning as well (Sharma et al. 2020). Assessing requirements and case studies to evaluate the practicability and usability of these systems for local stakeholders and societal needs is subject to current analysis among atmospheric and climate communities (Halbig et al. 2019).

In addition to process-based models, statistical- and machine-learning-based models take a more empirical approach towards understanding how data obtained in urban networks and studies may be combined. These types of models are typically simpler and faster than process-based models (Zhang et al. 2018). When applying data of sufficient temporal coverage, such approaches also allow descriptions of temporal evolutions in urbans contexts, e.g. urban growth or densification (Gómez et al. 2020). The lower complexity and computational expense of these models, however, come with a cost. Such models need to be specifically created for each urban area individually and have a large dependence on existing data, hence a limited ability to make predictions or produce transferrable results.

As presented in the paragraphs above, modelling approaches in the past decades have seen a steady trend towards smaller scales and increased complexity.

Research and innovation like increasing availability of computational resources as well as higher coverage and resolution of observational data (e.g. satellites) and the development of modern data processing techniques have played a large role in enabling more simulations at finer scales and for longer time periods. Collaborations like the transition towards open-source community development of modelling systems (e.g. WRF and PALM) and efforts towards facilitating transdisciplinary research have pushed the complexity of modern numerical model systems.

To summarise, the evolution of complexity as well as spatial and temporal scales at which numerical models are applied in atmospheric research is still moving forward. The resulting challenge is not necessarily of a technical nature: How to facilitate aforementioned tools and experiences for decision-making for sustainable planning and development?

8.6 Conclusion

This review offers a starting point for bringing together impulses from research and application from different disciplines thus fostering collaborations among different communities and strengthening argumentations with stakeholders. For example, after Superstorm Sandy in 2012, many cities in the United States and globally have installed dedicated climate resilience officers who transfer information into the generation of resilience plans. With the constraint to reduce carbon emission to zero by 2050, many cities in Europe also have followed with the installation of climate managers in urban planning. Model approaches such as those mentioned briefly in Section 8.5 will help climate resilience officers and urban climate managers to fulfil their tasks in the future. These positions are particularly important when official directives such as the announcement of the climate emergency are to be transferred from a symbolic statement to applied action.

However, current legislative regulations and also architectural views on the representativeness of buildings still limit the implications of certain measures. In our opinion, the financial barrier for private households is still too high and more financial benefits and support have to be provided, e.g. through national programmes (e.g. as already in place for Berlin, Germany). Furthermore, we believe that the perception of public space has to change in a way and that greening has to be considered in the pre-planning phase as we find especially new city quarters are still lacking green space. That aspect should be manifested in dedicated legislative regulations. Another reason for the lack of implementation might as well be the technical standards as being discussed in Emeis and Fallmann (2022).

It is imperative to evaluate climate-sensitive urban planning measures with regard to their cost–benefit factors both environmentally and economically. The expected benefits of intended urban planning actions have to be communicated in a proper way and accompanied by actual numbers highlighting the mitigation, adaptation and risk reduction potential. The dialogue between science and application has, therefore, to be guided by official bodies such as the above-mentioned climate managers. Sustainable development has to cover many fields and span different spheres of interest, which makes holistic actions hard to construct in light of both climate change and population growth happening simultaneously. Extreme events often are followed by side effects and communities have to be best prepared against unforeseen feedback. Having been responsible for over 70,000 deaths in Europe, the 2003 heat wave has shown dramatically the expected impact of a changing climate on urban population. At about the same time, the East Coast of the United States suffered the most severe electrical outage in history affecting about 60 million households, which was due to overheating electrical circuits due to excess heat.

Sustainable urban planning has to cover different scales, from local to regional and from short term to climate scale. Cities have to be prepared for short-term effects, requiring immediate action but also long-term strategies in light of these extreme events happening more often in the future. In doing so, it has to be kept in mind that the impacts of different measures from Section 8.4 do not add linearly. Some measures even counteract others. Therefore, model studies with elaborated and sophisticated systemic models are necessary in order to assess the consequences of selected measures before realising them in reality. This document aims to fertilise the design of dedicated case studies and the transfer to action in order to allow for sustainable, climate-sensitive urban planning.

Funding Sources

This research did not receive any specific grant from funding agencies in the public, commercial or not-for-profit sectors. It fits into Topic 1 (Atmosphere under Change) within the Programme 'Changing Earth' of the Fourth Phase of the National Programme-oriented Funding of the Helmholtz Association (PoF IV) in Germany.

Conflict of Interests

The authors share no conflicts of interest.

References

Ablain, M., Legeais, J.F., Prandi, P. et al. (2017). Satellite altimetry-based sea level at global and regional scales. In: *Integrative Study of the Mean Sea Level and Its Components* (ed. A. Cazenave, N. Champollion, F. Paul, and J. Benveniste), 9–33. Cham: Springer International Publishing.

Akbari, H., Bretz, S., Kurn, D.M., and Hanford, J. (1997). Peak power and cooling energy savings of high-albedo roofs. *Energy and Buildings* 25 (2): 117–126. https://doi.org/10.1016/S0378-7788(96)01001-8.

Akbari, H., Menon, S., and Rosenfeld, A. (2009). Global cooling: increasing world-wide urban albedos to offset CO2. *Climatic Change* 94 (3–4): 275–286. https://doi.org/10.1007/s10584-008-9515-9.

Akbari, H., Cartalis, C., Kolokotsa, D. et al. (2016). Local climate change and urban heat island mitigation techniques–the state of the art. *Journal of Civil Engineering and Management* 22 (1): 1–16.

Amado, M., Poggi, F., and Amado, A.R. (2016). Energy efficient city: a model for urban planning. *Sustainable Cities and Society* 26: 476–485.

Amouei Torkmahalleh, M., Akhmetvaliyeva, Z., Omran, A.D. et al. (2021). Global air quality and covid-19 pandemic: do we breathe cleaner air? *Aerosol and Air Quality Research* 21 (4): 200567. https://doi.org/10.4209/aaqr.200567.

Andrade, M.D.F., Kumar, P., de Freitas, E.D. et al. (2017). Air quality in the megacity of São Paulo: evolution over the last 30 years and future perspectives. *Atmospheric Environment* 159: 66–82. https://doi.org/10.1016/j.atmosenv.2017.03.051.

Angelidou, M., Psaltoglou, A., Komninos, N. et al. (2018). Enhancing sustainable urban development through smart city applications. *Journal of Science and Technology Policy Management* 9 (2): 146–169.

Azami, A. (2005). Badgir in traditional Iranian architecture. *International Conference "Passive and Low Energy Cooling for the Built Environment"* (May 2005), Santorini, Greece. http://www.inive.org/members_area/medias/pdf/inive/palenc/2005/azami2.pdf.

Baker, C.J., Chapman, L., Quinn, A., and Dobney, K. (2010). Climate change and the railway industry: a review. *Proceedings of the Institution of Mechanical Engineers, Part C: Journal of Mechanical Engineering Science* 224 (3): 519–528. https://doi.org/10.1243/09544062JMES1558.

Bandarin, F. and van Oers, R. (2012). *The Urban Heritage Landscape: Managing Heritage in an Urban Century*. Oxford: Wiley-Blackwell.

Barlag, A.-B. and Kuttler, W. (1990). The significance of country breezes for urban planning. *Energy and Buildings* 15 (3–4): 291–297. https://doi.org/10.1016/0378-7788(90)90001-y.

Bibri, S.E. and Krogstie, J. (2017). Smart sustainable cities of the future: an extensive interdisciplinary literature review. *Sustainable Cities and Society* 31: 183–212. https://doi.org/10.1016/j.scs.2017.02.016.

Blocken, B., Stathopoulos, T., and van Beeck, J. (2016). Pedestrian-level wind conditions around buildings: review of wind-tunnel and CFD techniques and their accuracy for wind comfort assessment. *Building and Environment* 100: 50–81. https://doi.org/10.1016/j.buildenv.2016.02.004.

Bonan, G.B. and Doney, S.C. (2018). Climate, ecosystems, and planetary futures: the challenge to predict life in Earth system models. *Science (New York, N.Y.)* 359 (6375): https://doi.org/10.1126/science.aam8328.

Brown, J.B., Hanson, M.E., Liverman, D.M., and Merideth, R.W. (1987). Global sustainability: toward definition. *Environmental Management* 11: 713–719.

Brundtland, G.H. (1987). Our common future. *Report of the World Commission on Environment and Development.*

Buffi, A. and Angelini, G.M. (2019). Adaptive timber towers. An evolutionary prototype for the 21st century skyscraper. In: *Digital Wood Design*, Part of the Lecture Notes in Civil Engineering Book Series (LNCE, volume 24), 971–987. Springer.

Bunning, J. (2014). Governance for regenerative and decarbonised eco-city regions. *Renewable Energy* 67: 73–79. https://doi.org/10.1016/j.renene.2013.11.041.

Cascone, S., Coma, J., Gagliano, A., and Pérez, G. (2019). The evapotranspiration process in green roofs: a review. *Building and Environment* 147: 337–355. https://doi.org/10.1016/j.buildenv.2018.10.024.

Chandler, T.J. (1970). Selected bibliography on urban climate (No. 276). Secretariat of the World Meteorological Organization.

Chatzidimitriou, A. and Yannas, S. (2016). Microclimate design for open spaces: ranking urban design effects on pedestrian thermal comfort in summer. *Sustainable Cities and Society* 26: 27–47. https://doi.org/10.1016/j.scs.2016.05.004.

Chen, F., Kusaka, H., Bornstein, R. et al. (2011). The integrated WRF/urban modelling system: development, evaluation, and applications to urban environmental problems. *International Journal of Climatology* 31 (2): 273–288. https://doi.org/10.1002/joc.2158.

Churkina, G., Organschi, A., Reyer, C.P.O. et al. (2020). Buildings as a global carbon sink. *Nature Sustainability* 3 (4): 269–276. https://doi.org/10.1038/s41893-019-0462-4.

Claussen, M., Mysak, L., Weaver, A. et al. (2002). Earth system models of intermediate complexity: closing the gap in the spectrum of climate system models. *Climate Dynamics* 18 (7): 579–586.

Cleugh, H. and Grimmond, S. (2012). Urban climates and global climate change. In: *The Future of the World's Climate*, 47–76. Elsevier.

Dimassi, N. and Dehmani, L. (2012, 2012). Thermal efficiency of a solar wall in Tunisia. *ISRN Renewable Energy* (4–5): 1–8. https://doi.org/10.5402/2012/465249.

Eliasson, I. (2000). The use of climate knowledge in urban planning. *Landscape and Urban Planning* 48 (1–2): 31–44. https://doi.org/10.1016/S0169-2046(00)00034-7.

Eltges, M. (2010). Leipzig charter on sustainable european cites – a work in progress. *European Spatial Research and Policy* 16 (2): 63–78. https://doi.org/10.2478/v10105-009-0013-5.

Emeis, S. and Fallmann, J. (2022). Unsatisfying transfer of climate research to urban planning: the regulatory trap in the triple helix. *Triple Helix* 9 (3): 275–295.

Fallmann, J. and Emeis, S. (2020). How to bring urban and global climate studies together with urban planning and architecture? *Developments in the Built Environment* 4: 100023. https://doi.org/10.1016/j.dibe.2020.100023.

Fallmann, J., Emeis, S., and Suppan, P. (2013). Mitigation of urban heat stress–a modelling case study for the area of Stuttgart. *DIE ERDE–Journal of the Geographical Society of Berlin* 144 (3–4): 202–216.

Fallmann, J., Forkel, R., and Emeis, S. (2016). Secondary effects of urban heat island mitigation measures on air quality. *Atmospheric Environment* 125: 199–211. https://doi.org/10.1016/j.atmosenv.2015.10.094.

Fallmann, J., Lewis, H., Castillo, J.M. et al. (2017). Impact of sea surface temperature on stratiform cloud formation over the North Sea. *Geophysical Research Letters* 44 (9): 4296–4303. https://doi.org/10.1002/2017GL073105.

Fan, Y., Hunt, J.C.R., and Li, Y. (2017). Buoyancy and turbulence-driven atmospheric circulation over urban areas. *Journal of Environmental Sciences* 59: 63–71. https://doi.org/10.1016/j.jes.2017.01.009.

Feola, G., Sahakian, M., and Binder, C.R. (2020). *Sustainability Assessment of Urban Agriculture*. Cambridge University Press.

Ferrari, A., Kubilay, A., Derome, D., and Carmeliet, J. (2020). The use of permeable and reflective pavements as a potential strategy for urban heat island mitigation. *Urban Climate* 31: 100534. https://doi.org/10.1016/j.uclim.2019.100534.

Founda, D. and Santamouris, M. (2017). Synergies between urban heat island and heat waves in Athens (Greece), during an extremely hot summer (2012). *Scientific Reports* 7 (1): 769. https://doi.org/10.1038/s41598-017-11407-6.

Georgescu, M., Morefield, P.E., Bierwagen, B.G., and Weaver, C.P. (2014). Urban adaptation can roll back warming of emerging megapolitan regions, Proceedings of the National Academy of Sciences of the United States of America. 111 (8): 2909–2914. https://doi.org/10.1073/pnas.1322280111.

Geppert, A. (2020). Sorting the wheat from the chaff. *DisP - the Planning Review* 56 (3): 77–79. https://doi.org/10.1080/02513625.2020.1851911.

Gillott, M., Rodrigues, L.T., and Spataru, C. (2010). Low-carbon housing design informed by research. *Proceedings of the Institution of Civil Engineers - Engineering Sustainability* 163 (2): 77–87. https://doi.org/10.1680/ensu.2010.163.2.77.

Goetzberger, A. and Schmid, J. (2010). Review of components for passive solar energy utilization. *International Journal of Solar Energy* 3 (6): 309–328. https://doi.org/10.1080/01425918508914401.

Gold, J.R. (1998). Creating the charter of Athens: CIAM and the functional city, 1933-43. *Town Planning Review* 69 (3): 225.

Gómez, J.A., Patiño, J.E., Duque, J.C., and Passos, S. (2020). Spatiotemporal modeling of urban growth using machine learning. *Remote Sensing* 12 (1): 109. https://doi.org/10.3390/rs12010109.

Grant, A., Johnstone, C., and Kelly, N. (2008). Urban wind energy conversion: the potential of ducted turbines. *Renewable Energy* 33 (6): 1157–1163.

Gross, G. (2021). A numerical study on the effects of natural ventilation on summer nighttime indoor temperatures in an urban area. *Meteorologische Zeitschrift* 2021: 227–236. https://doi.org/10.1127/metz/2021/1066.

Haggard, K., Bainbridge, D.A., and Aljilani, R. (2016). *Passive Solar Architecture Pocket Reference*. Routledge.

Halbig, G., Steuri, B., Büter, B. et al. (2019). User requirements and case studies to evaluate the practicability and usability of the urban climate model PALM-4U. *Meteorologische Zeitschrift* 28 (2): 139–146. https://doi.org/10.1127/metz/2019/0914.

Hayles, C. and Aranda-Mena, G. (n.d.) Well-being in vertical cities: beyond the aesthetics of nature. *52nd International Conference of the Architectural Science Association,* 1 December 2018. The Architectural Science Association and RMIT University, Australia, pp. 331–338.

Hidalgo, J., Masson, V., Baklanov, A. et al. (2008). Advances in urban climate modeling. *Annals of the New York Academy of Sciences* 1146: 354–374. https://doi.org/10.1196/annals.1446.015.

Hidalgo, J., Lemonsu, A., and Masson, V. (2019). Between progress and obstacles in urban climate interdisciplinary studies and knowledge transfer to society. *Annals of the New York Academy of Sciences* 1436 (1): 5–18. https://doi.org/10.1111/nyas.13986.

Holst, C.C., Tam, C.-Y., and Chan, J.C.L. (2016). Sensitivity of urban rainfall to anthropogenic heat flux: a numerical experiment. *Geophysical Research Letters* 43 (5): 2240–2248. https://doi.org/10.1002/2015GL067628.

Howard, L. (1818–20). *The Climate of London: Deduced from Meteorological Observations, Made at Different Places in the Neighbourhood of the Metropolis*. Published by W. Phillips, George Yard, Lombard Street, sold also by J. and A. Arch, Cornhill; Baldwin, Cradock, and Joy, and W. Bent, Paternoster Row; and J. Hatchard. London: Picadilly.

Husain, A.A., Hasan, W.Z.W., Shafie, S. et al. (2018). A review of transparent solar photovoltaic technologies. *Renewable and Sustainable Energy Reviews* 94: 779–791. https://doi.org/10.1016/j.rser.2018.06.031.

Jandaghian, Z. and Berardi, U. (2020). Comparing urban canopy models for microclimate simulations in weather research and forecasting models. *Sustainable Cities and Society* 55: 102025. https://doi.org/10.1016/j.scs.2020.102025.

Javanroodi, K. and Nik, V.M. (2020). Interactions between extreme climate and urban morphology: investigating the evolution of extreme wind speeds from mesoscale to microscale. *Urban Climate* 31: 100544. https://doi.org/10.1016/j.uclim.2019.100544.

Kadasch, E., Sühring, M., Gronemeier, T., and Raasch, S. (2020). Mesoscale nesting interface of the PALM model system 6.0. *Geoscientific Model Development Discussion* 1–48. https://doi.org/10.5194/gmd-2020-285.

Kellert, S.R., Heerwagen, J., and Mador, M. (2011). *Biophilic Design: The Theory, Science and Practice of Bringing Buildings to Life*. New York, NY: Wiley.

Khan, B., Banzhaf, S., Chan, E.C. et al. (2021). Development of an atmospheric chemistry model coupled to the PALM model system 6.0: implementation and first applications. *Geoscientific Model Development* 14 (2): 1171–1193. https://doi.org/10.5194/gmd-14-1171-2021.

Kii, M. (2021). Projecting future populations of urban agglomerations around the world and through the 21st century. *Npj Urban Sustainability* 1 (1): 1–12. https://doi.org/10.1038/s42949-020-00007-5.

Kratzer, A. (1935). Das Klima der Städte. *Geographische Zeitschrift 41* (9.H): 321–339.

Krayenhoff, E.S., Jiang, T., Christen, A. et al. (2020). A multi-layer urban canopy meteorological model with trees (BEP-Tree): street tree impacts on pedestrian-level climate. *Urban Climate* 32: 100590. https://doi.org/10.1016/j.uclim.2020.100590.

Landsberg, H.E. (1981). *The Urban Climate*. New York, N.Y: Academic Press.

Levermore, G., Parkinson, J., Lee, K. et al. (2018). The increasing trend of the urban heat island intensity. *Urban Climate* 24: 360–368. https://doi.org/10.1016/j.uclim.2017.02.004.

Lewis, H.W., Castillo Sanchez, J.M., Graham, J. et al. (2018). The UKC2 regional coupled environmental prediction system. *Geoscientific Model Development* 11 (1): 1–42.

Lewis, H.W., Castillo Sanchez, J.M., Arnold, A. et al. (2019). The UKC3 regional coupled environmental prediction system. *Geoscientific Model Development* 12 (6): 2357–2400. https://doi.org/10.5194/gmd-12-2357-2019.

Li, D., Liao, W., Rigden, A.J. et al. (2019). Urban heat island: aerodynamics or imperviousness? *Science Advances* 5 (4): eaau4299. https://doi.org/10.1126/sciadv.aau4299.

Lin, D., Khan, B., Katurji, M. et al. (2020). WRF4PALM v1.0: a mesoscale dynamical driver for the microscale PALM model system 6.0. *Geoscientific Model Development Discussion* 1–37. https://doi.org/10.5194/gmd-2020-306.

Lindemann-Matthies, P. and Brieger, H. (2016). Does urban gardening increase aesthetic quality of urban areas? A case study from Germany. *Urban Forestry & Urban Greening* 17: 33–41. https://doi.org/10.1016/j.ufug.2016.03.010.

Linhares de Siqueira, G. (2015). Klimaadaptive Entwurfsmethodik: eine Untersuchung der Wechselwirkung zwischen Klimadaten, klimagerechten Entwurfsmethoden und dem adaptiven Komfortmodell. Hamburg, HafenCity Univ., Diss., 2015, Hamburg, HafenCity Universität Hamburg, Bibliothek.

Llaguno-Munitxa, M., Bou-Zeid, E., and Hultmark, M. (2017). The influence of building geometry on street canyon air flow: validation of large eddy simulations against wind tunnel experiments. *Journal of Wind Engineering and Industrial Aerodynamics* 165: 115–130. https://doi.org/10.1016/j.jweia.2017.03.007.

Maronga, B., Gross, G., Raasch, S. et al. (2019). Development of a new urban climate model based on the model PALM – project overview, planned work, and first achievements. *Meteorologische Zeitschrift* 28 (2): 105–119. https://doi.org/10.1127/metz/2019/0909.

Martin, M. and Molin, E. (2019). Environmental assessment of an urban vertical hydroponic farming system in Sweden. *Sustainability* 11 (15): 4124. https://doi.org/10.3390/su11154124.

Martos, A., Pacheco-Torres, R., Ordóñez, J., and Jadraque-Gago, E. (2016). Towards successful environmental performance of sustainable cities: intervening sectors. A review. *Renewable and Sustainable Energy Reviews* 57: 479–495. https://doi.org/10.1016/j.rser.2015.12.095.

Marzluff, J.M., Alberti, M., Bradley, G. et al. (2008). *Urban Ecology: An International Perspective on the Interaction Between Humans and Nature*. Boston, MA: Springer Science+Business Media LLC.

Masson, V., Heldens, W., Bocher, E. et al. (2020). City-descriptive input data for urban climate models: model requirements, data sources and challenges. *Urban Climate* 31: 100536. https://doi.org/10.1016/j.uclim.2019.100536.

McEvoy, S., Haasnoot, M., and Biesbroek, R. (2021). How are European countries planning for sea level rise? *Ocean and Coastal Management* 203: 105512. https://doi.org/10.1016/j.ocecoaman.2020.105512.

Meili, N., Manoli, G., Burlando, P. et al. (2020). An urban ecohydrological model to quantify the effect of vegetation on urban climate and hydrology (UT&C v1.0). *Geoscientific Model Development* 13 (1): 335–362.

Meyer, D., Schoetter, R., Riechert, M. et al. (2020). WRF-TEB: implementation and evaluation of the coupled weather research and forecasting (WRF) and Town Energy Balance (TEB) model. *Journal of Advances in Modeling Earth Systems* 12 (8): e2019MS001961. https://doi.org/10.1029/2019MS001961.

Mills, G., Cleugh, H., Emmanuel, R. et al. (2010). Climate Information for Improved Planning and Management of Mega Cities (Needs Perspective). *Procedia Environmental Sciences* 1: 228–246. https://doi.org/10.1016/j.proenv.2010.09.015.

Mussetti, G., Brunner, D., Allegrini, J. et al. (2020a). Simulating urban climate at sub-kilometre scale for representing the intra-urban variability of Zurich, Switzerland. *International Journal of Climatology* 40 (1): 458–476. https://doi.org/10.1002/joc.6221.

Mussetti, G., Brunner, D., Henne, S. et al. (2020b). COSMO-BEP-Tree v1.0: a coupled urban climate model with explicit representation of street trees. *Geoscientific Model Development* 13 (3): 1685–1710. https://doi.org/10.5194/gmd-13-1685-2020.

Naser, M.Z. (2019). Fire resistance evaluation through artificial intelligence - a case for timber structures. *Fire Safety Journal* 105: 1–18. https://doi.org/10.1016/j.firesaf.2019.02.002.

Nechyba, T.J. and Walsh, R.P. (2004). Urban Sprawl. *Journal of Economic Perspectives* 18 (4): 177–200. https://doi.org/10.1257/0895330042632681.

Nieuwenhuijsen, M.J. (2020). Urban and transport planning pathways to carbon neutral, liveable and healthy cities; a review of the current evidence. *Environment International* 140: 105661. https://doi.org/10.1016/j.envint.2020.105661.

Oke, T.R. (1974). Review of urban climatology. 1968–1973. *WMO Tech. Note* 134: 1–132.

Panagopoulos, T., González Duque, J.A., and Bostenaru Dan, M. (2016). Urban planning with respect to environmental quality and human well-being. *Environmental Pollution* 208: 137–144. https://doi.org/10.1016/j.envpol.2015.07.038.

Parrish, D.D., Singh, H.B., Molina, L., and Madronich, S. (2011). Air quality progress in North American megacities: a review. *Atmospheric Environment* 45 (39): 7015–7025. https://doi.org/10.1016/j.atmosenv.2011.09.039.

Parrish, D.D., Xu, J., Croes, B., and Shao, M. (2016). Air quality improvement in Los Angeles—perspectives for developing cities. *Frontiers of Environmental Science & Engineering* 10: 1–13.

Parsaee, M., Joybari, M.M., Mirzaei, P.A., and Haghighat, F. (2019). Urban heat island, urban climate maps and urban development policies and action plans. *Environmental Technology and Innovation* 14: 100341. https://doi.org/10.1016/j.eti.2019.100341.

Petit-Boix, A., Llorach-Massana, P., Sanjuan-Delmás, D. et al. (2017). Application of life cycle thinking towards sustainable cities: a review. *Journal of Cleaner Production* 166: 939–951.

Piroozmand, P., Mussetti, G., Allegrini, J. et al. (2020). Coupled CFD framework with mesoscale urban climate model: application to microscale urban flows with weak synoptic forcing. *Journal of Wind Engineering and Industrial Aerodynamics* 197: 104059. https://doi.org/10.1016/j.jweia.2019.104059.

Quigley, M.F. (2004). Street trees and rural conspecifics: will long-lived trees reach full size in urban conditions? *Urban Ecosystem* 7 (1): 29–39. https://doi.org/10.1023/B:UECO.0000020170.58404.e9.

Resler, J., Eben, K., Geletič, J. et al. (2020). Validation of the PALM model system 6.0 in real urban environment; case study of Prague-Dejvice, Czech Republic. *Geoscientific Model Development Discussion* 14: 4797–4842. https://doi.org/10.5194/gmd-2020-175.

Rossi, F., Castellani, B., Presciutti, A. et al. (2016). Experimental evaluation of urban heat island mitigation potential of retro-reflective pavement in urban canyons. *Energy and Buildings* 126: 340–352. https://doi.org/10.1016/j.enbuild.2016.05.036.

Salmond, J., Sabel, C., and Vardoulakis, S. (2018). Towards the integrated study of urban climate, air pollution, and public health. *Climate* 6 (1): 14. https://doi.org/10.3390/cli6010014.

Santamouris, M. (2013). Using cool pavements as a mitigation strategy to fight urban heat island—a review of the actual developments. *Renewable and Sustainable Energy Reviews* 26: 224–240. https://doi.org/10.1016/j.rser.2013.05.047.

Satterthwaite, D. (2008). Cities' contribution to global warming: notes on the allocation of greenhouse gas emissions. *Environment and Urbanization* 20 (2): 539–549.

Schau-Noppel, H., Kossmann, M., and Buchholz, S. (2020). Meteorological information for climate-proof urban planning - the example of KLIMPRAX. *Urban Climate* 32: 100614. https://doi.org/10.1016/j.uclim.2020.100614.

Scherer, D., Antretter, F., Bender, S. et al. (2019). Urban climate under change [UC]2 – a national research programme for developing a building-resolving atmospheric model for entire city regions. *Meteorologische Zeitschrift* 28 (2): 95–104. https://doi.org/10.1127/metz/2019/0913.

Searchinger, T.D., Beringer, T., Holtsmark, B. et al. (2018). Europe's renewable energy directive poised to harm global forests. *Nature Communications* 9 (1): 3741. https://doi.org/10.1038/s41467-018-06175-4.

Sharifi, A. and Yamagata, Y. (2015). Roof ponds as passive heating and cooling systems: a systematic review. *Applied Energy* 160: 336–357. https://doi.org/10.1016/j.apenergy.2015.09.061.

Sharma, A., Wuebbles, D.J., Kotamarthi, R. et al. (2020). Urban scale processes in high spatial resolution earth system models (ESMs). *Bulletin of the American Meteorological Society* 101 (9): E1555–E1561. https://doi.org/10.1175/BAMS-D-20-0114.1.

Shirzadi, M., Tominaga, Y., and Mirzaei, P.A. (2020). Experimental and steady-RANS CFD modelling of cross-ventilation in moderately-dense urban areas. *Sustainable Cities and Society* 52: 101849. https://doi.org/10.1016/j.scs.2019.101849.

Siegert, M., Alley, R.B., Rignot, E. et al. (2020). Twenty-first century sea-level rise could exceed IPCC projections for strong-warming futures. *One Earth* 3 (6): 691–703. https://doi.org/10.1016/j.oneear.2020.11.002.

Silva, B.N., Khan, M., and Han, K. (2018). Towards sustainable smart cities: a review of trends, architectures, components, and open challenges in smart cities. *Sustainable Cities and Society* 38: 697–713. https://doi.org/10.1016/j.scs.2018.01.053.

Simon, H., Fallmann, J., Kropp, T. et al. (2019). Urban trees and their impact on local ozone concentration—a microclimate modeling study. *Atmosphere* 10 (3): 154. https://doi.org/10.3390/atmos10030154.

Stathopoulos, T., Wu, H., and Zacharias, J. (2004). Outdoor human comfort in an urban climate. *Building and Environment* 39 (3): 297–305. https://doi.org/10.1016/j.buildenv.2003.09.001.

Steeneveld, G.-J., Klompmaker, J.O., Groen, R.J., and Holtslag, A.A. (2018). An urban climate assessment and management tool for combined heat and air quality judgements at neighbourhood scales. *Resources, Conservation and Recycling* 132: 204–217. https://doi.org/10.1016/j.resconrec.2016.12.002.

Takano, Y. and Moonen, P. (2013). On the influence of roof shape on flow and dispersion in an urban street canyon. *Journal of Wind Engineering and Industrial Aerodynamics* 123: 107–120. https://doi.org/10.1016/j.jweia.2013.10.006.

Tan, J., Zheng, Y., Tang, X. et al. (2010). The urban heat island and its impact on heat waves and human health in Shanghai. *International Journal of Biometeorology* 54 (1): 75–84. https://doi.org/10.1007/s00484-009-0256-x.

Tappert, S., Klöti, T., and Drilling, M. (2018). Contested urban green spaces in the compact city: the (re-)negotiation of urban gardening in Swiss cities. *Landscape and Urban Planning* 170: 69–78. https://doi.org/10.1016/j.landurbplan.2017.08.016.

Torres, A., Brandt, J., Lear, K., and Liu, J. (2017). A looming tragedy of the sand commons. *Science (New York, N.Y.)* 357 (6355): 970–971. https://doi.org/10.1126/science.aao0503.

Trusilova, K., Schubert, S., Wouters, H. et al. (2016). The urban land use in the COSMO-CLM model: a comparison of three parameterizations for Berlin. *Meteorologische Zeitschrift* 25 (2): 231–244. https://doi.org/10.1127/metz/2015/0587.

Tupėnaitė, L., Žilėnaitė, V., Kanapeckienė, L. et al. (2019). Multiple criteria assessment of high-rise timber buildings. *Engineering Structures and Technologies* 11 (3): 87–94. https://doi.org/10.3846/est.2019.11952.

Tupėnaitė, L., Žilėnaitė, V., Kanapeckienė, L. et al. (2020). Multiple criteria assessment of high-rise timber buildings. *Engineering Structures and Technologies* 11 (3): 87–94. https://doi.org/10.3846/est.2019.11952.

Valley, W. and Wittman, H. (2019). Beyond feeding the city: the multifunctionality of urban farming in Vancouver, BC. *City, Culture and Society* 16: 36–44. https://doi.org/10.1016/j.ccs.2018.03.004.

Veratti, G., Fabbi, S., Bigi, A. et al. (2020). Towards the coupling of a chemical transport model with a micro-scale Lagrangian modelling system for evaluation of urban NOx levels in a European hotspot. *Atmospheric Environment* 223: 117285. https://doi.org/10.1016/j.atmosenv.2020.117285.

Visser, M. (2019). The geography of vertical forests: exploring the green city. Master's thesis. Utrecht University. https://studenttheses.uu.nl/handle/20.500.12932/35208

Vranckx, S., Vos, P., Maiheu, B., and Janssen, S. (2015). Impact of trees on pollutant dispersion in street canyons: a numerical study of the annual average effects in Antwerp, Belgium. *The Science of the Total Environment* 532: 474–483. https://doi.org/10.1016/j.scitotenv.2015.06.032.

Wallbaum, H., Ostermeyer, Y., and Goto, Y. (2011). Nutzer-und klimaangepasstes Bauen—Leitbild und Chance für die globale Verbreitung von nachhaltigen Gebäudelösungen. *Bautechnik* 88 (1): 3–16.

Walloth, C., Gebetsroither-Geringer, E., Atun, F., and Werner, L.C. (2016). *Understanding Complex Urban Systems: Integrating Multidisciplinary Data in Urban Models*. Cham: Springer.

Wang, Y., Bakker, F., de Groot, R., and Wörtche, H. (2014). Effect of ecosystem services provided by urban green infrastructure on indoor environment: a literature review. *Building and Environment* 77: 88–100. https://doi.org/10.1016/j.buildenv.2014.03.021.

Ward, K., Lauf, S., Kleinschmit, B., and Endlicher, W. (2016). Heat waves and urban heat islands in Europe: a review of relevant drivers. *The Science of the Total Environment* 569–570: 527–539. https://doi.org/10.1016/j.scitotenv.2016.06.119.

William, R., Goodwell, A., Richardson, M. et al. (2016). An environmental cost-benefit analysis of alternative green roofing strategies. *Ecological Engineering* 95: 1–9.

Wouters, H., Demuzere, M., de Ridder, K., and van Lipzig, N.P. (2015). The impact of impervious water-storage parametrization on urban climate modelling. *Urban Climate* 11: 24–50. https://doi.org/10.1016/j.uclim.2014.11.005.

Ye, Y., Wang, C., Zhang, Y. et al. (2018). Low-carbon transportation oriented urban spatial structure: theory, model and case study. *Sustainability* 10 (2): 19. https://doi.org/10.3390/su10010019.

Zeng, F., Lei, C., Liu, J. et al. (2020). CFD simulation of the drag effect of urban trees: source term modification method revisited at the tree scale. *Sustainable Cities and Society* 56: 102079. https://doi.org/10.1016/j.scs.2020.102079.

Zhang, X., Shen, L., Tam, V.W.Y., and Lee, W.W.Y. (2012). Barriers to implement extensive green roof systems: a Hong Kong study. *Renewable and Sustainable Energy Reviews* 16 (1): 314–319.

Zhang, X., Huang, T., Zhang, L. et al. (2016). Three-north shelter forest program contribution to long-term increasing trends of biogenic isoprene emissions in northern China. *Atmospheric Chemistry and Physics* 16 (11): 6949–6960. https://doi.org/10.5194/acp-16-6949-2016.

Zhang, Y., Grignard, A., Lyons, K. et al. (2018). Real-time machine learning prediction of an agent-based model for urban decision-making. *Proceedings of the 17th International Conference on Autonomous Agents and MultiAgent Systems*. AAMAS '18: Autonomous Agents and MultiAgent Systems Stockholm Sweden, 10–15 July 2018. Association for Computing Machinery (pp. 2171–2173).

Zhu, K., Blum, P., Ferguson, G. et al. (2011). The geothermal potential of urban heat islands. *Environmental Research Letters* 6 (1): 19501. https://doi.org/10.1088/1748-9326/6/1/019501.

9

Berlin's Pathway to Climate Neutrality: Scenarios and Measures for a European Metropole

Bernd Hirschl[1,2]

[1] *Sustainable Energy and Climate Protection, Institute for Ecological Economy Research, Berlin, Germany*
[2] *Management of Regional Energy Systems, Brandenburg University of Technology Cottbus-Senftenberg (BTU), Germany*

9.1 Introduction

On 12 December 2015, after many years of intensive negotiations on the so-called Paris Agreement, a historic outcome was finally achieved at the 21st International Climate Protection Conference (Conference of the Parties – COP 21). For the first time, almost all countries of the world were able to broadly agree on a legally binding, global climate protection agreement (UNFCCC 2015). At the core of the Paris Agreement is a global action plan to counteract the dangerous effects of man-made climate change by limiting global warming to well below 2°C, and if possible, no more than 1.5°C, over pre-industrial levels. The adoption of this treaty fundamentally altered the foundation of climate protection concepts, programmes and targets around the world, in all countries and communities.

This article will explore the development of climate policy in the city of Berlin, including current scenarios and strategies for climate neutrality. With roughly 3.7 million inhabitants, Berlin is the largest city in the Federal Republic of Germany, and also its capital. At the same time, Berlin is also a city state – 1 of the 16 states of the Federal Republic of Germany.[1] The state of Berlin committed itself to the long-term goal of climate neutrality already in 2011. The government elected that year a coalition of social democrats (Social Democratic Party of Germany, SPD) and conservatives (Christian Democratic Party, CDU) and formulated the goal of a climate-neutral Berlin by 2050 in the governing coalition agreement (SPD Berlin and CDU Berlin 2011). Accordingly, a scientific feasibility study, Climate-Neutral Berlin 2050, was commissioned in 2012 to investigate whether and how Berlin could achieve this goal (Reusswig et al. 2014). Climate neutrality was understood in

1 Further data and facts on Berlin and the Berlin climate protection strategy can be found under the Senate Department for the Environment, Transport and Climate Protection (SenUVK n.d.).

Smart Cities, Energy and Climate: Governing Cities for a Low-Carbon Future, First Edition.
Edited by Oleg Golubchikov and Komali Yenneti.

terms of the two-degree target, which was viewed by some influential figures in climate science as suitable for averting the danger of climate change (Randalls 2010; WBGU 2009). The two-degree target was then recognised as the lowest common denominator after the failure of the international climate conference in Copenhagen (COP 15) to formulate a successor to the Kyoto Protocol.[2]

On this basis, long-term CO_2 emissions reduction targets on the order of an 80–95% reduction over 1990 were established in many international, national and communal climate protection concepts – targets that the current science tells us were still far from true climate neutrality. This was also the case in Berlin, which committed itself to reduce CO_2 levels by 85% by 2050, thus a per capita emission level below 2 t CO_2, in accordance with the long-term worldwide per capita goal associated with the two-degree target (WBGU 2009; Reusswig et al. 2014).[3] The state of Berlin subsequently passed the Berlin Energy Transition Act (Energiewendegesetz Berlin) in 2016, in which, in accordance with the above study, achievement of an 85% reduction in CO_2 over 1990 levels by the year 2050 was set as a binding climate protection target, as well as intermediate targets of a 40% reduction by 2020 and 60% by 2030. Additionally, key components of the law include the leadership role of the public sector and the reference to a comprehensive energy and climate protection programme (BEK 2030; SenUVK 2019) consisting of more than 100 measures, which are to be continually updated in accordance with an ongoing monitoring programme.[4]

With the formation of a new governing coalition made up of the SPD, DIE LINKE and Bündnis90/Die Grünen in 2016, it was formally agreed that from then on, the government would be guided by the new targets of the Paris Agreement (SPD Berlin et al. 2016). The coalition partners also agreed to phase out the use of coal in the state of Berlin by 2030 at the latest, which was finally included in the Berlin Energy Transition Act in November 2017 in the course of its first amendment. The programme of measures prescribed by law since 2016 (BEK 2030, see above), after some delays, was finally adopted in 2018 (SenUVK 2020).[5] In addition to measures for climate protection and in support of the energy transition, the BEK 2030 also includes measures addressing climate change adaptation. In 2019, through a Senate resolution, Berlin also became the first federal state in

---

2 The so-called two-degree target as a political goal goes back to a proposal by the European Union in 1996 that was later taken up and further developed by the German Advisory Council on Global Change (WBGU 2009). The goal was included in the final document of the UN Climate Change Conference in Copenhagen but remained non-binding until the 2015 Paris Agreement.

3 The consideration here, as in many other cities and regions as well as with the IPCC, is not with respect to CO_2 equivalents, which include all greenhouse gases, but specifically only CO_2 emissions, which in a city like Berlin comprise around 98% of all greenhouse gases (Reusswig et al. 2014).

4 The BEK 2030 (Berlin Energy and Climate Protection Program) is based on an extensive, participatory study on the basis of which the measures adopted by the Berlin government were significantly developed (Hirschl et al. 2015).

5 Both the digital monitoring and information system diBEK at https://dibek.berlin.de/?lang=de and the Senate's annual monitoring reports (ibid, on diBEK homepage) provide information on the implementation progress being made with the BEK 2030.

Germany to formally recognise the existence of a climate emergency (SenUVK 2021).[6] Also included in the resolution was the call for the study (Hirschl et al. 2021) whose relevant details are presented here in excerpt form. In early 2020, the contract was finally issued to ascertain, in the course of a revision of the original Climate-Neutral Berlin 2050 feasibility study (see above), the feasibility of more rapid attainment of climate neutrality, in accordance with the stricter requirements of the Paris Accord and the measures that would be necessary.

Which climate policy goals should a city such as Berlin be pursuing? The following text considers this question, followed by a look at previous and current developments in order to demonstrate the city's need for action on the pathway to climate neutrality. A restrictions-based scenario approach for the target years 2050, 2030 and 2040 is then applied; based on their results, specific recommendations for the state of Berlin and an urban energy transition, in general, are offered.

9.2 The Search for a Climate Policy Target for Berlin

9.2.1 The Climate Policy Framework: No Standards for Cities or Municipalities

As apparently clear as the goals and agreements of the Paris Agreement seem to be, neither the document itself nor the subsequent regulations offer any precise specifications and standards – neither for countries nor sub-state entities such as cities and municipalities. Cities and regional authorities are counted among the 'important stakeholders' who, as 'non-contractual partners', are also intended to comply with and support the Paris Goals (United Nations 2015, 2016); however, no direct specifications or benchmarks for a federal state or a city such as Berlin can currently be derived from the international documents – nor from the national regulatory framework in Germany.

According to Art. 4, Para. 1, 'Parties aim to reach global peaking of greenhouse gas emissions as soon as possible, recognizing that peaking will take longer for developing country Parties, and to undertake rapid reductions thereafter in accordance with best available science, so as to achieve a balance between anthropogenic emissions by sources and removals by sinks of greenhouse gases in the second half of this century [. . .]' (United Nations 2015, p. 4). This equilibrium between sources and sinks is the basis behind the principle of climate neutrality or greenhouse gas neutrality, although the term itself is not mentioned directly in the text of the agreement. At the same time, achieving this state of equilibrium must satisfy the core requirement of the Paris Agreement according to Art. 2, Para. 1 (i): 'Holding the increase in the global average temperature to well below 2°C above pre-industrial levels and pursuing efforts to limit the temperature increase to 1.5°C above pre-industrial levels [. . .]' (United Nations 2015, p. 3).

---

6 The state of Berlin is thus following a larger international movement of many cities and countries relying on the concept of a climate emergency and striving by means of ambitious climate protection measures to achieve the 1.5°C goal of the Paris Agreement. For an assessment of the climate emergency movement in German municipalities (in German), see Hirschl and Pfeifer (2020).

In order to achieve this temperature corridor, it is necessary to determine the amount of greenhouse gases that can still be emitted up to this point in time. In the course of COP 21, the Intergovernmental Panel on Climate Change (IPCC) was asked to determine the relevant global emissions budget. In its special report *Global Warming of 1.5°C,* the IPCC (2018) ascertained a global CO_2 emissions budget – initially without consideration of other greenhouse gases – available worldwide until average annual global temperatures have increased, respectively, by 1.5°C and 2°C, as well as the differences in the respective climate impacts. The respective total carbon budgets were in turn calculated for several IPCC scenarios with varying probabilities of occurrence (see IPCC 2018, Table 2.2). The report provides the following key findings: Limiting warming to 1.5°C can be achieved within the framework of assumptions with a likely probability[7] of 67% by limiting the global emissions budget to 420 metric gigatons (Gt). Limiting global warming to 1.5°C with a probability of 50% would yield an emissions budget of approx. 580 Gt. Considering a target temperature of 1.75°C – thus the mean of 1.5°C and 2°C – and an occurrence probability of 67%, the budget remaining increases to 800 Gt.

The IPCC special report thus provides various ranges of global CO_2 emissions budgets but does not break them down by country group (e.g. industrialised vs. developing countries) or sector. This question of distribution, however, is extremely relevant for a fair and thus politically successful implementation in terms of achieving the overall global goal. In this context, various distribution principles have been proposed in debates about so-called climate or carbon justice.[8] Beginning with the United Nations Framework Convention on Climate Change (UNFCCC), for example, there has been a recognition of varying levels of responsibility and (economic) opportunities for the industrialised, developing and emerging countries with respect to introducing climate protection and climate change adaptation measures.[9] Individual countries, furthermore, will be affected to a varying degree by the consequences of climate change, which in turn influences the possibilities of making effective contributions to climate protection and adapting to the consequences of climate change. With regard to these varying responsibilities, not only historical emission burdens play a decisive role[10] but also reductions in emission levels already achieved. Moreover, with respect to the viewpoint of justice and fairness, each and every person is fundamentally entitled to the same carbon budget (WBGU 2009).

The present lack of standardisation means that each and every country and municipality can establish its own carbon budget by means of the most advantageous redistribution method, which means that comparability, fairness or the global target of climate neutrality can be achieved with any certainty. In short, the Paris Agreement and the IPCC allow for municipalities to play an important role, but offer no possibility for the derivation of

7 An occurrence probability of 50%, for example, means that the corresponding maximum temperature increase (here: 1.5°C) can be maintained in 50% of the scenarios with the global emissions budget specified by the IPCC (2018).

8 See, for example, German Advisory Council on the Environment (SRU 2020).

9 See also the formulation of the above-cited Art. 4. Para. 1 of the Paris Agreement (UNFCCC 2015).

10 According to the World Resource Institute, United States with 27% and European Union with 25% are responsible for more than 50% of the cumulative CO_2 emissions since 1850, followed by China with 11% and Russia with 8% (Ge et al. 2014).

specific, uniform objectives or targets beyond the general requirement of achieving climate neutrality.

9.2.2 What Other Cities are Doing: Urban Climate Neutrality in Theory and Practice

Cities and their inhabitants are already contributing substantially to climate change – thus it is essential that they become an essential part of the solution to the global climate crisis. Although today only a bit more than 50% of the world lives in urban communities, roughly 80% of the global GDP is associated with urban regions. Cities also account for almost 70% of the world's energy consumption and emit around 80% of global greenhouse gas emissions. And the forecasts assume that these developments may worsen if by 2050 more than 70% of the world population is living in cities, as this will further increase the energy demand and thus also the pressure on the environment and the climate unless we change the direction (IPCC 2018; IEA 2021).

Against this background, many cities have for some time now been developing their own climate protection goals and measures and actively participating in climate policy debates, for example, within the framework of international city networks (Bansard et al. 2017; C40 and Arup 2019).[11] The extent of their success in these efforts is increasingly being discussed and analysed in the literature (e.g. in Johnson et al. 2017; Johnson 2018; Fuhr et al. 2019; van der Heijden et al. 2019). In many countries, the creation of a municipal energy or climate plan has been prescribed for a number of years already, e.g. in France, Sweden or Italy; in other countries, such as Germany, for example, the creation of such plans is being promoted at the national level. All in all, a number of studies indicate that many of the current municipal climate protection targets are still a long way off from the objectives of the Paris Agreement. Reckien et al. (2018) and Salvia et al. (2021) have worked out in their comparative analyses of a large number of urban climate protection concepts that the level of ambition of these plans often also correlates with the size of the cities, since larger, more populous municipalities, for example, can usually more easily accommodate the administrative capacity called for. Kona et al. (2018) see some of these concepts as the 'path to the 1.5° target' already taken; however, the vast majority of cities have not yet updated their plans with respect to the Paris Agreement. In the case of those cities that have committed themselves to achieving climate neutrality, a closer look reveals very heterogeneous formulations of the target objectives and level, as well as requirements for achievement (C40 and Arup 2019; Kona et al. 2018; Reckien et al. 2018; Salvia et al. 2021). Thus, the evidence shows (as do the documents and information from the urban climate protection networks) that even among the various directly involved stakeholders themselves there is currently no standardised approach to calculating reduction targets or remaining emissions budgets for achieving climate neutrality in accord with the global targets (e.g. 1.5°C).[12]

---

11 Examples of such climate city networks are the C40 Initiative, the EU Covenant of Mayors (CoM) and the Climate Alliance.

12 In the above-mentioned comparative analyses to date, neither the (in some cases very different) prerequisites for achieving the targets nor previous compliance has been taken into account or the cities accordingly clustered. This is reserved for future investigations.

The pioneers with the most ambitious climate neutrality goals in Europe include, for example, cities such as Copenhagen (DK, target year 2025); Aarhus (DK, 2030); Linköping (SE, 2025); Turku (FI, 2030); and Birmingham, Exeter, Glasgow, Leicester and Newcastle (GB, 2030). Most of these cities with very early target years have access to a greater-than-average supply of renewable energy sources (e.g. Copenhagen) and are only of medium size, which makes it easier to become 'climate-neutral' on the balance sheet and offset fossil fuel consumption with credits for renewable energy exports (virtual or otherwise) or grey energy (if taken into account). In those urban regions without sufficient renewable energy resources, some municipalities are explicitly planning carbon offsets, e.g. in the surrounding region or even abroad. Helsinki has set itself the goal of becoming CO_2 neutral by 2035 by reducing its own emissions by 80% and offsetting the remainder through compensation measures (City of Helsinki 2018).

At present, the use of a carbon budget to manage emissions reductions is only being discussed in very few cities and practiced in even fewer. The city of Oslo is pursuing the goal of reducing CO_2 emissions by 95% by 2030 (vs. 1990 levels) but is also pursuing sector-specific climate budgets as a governance instrument that in turn should influence the choice of measures necessary (City of Oslo 2016). London also has some experience with the use of a carbon budget as a steering and controlling instrument (Greater London Authority 2018), and the city of Vienna is currently examining and preparing such an instrument (Stadt Wien 2019; Köppl et al. 2020). The city of Manchester must be viewed as a vanguard for its role in establishing a fixed permissible remaining CO_2 emissions budget derived from the IPCC special report of 2018. A study for Manchester ascertained a mid-level 'Paris-compliant' budget of 15 Mt CO_2 from 2018 until climate neutrality (with a range of 8–24 Mt CO_2). If historical emissions levels were to continue unchanged, this budget range would be exhausted within 4–10 years (Kuriakose et al. 2018). The study, therefore, recommends a reduction path based on a uniform 13% p.a. rate of reduction. In addition, in line with the Climate Change Act, the total budget is divided into five-year segments. The City Council of Manchester has approved the recommendations and currently is set on becoming CO_2 neutral by 2038 (Manchester City Council 2019).

The analysis shows that at present comparatively few cities are orienting their efforts on the Paris-compliant climate-neutrality targets and, that in most cases, the target levels are being chosen in combination with other target years and requirements. Moreover, the climate budget approach still remains the absolute exception and, despite the ranks of active cities and urban networks, there are still no agreed-upon methodologies or standards.

9.2.3 Range of Possible CO_2 Budgets for Berlin

The above findings suggest that the preparation of a CO_2 budget for a large city like Berlin is at present largely uncharted territory. With this in mind, the study by Hirschl et al. (2021) was an attempt to ascertain possible emission budgets for Berlin in order to be able to offer plausible recommendations for policy objectives and targets.

The starting point for Berlin is to be found in a report by the German Advisory Council on the Environment (SRU) (2020). Using the CO_2 budgets for Germany were calculated on the basis of the IPCC special report from 2018. The IPCC climate scenarios were calculated on the basis of a 67% likelihood of achievement for both 1.5°C and 1.75°C. Numbers for the country as a whole were obtained by proportional conversion on the basis of population without taking other factors into account. For the scenarios under consideration here, this yields a CO_2 budget of 5.5 Gt (1.5°C, 67% likelihood) and 9.7 Gt CO_2 (1.75°C, 67%).

Potential figures for Berlin were then determined on the basis of three proportionally based distribution schemes: share of population, share of CO_2 emissions and, finally, economic power as measured in terms of the share of GDP relative to the country as a whole. The results, shown in Table 9.1 below, are widely divergent. In a city such as Berlin, with a strong service industry and below-average industrial production, an apportionment on the basis of population, for example, leads to a highly permissible CO_2 budget due to the very low primary energy consumption. If apportionment is made on the basis of CO_2 emissions, the picture is reversed: Berlin's already comparatively low CO_2 emissions mean that only a very small proportion can be allotted to Berlin. The third distribution approach follows the logic established in international climate policy of the common but differentiated responsibilities of the various nations according to their economic strength. This principle, which essentially stands for the distinction with respect to obligations and possibilities between the industrialised countries (polluter states) and developing and emerging countries, can logically also be applied to a distinction between sub-national regions such as federal states and municipalities. Accordingly, distribution can be based on economic performance, e.g. on the basis of the statistical indicator gross domestic product (GDP). The results for Berlin are roughly in line with the magnitude of the population-based CO_2 budget.

Altogether, the three methods chosen yield an enormous range of possibilities and 'fair' CO_2 budgets: If Berlin wants to do its share in meeting the Paris climate protection targets for a target temperature corridor of between 1.5° and 1.75°C (with a 67% likelihood according to the IPCC), an emissions budget margin of somewhere between 60 and 290 Mt CO_2 is still available from 2020 onwards. It quickly becomes clear that such a wide range is not suitable for operational planning and that the carbon budget apportionments by the various municipalities and states should most definitely be made in accordance with uniform distribution procedures. Until the issue of standardisation is decisively resolved, the averaging of values could provide direction: If the methods used here are weighted equally, a fourth value results for each scenario, which is also shown in Table 9.1.

Table 9.1 CO_2 emission budgets for Germany and Berlin on the basis of various apportionment methods.

Target temperature	**1.5°C (67%)**	**1.75°C (67%)**
Emissions Budget Germany from 2020[a]	*2.5 Gt CO_2*	*6.7 Gt CO_2*
Emissions Budget Berlin from 2020		
... based on share of population (4.35%)	107.8 Mt CO_2	289.8 Mt CO_2
... according to share of CO_2 emissions (2.24%)	55.6 Mt CO_2	149.5 Mt CO_2
... based on GDP (4.25%)	105.4 Mt CO_2	283.2 Mt CO_2
... average of the above three distribution methods (1/3 each)	89.5 Mt CO_2	240.5 Mt CO_2

[a] Share of the global budget in accordance with IPCC (2018) calculated on the basis of population share.
Source: Author; data from destatis (n.d.), AfS BB (2020a,b) and UN (2019).

If current CO_2 emissions in Berlin in 2020 were to remain constant at this level, the emission budgets shown here will be exhausted, in the worst case, in less than four years (3.6), or at most, after 17.5 years. Apportionment of the budgets shown here, in accordance with a linear rate of consumption decreasing from 2020 onwards, would achieve zero emissions (climate neutrality) in the following target years: The more ambitious budgets based on a 1.5°C (67%) limit would lead to 'zero emissions target years' somewhere between 2026 and 2032, while those based on a maximum increase of 1.75°C (67%) would lead to net zero between 2039 and 2054. These interim results show that the question of budgeting at the state and especially the sub-state level is urgently in need of international agreement. Until then, viable solutions must be found that can be equally applied by the majority of countries, states and municipalities alike, for example, in line with uniform specifications via the European Commission. As a result, for the present study and the case of Berlin, the scenarios and policy recommendations were not developed on the basis of a CO_2 emissions budget, but rather in cooperation with the Berlin city administration on the 'conventional' basis of a reduction target. For the subsequent investigation and the sake of simplicity, it was therefore stipulated that climate neutrality be considered achievable through a reduction in CO_2 emissions[13] by 95% over 1990 levels.

9.3 The Status Quo and Current Trends: Berlin Not Yet on the Path to Climate Neutrality

As in other cities, Berlin's official energy use and CO_2 emissions statistics typically lag several years behind the present. At the time of this study, the most recent official statistical data for Berlin were for the year 2017, making it necessary to rely on our own estimates for the relevant base year 2020. We drew where possible on available current data (e.g. the growth of renewable energies) as well as the current trends and parameters for each sector in order to derive plausible values for the year 2020. The energy supply, building, transport, business and private household sectors were each analysed separately (final and primary energy) to produce a consolidated balance sheet for the city of Berlin as a whole. This required a partial restructuring of the data, as the building sector, for example, despite being of the greatest relevance in a service-oriented city with little industry such as Berlin, is currently not included in the official statistics of the state of Berlin. From this energy consumption data, CO_2 emissions were then calculated on the basis of both a source-based (producers) and a polluter-based (end use) means of allocation. In order to better illuminate the need for action on the basis of a simplified forward projection, effects with respect to the COVID-19 crisis for the year 2020 were not considered as this would have too strongly skewed the projection, despite being a one-time factor.

The result, looking at the primary energy source balance, suggests that in the year 2020, 92% of Berlin's energy use will still come from fossil fuels – this illustrates the enormity of the task of the energy transition to climate neutrality. Natural gas dominates with 41%,

13 In Berlin, CO_2 emissions make up around 98% of all greenhouse gases. Given the current lack of climate balance sheet data on sink and other climate-relevant effects, the (gross) reduction in CO_2 emissions is used as the central benchmark for the following scenarios in this study. Nonetheless, further measures to achieve climate neutrality are also identified.

followed by petroleum (liquid fuels and heating oil) with 35% and a hard coal share of 11%. Another 7% is accounted for by imported electrical power, 45% of which in Germany was already being derived from renewable energy supplies in 2020. Accordingly, the Berlin share of renewable energies is still quite low and mainly attributable to wood combustion in power plants and biofuel admixture quotas. Solar energy currently barely registers in absolute terms but is showing dynamic growth rates.

Breaking these figures down according to the above-mentioned relevant economic sectors in Berlin confirms that the building sector, with 44% of primary energy, is by far the largest contributor to the city's energy consumption. Almost half of Berlin's building energy needs are provided by natural gas, one-third is via district heating and roughly a sixth is from heating oil. The transport sector, in second place, is responsible for around 30% of Berlin's primary energy consumption, with diesel dominating over gasoline and kerosene consumption. Of the 51 PJ of electricity consumed in Berlin in 2020, around 50% was produced in central natural gas power plants with heat extraction, and almost 40% of the city's electrical power is imported. Here, too, only a very small proportion (3%) comes from renewable energies.

The development of CO_2 emissions in Berlin presents a somewhat better picture. However, this is largely due to efficiency gains implemented after German re-unification, a period in the 1990s during which many buildings were renovated and inefficient factories and businesses shut down or modernised. Together with the likewise existing or introduced climate protection measures at both the federal and local levels, by 2020, roughly 38% of CO_2 emissions could be saved (excluding COVID-19 effects, polluter-based climate balance sheet). This results in a value of 18.1 Mt CO_2 for 2020 according to the polluter-based methodology (climate adjusted). Electrical power dominates with 28%, followed by decentralised natural gas (18%) and district heating (15%). On the basis of the source-based climate balance sheet, 16.5 Mt CO_2 was emitted in 2020. Here too, natural gas dominates with a 40% share, followed by coal (17%), diesel (15%), heating oil (11%) and gasoline (10%). In recent years, the exit from lignite-based power generation has become particularly noticeable in the source-based balance sheet, likewise, the phase-out of coal, which has already begun at the federal level, as well as the general increase in renewable energies in the power grid.

A simplified forward projection of the city's CO_2 emissions from the years 2010 to 2020 suggests that, despite the downward trend and regardless of whether we use a source-based or end user (polluter)-based accounting methodology, Berlin will clearly fall short of achieving climate neutrality by 2050 (see Figure 9.1). This underscores the great need for political and social action to further reduce emissions as comprehensively as possible in the short and long term in order to reach a path to climate neutrality. This analysis of the status quo and the demonstrated need for action comprised the starting point for the following scenarios and the choice of methodologies used.

9.4 Scenarios for 2050, 2030 and 2040 – A Restrictions-Based Approach

In order to develop recommendations for Berlin energy and climate policy and, at the same time, realistic and plausible scenarios for the city, a so-called restrictions-based scenario was developed that, particularly in the short-to-medium term, illuminates the primary limiting factors. This differs from a large number of currently available scenario studies that

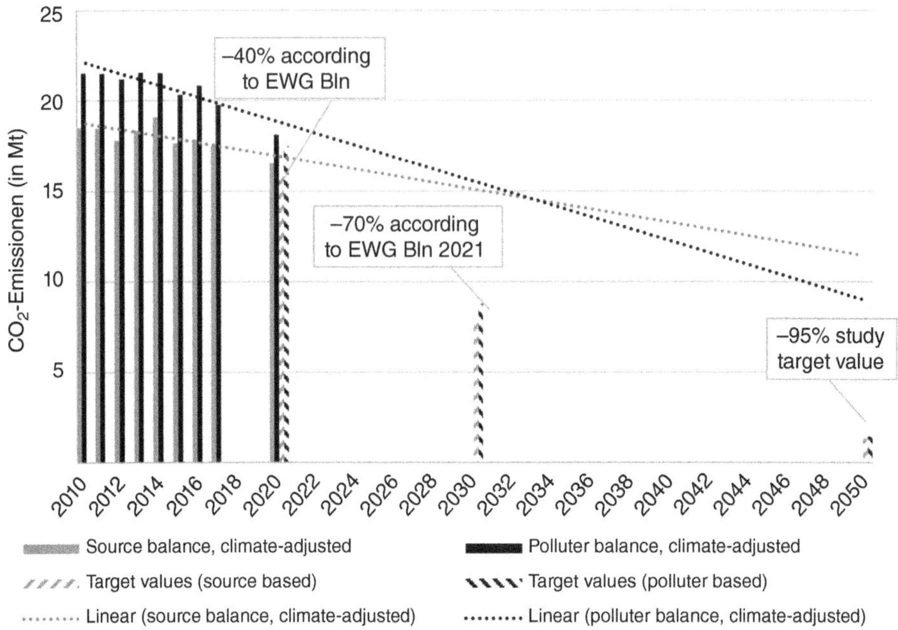

Figure 9.1 CO_2 emissions development for the years 2010–2017 and 2020, linear extrapolations, political target values for the state of Berlin in accordance with the 2021 Energy Transition Act and the target value of the study. *Source:* Adapted from Hirschl et al. (2021).

for the most part normatively model climate-neutral futures for an arbitrary number of early target years, but without taking a closer look at how the target scenarios are to be achieved. Often overlooked is the fact that to date, despite numerous subsidies and a conducive framework for climate protection measures in the key sectors, the necessary progress has not been achieved – neither in energy-conservation refurbishment nor in the expansion of public transport, bicycle transport or electromobility; nor in terms of growth in solar energy, heat pumps or geothermal energy. While some developments are similar to those nationwide, others are specific to Berlin. These include, among others, a particularly low rate of energy efficiency improvements to existing buildings, for example, and the conflict of goals between geothermal energy and groundwater protection. Given that Berlin's drinking water is drawn entirely from groundwater, the role of geothermal energy has been dealt with very cautiously so far (although it will very likely have to play a more important role in Berlin's energy transition in the future). A look at such conflicts and obstacles thus is important when it comes to earlier attainment - moving forward the goal of climate neutrality to an earlier target year.

In the first step, a long-term scenario for 2050 was developed, in which a largely restriction-free situation was assumed and modelled. Until then, the necessary energy savings can be achieved with a comparatively moderate rate of change and renewable energies can be expanded both within and outside Berlin to the necessary extent. Overall, we, therefore, assume that by 2050, a full reduction in CO_2 emissions can be achieved in Berlin (–100% CO_2 emissions vs. 1990).

- The drastic changes that will have to be implemented in Berlin's energy system by then can be seen in the example of electrical power: We assume a 140% increase in demand for electrical power, which can be achieved in the long term with 37% photovoltaics and 51% imported energy – mainly wind power from the surrounding region. Only 9% will come from power cogeneration plants, with the heat output being fed almost entirely into district heating. The efficient use of heat in urban district heating is the main reason why backup power stations, which will remain necessary in the future to ensure the stability of the country's power grid, should be sited in urban areas. We also assume that electrolysis and Power-to-X systems can be operated efficiently in Berlin, here again, primarily due to the efficient use of the heat generated in these systems. Berlin power plants will have to generate electrical power and heat as efficiently as possible, for which fuel cells, for example, are ideal.
- Gas consumption sinks to 36% relative to 2020, with five-sixths coming directly from a hydrogen pipeline supplying the relevant large-scale consumers such as power plants. In the long-term, the existing gas distribution network in Berlin can probably only accommodate a 50% share by volume of hydrogen. Therefore, Berlin will need to rely on methanated renewable energy gas for the long-term operation of the remaining share of the gas network.
- In order to be able to use the comparatively few emission-free urban energy sources efficiently, the use of primary energy must be drastically reduced overall, with the most extensive reductions, nearly 50%, to be sought in the areas of electrical power and in the building sector. In district heating, too, volume will drop by about 30%; however, market share and number of connections will increase by about 30% by the year 2050. In terms of supply, the importance of heat pumps also increases, with the share of gas drastically decreasing. In the case of energy efficiency refurbishments in the building sector, the current rate of renovation – far too low – must continually increase to an annual rate of more than 3% by 2030, and then be maintained until 2050. In addition, if we are to reach the desired goals, a greater effort must be made to implement climate-neutral efficiency standards – not only in new construction but also in the refurbishment of older buildings.
- In the transport sector, too, it is assumed for the long-term scenario of 2050 that a thorough turnaround can largely be achieved. Ninety per cent of our transport needs will be covered by environmentally friendly means of transport – the so-called environmental network. Local public transport (33%) and foot traffic (32%) will make up the lion's share. A quarter of all journeys, on average, will be by bicycle. The remaining automobile traffic, which will drop from more than 6 km per person per day to about 2.5 km, will be almost entirely electric. Freight transport will increase in all transport modes, with rail freight recording the strongest growth. In the shipping sector, passenger shipping is fully electrified, likewise sports and leisure boats. Freight shipping will be powered by various electric power-based fuels.

In the second step of our study, we considered the question of how this long-term, largely restriction-free scenario might be carried forward to the year 2030: How will the energy system look in the various sectors if, on the one hand, we were to target 2030 and accordingly accelerate long-term developments to the fullest extent possible, but at the same time take into account the limiting factors, obstacles and conflicting goals in many areas that objectively stand in the way of more quickly ramping up climate protection measures? These include, for example, the role of investment and modernisation cycles, social and architectural or urban cultural problems, shortages of skilled workers and the unresolved conflict between geothermal energy and drinking water protection mentioned above. A closer look at these restrictions does not mean that they cannot be modified or in part fully overcome; however, a shortage of skilled workers, the need for infrastructure modifications and slow rates of energy-efficiency refurbishment in the housing sector cannot be drastically overcome in a few years – a situation that many model-based studies do not take into account. A look at the relevant restrictions enables a more plausible and realistic assessment of the short- to medium-term situation on the one hand, but on the other it also allows us to shift the focus to the urgent measures necessary to reduce or overcome them.

In the case of the Berlin year 2030 scenario, it was furthermore assumed that developments at the federal level would proceed in a more progressive manner, both in terms of the overall framework and implementation.

- We assume, for example, that the national phase-out of coal will be completed by 2030, although in accord with current regulations, this is not called for until 2038. Such an assumption strongly influences the calculation of overall CO_2 emissions: roughly a 5% lower reduction in the Berlin climate balance sheet (polluter based) for 2030 if the nationwide coal phase-out is not completed by then. In addition, a 65% share of renewable energies in the electrical power mix by 2030 was assumed, which means that any direct use of electricity, e.g. for heating purposes or in electromobility, will have a significant emission-reducing effect.
- On the other hand, we assume that by the year 2030, it will not be possible to significantly increase the previously planned share of 20% by volume of green hydrogen in the gas network, as not enough renewable energy capacity will be available, either in Germany or elsewhere. Hence, it is unlikely that by 2030, Berlin will be connected to a pure hydrogen pipeline as part of the planned European hydrogen network. This means that hydrogen will only have a very slight effect on reducing emissions in 2030, and the gas distribution network will remain 80% natural gas. Even if we extend our assumption of a substantial hydrogen electrolysis plant with a volume of 300 MWel for the 2030 scenario to include further Power-to-X uses of green power, a comparatively high proportion of fossil fuels remains.
- In the building sector, this is also due to the fact that a ramp-up of energy-efficiency renovations by 2030 will take some time due to the various social, architectural, urban-cultural and labour-related factors involved. The implementation of energy-efficiency improvements and refurbishments of existing structures is currently competing not only with a very strong demand for new construction in Berlin but also with acceptance and economic viability issues; additionally, there is the problem of rising construction prices and the already noted shortage of skilled labour. In the 2030 scenario, despite higher assumptions, we thus arrive at an average rate of building refurbishment of just under 2% – which nonetheless implies renovation of almost 20% of the building stock by 2030. In the case of heating, natural gas still dominates despite the assumption of a substantial dynamic for change; due to their technical lifespan, the number of oil heating systems operating in 2030 will also be greater.

– In the transport sector, the lengthy planning and construction phases for the implementation of infrastructure projects (subway and tram, road construction, parking garages, charging station infrastructure, etc.) will show a particularly strong impact in the years leading up to 2030. Added to this is motor vehicle longevity, which tends to hinder the rapid modernisation of vehicle fleets. These obstacles suggest that by 2030, private car use will decrease to an 18% use share – a reduction of almost a third compared to the current figure (26%). The proportion of bicycle use is 23%, roughly the same as public transport (29%) and foot traffic (30%). In the case of air traffic, which in our scenarios is still allocated proportionately to the state of Berlin despite relocation (after the closure of Berlin Tegel and new opening of Berlin Brandenburg Airport) to the neighbouring state of Brandenburg, following a recovery we assume a very moderate increase leading up to 2030.

The scenarios suggest that with a very substantial effort, a 65% reduction in CO_2 emissions in Berlin (source-based climate balance sheet) or −67% (polluter-based) can be achieved by 2030 as compared to 1990.

For the year 2040 scenario, some non-linear relationships between the two previous scenarios were ultimately assumed. For one thing, it is possible that individual developments from the 2050 scenario can be moved ten years forward more or less without any restrictions, and thus entirely into the year 2040. At the same time, however, there are also much slower developments that will only begin to show an effect in the 2030s. This includes the construction or conversion of infrastructure for hydrogen production and transport, for example, as well as other Power-to-X products. Both cases, the gradual ramp-up and as well as the unrestricted earlier target date, suggest that we cannot necessarily assume the possibility of a simple 'middle' path for 2040 scenario. In the area of energy supply, we assume, for example, that district heating cannot be entirely converted and expanded by the year 2040. And the number of full-load hours of the power plants, the share of hydrogen gas and the expansion of photovoltaics are also lower compared to 2050. In the area of heat generation, the switch to environmentally friendly heating could largely be achieved by 2040, but in the case of end-use energy savings through energy conservation refurbishment measures, we do not assume that this can be fully moved forward; here, we assume a 42% reduction compared to 1990 – 10% less than in the long-term scenario. In the transport sector, the environmental network transport share for all journeys would be an already remarkable 86%. Nevertheless, neither the necessary expansion of the rail network nor the replacement of fossil vehicle fleets can be fully realised in this time frame. The rate of automobile electrification is already assumed to be 100%, but around a quarter of these are still plug-in hybrids. In total, the 2040 scenario achieves a bottom-line reduction of 87% according to both the source-based and the polluter-based climate balance sheets.

All said, it becomes clear that even with a very substantial effort, Berlin in the year 2030 will still be comparatively far away from achieving climate neutrality. It should again be noted that the most ambitious assumptions of all the scenarios were made for the 2030 scenario, including a nationwide coal phase-out and the largest annual increase in energy conservation measures in building renovation. The 2040 scenario comes much closer to climate neutrality, but still falls far short of it. If, for the purposes of simplification, we carry forward in a linear manner the results for the 2030 and 2040 scenarios on the basis of 2020, the goal of 95% CO_2 emissions reductions (over 1990) will be achieved by the 2030 scenario

Table 9.2 Key results of the scenarios for the years 2050, 2040 and 2030.

	2020	Scenario 2050	Scenario 2040	Scenario 2030
End energy (in PJ)	249	147	160	198
Primary energy (in PJ)	277	154	171	217
CO_2, polluter-based allocation (in Mt)	18.1	0	3.7	9.5
CO_2, source-based allocation (in Mt)	16.5	0	3.6	9.5
Reduction (source-based allocation) versus 1990 (%)	−38.3	−100	−86.5	−64.6
Year climate neutrality achieved (−95% CO_2 emissions vs. 1990, with linear extrapolation)	—	2048	2044	2042

Source: Adapted from Hirschl et al. (2021).

in 2042 and by the 2040 scenario in 2044. The scenario for 2050, in which a 100% reduction is already assumed, will reach −95% in 2048. Table 9.2 shows relevant results at a glance.

9.5 Strategy Recommendations and Measures for a Climate-Neutral Berlin

The analysis of the developmental dynamics of CO_2 emissions to date clearly shows that the state of Berlin must more consistently promote climate protection and the energy transition if it intends to become climate neutral by 2050 at the latest. If the intent is to meet this goal even sooner, then more ambitious targets must be pursued in all sectors. The scenarios clearly show that achievement of climate neutrality is scarcely feasible in the 2030s, but can be achieved in the 2040s. Additionally, in accordance with the concept of the carbon budget discussed initially, an effort should be made now, in the short-term, to achieve the greatest possible emission reductions, such that the total remaining budget – no matter how high –can be spared in the interest of intergenerational fairness.[14] Based on these relationships, the assumptions and sector target values in the 2030 scenario should dictate the path Berlin will need to follow. In doing so, it will be necessary, as shown, to minimise or eliminate the key obstacles and conflicts between climate protection and other social obligations that currently stand in the way of the path to climate neutrality.

This will require a number of individual measures at the sector level, as well as a new climate governance structure, which will enable the attainment of goals through appropriately effective governance and binding measures while also ensuring adequate means for participation. Relevant elements of this new **climate governance architecture** include, for example:

- situating political responsibility for climate issues at the highest level (in Berlin with the governing mayor)

14 This line of reasoning also follows the much-noticed decision of the German Federal Constitutional Court of 29 April 2021, in which the German Climate Protection Act was determined in part to be non-constitutional (BVerfG 2021).

- mainstreaming the responsibility in all administrations, establish a high-level committee with the leaders of climate-relevant administrations (in Berlin: climate senate), establish binding sector goals.
- ongoing, preferably annual, monitoring of both sectoral and cross-sectoral developments in order to initiate immediate countermeasures as necessary
- a consistent approach and solution to the conflicting goals and obstacles identified, with suitable participation formats for stakeholders and citizens

The federal commissions that successfully stewarded the nuclear and coal phase-outs could serve as models for the last-mentioned point (Commission on Growth, Structural Change and Employment 2019; Ethics Commission for a Safe Energy Supply 2011). In both cases, the political situation had been such that social conflicts remained unresolved for many years; here, the stakeholder commissions were able to successfully take on this task with a clear mandate and within a limited time. In Berlin, when it comes to climate protection, there are currently conflicts and issues involving tenant rights, limited road space versus the desire for automobility, historical preservation and urban culture issues and drinking water protection. The shortage of skilled workers is also one of the obstacles that can scarcely be resolved without targeted initiatives. Since these and other issues involve a number of ministries, departments and stakeholders, some of these conflicts can certainly be dealt with in parallel. The involvement and participation of the public is also important aspect that can increase acceptance and provide legitimation; however, many of the restrictions and conflicting goals mentioned here cannot be resolved without the involvement of critical stakeholders.

The necessary growth in many areas also requires an increase in human resources in administration with regard to all aspects of climate protection and climate adaptation. At the same time, the financial means must be found to enable the public sector to better serve as a role model because only then can it credibly issue stricter guidelines and demand more from the private sector. Berlin must also work together with the surrounding region – primarily the neighbouring state of Brandenburg, which entirely encircles the city – in order to develop joint energy supply projects and business models, joint hydrogen, bioeconomy, circular economy concepts, etc. to achieve climate neutrality and strengthen regional added value. Finally, considerable measures must already be undertaken today to protect and if possible, expand the carbon sink effects of forests and moors as well as other urban green spaces, and concepts for further technical CO_2 sink projects must be developed.

In the study, we formulated more fifty separate measures for specific sectors. There is no single sector that can undertake significant reductions on behalf of the others; each and every sector will have to contribute to a similar extent to the achievement of the goals set out. At the same time, given the size of the task and the ever-tighter deadline, purely incentive-based measures will probably no longer be sufficient. Here regulatory efforts can have an orienting effect and be called upon when, e.g. a CO_2 price or subsidy does not have a sufficient effect. The individual sector measures include recommendations for socially compatible energy conservation refurbishment measures, the conversion of district heating to fossil fuel-free energy sources, a massive expansion of photovoltaics and the environmental network (public transport, cycling and walking) and electrification of the majority of all vehicles. In the medium and long term, emission-free gases such as hydrogen will support the complete energy transition, but green hydrogen will remain scarce in the long term and will only be available in small quantities until 2030.

The path to climate neutrality – the scenarios show this – is in principle technically feasible, but with respect to social issues, it will be particularly challenging. It is, therefore, important that the costs and benefits be distributed in a balanced way. In the process, the benefits should not be underestimated, since failure to achieve climate neutrality will lead to significantly higher costs in the long term. Additionally, care must be taken with the public in order to avoid provoking acceptance problems such as those that have taken place in France with the yellow vest protests (Douenne and Fabre 2020).

Given that Berlin cannot achieve climate neutrality independent of or any faster than the federal level, the study ultimately included numerous recommendations at the national level in order to improve the policy framework (i.e. federal legislation) for an urban energy transition. This includes the following:

- an earlier phase-out of coal at federal level, possibly by 2030, as this has by far the greatest impact on CO_2 emissions
- the creation of a legal framework for stricter energy-efficiency standards for buildings sufficient to meet the goal of climate neutrality
- socially acceptable assessments and a fair distribution of costs between tenants and landlords in the energy-efficiency retrofitting of existing buildings
- better opportunities for the use of self-generated energy (prosuming), both in general and in support of the power grid
- the introduction of so-called energy communities, which have been stipulated for several years already in accordance with the current EU Renewable Energy Directive (European Parliament and Council 2018), as well as citizens' energy in general, as these will play a key role in public acceptance
- the creation of effective incentives for regional flexibility

In view of the latest resolutions of the so-called 'Fit for 55' package at the EU level (European Commission 2021), a new federal government will anyway have to adopt stricter and more ambitious laws – in the process, cooperation with respect to climate protection and the energy transition between the federal states and local municipalities must also improve and a better framework for an urban energy transition needs to be established. At the international climate policy level, the rules on the distribution of the global CO_2 budget need to be specified more precisely such that they can be directed in a more targeted and internationally uniform manner.

9.6 Conclusions

The case study presented here examined the question of when, and more particularly how, Berlin will achieve climate neutrality in light of the modified regional, national and international requirements, and the requirements that will have to be met to achieve this. On the basis of an analysis of the concept of the so-called climate budget in both theory and practice, it was initially determined that the state of Berlin presently is not yet capable of bringing its climate policy into alignment with such a climate or carbon budget. Calculations using differing distribution methods resulted in a very wide range of possible target years for achieving CO_2 neutrality – from 60 to 290 Mt CO_2 for a target temperature corridor of 1.5–1.75°C. This suggests that there is a need for international coordination, such that management on the basis of an operative carbon budget at the national and subnational levels can be carried out in a uniform, standardised manner.

Given this background, the study considered the extent to which CO_2 emissions could be reduced by the greatest possible means for each of the target years: 2050, 2030 and 2040. Sink effects for CO_2 and other greenhouse gases could not be taken into account due to the current lack of available data. Calculations were initially made for the reference year 2020, but under the exclusion of one-off COVID-19 effects, which would have too severely skewed future projections. A forward projection of the current time series clearly shows that Berlin presently is nowhere near the path to climate neutrality. This is true of the consolidated Berlin source and polluter climate balance statement, as well as the quantitatively most significant three sectors: energy, building and transport, with the latter even showing an upward emission trend in recent years.

A restrictions-based approach was developed for the scenarios. Initially, a largely restriction-free situation was modelled for the long-term scenario 2050, in which sufficient renewable energies are available and a full reduction in emissions can be achieved. Subsequently, a closer look was taken at the limiting factors for the 2030 scenario. These are the obstacles and conflicting goals that are already obstructing climate protection and energy transition efforts today that thus stand in the way of efforts to achieve climate neutrality by an earlier date. They include, for example, the timing of investment and modernisation cycles, social and architectural heritage problems, a shortage of skilled workers and the still unresolved conflict between geothermal energy and drinking water protection.

The results of the scenarios: Achieving climate neutrality in the sense of reducing CO_2 emissions by at least 95% compared to 1990 appears achievable for the state of Berlin in the 2040s; a significantly earlier target in the 2030s, on the other hand, is rather unlikely. With a substantial, immediate effort, a 65% reduction in CO_2 is possible by 2030 but only with a much more ambitious roll-out at the national level (e.g. a nationwide phase-out of coal by 2030). Socially sustainable energy conservation refurbishment measures, the conversion of district heating to fossil-free energy, a massive expansion of photovoltaics, expansion of the environmentally friendly transport network and the electrification of the majority of all vehicles are key elements. In the medium and long term, emission-free gases such as hydrogen will support the complete energy transition, but green hydrogen will remain scarce in the long term; until 2030, only small quantities will be available to a city such as Berlin.

In order to achieve the climate protection goals, we recommend a new climate governance architecture to ensure the mainstreaming of the subject in all departments, efficient management and binding implementation. The introduction of a climate senate and individual sector goals as well as a consistent approach and resolution of conflicting goals and obstacles can contribute to this. In addition, the study includes approx. 50 sector-specific proposals for measures in the state of Berlin that are intended to complement existing instruments and support the path to climate neutrality in all areas of activity. Last but not least, the study contains recommendations for the federal level, where it is necessary to formulate a more conducive legislative framework for the urban energy transition and the path to climate neutrality.

Acknowledgement

Parts of this Chapter are based on a study that was carried out on behalf of the state of Berlin, see Hirschl et al. 2021.

Literature/References

Amt für Statistik Berlin-Brandenburg (AfS BB) (2020a). Bevölkerungsstand – Zeitreihen. https://www.statistik-berlin-brandenburg.de/bevoelkerung/demografie/bevoelkerungsstand (accessed 01 September 2021).

Amt für Statistik Berlin-Brandenburg (AfS BB) (2020b). Energie- und CO2-Bilanz im Land Berlin und Brandenburg 2012–2017 – Statistische Berichte. https://daten.berlin.de/datensaetze/energie-und-co2-bilanz-im-land-berlin-und-brandenburg (accessed 01 September 2021).

Bansard, J.S., Pattberg, P.H., and Widerberg, O. (2017). Cities to the rescue? Assessing the performance of transnational municipal networks in global climate governance. *International Environmental Agreements: Politics, Law and Economics* 17: 229–246.

C40 and Arup (2019). Deadline 2020 - How Cities Will Get the Job Done. London: C40 and Arup. https://www.c40.org/researches/deadline-2020 (accessed 30 July 2021).

City of Helsinki (2018). *The Carbon-neutral Helsinki 2035 Action Plan*. Helsinki: City of Helsinki.

City of Oslo (2016). *Climate and Energy Strategy for Oslo*. Oslo: City of Oslo.

Commission on Growth, Structural Change and Employment (2019). *Final Report*. Berlin: Federal Ministry for Economic Affairs and Energy (BMWi).

destatis (n.d.). Volkswirtschaftliche Gesamtrechnungen der Länder VGRdL. https://www.statistikportal.de/de/vgrdl (accessed 24 September 2020).

Douenne, T. and Fabre, A. (2020). French attitudes on climate change, carbon taxation and other climate policies. *Ecological Economics* 169 (C): 106496.

Ethics Commission for a Safe Energy Supply (2011). *Germany's Energy Transition – A Collective Project for the Future*. Berlin: Ethics Commission for a Safe Energy Supply.

European Commission (2021). "Fit for 55": delivering the EU's 2030 climate target on the way to climate neutrality - communication from the commission to the European Parliament, the Council, the European Economic and Social Committee and the Committee of the Regions COM(2021) 550 final.

European Parliament and Council (2018). Directive (EU) 2018/2001 of the European Parliament and of the Council of 11 December 2018 on the promotion of the use of energy from renewable sources.

Federal Constitutional Court (BVerfG) (2021). Order of the First Senate of 24 March 2021 - 1 BvR 2656/18 -, paras. 1-270.

Fuhr, H., Hickmann, T., and Kern, K. (2019). Cities as leaders in EU multilevel climate governance: embedded upscaling of local experiments in Europe. *Environmental Politics* 28: 125–145.

Ge, M., Friedrich, J., and Damassa, T. (2014). 6 graphs explain the world's top 10 emitters. https://www.wri.org/blog/2014/11/6-graphs-explain-world-s-top-10-emitters (accessed 27 November 2020).

German Advisory Council on Global Change (WBGU) (2009). Solving the climate dilemma: the budget approach. Special Report. German Advisory Council on Global Change, Berlin. https://www.wbgu.de/fileadmin/user_upload/wbgu/publikationen/sondergutachten/sg2009/pdf/wbgu_sn2009_en.pdf (accessed 30 July 2021).

German Advisory Council on the Environment (SRU) (2020). Using the CO2 budget to meet the Paris climate targets – Environmental Report 2020 Chapter 2. German Advisory Council on the Environment, Berlin. https://www.umweltrat.de/SharedDocs/Downloads/EN/01_Environmental_Reports/2020_08_environmental_report_chapter_02.pdf;jsessionid=83E1C4DD2B7E98DB8A8E45E9E20064B6.2_cid321?__blob=publicationFile&v=6 (accessed 30 July 2021)

Greater London Authority (2018). *Zero Carbon London: A 1.5°C Compatible Plan*. London: Greater London Authority.

van der Heijden, J., Patterson, J., Juhola, S. et al. (2019). Special section: advancing the role of cities in climate governance – promise, limits, politics. *Journal of Environmental Planning and Management* 62: 365–373.

Hirschl, B. and Pfeifer, L. (2020). Kommunen im Klimanotstand: Wichtige Akteure für kommunalen Klimaschutz. Discussion Paper (No. 71/20). Berlin: Institute for Ecological Economy Research (IÖW).

Hirschl, B., Reusswig, F., Weiß, J. et al. (2015). *Entwurf für ein Berliner Energie- und Klimaschutzprogramm (BEK) (Endbericht im Auftrag des Landes Berlin, Senatsverwaltung für Stadtentwicklung und Umwelt)*. Berlin: Institute for Ecological Economy Research (IÖW).

Hirschl, B., Schwarz, U., Weiß, J. et al. (2021). Berlin Paris-konform machen. Eine Aktualisierung der Machbarkeitsstudie "Klimaneutrales Berlin 2050". Bezug auf die Anforderungen aus dem Übereinkommen von Paris 2015.

Intergovernmental Panel on Climate Change (IPCC) (2018). Global warming of 1.5°C. An IPCC Special Report on the impacts of global warming of 1.5°C above pre-industrial levels and related global greenhouse gas emission pathways, in the context of strengthening the global response to the threat of climate change, sustainable development, and efforts to eradicate poverty.

International Energy Agency (IEA) (2021). *Empowering Cities for a Net Zero Future: Unlocking Resilient, Smart, Sustainable Urban Energy Systems*. Paris: International Energy Agency.

Johnson, C.A. (2018). *The Power of Cities in Global Climate Politics – Saviours, Supplicants or Agents of Change?* London: Palgrave Macmillan.

Johnson, C., Toly, N., and Schroeder, H. (ed.) (2017). *The Urban Climate Challenge: Rethinking the Role of Cities in the Global Climate Regime*. Routledge.

Kona, A., Bertoldi, P., Monforti-Ferrario, F. et al. (2018). Covenant of mayors signatories leading the way towards 1.5 degree global warming pathway. *Sustainable Cities and Society* 41: 568–575.

Köppl, A., Muehlberger, M., and Steininger, K.W. (2020). *Klimabudget Wien – Klimaindikatoren im Rahmen eines Klimabudgets*. Wien: Österreichisches Institut für Wirtschaftsforschung.

Kuriakose, J., Anderson, K., Broderick, J. et al. (2018). *Quantifying the Implications of the Paris Agreement for Greater Manchester*. Manchester: University of Manchester.

Manchester City Council (2019). *State of the City Report 2019*. Manchester: Manchester City Council.

Randalls, S. (2010). History of the 2°C climate target. *WIREs Climate Change* 1 (4): 598–605.

Reckien, D., Salvia, M., Heidrich, O. et al. (2018). How are cities planning to respond to climate change? Assessment of local climate plans from 885 cities in the EU-28. *Journal of Cleaner Production* 191: 207–219.

Reusswig, F., Hirschl, B., and Lass, W. (2014). *Climate-Neutral Berlin 2050 - Results of a Feasibility Study. Brochure*. Berlin: Senate Department for Urban Development and the Environment.

Salvia, M., Reckien, D., Pietrapertosa, F. et al. (2021). Will climate mitigation ambitions lead to carbon neutrality? An analysis of the local-level plans of 327 cities in the EU. *Renewable and Sustainable Energy Reviews* 135: 110253.

Senate Department for the Environment, Transport and Climate Protection (SenUVK) (2019). *BEK 2030 - Berlin Energy and Climate Protection Programme 2030. Brochure*. Berlin: Senate Department for the Environment, Transport and Climate Protection.

Senate Department for the Environment, Transport and Climate Protection (SenUVK) (2020). *Monitoringbericht zur Umsetzung des Berliner Energie- und Klimaschutzprogramms (BEK 2030) - Berichtsjahr 2019*. Berlin: Senate Department for the Environment, Transport and Climate Protection.

Senate Department for the Environment, Transport and Climate Protection (SenUVK) (2021). *Decision of the Berlin Senate: Increased action in recognition of a climate emergency (in German)*. Berlin: Senate Department for the Environment, Transport and Climate Protection.

Senate Department for the Environment, Transport and Climate Protection (SenUVK) (n.d.). Berlin – Facts and Figures. https://dibek.berlin.de/?lang=en#caption_c2c8 (accessed 30 July 2021).

SPD Berlin and CDU Berlin (2011). Berliner Perspektiven für starke Wirtschaft, gute Arbeit und sozialen Zusammenhalt. Koalitionsvereinbarung zwischen Sozialdemokratische Partei Deutschlands (SPD), Landesverband Berlin und Christlich Demokratische Union Deutschlands (CDU), Landesverband Berlin. https://www.cdu-fraktion.berlin.de/image/inhalte/file/SPD_CDU_koa_vertrag.pdf (accessed 4 November 2020).

SPD Berlin, DIE LINKE Berlin, BÜNDNIS 90/DIE GRÜNEN Berlin (2016). Berlin gemeinsam gestalten. Solidarisch. Nachhaltig. Weltoffen. Koalitionsvereinbarung zwischen Sozialdemokratische Partei Deutschlands (SPD), Landesverband Berlin und DIE LINKE, Landesverband Berlin und BÜNDNIS 90/DIE GRÜNEN, Landesverband Berlin für die Legislaturperiode 2016–2021. http://www.die-linke-berlin.de/fileadmin/download/2016/161116_Koalitionsvertrag_finale_Fassung.pdf (accessed 30 July 2021).

Stadt Wien (2019). *Smart City Wien Rahmenstrategie 2019-2050*. Wien: Stadt Wien.

United Nations, Department of Economic and Social Affairs, Population Divisio (UN) (2019). *World Population Prospects 2019, Volume I: Comprehensive Tables*. New York: United Nations.

United Nations Framework Convention on Climate Change (UNFCCC) (2015). *Paris Agreement*. UNFCCC.

United Nations Framework Convention on Climate Change (UNFCCC) (2016). Report of the conference of the parties on its twenty-first session. Paris, France (30 November – 13 December 2015).

10

City, Neighbourhood and Citizens: Putting the '20-Minute' Idea to Work in Edinburgh

Alice Creasy[1], *Matthew Lane*[2], *and Dan van der Horst*[2]

[1] *Net Zero Cities consultant at WSP, Edinburgh, UK*
[2] *School of Geosciences, University of Edinburgh, Edinburgh, UK*

10.1 Introduction

The city of Edinburgh, like so many around the world, has embarked on a strategy to think about (and ultimately deliver on) the idea of being a '20-minute' place – somewhere where citizens can access their daily needs and services within a 20-minute walk from home. Concomitantly stimulated by the Scottish Governments' aspirations to embrace this globally circulating concept at a national level and by the structural shifts in governance brought on by the COVID-19 pandemic, City of Edinburgh Council is working to incorporate the idea across all levels of its strategic operations (City of Edinburgh Council 2021a). Building on the notion of a 'smart city' (Moreno et al. 2021), the 20-minute neighbourhood is a contemporary urban planning strategy that focuses on the decentralisation of cities by bringing essential services into neighbourhoods. Echoing what Jane Jacobs describes as the 'intricate mingling' of different services (1961, p. 222), the 20-minute concept aims to guide the development of lively, low-carbon and safe communities. Given the growing popularity of this concept, governments across the world are faced with the question of how to strategically implement the 20-minute neighbourhood within local geographies. However, as this concept transitions into a formalised policy strategy, we might question whether it may serve to stifle the dynamism of a less controlled urban morphology that seeks to embrace further exacerbating entrenched urban inequalities.

Over the course of the COVID-19 pandemic, the concept of the 20-minute neighbourhood has risen in prominence as a potential organising principle for how governments, organisations and communities across the world manage their interaction with the 'local'. From Ottawa and Bogota to Seoul and Paris, this concept has captured imaginations as people were forced to adapt their lifestyles and maximise use of the amenities on their doorsteps. However, while the 20-minute neighbourhood has hit the headlines over the last few years, the principles that this concept embodies are by no means new.

Smart Cities, Energy and Climate: Governing Cities for a Low-Carbon Future, First Edition.
Edited by Oleg Golubchikov and Komali Yenneti.

In fact, many cities already possess much of the pre-industrial infrastructure to support 20-minute neighbourhoods. Planned and built before the automobile began to shape urban design, the core of many older cities is made up of a collection of neighbourhoods, each with its own local centre, transport links, services and leisure spaces. London, for example, has over 600 high streets (Talk London 2020) while cities such as Barcelona have hundreds of local squares and plazas. For centuries, these places have been the centre of local economies, places of retail, work and social activity. However, while much of the infrastructure exists, after decades of funding cuts and changing consumer habits, many of these local services have fallen into disrepair. This has been a particular feature in the United Kingdom where the declining high street has been a central feature of planning narratives for decades (Dolega and Lord 2020; Eichler 2018; Oxford Analytica 2018; Turner and Gardner 2014). As a result, aspirations, particularly from older cities (such as Edinburgh), to embrace the contemporary popularity of the 20-minute idea offer a fascinating laboratory for thinking through the relative 'smartness' associated with this new governance mandate.

In an increasingly dynamic and uncertain urban design arena, one reason behind the popularity of this concept perhaps lies in its perceived simplicity. Although many of the themes that the 20-minute idea embodies have been around for centuries, the quantifiable and appealing 'neatness' of using a unit of time to orientate questions of service accessibility, neighbourhood walkability and intra-city traversal has hooked into the hope-filled, locally focussed, post-pandemic narrative. The shift to working from home and the re-orientation to local geographies, catalysed by the COVID-19 pandemic has sparked a renewed interest in the role of local centres in local social and economic resilience. With this in mind, the need for accessible, sustainable neighbourhoods that promote physical and mental well-being is being recognised across the planet. The 20-minute idea has thus emerged as a popular concept through which to assess urban liveability and frame the implementation of a new wave of urban planning strategies.

From an academic perspective, the 20-minute concept aligns closely with other previous sustainability planning ideas which gained traction worldwide. Concepts such as transit-orientated development, smart growth and new urbanism share a great deal with the principles of the 20-minute place. The principal difference between these movements and the 20-minute concept is the attribution of a time dimension to what constitutes cohesive, walkable and accessible neighbourhoods. In recent years much of the literature and practice around the 20-minute concept has been practitioner led and focussed around the cities of Portland (Portland City Council 2012) and Melbourne (Victoria State Government 2016), both of which have pioneered the implementation of this concept. In particular, much of this recent grey and academic literature on 20-minute neighbourhoods has focussed on Melbourne with research on issues such as the economic benefits of 20-minute neighbourhoods (Angelopoulos et al. 2019), designing healthy communities (Gunn et al. 2017), transit-orientated urban design (Dovey and Woodcock 2014; Stanley et al. 2015) and building equity into 20-minute plans (Clark 2019). Beyond the geographical focus of Melbourne or Portland, in recent years there have been a growing number of papers explicitly exploring this concept as a tool with which to develop more inclusive, sustainable places (Moreno et al. 2021; Pozoukidou and Chatziyiannaki 2021; Weng et al. 2019).

Steered by the terminology drawn upon by the City of Edinburgh, for the purposes of this chapter, we will refer principally to the '20-minute' idea; however, it is important to recognise that this concept has been adapted to different geographies and as a result there exists not only 20-minute plans but 15 and even 10-minute neighbourhoods (Moreno et al. 2021). Essentially, the difference between these interpretations is the travel time within which essential services should be located. However, what this means in practice varies between geographies and is often not clearly operationalised, e.g. does the unit of time correspond to the round trip or just the amount of time it takes to reach the service? At what uniform pace would you have to walk (or could you cycle?) to reach your destination in that time? How can this concept be applied in more rural geographies? What services count as 'essential'? While the snappiness of the 10-, 15- or 20-minute idea has captured people's imaginations, these practical questions expose the vagueness of this concept, posing a number of challenges for local decision-makers in urban planning and design. Challenges that we will unpack in greater detail over the course of this chapter.

This chapter poses the question of how the 20-minute 'idea' might be put to work by city government's eager to draw on the concept as an organising principle for 'smart' governance. In doing so the chapter has two core purposes. Firstly, to review the existing work done to date (within both grey and academic literature) in order to identify and conceptualise the different ways in which the 20-minute concept has been put to work as both an evaluative empirical (what is our city currently like) and a normative strategic (what should our city be like) tool. The second purpose is to present and subsequently reflect upon a case study of the city of Edinburgh drawn from our experiences with an ongoing research partnership between the University of Edinburgh and the City of Edinburgh Council exploring the potential for the 20-minute idea to usher in a new data-driven approach to urban planning and design governance in the city. Building on the work of Moreno et al. (2021) to conceptualise a framework for 20-minute governance at city, neighbourhood and individual citizen levels, and subsequently putting this framework into operation alongside the city government, we ultimately use the chapter to reflect on the different scales at which city 'smartness' can (and should) be pursued.

10.2 The 20-Minute Idea

The 20-minute neighbourhood has a complex global geography. While many tenets of this idea have influenced urban planning for centuries, it is possible to identify a transition of sorts from a status of planning *principle* (an ethos that was viewed as common sense) to a planning *policy* (a visionary agenda pro-actively incorporated into city strategies by politicians and managers). This transition from principle to policy has significant implications for urban areas across the world; areas which, over the last half a century, have come to be shaped by and designed around the use of private vehicles. This interim period in which urban morphology was dominated by the presence of automobiles helps explain the disappearance, and now resurgence, of an idea with a considerable history.

In contrast to the pre-industrial beginnings of many European cities, the original design and subsequent expansion of many North American and Australian cities in the post-industrial era were explicitly delivered with the automobile in mind (Gandy 2002; Harris and Lewis 2001). In reaction to this, some twentieth-century urban planners began advocating for more accessible urban design that recognises the importance of neighbourhoods as key building blocks of the city. The work of practitioners and writers such as Clarence Perry (1929) and Jane Jacobs (1961) pushed back against the centralisation of cities which happened during the twentieth century and advocated for the creation of liveable, accessible and thriving local neighbourhoods. In particular, Jacob's book, *The Death and Life of Great American Cities* (1961) in which she coined the terms 'social capital', 'eyes on the street' and 'mixed primary uses', remains one of the most influential books in the history of American urbanism and is regularly cited as an influential inspiration behind the contemporary 20-minute neighbourhood concept (Moreno et al. 2021; Pozoukidou and Chatziyiannaki 2021).

In Europe, the pre-industrial architecture of many urban centres has formed the foundation not only for the development of existing local centres but also the creation of new, accessible neighbourhoods. Evidence of this can be seen in cities across the continent, from new neighbourhoods such as Frauen-Werk-Stadt in Vienna and Nordhavn in Copenhagen, which are explicitly designed around local accessibility, to wider planning strategies such as Barcelona's 'Superblocks' and active travel infrastructure like that in Oulu that facilitates cycling throughout the Scandinavian winter. However, while the pre-industrial architecture of European cities has certainly helped to maintain a cultural value for well-serviced, accessible local neighbourhoods, rapid urban expansion, rising use of private cars and changing consumer habits have all served to fragment the way in which these cities have grown (EEA 2006; Evans 2019). As a result, the 20-minute concept has been picked up by a number of European local and national governments in a way that has seen it evolve from a planning and design principle to an overarching policy for future urban governance. Catalysed by the COVID-19 pandemic, this concept is now being pioneered by leading government officials and subsequently engaged with by academics, across the continent (Moreno 2020; O'Sullivan 2020).

One example of this rapid adoption of the concept is in Paris where, spurred on by the COVID-19 crisis, Mayor Anne Hidalgo made the '15-minute neighbourhood' a central feature of her successful Mayoral campaign in 2020 and is currently building on the city's existing work using the 15-minute neighbourhood as a way of creating a 'city of proximities' (Moreno et al. 2021; Yeung 2021). This vision takes a holistic approach built around the concept of 'hyper proximity' which focuses on ease of travel, walkability and public services as well as considering changing workplaces, cultural activities and social connections (O'Gorman and Dillon-Robinson 2021). Hidalgo aims to create a city for people rather than cars by turning over 70% of on-street car parking space to other uses, increasing the provision of offices and co-working spaces in neighbourhoods, expanding the uses of infrastructure and buildings outside of standard hours, encouraging people to use their local shops and creating small parks in school playgrounds that would be open to local people outside of school hours to combat the city's lack of public green space (ibid).

Hidalgo's work in Paris has been inspired by that of Professor Carlos Moreno who, over the course of the pandemic, has worked closely with the city to integrate the 15-minute concept into policy-making in Paris. Influenced by the idea of chrono-urbanism, which advocates for multiple uses of one place as a way of offering greater flexibility and improved quality of life in cities (Gwiazdzinsk 2015; Gwiazdzinski 2014), Moreno first developed the

15-minute concept in 2016 (Moreno 2016). Since then, this idea has been picked up in academic texts from China (Weng et al. 2019) to the United States (Capasso Da Silva et al. 2020) and has come to take an increasingly prominent role in steering both national and local policy-making (Comune di Milano 2019; Ottawa City Council 2019; Scottish Government 2020; Victoria State Government 2017).

As the concept has transitioned from a planning principle to a planning policy, there have been numerous different interpretations of the *scale* at which this idea is most pertinent as a governance concept and a policy idea. For example, while the 'neighbourhood' is the scale that is most colloquially talked about in '20 minute' terms, as illustrated above, it is often cities, rather than specific neighbourhoods, that are given the 'exemplar' status in how they are enacting the idea. Furthermore, with the convergence of the COVID pandemic and the sustainability agenda, conversations are growing about the potential for this concept to change the behaviours of individual citizens and empower them to take more control over how they manage their day-to-day traversal of the urban space.

Having introduced the 20-minute idea's resurgence as a governance ideal in the post-COVID-19 world, in what follows, we sketch out what we see as three distinct *scales* through which the 20-minute idea is being operationalised as a tool for governance. In doing so we build directly on the work of Moreno et al. (2021) and align our three typologies with the themes of 'proximity', 'density' and 'diversity' captured by their recently developed framework (Figure 10.1). Describing these as the 20-minute *city* (density), the 20-minute *neighbourhood* (proximity) and the 20-minute *citizen* (diversity), in the subsequent section of the

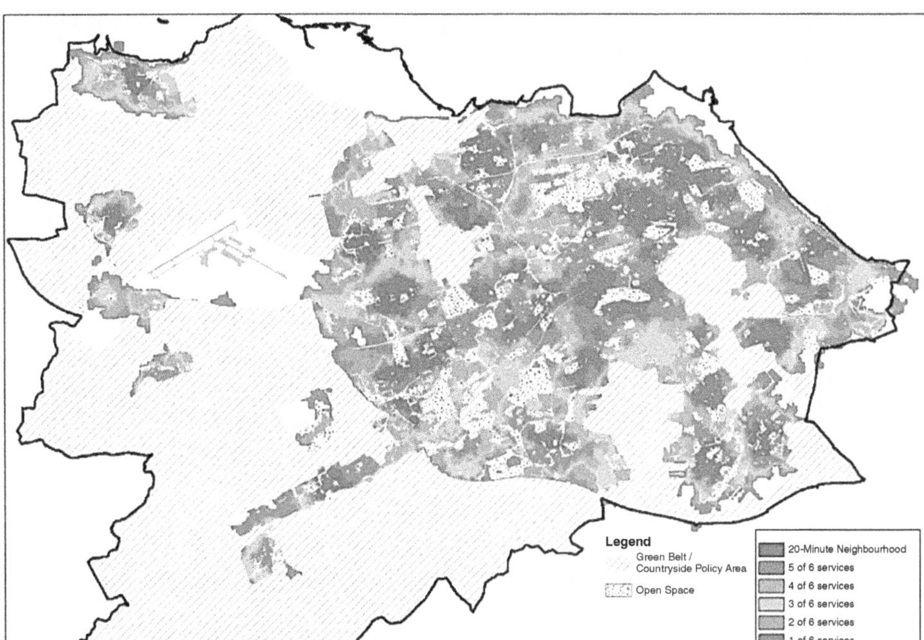

Figure 10.1 The council's heat map indicating access to key services. This was created by applying 'as the crow flies' buffer zones around locations of a particular service, and then overlaying (six of) these service-specific maps. *Source:* Adapted from Edinburgh City Council.

chapter (Section 10.3), we then turn our attention to the case of Edinburgh and to the fourth and final element of Moreno et al.'s framework: digitalisation.

10.2.1 The 20-Minute City

Over the last decade, a number of cities have driven the implementation of the 20-minute idea, with each taking a unique approach to its development across their urban geographies. One of these approaches is mapping the existing spread of key services across the entirety of the urban landscape in order to identify those areas where service access is most localised and where carrying out one's day-to-day activity most likely involves the use of some form of transportation.

In 2010, Portland became the first city to officially coin the idea of a 20-minute city (Portland City Council 2012). One of Portland's approaches to this idea takes a city-scale 20-minute analysis in order to visualise (and augment) catchment areas for existing key services and their spread across the city. This birds-eye view has been represented as a city heat map which visualises the city's 20-minute or 'complete' neighbourhood index. This map is broken down into 24 districts which allows planners to identify gaps at both a city and neighbourhood level. If a neighbourhood achieves a score of 70 or higher, on a scale of 0–100, it is considered a relatively complete neighbourhood in terms of local (and walkable) service provision (Portland City Council 2012). The analysis also indicates the per cent of population within 1/2 mile of grocery store, 1/2 mile of a park, 3 miles of a full-service community centre, 1/2 mile of elementary school and 1/4 mile of frequent transit (Portland City Council 2012). Analysis found that less than half of Portland residents (45%) live in areas with good access to schools, parks, grocery stores, sidewalks and transit (Portland City Council 2012).

Portland is widely recognised as a cosmopolitan centre of progressive politics, creativity and sustainability. It is often held up as a leading case study of a sustainable city due to its abundant green space, extensive public transport system and biking infrastructure. The city became one of the first municipal pioneers of urban sustainable development when in the 1970s it began to re-orientate city planning away from suburbanisation and urban sprawl and began to prioritise urban densification (Allison 2020; Goodling et al. 2015; Jun 2004; Resnik 2010). In 2012, Portland City Council set the goal of increasing the percentage of Portlanders with safe, walkable access to goods and services to 80%. Given the existing challenges of unequal service distribution identified in the city-wide 20-minuite neighbourhood analysis, the Portland Plan sets out four key objectives to meet this goal: increase housing in areas with services, support economic development, bring more services (including transit) into areas with poor accessibility and retain and attract grocery stores and markets in areas that are currently underserved.

Despite the early adoption of densification policies and Portland's world-renown status as a leading sustainable city, the implementation of sustainable development policies in Portland has produced severe spatial inequality and environmental racism which continue to impact people's lives to this day (Allison 2020; Goodling et al. 2015). In their paper on Greening Inequality in Portland, Allison writes that this inequality *'is largely due to the capitalist mindset and neoliberal economics locked into the city's development plans. Urban development in Portland historically correlates to racism, environmentalist branding, and profit. Portland has shaped its imaginary as a progressive, sustainable, creative city. It touts a culture of social equality and ecological awareness, but after exploring the margins of the city, it is easy to see that this is a façade'* (2020, p. 22). The historical rejection of urban sprawl and commercialisation of sustainability in Portland has led to a pattern of green gentrification and the displacement, and subsequent segregation, of low-income, often non-white residents, from commercially valuable urban land within the city (Allison 2020; Gould and Lewis 2017; Habitat for Humanity 2020). Consequently, Goodling et al. observe that '*the sustainability fix in Portland is very much a spatial one*' (2015, p. 14) and while certain areas enjoy the benefits of green amenities, districts such as East Portland have accumulated much of the city's poverty and environmental hazards. As Allison writes, 'East Portland has become a dumping zone for the city's burdens to keep them out of sight and out of mind, and to maintain the crisp image of its sustainability brand' (2020, p. 33).

The history of spatial segregation in Portland as a result of the neo-liberalisation of sustainability development throws up a plethora of challenges for planners working to create 20-minute neighbourhoods. The historic disinvestment in certain communities has left many neighbourhoods facing a range of social and economic challenges within physical neighbourhood environments that often lack the infrastructure found in wealthier parts of the city (Allison 2020). This means that those wishing to develop 20-minute neighbourhoods face not only an infrastructural gap but must also confront historic spatial and systemic environmental, economic and social justice issues. Moreover, these challenges must be addressed without triggering the cycle of green gentrification and displacement that has happened in other parts of the city (Gould and Lewis 2017; Pitter 2021). In sum, Portland's application of the 20-miunte idea at the city scale severs to lay bare the material fragmentation of urban space left in the wake of decades of green gentrification and displacement. Overcoming such challenges will likely require an altogether more holistic application of the 20-minute idea which engages more directly with the fragmented communities that make up the wider urbanity.

10.2.2 The 20-Minute Neighbourhood

While relatively compact cities such as Portland have been able to apply the 20-minute concept to a city-scale analysis, this approach is more complex in larger urban centres. Over the last hundred years, the growth of cities across the world has been guided by rising levels of private car ownership. In many instances, this has led to the rapid suburbanisation of cities and the creation of peripheral neighbourhoods with limited transport, retail, health and education infrastructure. Although Portland has actively resisted suburban expansion, this pattern of urban sprawl is evidenced in cities across the world and is particularly prevalent in the cities of north America and Australia (Randolph and Tice 2014; Davison 1997, 2006; Gandy 2002; Harris and Lewis 2001). With this in mind, rather than mapping service provision at the city-scale, the 20-minute neighbourhood has been used as a means to devolve decision-making around service delivery to neighbourhood hubs spread across the city.

Over the last decade or so, much of both the grey and academic literature on the subject of 20-minute planning has been focussed on Melbourne, whose local and regional governments have been working to implement the 20-minute concept since 2017. While Portland has developed a city-wide heat map, in Melbourne a key element in the development of this approach has been collaboration between stakeholders to align decision-making on local assets around key 'activity centres' (Victoria State Government 2017). These activity centres act as co-location hubs which provide services such as entertainment facilities, retail services and health centres.

Emphasising the importance of local density, this approach takes the neighbourhood as its starting geography and sees the city as being comprised of many walkable neighbourhoods, focussing city development strategies at the sub-city scale. In doing so, it involves collaboration among local authorities, community groups and public to align decision-making about local assets around community hubs (Victoria State Government 2018). Using existing inner-city neighbourhood boundaries as a starting point, the 20-minute neighbourhood approach poses questions of both the walkability of these areas and whether or not key services are located within their reach.

To support the implementation of the 20-minute neighbourhood concept, the State Government has run four pilot projects in neighbourhoods in the city (Victoria State Government 2018). For each of these pilots' walkability, movement, place, housing, vegetation and heat vulnerability assessments have been carried out. Additionally, Victoria State Government has commissioned and collaborated on a number of pieces of research into the 20-minute neighbourhood including the creation of evidence-based metrics for the implementation of this concept (Gunn et al. 2017), the economic benefits of 20-minute neighbourhoods (Angelopoulos et al. 2019), community infrastructure (Grodach et al. 2020) and an assessment of retail model in greenfield development settings (Goodman and Kroen 2019; Resilient Melbourne 2020).

Governance of the 20-minute concept must take into account historical design and socio-economic differences between neighbourhoods. While many suburban areas are associated with the middle class, over recent decades there has been sustained suburbanisation of disadvantage in many Australian cities (Baum et al. 2006; Hulse and Reynolds 2018; Pawson et al. 2015; Randolph and Tice 2014, 2017). As Randolph and Tice write, the neoliberal economic policies which have dominated urban development since the 1980s have led to a dramatic locational shift in concentrations of disadvantage in Australian cities where a largely deregulated housing market led to the displacement of inner-city communities and their re-location to suburban neighbourhoods. This 'neoliberal turn' has led to the fragmentation of many urban areas and the emergence of new geographies of underprivilege (Randolph and Tice 2014).

The suburbanisation of disadvantage presents a number of social, economic and infrastructural challenges to the development of 20-minute neighbourhoods. This is highlighted by Stanley (2020) who has observed the stark lack of basic infrastructure in disadvantaged suburban neighbourhoods in Melbourne. The lack of infrastructure, argues Stanley, is a significant driver of social exclusion and low levels of well-being. However, while this presents a key challenge to urban planners Stanley suggests that the 20-minute concept is an important part of the solution as it would improve not only accessibility but also the acquisition of social capital, health and employment opportunities.

Beyond the challenges of urban fragmentation, displacement and suburban sprawl, there is also the issue of governance. While de-centralising cities has the potential to achieve a range of social and economic benefits, it is important to ensure that the governance of these services is joined up, appropriately resourced and robust to ensure the longevity of these developments. This is an important point to make as there is a well-recognised pattern of national governments devolving responsibility to local councils without the accompanying power to make change (Broto 2019; Davidson and Iveson 2015; Mouffe 2005; Barnett 2003; Harvey 1989). This devolution has, in many instances, led to the fracturing of urban governance regimes as under-resourced local authorities further devolve responsibility to a series of unelected actors including community groups, private and third-sector partners and individual 'experts' (Bentley et al. 2017; Hodson and Marvin 2010; Howell 2015; Swyngedouw 2009). Given the spatial de-centralisation necessary to achieve 20-minute neighbourhoods, it is vital that any accompanying devolution of governance responsibilities to local neighbourhoods is appropriately resourced and supported.

10.2.3 The 20-Minute Citizen

At the heart of the 20-minute neighbourhood lies the notion of equality in relation to space and mobility; all residents across the urban space should have the right and opportunity to access key services locally. In doing so, this concept opens the door to wider considerations of inclusivity and justice within urban environments and offers a tool through which those conversations can be embedded into urban decision-making and used to engage citizens in place-making. As Moreno et al. (2021) observe, the 20-minute neighbourhood aligns with the concept of chrono-urbanism and research highlighting the importance of urban rhythms for understanding quality of life in cities (Mulícek et al. 2015). Given the importance placed on social diversity by Moreno et al. (2021), in order to realise the potential of the 20-minute concept, it is important to first recognise the ways that different groups use urban space.

Despite the clear disparities in the way that different people use urban space, for decades interventions (such as investment in road infrastructure and cutting public transport) have benefited the movement of certain groups (Criado-Perez 2019). With this in mind, for many the 20-minute neighbourhood represents a new wave of feminist urban planning initiatives that push back against traditional male-dominated urban design to create more liveable and accessible spaces (Kern 2020). Others, however, have voiced strong criticisms of this concept which is increasingly seen by some communities as a top-down policy intervention which does not take into account histories of marginalisation, segregation and injustice. At a 2021 CityLab conference, for example, Toronto-based urban designer and thinker Jay Pitter warned against the application of this concept stating that '*It doesn't take into account the histories of urban inequity, intentionally imposed by technocratic and colonial planning approaches, such as segregated neighbourhoods, deep amenity inequity and discriminatory policing of our public spaces*' (Pitter 2021). Pitter goes on to raise concerns about the prospect of green gentrification and further displacement that is likely to happen as a result of 20-minute policies.

Other commentators echo Pitter's concern about the potential for this concept to exclude already marginalised communities. In one article, Anna Zivarts warns that '*in addition to the racialized and colonial assumptions at play, the 15-minute city model also erases disabled bodies and movement*' (2021). The physical ability of an individual to negotiate public space varies hugely. For example, the walking pace of a healthy, able-bodied adult may be different from that of a child, an older person, someone in a wheelchair, who is partially sighted or pushing a buggy. Additionally, stopping at intersections, walking uphill and negotiating street furniture, kerbs or uneven surfaces may also impact the speed at which different people can travel from a to b. Many of these factors are not commonly incorporated into mainstream walkability ratios or planning strategies (Criado-Perez 2019) and are, by definition, not accounted for in the attention-grabbing 20-minute concept.

In reaction to the top-down nature of the 20-minute concept, researchers and practitioners in Sweden have designed an alternative urban design concept rooted in the practice of community engagement. Termed the 'one-minute city', this approach engages residents, through workshops and consultations, with the place just outside their front doors (ArkDes 2021). By allowing residents to be the architects of their own streets, this strategy challenges top-down policy approach associated with the 20-minute neighbourhood. Importantly, the hyper-local one-minute city rejects the deceptively complex and socially exclusive '20-minute' neighbourhood and instead focuses empowering and understanding the needs of individual citizens. Designers pioneering this concept have developed a suit of street furniture – including parklets, small play areas, public seating and EV charging points – which are able to fit into the dimensions of a standard parking space. Residents are able to collectively decide the pieces they would like to be installed on their street. These pieces of furniture are designed to be conversation starters, ways of engaging with and connecting people with their neighbours and the places they live (O'Sullivan 2021).

In doing so, it challenges the need to engage with the 10, 15 and 20-minute time frame, and re-enforces the need for grounded, inclusive planning practices. While many smart city strategies highlight the importance of technological innovation, the one-minute city acts as a reminder to recognise diversity and to anchor decision-making in meaningful, ongoing community engagement practices including the need for co-design.

10.3 Case Study: Putting the 20-Minute Concept to Work in Edinburgh

10.3.1 Background

In Autumn of 2020, the idea of 20-minute neighbourhoods was captured in the Scottish Government's 'Programme for Government', which pledged to work with local government to implement this concept across the country (Scottish Government 2020). In Edinburgh, the call for 20-minute neighbourhoods has been echoed by both the Poverty and Climate 'Commissions' who have highlighted the potential for this concept to underpin not only infrastructure design and implementation but also poverty prevention, well-being and sustainability. This increasingly popular concept has become a new boundary object around which to orientate goals and to help move strategies forward.

Through statutory policies such as Edinburgh's City Plan 2030, City Mobility Plan, and Active Travel Action Plan, the Council has recognised the need for developing well-connected, sustainable, diverse and affordable neighbourhoods, bridging the gap between land use and transportation governance in the process. The current draft versions of these plans already align with many of the principles of the '20-minute' concept, particularly on the issues of mobility and active travel. With this in mind, the aspirations of the research project to be reflected on here (and involving both University of Edinburgh researchers on this project and officers at the Council) were to think through how this then translates into decision-making at the neighbourhood scale, how some of the impacts of these decisions on the city's '20-minute' status might be modelled and forecasted and how the concept might bring together different elements of the Council' portfolio of responsibilities. Aligning here with the importance of 'digitalisation' to the 20-minute idea's application (the fourth and final aspect of Moreno et al.'s (2021) framework), the project sought to draw on available spatial datasets and GIS capabilities in order to explore how the different iterations of the 20-minute idea presented above could be put to work in Edinburgh. As Moreno points out, digitalisation aligns closely with the concept of smart cities, a key inspiration for the 20-minute idea. Within the smart city concept, digital platforms are a key tool through which to engage citizens, better understand urban geographies and model future scenarios (Moreno et al. 2021).

By reflecting on our experiences through this case study, we intend to explore if, when and how the different iterations outlined in the previous sections are able to be put to work by local government officials; how the different approaches complement and at times contradict one another when applied to a city such as Edinburgh; and discuss the implications this has for the potential to usher in a data-driven approach to '20-minute' urban governance. As the research is a co-produced piece of work alongside City of Edinburgh Council, we do not draw on direct quotes from participants here to illustrate the points being made but, instead, reflect more broadly on how the ideas discussed above have been taken up in the city through the research project.

10.3.2 Twenty-Minute Edinburgh

Service distribution and accessibility are key elements of the 20-minute concept. The inevitable first step towards embracing the 20-minute idea is therefore an evaluative exercise to better understand the distribution of key services across the city. Prior to the initiation of the collaborative research involving the university and the city, council officers in the GIS department had begun to develop a city-scale heat map (Figure 10.2) to act as a baseline from which to think about current and future geographies of service delivery across Edinburgh. Also, pre-dating the formalised adoption of the 20-minute idea by the City of Edinburgh Council, the development of this heat map took on a somewhat experimental form, with officers themselves deciding on what might ultimately come to be deemed a 'key' service, for which all residents can expect to reach within a certain amount of walking time. These two variables: how long a 20-minute walk actually *is* and the services which should be rendered as important enough to be accessible for all within this time would come to pose challenges both for the research project's adopted methodologies and for any strategic policy agendas that seek to build upon it.

The six selected key services to use in running this initial analysis of Edinburgh's status as a 20-minute city were as follows: a health centre (general practitioner); a pharmacy; a food shop; a public library; a play park; and a primary school. As Figure 10.1 demonstrates,

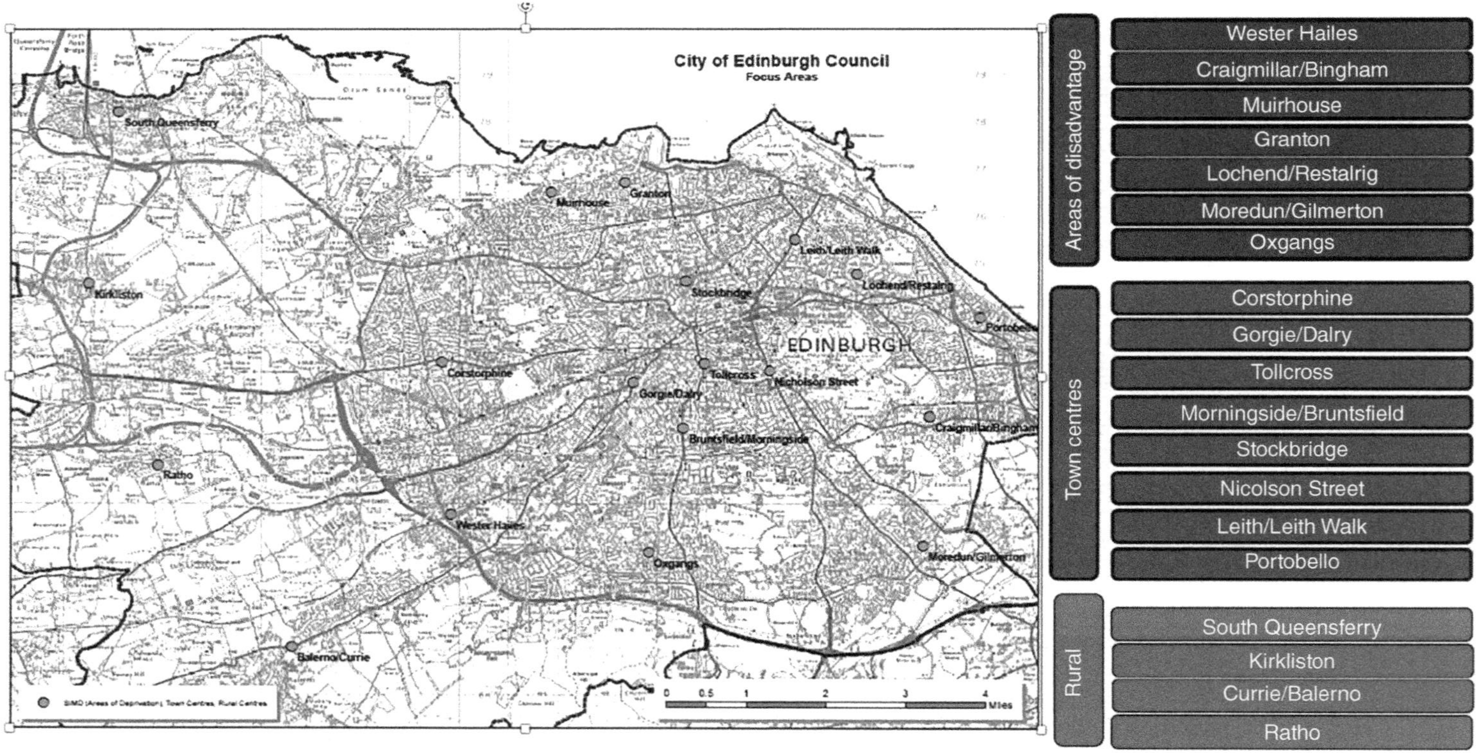

Figure 10.2 The council's 19 focus areas for neighbourhood multi-service provision. *Source:* Data from City of Edinburgh Council.

mapping the spread of these services across the city space shows Edinburgh to have relatively high levels of accessibility with large swathes of the city living in close proximity to these six services. For Edinburgh, then, this heat map became as much an important device for illustrating where the status quo is performing well as one for identifying parts of the city that are underperforming with regards to the idea of being a 20-minute city. Indeed, despite starting life as an experimental mapping exercise in the backrooms of the council's GIS department, the heat map has subsequently made its way into the City Mobility Plan and City Plan 2030. Given that these visions take the form of long-term strategy documents, the issue of how the 20-minute city concept might be drawn upon in the short and medium term remains an important question.

In this vein, as an object for starting conversations and orientating stakeholders to this idea of a '20-minute city', the map raised as many questions about how to do urban governance as it has answered questions about the existing urban morphology. These questions inevitably focussed on the issue of which services had been chosen and designated as 'key' and whether 20 minutes meant there and back or one way (both versions of the map are included in the city's recently adopted City Mobility Strategy 2030). However, in response to such questions, concerns were raised by those in the planning team about turning the question of what counts as a 'key' service into something prescriptive. Such a scenario might result in a considerable loss of power on the part of the council when it comes to negotiation with developers and landowners who would be able to point to 'missing' services in different parts of the city as justification for what they were offering. In other words, it is only by keeping the question of what counts as a 'key' service context dependent and variable that authority over the future of urban space can be retained by local governments. As highlighted at the outset of the chapter, by becoming too prescriptive about what is and is not included within a 20-minute neighbourhood, and the geographical extremities of this neighbourhood, the very dynamism and 'intermingling' potential of urban space that the 20-miunte idea was spawned out of becomes curtailed.

The heat map in Figure 10.1 provides a useful baseline planning tool and is something that through the project we have taken down to the granular level of assigning every building across the city an accessibility 'level' based on the number of key services that can be accessed in a certain period of walking time. When taking down this level of granularity, one area where there has been enthusiasm for the potential of service accessibility mapping to aid local decision-making is around housing allocations. By mapping the proximity of services, it becomes conceivable that future home construction might be delivered in a manner that ensures additions to the city's housing stock do not impinge on its 20-minute status by prioritising (re)-development in already serviced areas. This possibility, of course, in no way mitigates the powerful forces of politics and economics that currently shape decision-making around approving where new housing developments can be located.

It is also important to recognise that different services will likely have different accessibility requirements across different timescales. For example, while it may be the case that people are willing to walk a 40-minute round trip to visit the dentist once every 6 months, it is unlikely that this length of trip would be what one person would take on a daily basis in order to buy groceries. This presents a question of how to weight services based on how often the services are needed. Similarly, its crucial to take into account the quality of the services on offer, e.g. the price and range of goods available in shops, opening hours and demand. Assessing these factors requires not only a large number of up-to-date datasets but, importantly, a significant degree of 'ground truthing' including community engagement and consultation. With this in mind analysis such as the 20-minute heat map needs to be balanced with the needs of citizen groups across the different geographies of the city.

10.3.3 Twenty-Minute Neighbourhoods

One of the challenges of thinking about governance at the city scale is the diverse and heterogenous set of stakeholders and actors who are involved with shaping the landscape and fabric of urban space. The number of historically and culturally rendered variables at play in shaping urban morphology means that city governments are not in a position to comprehensively operationalise new mandates (such as the 20-minute city idea), from the top down. In contrast to a city-scale evaluation of service proximity and accessibility, the idea of the 20-minute *neighbourhood* engages with this governance challenge head-on, prioritising the delivery of densely co-located services in easily accessible locations spread across the city. It does so in two important ways. First, by focussing attention on the areas of urban governance where local councils have most authority and power to intervene. And secondly by thinking through the viability of devolving questions of governance to smaller constituent geographies across the city.

The principal implication of the 20-minute idea is its emphasis on the *neighbourhood* scale and, therefore, the re-orientation of land-use and transportation decision-making away from the city scale and towards a network of service 'activity centres'. Led by the council's estate management team and drawn upon to re-evaluate *how* service delivery should be coordinated across the city, a process is underway in Edinburgh to prioritise attention to key areas where services are (i) already densely present and can be further strengthened with maximum impact; (ii) currently lacking in areas where they are most needed; or (iii) have unique characteristics by virtue of the city council's unusual portfolio comprising both heavily urbanised as well as rural space. Totalling nineteen sites spread across the geography of Edinburgh, these strategic 'focus areas' are made up of (i) eight historic town centres; (ii) seven areas of recognised disadvantage (according to the Scottish Indices of Multiple Deprivation, SIMD); and (iii) four 'rural centres' located in the city's hinterland but under the authority of the city government. Figure 10.2 shows the locations of these places.

Representing the City of Edinburgh Council's strategy towards the delivery of 20-minute neighbourhoods, the aspiration is to pursue the development of multi-service 'hubs' located within each of these focus areas. Such co-location hubs are becoming an increasingly popular idea across local authorities in Scotland and the United Kingdom. For local councils, the co-location of services in one central neighbourhood hub provides a cost-effective method of service delivery as not only does it reduce the cost of utilities and maintenance but also it would allow the council to consolidate its existing assets, prioritising the delivery of densely organised buildings, serviced by well-established and integrated active travel networks – something which is seen as increasingly necessary given current financial pressures.

Having taken the lead in identifying *where* co-located service hubs should be delivered across the city space, attention for both the council and the research team turned towards ensuring that these locations are highly 'walkable', allowing as many citizens as possible residing in the areas immediately surrounding these hubs, to access them. While a great deal of academic research has been done to develop sophisticated means of generating walkability 'indicators', once again the 20-minute idea offers a simplistic means of interpreting this in practice and, in doing so, facilitates comparison at the neighbourhood level (between the different identified hub locations) and even down to the individual building level (with regards to how assets may be managed *within* a particular hub location). This can be done by contrasting the 'theoretical' distance one might be able to travel from a particular location in 10 minutes with the actual distance it is possible to travel using the existing pedestrian infrastructure network. Referred to as a 'Neighbourhood Walkability Ratio' (NWR), this is shown below in Figure 10.3.

Starting with the neighbourhood scale, the research developed a means of applying this NWR approach to the 19 identified focus areas across the city. Figure 10.4 shows this for the area of 'Corstorphine', located to the West of the city space in Figure 10.2. As shown in the image, an overall 'theoretical' 20-minute walking space is captured by the dark line around the Corstorphine town centre, while the shaded inset illustrates the space from which one can currently access the centre using the existing footpath network. An NWR of 0.667 is generated for this location and can be used to compare walkability ratios across the different hubs.

While comparison between different areas of the city in terms of relative NWR offers some insight into the viability of these locations as priority service delivery hubs for the city, this alone offers limited insight into how decision-making around the governance *of* these new 20-minute neighbourhoods might make best use of the available data. This is where, upon zooming in to the neighbourhood scale, the application of walkability ratios and service accessibility levels mutually benefit from being brought into conversation with one another, rather than treated as abstract, academic evaluation indicators. By generating a granular, building-level service accessibility heat map and overlaying the NWR map on top of this (Figure 10.5), it is possible to identify areas around the proposed focus centres where improvements to the pedestrian infrastructure would result in a greater number of buildings being able to access the town centre and, vice versa, where areas of the city

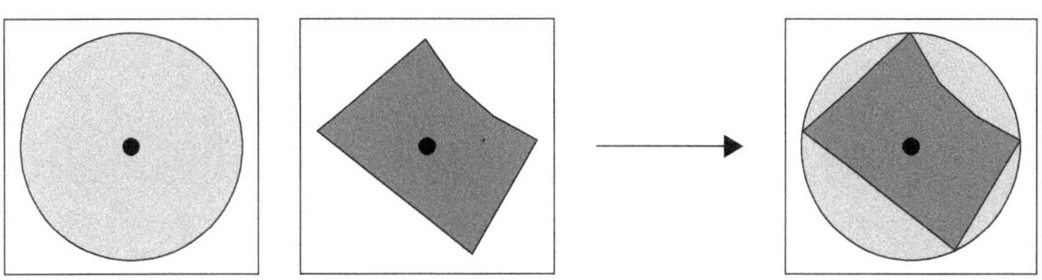

Figure 10.3 A visual representation of the neighbourhood walkability ratio, i.e. dividing the size of the area you can reach from the centroid while covering distance X along existing roads by the area you could theoretically reach by covering distance X 'as the crow flies'. *Source:* Schlossberg (2006)/with permission of SAGE Publications.

Figure 10.4 Map of 20-minute walkability around Corstorphine on the west side of Edinburgh yielding a walkability ratio of 0.667. Rather than using a single centroid (as in Figure 10.3), our GIS analysis used multiple 'town centre access points' around the 'town centre' identified by the council as the focus area for multi-service location. *Source:* Authors.

Figure 10.5 Mapping key service accessibility at the building level in Corstorphine. *Source:* Authors.

suffering from poor service access levels lie outwith even the theoretical 20-minute catchment areas of these new service delivery hubs.

Turning attention to the question of governance *within* the different focus area hubs, the addition of demographic information about the number of people living within a 20-minute walk of different builds (accessed by the research team from National Records of Scotland data from 2017) allows for analysis to be undertaken which examines the levels of walkability and accessibility associated with specific buildings. Located in the area of the city directly north of Corstorphine, and therefore beyond the reach of any of the 19 focus areas, the Drumbrae area contains a number of assets owned and operated by City of Edinburgh Council (Table 10.1). By drawing together indicators of walkability and service accessibility, along with the populations serviced by particular buildings, it becomes possible for the council to not only *identify* those assets that are most difficult to access on the part of the population but also to *explain* why this is the case, whether it be because the service is isolated from other services (lacking in urban density) or due to inefficiencies in the existing pedestrian network.

For example, while a pattern can be identified that correlates a strong (>0.5) NWR with a high service accessibility level and ultimately with a significant number of people able to access the building in question within 10 minutes, anomalies to this pattern point to potential places where council intervention would be well valued. The example of Clermiston Primary School illustrates how strategic investment in improving the pedestrian infrastructure around this site could potentially grant access to a whole host of services for a much greater number of people.

Alongside an analysis of each of the 19 town centres, we have also taken a 'deep dive' into the neighbourhood of Granton in the north of Edinburgh to gain a better understanding of the specific characteristics and needs of a neighbourhood. Granton is undergoing significant redevelopment driven by the city's housing shortage and desire to address levels of inequality in the neighbourhood and the surrounding area (City of Edinburgh Council 2021b). The council's ability to take a leading role in this redevelopment is strongly linked to land ownership; in 2018, the council bought a 66-acre site from the National Grid

Table 10.1 How GIS analysis can be used to rate existing council assets in terms of accessibility; current urban morphology (NWR), catchment population and inter-service accessibility.

Council asset name	Neighbourhood walkability ratio (NWR)	Number of key services accessible	Estimated '10-minute' population served
Drumbrae Library	0.53	7	5,313
Drumbrae Care Home	0.54	7	4,872
Fox Covert R.C. Primary School	0.37	2	2,648
East Craigs Nursery/ Primary School	0.55	7	6,335
Clermiston Primary School	0.17	6	1,301

Source: Authors.

and will be adding adjacent sites through compulsory purchase order. Through the development of this site, the council plans to make the Granton waterfront area a key 'destination' for residents and visitors to the city. An important part of this plan involves the construction of over 450 affordable and mid-market homes (City of Edinburgh Council 2020). For this reason, the Council has asked us to take a closer look at this neighbourhood to explore how the 20-minute concept might be used to guide the plans for development. In addition to the opportunities afforded by Granton's status as somewhat of a 'blank slate' redevelopment space, this steer was also based on the fact that Granton is seen as somewhat of a testbed for thinking through how ideas around service co-location (aspired towards across the city as a whole) might be put to work in practice.

10.3.4 20-Minute Edinburghers

Although the 20-minute data analytics can inform policy at the city level and shape the design of neighbourhood-scale redevelopments, citizens' experiences and perspectives are the ultimate litmus test for the success of any intervention measure. In practice, intervention options tend to be constrained (e.g. by finance, competencies, ownership of assets and legal rules), while public perceptions and preferences are known to be diverse and dynamic. People may not only have different perspectives on what types of services are essential or what constitutes good access but they may also have personal preferences that are harmful to themselves and/or others.

As a consequence, 'success' should not be perceived as a state or end point but rather as an inclusive process of public and stakeholder engagement and a dynamic alignment of service provision that is feasible and citizens' experiences that are acceptable. Observed behaviour (e.g. travel patterns and footfall) can yield important empirical evidence of the aggregate effectiveness of interventions, but the value of such technical, ex post monitoring can be hugely improved through active citizen participation in design, data collection and analysis. Ensuring citizen participation that is sufficiently proactive and diverse is an ongoing challenge in democratic governance.

While the practical limitations of our project meant we were not able to engage directly with citizens about the 20-minute idea, our conversations have yielded three observations that are worth sharing.

First of all, while our council counterparts were interested in the 20-minute GIS analysis, we showed them they were also willing and able to engage critically with the limitations from a resident perspective. Some expressed concerns about inclusivity, wishing the analysis to be made more sophisticated to account for the mobility needs of particular groups, e.g. parents with prams or people in a wheelchair. Others asked if we could take account of car traffic, as it slowed down pedestrians or increased their risk of exposure to air pollution and traffic accidents. As our GIS analysts on the video call shared their screen to show the GIS maps, civil servants expressed desire to see their own part of town, and some were happy to express personal preferences with regard to key services. For example, one person pointed out that as someone without children, they would prefer NOT to live close to schools, as these bring the disamenities of noise and car traffic to pick up or drop off the pupils. It made us realise that 20-minute GIS analysis has direct value for people moving home if it is a dynamic online system where they can select the criteria (amenities and disamenities) that are important for them. If such online, open-access GIS facilities are made available, how does this influence the debate on 20-minute citizens? To what extent would or should '20 minute informed' citizens who have the freedom (financial and otherwise) to make a residence choice have a say about changes in service provision? Under what kind of conditions could increased residential mobility provide a (or an alternative) route to mitigate some of the inequalities arising from the uneven spatial patterns of existing services?

Second, it is important to acknowledge the existence of citizen-oriented local urban governance projects running in Edinburgh with input or support from various city stakeholders, including the council. These include both city-level mapping of assets valued by citizens, including many cultural assets that might not feature on the City Council's GIS database and community-level mapping of problems, assets and desired improvements (Currie and Correa 2021).

Finally, we noticed an interest from data scientists to develop more sophisticated assessments of accessibility, to bring it closer to the lived experience of residents. These improvements can consist of tailoring the key services (e.g. frequency of use and opportunities for multi-purpose trips), adding more relevant information about the routes (e.g. width, steepness, quality of the surface and amenities and disamenities along the route) and taking account of more dynamic travel factors (e.g. weather conditions, traffic jams, crowdedness and sense of safety after dark). It is beyond the scope of this chapter to explore access modelling in more detail, but in terms of citizen involvement, it may be useful to crudely distinguish between two types of dynamic accessibility models. First of all, there are models that are informed by observed behaviour, which can predict aggregate mobility patterns within certain set parameters. Within those parameters that are set by the empirical data, these models can be highly accurate and reliable. They are suited to identify bottlenecks and quantify and rank the expected improvements in aggregate mobility arising from relatively minor and specific infrastructural interventions (e.g. moving a particular key service from location A to location B; building a footbridge at location C). The other type is characterised by a more extensive input of personal preferences. It allows the tailoring of accessibility modelling to reflect the needs and wants of particular individuals and groups. These models have a greater potential for use in citizen engagement and if adopted more widely, they can yield novel insights into potentially conflicting demands for accessibility to services and urban mobility more broadly.

10.4 Discussion

While the underlying ideas are not new, the '20-minute' concept has captured people's imaginations and has the potential to act as a unifying and accessible vision around which to plan holistically and exchange information. Against this backdrop, this chapter had two core purposes. First, to broach the existing grey and academic literature on the concept of 10-/15-/20-minute neighbourhoods and, in doing so, set out a conceptual framework capable of thinking through how this idea is being (and might be in the future) interpreted to offer new insight and tools for urban governance at three different *scales*. Second, we have sought to apply this framework to the city of Edinburgh, via an ongoing research collaboration between the University of Edinburgh and the City of Edinburgh Council, targeting a data-driven approach to the 20-minute concept's application. In doing so, the chapter has built explicitly on the work of Moreno et al. (2021) and their aspirations for a 'digitalisation' of how we bring about 15- and 20-minute 'smart' cities. In this final section, we return to this work to reflect both conceptually and practically on how our experiences with the 20-minute concept in the context of Edinburgh help to further understandings of how (and with what potential challenges and opportunities) it can be transformed from a descriptive ideal used to evaluate urban morphology to a more concrete set of analytical capabilities, deployed to key datasets, that can aid future decision-making on urban design and the management of council assets and infrastructures of service provision.

10.4.1 Urban Governance in the Data-Driven Era

Perhaps *the* most important takeaway from the analysis carried out as part of our research alongside the city of Edinburgh is that questions of walkability, accessibility and demography must be used alongside one another, drawn together to illuminate not only the parts of the city that perform well as 20-minute neighbourhoods and the parts that do not but also to shed further light on how this might be addressed within the existing capacity of local governments to shape the development of urban spaces. Using Moreno et al.'s (2021) associated terminology, priority themes of 'proximity', 'density' and 'diversity' go hand in hand with one another, and it is the role of the fourth element of their framework – digitalisation – to act as the glue which can hold these other aspirational characteristics together by illuminating where the city is at today and how decision-making might take it on a different path to the future.

In particular, the process of digitalisation has been useful when identifying areas where pedestrian infrastructure upgrades would have the most 'impact', i.e. improving accessibility in those parts of the city that are currently not within the walkable catchment area of the 19 focus centres. The limited service provision may well encourage a higher level of car use by those who have cars. This might lead to more social inequality, especially between those who do not have cars and thus have to walk, cycle or use public transport, and those who have cars and are using them more to drive in the city to reach key services. By identifying these areas of the city, this analysis helps to inform council decision-making about sufficient, equitable and car-free access to services across urban space.

Identifying and subsequently targeting areas such as Ravelston offer the potential to reduce car dependency thus creating safer streets and lowering air pollution which has detrimental impacts on human and environmental health (Turner et al. 2020; Engemann et al. 2019). Meanwhile, as some commentators have pointed out, the ability to move about and switch locations is integral to innovation and advancement (Whittle 2020). With this in mind, maintaining affordable, efficient and frequent transport connections between urban neighbourhoods is vital to maintain social and cultural mixing across the city. Lastly, by increasing access to essential, affordable and well-maintained services both within and beyond people's neighbourhoods, this concept has the power to build social capital, increase well-being and reduce emissions (Ambrey 2016; Davern et al. 2017; Forrest and Kearns 2001; Stanley 2020; Ward et al. 2016).

10.4.2 Collaboration Among Place-Based Civic Actors

Beyond these concerns, the 20-minute concept must overcome a number of practical challenges. While this idea has captured imaginations, in practice its implementation in unique geographies across the world will not come without challenges, particularly during this uncertain and unstable period. In cities already working with this concept, governments have voiced the need to adapt to support increasing densities and intensity of activity, while also providing for high amenities and social inclusion (Victoria Planning Authority 2020). Perhaps most pressing questions of funding pose a significant challenge for local authorities across the world, many of who were already facing financial difficulties before the pandemic hit. In England, for example, the financial impact of COVID-19 on local government is estimated at £10.9 billion (Honeyben 2020).

The 20-minute concept has risen to prominence at a time when the COVID-19 pandemic has meant that the turn to digital technologies has been one of necessity rather than aspiration. A key reflection from the research collaboration presented in this chapter stems from the very uncertainty created by the pandemic. This is the relationship between research and policy practice, represented in this project by stakeholders from the university and the Council. Relationship building between these two groups undoubtedly benefited from the move to online meetings precipitated by the pandemic as this allowed different combinations of stakeholders to meet more frequently. In part, the organic nature of this relationship was also helped by the lack of clarity about the 20-minute concept on both sides. Although frustrating at times, this lack of direction allowed for more open discussions and presented the opportunity to forge new relations which draw on the respective resource bases of key civic actors within cities to help build up clearer pictures of how the city is used, how this is shifting and how it could potentially be used in the future.

From a practical perspective, there is also the challenge of data collection and standardisation. The assessment, implementation and monitoring of 20-minute neighbourhood require the collection and standardisation of a plethora of different datasets, from public transit GPS data and area demographics to environmental quality and the availability of affordable and healthy food. Not only will much of this data come in different formats but it will likely be owned and curated by different council departments, state agencies and the private sector.

10.4.3 Looking Forward

Although local neighbourhoods have often borne the brunt of funding cuts and changing patterns of work and consumption, made worse by austerity policies after the 2008 financial crash, the shift to working from home and re-orientation to local geographies, catalysed by the COVID-19 pandemic, has sparked a renewed interest in the role of local centres in local economies.

Beyond the issue of finance, the challenge of monitoring and reporting has been highlighted by a number of 20-minute city plans which fail to specify what 'access' means and how it is measured (Whizman 2017). This is challenging given the fact that, as we have explored, different people move around urban space in different ways and at different speeds so what is a 10-minute walk to the shops for one person might be much longer for someone else. Furthermore, it is important to recognise that the term 'essential services' will vary from person to person depending on individual needs. With this in mind, however, it is also important to also consider the quality of those services such as the affordability of food, environmental quality of green space and the frequency of public transport as well as the qualities (however perceived) of the paths, pavements and streets people take to get to them. Furthermore, people will often seek to undertake multi-purpose trips for the sake of time management or may seek the opposite, i.e. longer walks for leisure or health.

The extent to which the 20-minute concept can work its way into the more granular aspects of urban governance (e.g. decision-making on housing development sites, where to prioritise upgrades to the active travel network and how to best utilise existing council assets) remains to be seen. What we can be sure of, however, is that the ultimate delivery of 20-minute neighbourhoods and the benefits that can be obtained through the embracing of digitalisation of the urban landscape will require a concerted effort on the part of local government and the active participation of its citizens.

Acknowledgements

The authors are grateful to acknowledge funding received from UK Research and Innovation (PCAN; Place-Based Climate Action Network, ES/S008381/1) and the Scottish Funding Council (through the DDI Data Platforms Innovation Programme).

References

Allison, J.R. (2020). Greening inequality: how urban sustainable development fails under neoliberalism. University Honors Theses. Paper 954. https://doi.org/10.15760/honors.977.

Ambrey, C. (2016). Urban greenspace, physical activity and wellbeing: the moderating role of perceptions of neighbourhood affability and incivility. *Land Use Policy* 57: 638–644. https://doi.org/10.1016/j.landusepol.2016.06.034.

Angelopoulos, S., Boymal, J., and de Silva, A. (2019). Identifying and valuing the economic benefits of 20-minute neighbourhoods higher density mixed use and walkability dimensions. RMIT University. https://www.planning.vic.gov.au/__data/assets/pdf_file/0018/450135/Economic-benefits-of-20-minute-neighbourhoods.pdf (accessed 10 May 2021).

ArkDes (2021). Street moves. https://arkdes.se/arkdes-play/nu-flyttar-streetmoves-fran-stockholm/ (accessed 27 May 2021).

Barnett, N. (2003). Local government, new labour and 'active welfare': a case of 'self responsibilisation'? *Public Policy and Administration* 18 (3): 25 38.

Baum, S., Haynes, M., van Gellecum, Y., and Han, J.H. (2006). Advantage and disadvantage across Australia's extended metropolitan regions: a typology of socioeconomic outcomes. *Urban Studies* 43 (9): 1549–1579. https://doi.org/10.1080/00420980600831759.

Bentley, G., Pugalis, L., and Shutt, J. (2017). Leadership and systems of governance: the constraints on the scope for leadership of place-based development in sub-national territories. *Regional Studies* 51 (2): 194–209.

Broto, V. (2019). Climate change politics and the urban contexts of messy governmentalities. *Territory, Politics, Governance* 8: 1–18.

Capasso Da Silva, D., King, D.A., and Lemar, S. (2020). Accessibility in practice: 20-minute city as a sustainability planning goal. *Sustainability* 12: 129.

City of Edinburgh Council (2020). £1.3bn sustainable regeneration at Granton Waterfront to begin. https://www.edinburgh.gov.uk/news/article/12873/13bn-sustainable-regeneration-at-granton-waterfront-to-begin (accessed 3 June 2021).

City of Edinburgh Council (2021a). 20-minute neighbourhood. https://www.edinburgh.gov.uk/adaptation-renewal-programme/20-minute-neighbourhood/1 (accessed 26 May 2021).

City of Edinburgh Council (2021b). Granton waterfront regeneration. https://www.edinburgh.gov.uk/homepage/10492/granton-waterfront-regeneration (accessed 3 June 2021).

Clark, C. (2019). Building equity into 20-minute city plans. https://www.eurekastreet.com.au/article/building-equity-into-20-minute-city-plans (accessed 14 June 2021).

Comune di Milano (2019). Milano 2030 - PGT VIGENTE. https://www.pgt.comune.milano.it/ (accessed 14 June 2021).

Criado-Perez, C. (2019). *Invisible Women: Exposing Data Bias in a World Designed for Men*. Vintage Publishing.

Currie, M. and Correa, M. (2021). Tangibles, intangibles and other tensions in the culture & communities mapping project. Cultural Trends. 10 May. https://doi.org/10.1080/09548963.2021.1910491.

Davern M, Gunn L, Whitzman C, Higgs C, Giles-Corti B, Simons K, Villanueva K, Mavoa S, Roberts R. & Badland H. (2017) Using spatial measures to test a conceptual model of social infrastructure that supports health and wellbeing, *Cities & Health*, 1:2, 194–209. https://doi.org/10.1080/23748834.2018.1443620

Davidson, M. and Iveson, K. (2015). Recovering the politics of the city: from the 'post-political city' to a 'method of equality' for critical urban geography. *Progress in Human Geography* 39 (5): 543–559.

Davison, G. (1997). The great Australian sprawl. Historic environment. Australia ICOMOS (International Council on Monuments and Sites). https://search.informit.org/doi/10.3316/informit.769402465333798 (accessed 14 June 2021).

Davison, A. (2006) Stuck in a Cul-de-Sac? Suburban history and urban sustainability in Australia *Urban Policy and Research*, 24:2, 201–216. https://doi.org/10.1080/08111140600704137

Dolega, L. and Lord, A. (2020). Exploring the geography of retail success and decline: a case study of the. Liverpool City Region. *Cities* 96: 102456.

Dovey, K. and Woodcock, I. (2014). Intensifying Melbourne. Transit-oriented urban design for resilient urban futures. https://msd.unimelb.edu.au/__data/assets/pdf_file/0007/2595616/Intensifying-Melbourne-2014_180dpi.pdf (accessed 14 June 2021).

Eichler, W. (2018). Councils slash £15m from parks budgets. Local Gov. https://www.localgov.co.uk/Councils-slash-15m-from-parks-budgets/45519 (accessed 12 May 2021).

Engemann, K., Pedersen, C., Arge, L. et al. (2019). Residential green space in childhood is associated with lower risk of psychiatric disorders from adolescence into adulthood. *Proceedings of the National Academy of Sciences* 116 (11): 5188–5193. https://doi.org/10.1073/pnas.1807504116.

European Environment Agency (2006). Urban Sprawl in Europe, the ignored challenge. The European Commission.

Evans, J. (2019). Death of the high street weighs on landlords across the world. Financial Times. https://www.ft.com/content/4e5e1022-8df3-11e9-a1c1-51bf8f989972 (accessed 14 June 2021).

Forrest, R. and Kearns, A. (2001). Social cohesion, social capital and the neighbourhood. *Urban Studies* 38 (12): 2125–2143. https://doi.org/10.1080/00420980120087081.

Gandy, M. (2002). *Concrete and Clay Reworking Nature in New York City*. Cambridge, MA: MIT Press.

Goodling, E., Green, J., and McClintock, N. (2015). Uneven development of the sustainable city: shifting capital in Portland, Oregon. *Urban Geography* 36: 504–527.

Goodman, R. and Kroen, A. (2019). Assessment of retail model in greenfield development settings: the social and health impacts of the form of shopping centres in new suburbs. RMIT University. https://www.planning.vic.gov.au/__data/assets/pdf_file/0017/450134/Assessment-of-retail-model-in-greenfield-development-settings.pdf (accessed 14 June 2021).

Gould, K. and Lewis, T.L. (2017). *Green Gentrification: Urban Sustainability and the Struggle for Environmental Justice (Routledge, Equity, Justice, and the Sustainable City Series)*. Abingdon, Oxon; New York, NY: Routledge.

Grodach, C., Kamruzzaman, L., and Harper, L. (2020). 20-minute neighbourhood - living locally research. Monash University. https://www.planning.vic.gov.au/__data/assets/pdf_file/0023/450275/Mambourin-Report-Staging-Community-Infrastructure.pdf (accessed 14 June 2021).

Gunn, L.D., Mavoa, S., Boulangé, C. et al. (2017). Designing healthy communities: creating evidence on metrics for built environment features associated with walkable neighbourhood activity centres. *International Journal of Behavioral Nutrition and Physical Activity* 14: 164. https://doi.org/10.1186/s12966-017-0621-9.

Gwiazdzinsk, L. (2015). Space-time design for chrono-urbanism in smart city. *Smart Sustainable City and Big Data, The 3rd International Conference*, Shangaï University. Shangai China, July 2015.

Gwiazdzinski, L. (2014). Adaptable cities and chrono-urbanism. Adaptable Cities and Chrono-Urbanism, Sweden.

Habitat for Humanity (2020). The early history of Portland's racist housing strategies: part one. Habitat for Humanity: Portland/Metro East. https://habitatportlandmetro.org/the-early-history-ofportlands-racist-housing-strategies-part-one/ (accessed 26 June 2020).

Harris, R. and Lewis, R. (2001). The geography of North American Cities and suburbs, 1900–1950: a new synthesis. *Journal of Urban History* 27 (3): 262–292. https://doi.org/10.1177/009614420102700302.

Harvey, D. (1989). From managerialism to entrepreneurialism. *Geografiska Annaler. Series B. Human Geography* 71: 3–17.

Hodson, M. and Marvin, S. (2010). Urbanism in the anthropocene: ecological urbanism or premium ecological enclaves? *City* 14: 298–313.

Honeyben, P. (2020). Impact of Covid-19 on local government finance. London Councils. https://www.londoncouncils.gov.uk/members-area/member-briefings/local-government-finance/impact-covid-19-local-government-finance (accessed 14 June 2021).

Howell, A. (2015). Resilience as enhancement: governmentality and political economy beyond 'responsibilisation. *Politics* 35 (1): 67–71.

Hulse, K. and Reynolds, M. (2018). Investification: financialisation of housing markets and persistence of suburban socio-economic disadvantage. *Urban Studies* 55 (8): 1655–1671. https://doi.org/10.1177/0042098017734995.

Jun, M.-J. (2004). The effects of Portland's urban growth boundary on urban development patterns and commuting. *Urban Studies* 41 (7): 1333–1348.

Kern, L. (2020). *Feminist City Claiming Space in a Man-Made World.* New York: Verso.

Moreno, C. (2016). La ville du quart d'heure: pour un nouveau chrono-urbanisme. La Tribune. https://www.latribune.fr/regions/smart-cities/la-tribune-de-carlos-moreno/la-ville-du-quart-d-heure-pour-un-nouveau-chrono-urbanisme-604358.html (accessed 26 May 2021).

Moreno, C. (2020). Carlos Moreno: the 15-min city. C40 Knowledge. https://www.c40knowledgehub.org/s/article/Carlos-Moreno-The-15-minute-city?language=en_US (accessed 14 June 2021).

Moreno, C.; Allam, Z.; Chabaud, D.; Gall, C.; Pratlong, F. Introducing the "15-minute city": sustainability, resilience and place identity in future post-pandemic cities. *Smart Cities* 2021, *4*, 93–111. https://doi.org/10.3390/smartcities4010006

Mouffe, C. (2005). *On the Political.* Routledge.

O'Gorman, S. and Dillon-Robinson, R. (2021). *20 Minute Neighbourhoods in a Scottish Context.* Ramboll https://doi.org/10.7488/era/808.

Mulíček, O., Osman, R., and Seidenglanz, D. (2015). Urban rhythms: a chronotopic approach to urban timespace. *Time & Society* 24 (3): 304–325.

O'Sullivan, F. (2020). Paris Mayor: it's time for a '15-minute city'. https://www.bloomberg.com/news/articles/2020-02-18/paris-mayor-pledges-a-greener-15-minute-city (accessed 14 June 2021).

O'Sullivan, F. (2021). Make way for the 'one-minute city'. Bloomberg City Lab. https://www.bloomberg.com/news/features/2021-01-05/a-tiny-twist-on-street-design-the-one-minute-city (accessed 27 May 2021).

Ottawa City Council (2019). The 5 big moves. https://engage.ottawa.ca/the-new-official-plan/news_feed/the-5-big-moves (accessed 14 June 2021).

Oxford Analytica (2018). Rising living costs will hasten 'high streets' decline. Expert Briefings. https://doi.org/10.1108/OXAN-DB235918.

Pawson, H., Hulse, K., and Cheshire, L. (2015). Addressing concentrations of disadvantage in urban Australia, AHURI Final Report No.247. Australian Housing and Urban Research Institute, Melbourne. http://www.ahuri.edu.au/publications/projects/myrp704 (accessed 14 June 2021).

Perry, C. (1929). The neighbourhood unit. Volume VII, regional New York and its environs, Monograph I. New York. *National Municipal Review* 18: 636–637.

Pitter, J. (2021). CityLab 2021 conference. Cited in O'Sullivan, F. (2021). Where the 15-minute city falls short. Bloomberg CityLab. https://www.bloomberg.com/news/articles/2021-03-02/the-downsides-of-a-15-minute-city (accessed 14 June 2021).

Portland City Council (2012). The Portland plan. https://www.portlandonline.com/portlandplan/index.cfm?c=58776 (accessed 12 May 2021).

Pozoukidou, G. and Chatziyiannaki, Z. (2021). 15-minute city: decomposing the new urban planning Eutopia. *Sustainability* 13 (2): 928. https://doi.org/10.3390/su13020928.

Randolph, Bill & Tice, Andrew. (2014). Suburbanizing disadvantage in Australian cities: sociospatial change in an era of neoliberalism. *Journal of Urban Affairs* 36 384–399. https://doi.org/10.1111/juaf.12108.

Bill Randolph & Andrew Tice (2017) Relocating disadvantage in five Australian cities: socio-spatial polarisation under neo-liberalism, *Urban Policy and Research*, 35:2, 103–121. https://doi.org/10.1080/08111146.2016.1221337

Resilient Melbourne (2020). Living locally: creating resilient 20-minute neighbourhoods in greenfield growth areas. https://www.planning.vic.gov.au/__data/assets/pdf_file/0033/487509/Living-Locally-20MN-in-Greenfield-Growth-Areas.pdf (accessed 14 June 2021).

Resnik, D.B. (2010). Urban sprawl, smart growth, and deliberative democracy. *American Journal of Public Health 100* (10): 1852–1856. https://doi.org/10.2105/AJPH.2009.182501.

Schlossberg, M. (2006). From TIGER to audit instruments measuring neighborhood walkability with street data based on geographic information systems. *Transportation Research Record* 1982: 48–56. https://doi.org/10.3141/1982-08.

Scottish Government (2020). Protecting Scotland, renewing Scotland: the government's programme for Scotland 2020–2021. https://www.gov.scot/publications/protecting-scotland-renewing-scotland-governments-programme-scotland-2020-2021/ (accessed 14 June 2021).

Stanley, J. (2020). Chapter 12: The interface between family well-being and government policy in Australia. In: *Cross-Cultural Family Research and Practice* (ed. W.K. Halford and F. van de Vijver), 373–387. Academic Press ISBN 9780128154939. https://doi.org/10.1016/B978-0-12-815493-9.00012-0.

Stanley, J., Stanley, J., and Davis, S. (2015). Moving people: connecting neighbourhoods: the 20-minute city, bus industry confederation inc. https://www.researchgate.net/publication/274064812_Moving_People_Connecting_Neighbourhoods_The_20_minute_city_Bus_Industry_Confederation_Inc (accessed 14 June 2021).

Swyngedouw, E. (2009). The antinomies of the postpolitical city: in search of a democratic politics of environmental production. *International Journal of Urban and Regional Research* 33: 601–620.

Talk London (2020). 15 minute cities – the city on your doorstep. https://www.london.gov.uk/talk-london/communities-regeneration/15-minute-cities-%E2%80%93-city-your-doorstep (accessed 12 May 2021).

Turner, J.J. and Gardner, T. (2014). Critical reflections on the decline of the UK high street: exploratory conceptual research into the role of the service encounter. In: *Handbook of Research on Retailer-Consumer Relationship Development* (ed. F. Musso and E. Druica), 127–151. IGI Global. https://doi.org/10.4018/978-1-4666-6074-8.ch008.

Turner, M.C., Andersen, Z.J., Baccarelli, A. et al. (2020). Outdoor air pollution and cancer: an overview of the current evidence and public health recommendations. *CA: a Cancer Journal for Clinicians* 2020 (70): 460–479. https://doi.org/10.3322/caac.21632.

Victoria Planning Authority (2020). Guidelines for precinct structure planning in Melbourne's greenfields draft for public engagement Victoria State Government. https://vpa-web.s3.amazonaws.com/wp-content/uploads/2020/09/Guidelines-for-Precinct-Structure-Planning-in-Melbournes-Greenfields-Draft-for-Public-Engagement-September-2020.pdf (accessed 14 June 2021).

Victoria State Government (2016). Plan Melbourne 2017–2015. https://planmelbourne.vic.gov.au/home (accessed 12 May 2021).

Victoria State Government (2017). 20-minute neighbourhoods Create more inclusive, vibrant and healthy neighbourhoods. https://www.planning.(2017)vic.gov.au/

policy-and-strategy/planning-for-melbourne/plan-melbourne/20-minute-neighbourhoods (accessed 14 June 2021).

Victoria State Government (2018). Living locally- 20-minute neighbourhoods. https://www.planmelbourne.vic.gov.au/__data/assets/pdf_file/0003/509736/Brochure-January-20-min-neighbourhood-2019.pdf (accessed 14 June 2021).

Ward, J., Duncan, S., Jarden, A., and Stewart, T. (2016). The impact of children's exposure to greenspace on physical activity, cognitive development, emotional wellbeing, and ability to appraise risk. *Health & Place* 40: 44–50, ISSN 1353-8292. 10.1016/j.healthplace.2016.04.015.

Weng, M., Ding, N., Li, J. et al. (2019). The 15-minute walkable neighborhoods: measurement, social inequalities and implications for building healthy communities in urban China. *Journal of Transport and Health* 13: 259–273.

Whittle, N. (2020). Welcome to the 15-minute city. Financial Times. https://www.ft.com/content/c1a53744-90d5-4560-9e3f-17ce06aba69a (accessed 14 June 2021).

Whizman, C. (2017). A 20-minute city sounds good, but becoming one is a huge challenge. The Conversation. https://theconversation.com/a-20-minute-city-sounds-good-but-becoming-one-is-a-huge-challenge-80082 (accessed 14 June 2021).

Yeung, P. (2021). How '15-minute cities' will change the way we socialise. BBC. https://www.bbc.com/worklife/article/20201214-how-15-minute-cities-will-change-the-way-we-socialise (accessed 14 June 2021).

Zivarts, A. (2021). The '15-minute city' isn't made for disabled bodies. CityLab Bloomberg. https://www.bloomberg.com/news/articles/2021-04-22/the-people-that-the-15-minute-city-leave-behind (accessed 14 June 2021).

11

From Smart Urbanism to Sustainable Urban Mobility Plan: A Critical Evaluation of the Case of Cagliari

Chiara Garau[1], *Giulia Desogus*[1], *and Vincenza Torrisi*[2]

[1] *Department of Civil and Environmental Engineering and Architecture (DICAAR), University of Cagliari, Cagliari, Italy*
[2] *Department of Civil Engineering and Architecture (DICAR), University of Catania, Catania, Italy*

11.1 Introduction: Sustainable Urban Mobility Plan (SUMP)

In the last century, urban systems have undergone a remarkable development, so that currently 55% of the world's population resides in urban areas. This percentage is set to become 68% by 2050 (United Nations 2018; World Urbanization Prospects 2018). This expansion can be attributable to various factors such as the concentration of activities and residences, the disposable income, the increase in urbanisation levels as well as the growing technological development. This phenomenon has produced tensions in the place-based system and is one of the reasons that prompted the European Union to implement urban investment policies to make cities more attractive and healthier through the introduction of technologies and innovative infrastructures (European Union 2020).

To respond to these challenges, the emerging philosophy of smart and resilient cities is to make cities 'green and healthy' by going far beyond the single city and the simple reduction in CO_2 emissions through efficiency and energy-saving measures or through sustainable urban transport. Indeed, the development of innovative solutions capable of improving urban environmental performance has led the most current research to give greater importance to the 'social' dimension with which these solutions are understood and adopted, giving value and benefits to urban populations (Heaton and Parlikad 2019; Tiboni et al. 2021). In addition, place-based governance is no longer limited to a single city (Azzari et al. 2018; Caprotti and Cowley 2019). It must be implemented through a network of technical-administrative tools that allow structured management of various problems through a connection of multiple systems of the entire area of competence.

In addition, the problems of a wider territory than a single city through the integration of multiple application sectors, in the theoretical field, are linked to the concept of smart

Smart Cities, Energy and Climate: Governing Cities for a Low-Carbon Future, First Edition.
Edited by Oleg Golubchikov and Komali Yenneti.

urbanism, understood in this research as an approach to urban planning and governance based on technological developments.

This combines urban planning and science by optimising and foreseeing forms of regional and urban planning that improve sustainability, efficiency, resilience and quality of life (Bibri 2019).

Smart sustainable urbanism also translates in the field of urban mobility as a series of implementation strategies supported by smart digital solutions for planning a more sustainable mobility in the environmental, social and economic fields.

Although research on transport sustainability began more than twenty years ago (Jordan and Horan 1997; Zawieska and Pieriegud 2018), implementation of sustainable mobility solutions is still a highly studied field (Torrisi et al. 2021a; Yigitcanlar 2018).

Starting from 2013 (Sustainable Urban Mobility Plans 2021; Urban Mobility Package 2021), the need for 'more sustainable and integrative planning processes as a way of dealing with the complexity of urban mobility' has been increasingly recognised (The SUMP Concept 2020, p. 1). This has led to new approaches to urban transport planning that are specified in the latest edition of the Sustainable Urban Mobility Plan (SUMP) guidelines defined as 'a strategic plan designed to satisfy the mobility needs of people and businesses in cities and their surroundings for a better quality of life. It builds on existing planning practices and takes due consideration of integration, participation, and evaluation principles' (Rupprecht et al. 2019, p. 9).

As described by the 'Guidelines for developing and implementing a Sustainable Urban Mobility Plan' (2019), the SUMPs are based on some principles that should guarantee the achievement of the 'EU goals for a competitive and resource efficient European transport system are met' (Rupprecht et al. 2019, p. 9). These principles show a strong connection to the concept of smart sustainable urbanism.

The Sustainable Urban Mobility Plans (SUMPs) are called in Italy '*Piani Urbani di Mobilità Sostenibile*' (PUMS). In Italy, the PUMS is established by Article 22 of Law no. 340 of 24 November 2000 as a strategic planning tool for achieving sustainability objectives through actions aiming to integrate the mobility system with urban and territorial developments (Teknoring 2017). In 2017 with Decree no. 4 of 4 August 2017, the Italian Ministry of Infrastructure and Transport (IMIT) introduced the Guidelines for Urban Sustainable Mobility Plans (Guidelines for Urban Sustainable Mobility Plans 2017; Minister of Economy and Finance 2017). The IMIT establishes that the 'regional and national planning scenario must be integrated and put into a system with the territorial and transport planning tools already existing at the local level, if the Administrations have them, acting as a superordinate plan to the plans sector' (Urban Sustainable Mobility Plan Guidelines 2017, p. 11). Since 2000, 46 PUMSs have been approved in Italy as planning tools at the municipal and/or metropolitan city level (Observatory 2021).

This chapter offers an overview of the transport policies, currently contained within the guidelines of the PUMS of the Metropolitan City of Cagliari (MCC), constituted by 17 Municipalities of South Sardinia, including Cagliari, the regional capital. These guidelines are compared with the national guidelines and discussed in relation to the real needs of the city. The authors add some implementative actions aimed at a more human-oriented approach, the use of technologies and the relationship of the MCC with water.

11.2 Comparison Between the Guidelines for Italian PUMS and the Guidelines of the MCC

The case of the MCC is particularly emblematic, first, because some of its 17 municipalities had already approved their own PUMS such as the municipalities of Capoterra and Sestu (Cagliari Observatory 2021), as shown in Figure 11.1. Second, because the MCC has a polycentric configuration in which the municipality of Cagliari has a central role for the whole of Sardinia (Rapporto di Analisi del Territorio 2020), and its transport system should develop better networks, according to a unitary, integrated and shared strategy (Garau et al. 2016; Rapporto di Analisi del Territorio 2020).

The infrastructural system of the MCC consists of a radial pattern with a single transversal element represented by state road 554 (Figure 11.2). This radial pattern – which consists of the following axes: state road 195; state road 195 connected with 554; state road 130 (*viale Elmas*); state road 131 (*viale Monastir*); state road 387 (Dolianova); state road 125 (Eastern Sardinia) and the provincial road 17 (*Viale Poetto*) – is characterised by unidirectional movements towards the centre of Cagliari (PUMS – Guidelines 2019). This structure causes congestion phenomena in the main urban-connecting arteries due to inappropriate road sections, insufficient to withstand the intense vehicular traffic that causes critical nodes (PUMS – Guidelines 2019).

However, experiments on car and bike sharing, electric mobility, the urban local public transport (LPT) system and soft and cycling mobility have been carried out and highlighted by (i) the inter-municipal spatial strategic plan (main urban planning tool of the MCC) and (ii) by two agreements in the plan relating to the governance of mobility in Cagliari entitled 'Preliminary design of a metropolitan system in the metropolitan area of Cagliari' and the 'Integrated program for urban development and cycling, pedestrian and commuter in the metropolitan area of Cagliari' (Garau et al. 2016).

In this regard, the Agreement for the Development of the MCC, signed in 2016, identifies as priority interventions those aimed at promoting sustainable mobility. However, a sustainable mobility plan for the entire Metropolitan city has not yet been defined. The MCC rather drafted the PUMS guidelines (PUMS – Guidelines 2019) which 'constitute the strategic document having the function of systemising policies for mobility and interventions on infrastructures with the economic, social, urban planning and environmental protection strategies, in line with the provisions of the guidelines of the European Commission and Italian legislation' (PUMS – Guidelines 2019, p. 3).

Table 11.1 shows the state of the art relating to the drafting of PUMS in Italy.

In particular, the PUMSs classified as 'being drafted' are those for which the planning process has been officially launched through the drafting of the guidelines or publication of the notice for the assignment of the drafting task. MCC falls into this category.

As mentioned above, the MCC does not present a definitive PUMS that would include all the municipalities unified in 2016 (Regional Law No. 2 2016), but only the guidelines (PUMS – Guidelines 2019). These guidelines define, preliminarily, the priorities of the sustainable mobility system and the interventions on infrastructures with economic, social, urban planning and environmental protection strategies. These guidelines define

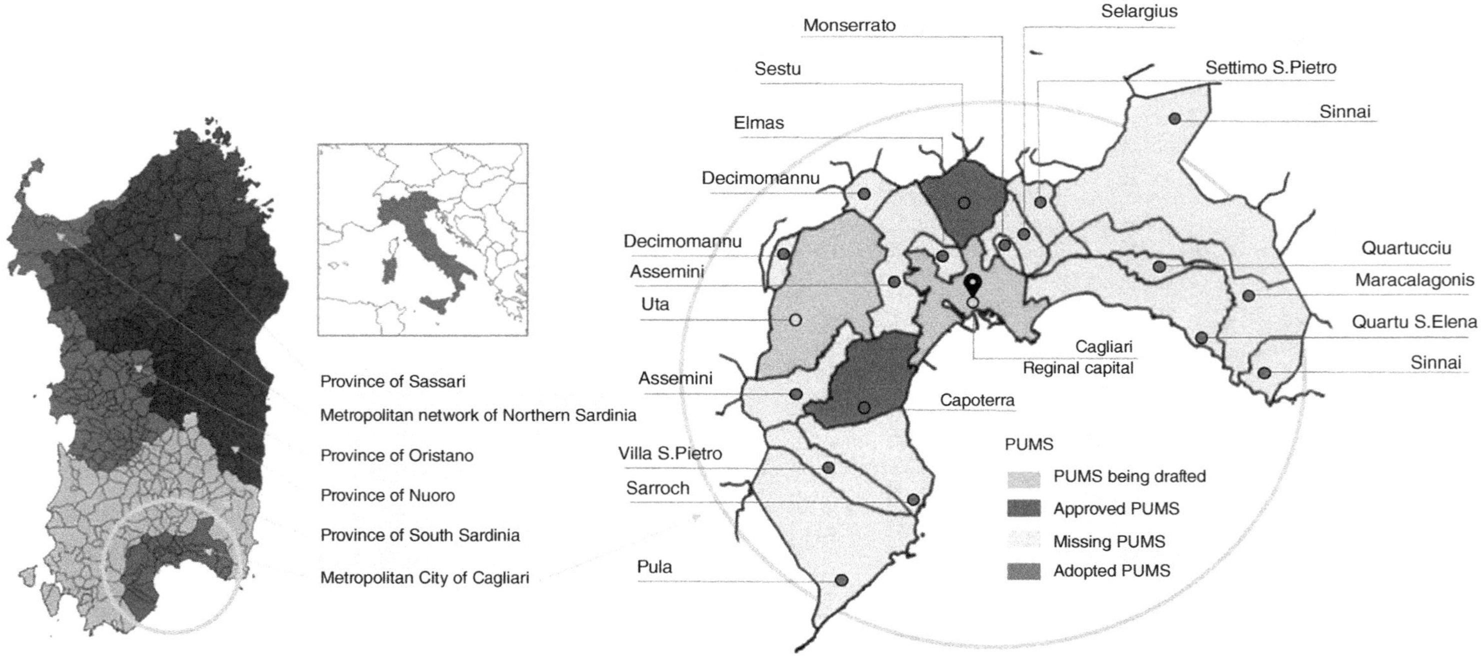

Figure 11.1 Approved, adopted and in-progress PUMS in the MCC. *Source:* Adapted from Cagliari Observatory (2021).

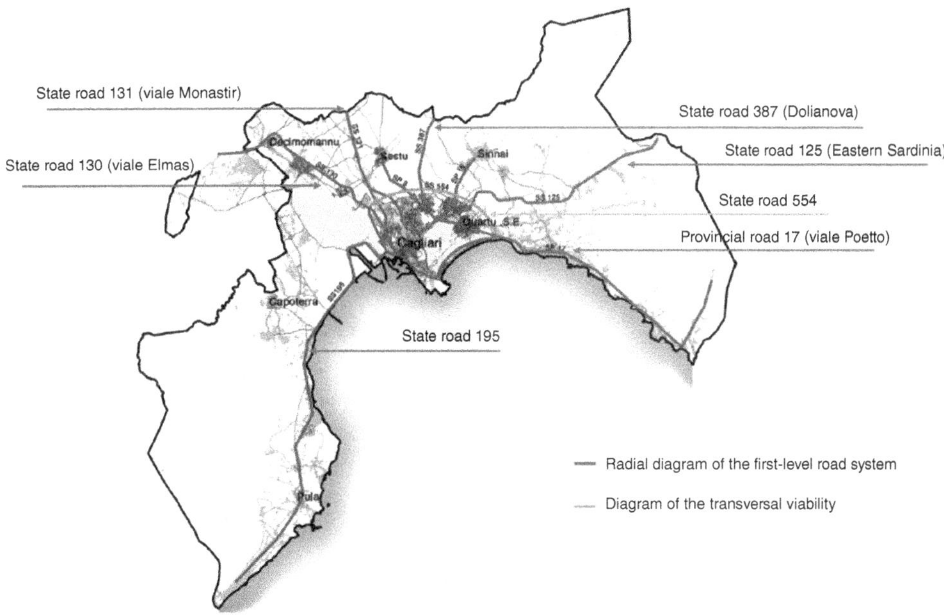

Figure 11.2 Viability scheme of the MCC. *Source:* PUMS – Guidelines (2019).

Table 11.1 The number of PUMS in Italy.

State of drafting of the PUMS in Italy	Numbers
Approved PUMS	46
Adopted PUMS	46
Being drafted PUMS	97
Total PUMS	189

Source: Author elaboration from National Observatory of PUMS (2021).

four macro-objectives (Table 11.2, column 1) with related areas of interest (Table 11.2, column 2) and 15 specific objectives (Table 11.2, column 3) that the metropolitan city should achieve within 10 years of the approval of the Plan.

In general, these objectives promote the mobility of residents and city users, favouring travel on foot, by bicycle, public transport or private transport with low environmental impact (PUMS – Guidelines 2019).

In order to achieve these objectives, the MCC guidelines identify various strategies and actions, on which the MCC will be able to plan, create and organise the infrastructures. These strategies are in line with the decree of 4 August 2017 of the IMIT, which contains the National PUMS's guidelines (Teknoring 2017), on which local people will have to define the criteria of sustainable mobility to access state funding. In fact, the decree identifies not only the procedure for the preparation and approval of PUMS but also some actions (Table 11.3, column 2) that contribute to the concrete implementation of good strategies (Table 11.3, column 1).

Table 11.2 Objectives outlined in the guidelines of the MCC's PUMS.

Macro-objectives	Areas of interest	Specific objectives
A. Effectiveness and efficiency of the mobility system	A1. Improvement of the LPT A2. Modal rebalancing of mobility A3. Congestion reduction A4. Improvement of the accessibility of people and goods A5. Improvement of the integration between the development of the mobility system and the structure and development of the territory A6. Improvement of the quality of street and urban space	1 Improving the attractiveness of collective transport 2 Improving the attractiveness of shared transport 3 Improving the economic performance of LPT 4 Improving the attractiveness of cycle–pedestrian transport 5 Reducing road congestion 6 Promoting the introduction of low-polluting means 7 Reducing irregular stops 8 Making urban logistics efficient
B. Energy and environmental sustainability	B1. Reduction in the consumption of traditional fuels other than alternative fuels B2. Improvement of air quality B3. Reduction in noise pollution B4. Pedestrianisation of historic centres and the most delicate areas (environmental islands, pedestrian traffic and smart parking)	9 Improving the energy and environmental performance of the passenger and freight vehicle fleet 10 Ensuring accessibility for people with reduced mobility 11 Ensuring mobility for people with low incomes 12 Ensuring mobility for older people 13 Improving the safety of vehicular traffic
C. Road mobility safety	C1. Reduction in road accidents C2. Significant decrease in the general number of accidents with deaths and injuries C3. Significant reduction in the social costs deriving from accidents C4. Significant decrease in the number of accidents with deaths and injuries among weak users	14 Improving the safety of pedestrians and cyclists 15 Increasing modal choice alternatives for citizens
D. Socio-economic sustainability	D1. Improvement of social inclusion D2. Increased citizenship satisfaction D3. Increase in the employment rate D4. Reduction in mobility costs (related to the need to use a private vehicle)	

Source: PUMS – Guidelines (2019).

Table 11.3 Strategies and actions of the IMIT and the guidelines of the MCC.

Strategies and actions of the SUMPs' guidelines (IMIT 2017)		Strategies and actions of the guidelines of the PUMS of MCC (PUMS – Guidelines 2019)	
Strategies	**Actions**	**Strategies**	**Actions**
1. Integration between transport systems, which also include rapid mass transport systems, where economically and financially sustainable	a. The redistribution of the transport network in a hierarchical and synergistic form and the recovery of shares of the road network and public spaces by integrating with new infrastructural interventions in favour of their better usability and sharing by pedestrians, cyclists, LPT users and private mobility to low environmental impact	Establishment of a control room for an integrated unitary system of public transport	• Planning of all public transport services that allow the most effective use of the various modes
	b. The identification of possible forms of integration between the transport systems through the proper functioning of the existing interchange nodes to ensure appropriate additions to the primary and secondary network		• Strengthening of policies and intermodal hubs
	c. Make it possible to transport bicycles on public transport vehicles, trains and ferries by appropriately adjusting the spaces		
	d. The development of tariff integration also providing for the transport of bicycles on public transport vehicles, trains and ferries		
	e. Use of ITS and info-mobility systems to facilitate the integration of transport systems, for the provision of data on the priority urban network and the development of innovative mobility services		
	f. Activities conducted by the administrative mobility manager in collaboration with the individual company mobility managers with the aim of encouraging sustainability	Establishment of a control room for an integrated unitary system of public transport	• Definition of quality standards and their monitoring
	g. Develop integrated demand management policies		

(*Continued*)

Table 11.3 (Continued)

Strategies and actions of the SUMPs' guidelines (IMIT 2017)		Strategies and actions of the guidelines of the PUMS of MCC (PUMS – Guidelines 2019)	
Strategies	**Actions**	**Strategies**	**Actions**
2. Development of collective mobility to improve the quality of service and increase the commercial speed of public transport means	a. The creation of preferential lanes or lanes reserved for collective transport (buses or trams), which, in addition to having a positive impact on commercial speed, improves the reliability of the passages, safety and quality of service		
	b. The implementation of traffic light systems enslaved and preferential to LPT		
	c. The provision of interventions, including on infrastructures, for the fluidification of public transport routes (such as intersections, junctions and itineraries functional to the rectification of routes)	Light Rail	• Completion of the planned metro network • Connection of poles to create a network • Enhancement of the public transport service for evening/night/festive mobility
	d. increase accessibility to LPT for passengers with reduced mobility, increasing the number of equipped cars and carrying out interventions on the sidewalks at the stops		
	e. Use of ITS by public transport operators, through the increase in the provision of vehicles with systems for real-time monitoring of location and service (operations centre, automatic vehicle monitoring [AVM] and -automatic vehicle location [AVL]) aimed to adapt service timetables to actual passenger demand and also to intervene in real time for changes to operating plans		
	f. The detection of the number of passengers on board through the installation of devices on the vehicles, with the launch of specific experiments for the use of mobile telephony		

	g. The widespread use of the various communication channels to users: information on board and at stops; informative websites; social networks such as Facebook and Twitter; mobile phones, by means of SMS alerts; smartphone applications; screens and speakers in stations and at stops and inside the cars; touch screens and computers in strategic locations such as hospitals, shopping malls and universities; and variable message panels		
	h. Actions to improve the quality of the public transport service	Reorganisation of the metropolitan road system	• Realisation of the road works necessary to make the metropolitan traffic congruent to the inspiring principles of the PUMS • Priority of road works and compatibility with the objectives of the PUMS
3. Development of pedestrian and cycling mobility systems in order to consider cycle and pedestrian movements as an integral and fundamental part of urban mobility and not as a residual share	a. The implementation of bike-sharing services also for tourists and occasional users b. The improvement of the conditions of use of bicycles through the creation of cycle routes c. The improvement of pedestrian and cycling connections to the main places of public interest (schools, public offices and primary services) – dedicated bike-sharing, road services and dedicated routes (from stations to destinations of public interest)	Development of pedestrian and cycling mobility systems	• Improve the infrastructure and services for the use of bicycles: upgrading parking areas, safe parking systems and comfortable rentals • Provide the metropolitan city with an adequate network of paths and cycle paths, promote widespread cycling and also providing for the extension and mending of cycle and pedestrian paths already created and/or designed by individual municipalities • Promote safe home–school and home–work routes
		Enhancement of public road transport	• A supporting framework of inter-urban and metropolitan corridors on rail and road on which to offer rapid mobility services (train, metro and bus rapid transit [BRT]) • a significant equipment of collective road and individual services to be activated on foot and by bicycle through direct, interconnected, safe and comfortable routes (paths and cycle paths) and on shared cars (possibly electric)

(*Continued*)

Table 11.3 (Continued)

Strategies and actions of the SUMPs' guidelines (IMIT 2017)		Strategies and actions of the guidelines of the PUMS of MCC (PUMS – Guidelines 2019)	
Strategies	**Actions**	**Strategies**	**Actions**
			• New connections for the polarities of large metropolitan services (work and health) • More services on high-demand lines • Increase of preferential lanes or lanes reserved for collective transport, which, in addition to having a positive impact on commercial speed, improve the reliability of the passages and safety and quality of service • Redevelopment of public transport stops also in the modal rail–road–bike interchange points • Construction of a new road transport connection network to connect the existing and/or planned nodes of the surface metro
	d. The adoption of design solutions for specific areas of particular interest and/or particularly problem (such as area 30) e. The dissemination of services for cyclists, such as repair and storage services, public pumps and the construction of bicycle parking spaces		
	f. Creation of home–school routes for bicycles and on foot and promotion of forms of collective pedestrian mobility g. The implementation of promotion, awareness and marketing actions h. The spread of automated systems, way-finding signs and devices to aid in the mobility of weak users (traffic lights with acoustic signals, slides, tactile paths, etc.)	Development of pedestrian and cycling mobility systems	• Promote safe home–school and home–work routes • Promote information and training campaigns in favour of walking and cycling activities

4. Introduction of shared motorised mobility systems, such as car sharing, bike sharing, van sharing and car pooling	a. Equipment at metro/train stations, main bus stops and dedicated parking interchange nodes for the purpose of developing shared mobility with a view to enhancing accessibility to the public transport system		
	b. Use of ITS and software platforms capable of managing shared private transport and integrating it with LPT		
	c. Promotion of shared mobility in companies and public bodies		
	d. tariff policies in favour of car sharing, motorcycle sharing and carpooling		
	e. transit and parking facilities for vehicles with shared mobility		
5. Renewal of the car fleet with the introduction of vehicles with low polluting impact and high energy efficiency	a. Actions to favour the development of shared mobility with low polluting impact		
	b. Introduction of low-polluting vehicles in public and private company fleets		
	c. Introduction of low-polluting-impact vehicles for the urban distribution of goods and/or cargo bikes		
	d. Introduction of low-polluting tourist vehicles		
	e. Installation of columns for electric recharging and systems for the distribution of alternative fuels with low polluting impact	Development of electric mobility as a contribution to air quality	• Promote and encourage the creation of public charging infrastructures for electric vehicles accessible to the public

(*Continued*)

Table 11.3 (Continued)

Strategies and actions of the SUMPs' guidelines (IMIT 2017)		Strategies and actions of the guidelines of the PUMS of MCC (PUMS – Guidelines 2019)	
Strategies	**Actions**	**Strategies**	**Actions**
	f. Monitoring the composition and average age of the fleet of the local public transport fleet g. Reward systems for cargo bikes and low-polluting tricycles and quadricycles		
6. Rationalisation of urban logistics in order to reconcile the supply needs of the goods necessary to increase the vitality of the economic and social fabric of urban centres	a. Development of new governance models for efficient, effective and sustainable urban logistics that allows to optimise the process of collection and distribution of goods in the urban environment, contributing to the reduction in traffic and pollution b. Introduction of a reward system for vehicles with less impact from the point of view of dimensions (vans) c. Adoption of a comprehensive and integrated regulatory system (goods and passengers) to be implemented also through tariff policies for access to loading/unloading vehicles (paid access, articulation of discounts and/or subscriptions) that rewards a last eco-sustainable mile d. Rationalisation of the areas for the loading and unloading of goods by promoting and supervising, including through the use of electronic and IT tools, networks of areas (stalls) for loading/unloading goods.	Rationalisation of urban logistics	• Promotion of sustainability and optimisation of freight logistics • Management of freight transport in the last km and the limited traffic zones with low-impact vehicles

7. Dissemination of the culture related to the safety of mobility, with actions aimed at reducing the risk of accidents and others whose purpose is to reduce exposure to risk; with actions to protect weak users and others that aim to mitigate the consequences of accidents. Dissemination of culture and training on sustainable mobility in order to foster greater awareness and modal shift, especially for future generations	a. Infrastructural interventions for solving problems in the most at-risk points of the road network b. Introduction in the urban environment, on an experimental basis, of assessments, controls and safety inspections c. Increasing the safety of pedestrians and cyclists and LPT users, e.g. by creating and protecting 'island' stops and sidewalks at the stops, through the creation of protected cycle lanes, flow separation interventions, horizontal and vertical signs and through protected pedestrian lanes and creation of protected pedestrian paths home–school d. Awareness campaigns and road education e. Information campaigns and involvement in sustainable mobility, also through specific and widespread interventions in schools	Dissemination of culture related to mobility safety and sustainable mobility	• Introduction in the urban environment, on an experimental basis, of assessments, controls and safety inspections • Increase the safety of pedestrians and cyclists and LPT users, e.g. with the creation and protection of 'island' stops and sidewalks at the stops, through the creation of protected cycle lanes, flow separation interventions, horizontal and vertical signs and through protected pedestrian lanes and creation of protected pedestrian paths at school • Promote awareness campaigns on road education • Information and involvement campaigns on sustainable mobility, also through specific and widespread interventions in schools.
		Urban development and attraction poles in line with its main network of public transport	• Coordination between urban planning and mobility planning in the 17 municipalities of the MCC • Harmonisation between urban planning choices and the public mobility system in the 17 municipalities of the MCC • Quality urban planning in spaces for non-motorised users in the 17 municipalities of the MCC

Among these, various Italian municipalities identify strategies and actions to achieve the goal of sustainable mobility, through the analysis of their needs, their characteristics and the governance choices contained in the most recent regional and provincial planning tools. The choice of strategies and applicative actions is a fundamental procedural step (Mistretta and Garau 2013) because from these derives not only the construction of the scenario and therefore of a clear identification of the objectives of the whole PUMS (Torrisi et al. 2021b) but also the evaluation of current and future scenarios resulting from the identification of strategies and implementation of actions (Torrisi et al. 2021a,b).

Thus, the MCC will have to present a PUMS with not only the strengths and criticalities, by examining the environmental, social and economic impact, but also the relationships between all municipalities of the MCC. Furthermore, the specific settlement context and its urban configuration projected on the sea will be essential to analyse, for proposing a new network by sea between the secondary ports as well as through the restoration of the navigable canals for both tourism and commercial purposes. All this is clearly with a particular focus on sustainable urban mobility, by considering both the accessibility of LPT, by road and rail, and the inter-modality, as an alternative collective transport service towards low environmental impact and high tourist value (PUMS – Guidelines 2019).

The strategies of the MCC's PUMS guidelines respect almost all the strategies defined by the IMIT. However, the actions identified for the implementation of these strategies are minimal, not linked to the concept of a maritime city and, above all, do not consider alternatives useful for defining sustainable urban mobility if the residents show needs different from those described in the same guidelines. The next section analyses these assumptions in detail.

11.3 Results and Discussion

The comparative analysis above has outlined policies on sustainable urban mobility that the MCC intends to undertake once the SUMP has been drafted and implemented, in relation to the guidelines provided by the IMIT in 2017. In general, the analysis shows that almost all the strategies dictated by the IMIT have been included in the guidelines of the PUMS of MCC, even if not all the implementation actions were deemed necessary (Table 11.3). If the guidelines do not correctly foresee all the sustainable outcomes for mobility, there is a risk of having a plan that is not completely capable of satisfying the needs of the metropolitan city.

Only strategy No. 4 'Introduction of shared motorized mobility systems, such as car-sharing, bike-sharing, vans-haring, car-pooling' was not included in the guidelines' strategies, but is inserted on actions included in different strategies. The lack of attention towards the use of ITS and software platforms capable of managing shared private transport is also evident, although the macro-objectives (Table 11.2) are linked to improving the attractiveness of collective and shared transport. Nevertheless, the MCC's guidelines highlighted strong attention to the link between urban planning and mobility planning which, in the IMIT's strategies, is not present (last row, Table 11.3). This lack of IMIT was, however, filled with the second draft of the new guidelines realised in 2019, in which they place strong emphasis on the establishment of an interdisciplinary working group (urban planners, transporters, economists, environmentalists, etc.). This particular attention is due to the main problem of the MCC which, before 2016, the year of its establishment, had a particular settlement system of the 17 municipalities which was characterised by the absence of sharing planning and governance processes. In this sense, it appears even more important to prepare an initial cognitive framework in order to be able to launch the plan that certainly takes into account both urban planning and transport aspects and the MCC's guidelines found in the SUMP, a possible strategic tool that would define shared management of a capillary network among the 17 municipalities of the MCC.

In particular, the analysis frames some problems in the identification and formulation of the PUMS guidelines of the MCC, linked not only to the limited choice of strategies and actions compared to the guidelines of the IMIT (Table 11.3) but also to the same objectives that the PUMS of the MCC has given itself (Table 11.2). These problems can be summarised in (i) (in general) no comparison of the PUMS of the MCC with the new IMIT guidelines of 2019; (ii) little attention to the real needs of society; (iii) little attention to innovative and management systems; and (iv) little attention to the sea.

In general, the MCC's guidelines do not provide useful solutions to the needs of population. To integrate this lack, sustainable alternatives chosen on the basis of a collective process with a central role of stakeholders could be envisaged as an action. This is highlighted both in the European guidelines and in those of IMIT. These alternatives should already be present in the SUMP during the approval process and should facilitate: (i) urban redevelopment/regeneration processes; (ii) the increasing of the resilience of territories through infrastructures (including the green ones), aimed at safeguarding natural and human capital; and (iii) the creation of ecological-functional connections. In these processes, integrated systems for transport and logistics are essential for the development and efficiency of the sustainable mobility system.

The lack of attention to the real needs of society is highlighted by not including within the guidelines of the MCC's PUMS: (i) some implementative actions to link accessibility to LPT for passengers with reduced mobility (second action of the IMIT); (ii) the adoption of design solutions for specific areas (third action of the IMIT); (iii) the provision of shared motorised mobility systems and the intermodal tariff system (actions 4a, c, d and e of the IMIT); and (iv) urban logistics solutions (action 6c of the IMIT).

Furthermore, the MCC's guidelines show little attention to innovative and management systems. There is a complete lack of actions related to info-mobility systems (IMIT action 1e), real-time monitoring of the location and service (IMIT actions 2e–f and 4b), the widespread use of the various communication channels to users such as Facebook and Twitter (IMIT action 2g), the spread of automated systems (IMIT action 3h) and the renewal of the park with the introduction of vehicles with low polluting impact and high energy efficiency (IMIT strategy 5). Instead, development strategies for info-mobility services could be inserted in order to increase the use of intelligent transport systems (ITS) applications throughout the metropolitan area, e.g. through info-mobility systems. In fact, ITS not only improve the overall efficiency of the mobility system but also significantly reduce the costs of mobility without penalising accessibility (Coni et al. 2018; Pinna et al. 2020).

Finally, the authors of this Chapter believe that one of the most serious limitations of the analysed guidelines is the absence of the water–territory relationship [3] which instead should characterise the PUMS because the MCC develops, in large part, on the coast. In fact, the guidelines of the MCC hypothesise connections between its municipalities through a new network by sea between the secondary ports (PUMS – Guidelines 2019), and the strategies described do not expose implementation actions related to this context.

As other Italian maritime cities did (PUMS of Genoa 2019; PUMS of La Spezia 2018), a more precise improvement of accessibility by sea could be considered, also through the introduction of a limited traffic zone or pedestrian areas, as part of a more general strategy of environmental sustainability for preserving the attractiveness of the city, the access of residents/traders and the local economy. Therefore, actions could be defined, in accordance with the IMIT strategies 1c and 1d, which not only promote boat sharing, in the context of mobility and the sharing economy, but which define an organisational structure capable of efficiently and effectively managing entire process of maritime mobility.

11.4 Conclusions

This chapter has provided a critical framework on transport policies related to the concept of smart urbanism. The SUMPs or PUMS, as strategic tools for implementing these policies, are studied by focussing on the case study of the MCC. This has highlighted the limitations in its implementative actions and led the authors to understand what to implement in the final drafting of the MCC's PUMS.

Although the strategies of the MCC are characterised by an integrated approach between urban and transport planning on several levels, targeted actions are lacking, especially in relation to the needs of the population, integration of ITS and the relationship between city and sea. In this regard, the authors highlighted the need to implement the strategies adopted by proposing implementative actions aimed at a governance more focussed on (i) the needs of the population, (ii) the use of technologies and (iii) the relationship with water.

Furthermore, the case study leads to make some considerations regarding the relationship between smart urbanism and smart urban mobility. First, this study allows us to define guidelines for key points on the choice of actions to be adopted in a coastal metropolitan city, projected on the water: (i) the force that maritime mobility can have towards the conurbation of the metropolitan city makes possible to modify all the mobility of the same metropolitan city; (ii) the use of ITS appears essential to facilitate social and economic development.

Acknowledgements

This study was supported by the Ministry of Education and Merit (MIUR) through two projects: (i) 'WEAKI TRANSIT': WEAK-demand areas Innovative TRANsport Shared services for Italian Towns (Project protocol: 20174ARRHT_004; CUP Code: F74I19001290001), financed with the PRIN 2017 (Research Projects of National Relevance) programme; and (ii) 'CAGLIARI2020' (Project code: PON04a2_00381). We authorise the MIUR to reproduce and distribute reprints for governmental purposes, notwithstanding any copyright notations thereon. Any opinions, findings and conclusions or recommendations expressed in this material are those of the authors and do not necessarily reflect the views of the MIUR. This study was also supported by the project 'SUMPS and technologies, ITS systems, strategies and actions to promote sustainability in the key perspective of Smart Urbanism' and by the project 'Smart PlANning in Sardinia: place-oriented policies and technologies concerning urban environments – SPLAN', founded by the Foundation of Sardinia and Autonomous Region of Sardinia (Fondazione di Sardegna).

References

Azzari, M., Garau, C., Nesi, P. et al. (2018). Smart city governance strategies to better move towards a smart urbanism. Gervasi, O., Murgante, B., and Misra S. et al., ed. *Computational Science and Its Applications – ICCSA 2018*, Melbourne, Australia (2–5 July 2018). Springer International Publishing. https://doi.org/10.1007/978-3-319-95168-3_43.

Bibri, S.E. (2019). On the sustainability of smart and smarter cities in the era of big data: an interdisciplinary and transdisciplinary literature review. *Journal of Big Data* 6 (25): https://doi.org/10.1186/s40537-019-0182-7.

Bibri, S.I. (2020). Compact urbanism and the synergic potential of its integration with data-driven smart urbanism: an extensive interdisciplinary literature review. *Land Use Policy* 97: 104703. https://doi.org/10.1016/j.landusepol.2020.104703.

Caprotti, F. and Cowley, R. (2019). Varieties of smart urbanism in the UK: discursive logics, the state and local urban context. *Transactions of the Istitute of British Geographers* 44: 587–601. 10.1111/tran.12284.

Coni, M., Garau, C., and Pinna, F. (2018). How has Cagliari changed its citizens in smart citizens? Exploring the influence of ITS technology on urban social interactions. Gervasi, O. et al., ed. *Computational Science and Its Applications – ICCSA 2018*, Melbourne, Australia (2–5 July 2018). Springer International Publishing. https://doi.org/10.1007/978-3-319-95168-3_39

European Union (2020). What are smart cities?. https://ec.europa.eu/info/eu-regional-and-urban-development/topics/cities-and-urban-development/city-initiatives/smart-cities_en#what-are-smart-cities (accessed 26 April 2021).

Garau, C., Masala, F., and Pinna, F. (2016). Cagliari and smart urban mobility: analysis and comparison. *Cities* 56: 35–46. 10.1016/j.cities.2016.02.012.

Guidelines for Urban Sustainable Mobility Plans (2017). Individuazione delle linee guida per i Piani Urbani di Mobilita' Sostenibile. https://www.gazzettaufficiale.it/eli/id/2017/10/05/17A06675/sg (accessed 11 July 2023).

Heaton, J. and Parlikad, A.K. (2019). A conceptual framework for the alignment of infrastructure assets to citizen requirements within a smart cities framework. *Cities* 90: 32–41. 10.1016/j.cities.2019.01.041.

IMIT (2017). Guida alla Produzione 2017. https://www.imit.it/doc/GUIDA%20PRODUZIONE%202017%20ITA%20MAIL.pdf (accessed 11 July 2023).

Jordan, D.R. and Horan, T.A. (1997). Intelligent transportation systems and sustainable communities: findings of a national study. *Journal of the Transportation Research Board* 1588 (1): 70–76. 10.3141/1588-09.

Linee guida piani urbani di mobilità sostenibile (2017). Gazzetta Ufficiale della Repubblica Italiana, D.M. 04.08.2017. http://www.regioni.it/news/2017/10/06/d-m-04-08-2017-linee-guida-piani-urbani-di-mobilita-sostenibile-533192/ (accessed 26 April 2021).

Minister of Economy and Finance (2017). Connettere l'Italia: fabbisogni e progetti di infrastrutture. https://www.mit.gov.it/sites/default/files/media/notizia/2017-04/Allegato_MIT_AL_DEF_2017.pdf (accessed 26 April 2021).

Mistretta, P. and Garau, C. (2013). *Città e sfide. Conflitti e Utopie. Strategie di impresa e Politiche del territorio. Successi e criticità dei modelli di governance.* Cagliari: CUEC.

Murgante, B., Botonico, G., Graziadei, A. et al. (2019). Innovation, technologies, participation: new paradigms towards a 2.0 citizenship. *International Journal of Electronic Governance* 11 (1): 62–88. https://doi.org/10.1504/IJEG.2019.098814.

National Observatory of PUMS (2021). Osservatorio PUMS. https://www.osservatoriopums.it/ (accessed 11 July 2023).

OECD (2013). Working together for local integration of migrants and refugees in Barcelona. https://www.oecd-ilibrary.org/social-issues-migration-health/working-together-for-local-integration-of-migrants-and-refugees-in-barcelona/key-data-on-migrant-presence-and-integration-in-barcelona_9789264304062-3-en (accessed 26 April 2021).

Osservatorio (2021). I PUMS in Italia: stato dell'arte. https://www.osservatoriopums.it/osservatorio/pums#capoterra (accessed 26 April 2021).

Osservatorio Cagliari (2021). I PUMS in Italia: stato dell'arte. https://www.osservatoriopums.it/cagliari (accessed 26 April 2021).

Pinna, F., Garau, C., Maltinti, F. et al. (2020). Beyond architectural barriers: building a bridge between disability and universal design. Gervasi, O., Murgante, B., and Misra S. et al., ed. *Computational Science and Its Applications – ICCSA 2020*, Cagliari, Italy (1–4 July 2020). Springer International Publishing. https://doi.org/10.1007/978-3-030-58820-5_51.

PUMS – Guidelines (2019). Patto per lo sviluppo della Città Metropolitana di Cagliari. http://www.cittametropolitanacagliari.it/documents/36143/0/PUMS+Linee+Indirizzo.pdf/cfd6b031-a7f9-01e3-3410-01cfb3af81ae (accessed 26 April 2021).

PUMS of Genoa (2019). Piano urbano della Mobilità Sostenibile della Città Metropolitana di Genova. https://pums.cittametropolitana.genova.it/sites/default/files/Allegato_I%20-%20PUMS.pdf (accessed 29 April 2021).

PUMS of La Spezia (2018). Piano Urbano della Mobilità Sostenibile e Piano Urbano del Traffico della Città della Spezia. http://www.comune.laspezia.it/export/sites/SPEZIAnet/Aree_tematiche/Mobilita_Traffico/Documenti/PUMS_PUT-CC-01.08.2018.pdf (accessed 29 April 2021)

Rapporto di Analisi del Territorio (2020). La Città Metropolitana di Cagliari. https://www.cittametropolitanacagliari.it/documents/36143/0/Report+Analisi+del+Territorio.pdf/3b451db6-3e69-c6a4-9f50-87abd64c1e94 (accessed 26 April 2021).

Regional Law No. 2 (2016). Riordino del sistema delle Autonomie Locali della Sardegna. https://www.regione.sardegna.it/j/v/80?s=300929&v=2&c=13906&t=1 (accessed 28 April 2021).

Rupprecht, S., Brand, L., Baedeker, S.B. et al. (2019). Guidelines for developing and implementing a sustainable urban mobility plan second edition. https://www.eltis.org/sites/default/files/sump_guidelines_2019_interactive_document_1.pdf (accessed 26 April 2021).

Sustainable Urban Mobility Plans (2021). European commission clean transport, urban transport, mobility and transport. https://ec.europa.eu/transport/themes/urban/urban-mobility/urban-mobility-actions/sustainable-urban_en (accessed 6 April 2021).

Teknoring (2017). Linee guida per i Piani urbani di mobilità sostenibile (Pums). https://www.teknoring.com/news/trasporti-e-automotive/linee-guida-per-i-piani-urbani-di-mobilita-sostenibile-pums/ (accessed 26 April 2021).

The SUMP Concept (2020). The Urban Mobility Observatory, Eltis. https://www.eltis.org/mobility-plans/sump-concept (accessed 26 April 2021).

Tiboni, M., Rossetti, S., Vetturi, D. et al. (2021). Urban policies and planning approaches for a safer and climate friendlier mobility in cities: strategies, initiatives and some analysis. *Sustainability* 13: 1778. https://doi.org/10.3390/su13041778.

Torrisi, V., Garau, C., Inturri, G. et al. (2021a). Strategies and actions towards sustainability: encouraging good ITS practices in the SUMP vision. Simos, T., Kalagiratou, Z., and Monovasilis, T., ed. *International Conference of Computational Methods in Sciences and Engineering – ICCMSE 2020*, Crete, Greece (29 April–3 May 2020). AIP Conference Proceedings. https://doi.org/10.1007/978-3-030-58820-5_45.

Torrisi, V., Inturri, G., and Ignaccolo, M. (2021b). Introducing a mobility on demand system beyond COVID-19: evidences from users perspective. Simos, T., Kalagiratou, Z., and Monovasilis, T., ed. *International Conference of Computational Methods in Sciences and Engineering – ICCMSE 2020*, Crete, Greece (29 April–3 May 2020). AIP Conference Proceedings. https://doi.org/10.1063/5.0047889.

United Nations (2018). Department of economic and social affairs population division. https://www.un.org/development/desa/en/news/population/2018-revision-of-world-urbanization-prospects.html (accessed 26 April 2021).

Urban Mobility Package (2021). European commission, mobility and transport. https://ec.europa.eu/commission/presscorner/detail/en/MEMO_13_1160 (accessed 26 April 2021).

World Urbanization Prospects (2018). United Nations New York, department of economic and social affairs population division. ST/ESA/SER.A/420. https://population.un.org/wup/Publications/Files/WUP2018-Report.pdf (accessed 26 April 2021).

Yigitcanlar, T. (2018). Smart city policies revisited: considerations for a truly smart and sustainable urbanism practice. JAKO201809242559537.pdf (koreascience.or.kr) (accessed 26 April 2021).

Zawieska, A. and Pieriegud, J. (2018). Smart city as a tool for sustainable mobility and transport decarbonisation. *Transport Policy* 63: 39–50. https://doi.org/10.1016/j.tranpol.2017.11.004.

12

Analysing India's Smart Cities Mission from a Sustainability Perspective

Sarbeswar Praharaj

Department of Geography and Sustainable Development, University of Miami, FL, USA

12.1 Introduction

The urban landscape in the world's largest democracy and developing major economy – India – is changing dramatically, with the urban population increasing from 18% in 1961 to 31.2% by 2011 (Census of India 2011). By 2030, 590 million people will live in India's cities, with a net increase of 270 million working-age population (Sankhe et al. 2010). There will be 68 million-plus cities in the country by 2030, whereas all of Europe currently has 35 such urban agglomerations. Urban policymakers in India have somewhat failed to match the growing complexities of urban problems with adequate planning, governing instruments and physical infrastructure. The United Nations (UNI) placed India at 131st position among 182 countries according to a ranking of citizens' quality of life (UNDP 2020). In its pollution-level assessment, the World Health Organization (WHO) announced that half of the world's 20 most polluted cities are in India (Yenneti et al. 2019). McKinsey & Company rated the performance of Indian cities in the areas of urban planning, finance and governance as inferior – the lowest rating on a 4-point scale (Sankhe et al. 2010). Such poor standing across global benchmarks has somewhat forced the Government of India (GoI) to lay out a new strategy for local governments to think innovatively and transform service delivery strategies to avoid urban decay and ensure the competitiveness of cities, leading to the launch of Smart Cities Mission (SCM) in 2015.

The SCM is one of the world's most extensive smart city programmes engaging 100 cities and a US$14 billion investment. The GoI outlines that the programme aims to drive economic growth and improve the quality of life for people in 100 selected cities by enabling local development and harnessing technology to create smart solutions for citizens (Government of India 2015). The smart cities mission statement and guidelines (Government of India 2015) issued by GoI state that the concept of smart cities, relying on the foundations of urban innovation, creative policies and Information and Communication Technologies (ICT), attracts significant attention from the political class and provides policymakers with new technologies to address the rapid urbanisation challenges. The national

Smart Cities, Energy and Climate: Governing Cities for a Low-Carbon Future, First Edition.
Edited by Oleg Golubchikov and Komali Yenneti.

smart cities guideline encourages cities to focus both on area-based development (ABD) and pan-city infrastructure upgradation. The ABD approach is aimed at intensive use of ICT to develop selected parts of the city as 'smart precincts' or, more critically, referred to as 'smart enclaves' (Hollands 2008). In contrast, pan-city infrastructure is proposed to deliver smart solutions to city-wide challenges. The GoI's smart cities vision statement (Government of India 2015) asserts that digital technologies, modern infrastructure and a new form of urban governance will transform cities across the country, making them citizen friendly and sustainable.

This chapter queries whether smart cities by default lead towards sustainable development? It adopts a case study approach providing an on-ground picture to illustrate the widening gap between the promises and realities of the smart city. A classical sustainability assessment model is developed to measure the social, economic, environmental and institutional outcomes of smart city investments in Indian cities (Figure 12.1). The framework provides an opportunity to address existing gaps in urban studies where several researchers (Hollands 2008; Praharaj et al. 2018b; Yigitcanlar et al. 2019) highlight the need for critical analysis examining to what degree the increasing investment in urban technology is justified to address socio-economic inequalities and governance systems transformation. Such investigation is crucial in the global south, where the capacities and resources for urban development are limited (Praharaj et al. 2018c).

This chapter is laid out in five sections, beginning with this introduction. The second and third sections present an overview of the smart cities' mission and uncover the different development models adopted by Indian cities. The fourth section provides a critical analysis of the impact of smart city interventions on the four pillars of sustainability (social, economic, environmental and institutional). The concluding section highlights key lessons and provides recommendations for ensuring sustainable outcomes from smart city ventures.

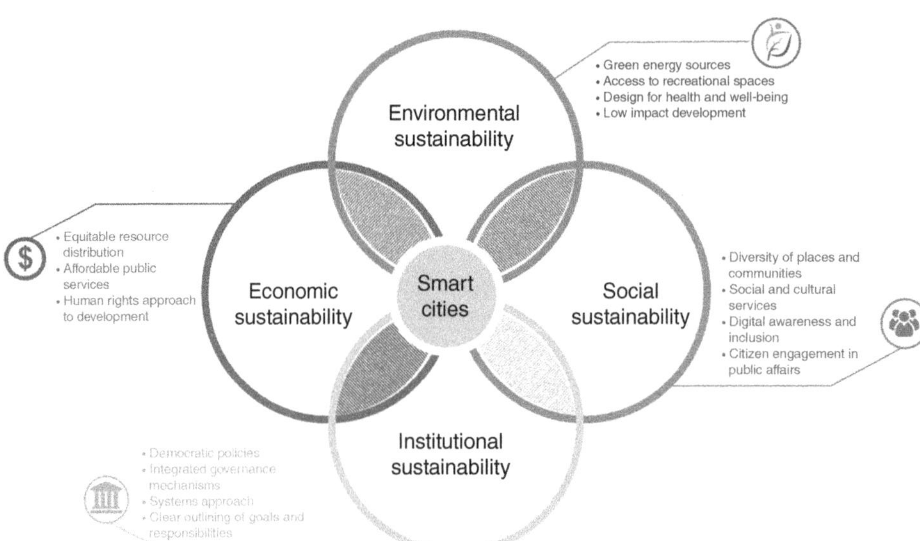

Figure 12.1 Sustainability assessment framework for smart cities.

12.2 Overview of the Smart Cities Mission in India

The initial idea of building smart cities in India was expressed in the pre-election manifesto of the Bhartiya Janta Party, led by India's current Prime Minister Narendra Modi, to address urbanisation challenges. The declaration stated, 'We will initiate building 100 new cities; enabled with the latest in technology and infrastructure – adhering to concepts like sustainability, walk to work, etc. and focused on specialized domains' (Bharatiya Janata Party 2014). After assuming office, the new government rolled out various urban policies to tackle growing issues in cities across the country. A close review of the SCM indicates that the initial idea of building smart cities evolved and, in fact, changed drastically when the newly elected government in August 2014 asked the state governments to identify existing towns for development under the SCM. This marked a shift in focus from greenfield to brownfield development and from the construction of '100 new smart cities' to 'making existing cities smart' (Bhattacharya et al. 2015). These discussions indicate that although the SCM appears as a strategic response to India's urbanisation challenges, it has a political angle to its genesis.

12.2.1 Smart City Selection Process

The distribution of cities under the 100 smart cities mission was based on the share of the urban population and the number of urban centres across the states in India. While the central government identified the overall number of proposed smart cities from each state, specific towns were left for the states to decide. Each state carried out a competitive merit criterion-based selection process to identify and nominate cities for participation in the mission. The state-level High Powered Steering Committee shortlisted the highest-scoring cities based on pre-established criteria. The winning cities in each state were added to the recommended shortlist of smart cities by the national government, which subsequently announced the list of 100 Smart Cities.

The GoI, in partnership with Bloomberg Philanthropies, conducted a Smart Cities Challenge to evaluate the capacity of cities and the robustness of the smart city proposals to prioritise investment in various phases. In the first phase of the competition, 20 cities were selected, followed by 13 cities in phase 2. A total of 50 cities were chosen for the implementation of smart city projects in phases 3 and 4. The rest of the ten cities received funding in the last phase of the SCM. An evaluation criterion was developed for the competitive selection of cities in the various rounds, which includes a 30% weighting on city-level assessment and 70% weighting on smart city proposal-level evaluation. The city-level assessment examined two aspects. The credibility and performance of the local government in the preceding three years (including operational efficiency improvement in transport scenario, administrative efficiency and user charges collection ratio) were considered. The city's vision and strategy, including clarity in overall future vision, articulation of how ICTs will be used and the potential impact of the vision on the economy, sustainability and inclusiveness, was also assessed.

The smart city proposal-level evaluation assessed the impact of the plan, its cost-effectiveness, innovation and scalability issues, and the process followed in drafting the plan. The plan's impact was measured through the depth and scale of citizen participation while identifying the development objectives and projects and the level of inclusiveness of the plan with common citizen aspirations, especially of the poor and disadvantaged. The plan's cost-effectiveness was assessed by evaluating the convergence of smart cities' projects with existing infrastructure initiatives and the scope for a diverse and reliable source of finance. Innovation and scalability of proposals were assessed by measuring their adaptiveness to global best practices, the number of smart solutions proposed and the scope for replicability of the solutions in other parts of the city and outside them. The extent of local consultation at each stage of the plan development, political participation and the intensity of social media outreach is considered for evaluating the robustness of the Smart City Plan development process. The evaluation was undertaken by a committee of national and international experts and agencies, including the World Bank, Asian Development Bank, London School of Economics and National Institute of Urban Affairs.

12.2.2 Smart City Implementation and Governance Framework

Cities under the SCM have set up special purpose vehicles (SPVs) as a nodal implementing agency for smart city projects. The SCM Guidelines cite the lack of resources and capacity of municipal corporations to raise revenue and manage large-scale development projects as justifying the move. These SPVs are conceived as independent and autonomous companies with 50:50 shareholding between state governments and municipal corporations. The argument behind having SPVs instead of municipal corporations for project planning and execution is that it will create a high-performance urban system and bring agility to strategic decision-making. The SPVs are conceived as a corporate body (functioning under the Companies Act 2013) that can take up processes and execute mechanisms that municipal governments are not empowered to perform as per prevailing laws and regulations. Local corporations require special approval from the state government to undertake activities such as raising large loans, entering joint venture arrangements with private organisations and selling off assets. However, such actions can be unilaterally performed by SPVs.

The GoI also approved partaking by private businesses in owning shares and management of SPVs. The smart city SPVs are headed by a chief executive officer (CEO) supported by a directorial board with representation from the central government, state government and local-level public utility-providing agencies. CEOs and smart city SPVs are entrusted with planning, appraisal, approval, releasing of funds, implementation, management, operation, monitoring and evaluating smart city projects.

12.2.3 Smart City Financing Mechanism

The national and state government provides US$140 million in funding to each smart city SPV, as an equal contribution of US$70 million each. The SPV companies have to raise additional funds from the financial market as debt or equity. Additionally, the municipal corporation provides funding to the smart city projects through convergence with other schemes. The mission guidelines make it mandatory for an equal amount of investment to be contributed jointly by the state and respective cities to be eligible to receive funds from the national government.

Cities were suggested to actively explore private sector funding via public–private partnership opportunities in developing and operating infrastructure to raise matching funds. Alternatively, they could seek convergence of smart city plans with other GoI-sponsored schemes such as Swachh Bharat Mission, Atal Mission for Rejuvenation and Urban Transformation (AMRUT) and Heritage City Development and Augmentation Yojana (HRIDAY) to bring in linkage grants. Revenues from the states and/or local authority resources through the collection of taxes, user fees, development charges and land monetisation are also being channelled into smart city development. The GoI also recommends that cities improve their credit rating to leverage borrowings from global financial institutions, including bilateral and multilateral institutions, to bridge the smart city funding gap.

12.3 Untangling the Indian Smart City Models

A review of the SCM guidelines (Government of India 2015) and smart city proposals (available at https://smartnet.niua.org) through this study reveal that GoI encouraged cities to adopt a 'smart city model' that focuses both on area-based development and pan-city infrastructure. The area-based development (ABD) approach (Praharaj and Han 2019b) aims at intensive use of ICT to develop certain parts of the city as 'smart precincts'. The GoI asserted that area-based development vis-a-vis new technologies acts as agents of transformation of existing areas through retrofitting or redevelopment strategies to improve the livability of urban precincts. These select areas will then act as lighthouses for other parts of the city as a best practice model to be replicated within and across regions. Overall, the national programme funds one ABD in each city that would showcase 100 smart precincts across the 100 cities. In contrast, pan-city development envisages improvements in the delivery and management of existing infrastructure or construction of new facilities to have a city-wide impact. The national government proposed that cities must leverage technology, information and data to create smart solutions to urban problems through the pan-city infrastructure projects. The SCM website (see https://smartcities.gov.in) identifies several solutions, including intelligent traffic management systems, smart water metering, video crime monitoring, smart parking, telemedicine, smart energy and sensor-based solid waste management.

The SCM outlines three strategic components of area-based smart cities development – city improvement (retrofitting), city renewal (redevelopment) and city extension (greenfield development). Retrofitting techniques are proposed for existing built-up areas to achieve the smart city objectives by enhancing the efficiency and livability neighbourhood. The size of the selected area shall be 500 acres or more for retrofitting model implementation. Rather than suggesting an enhanced land consolidation strategy, the guidelines expect existing structures to remain intact and call for some handpicked smart solutions injected into the identified area via the retrofitting approach. Examples of retrofitting projects include installing Wi-Fi hotspots, energy-efficient LED street lighting, CCTV surveillance, smart water metering and intelligent traffic and parking management solutions.

Redevelopment models will affect a more significant makeover of the existing built-up environment with new layouts, enhanced infrastructure, mixed land use and increased density strategies. As per the SCM guidelines, a redevelopment project shall cover more than 50 acres and emphasise a higher floor space index (FSI), high ground coverage over the selected land parcel and release more green spaces. The vision of this development model appears to have been influenced by the formerly town-planning schemes of Gujarat and Maharashtra, considered the best available land development models in India. Smart city redevelopment precincts engage in projects such as road widening, pedestrian and bike infrastructure development, design–implementation of waterfront public spaces and high-intensity mixed-land use development around public transit stations.

Greenfield development seeks to introduce advanced infrastructure solutions in a greenfield area (more than 250 acres) using modern planning strategies, green technologies and innovative financing and implementation tools. The MoHUA explicitly suggests that greenfield developments should address the needs of the expanding population in urban peripheries to tackle future growth while also providing high-end infrastructure for attracting businesses. While the retrofitting and redevelopment strategies are implemented within the limits of the core city area, greenfield development ideally takes place beyond them but within the jurisdiction of the metropolitan authority. Key features and projects of greenfield city extension are green buildings, adequate provision for affordable housing, quality health and recreation facilities, trade facilitation and startup incubation centres (Figure 12.2).

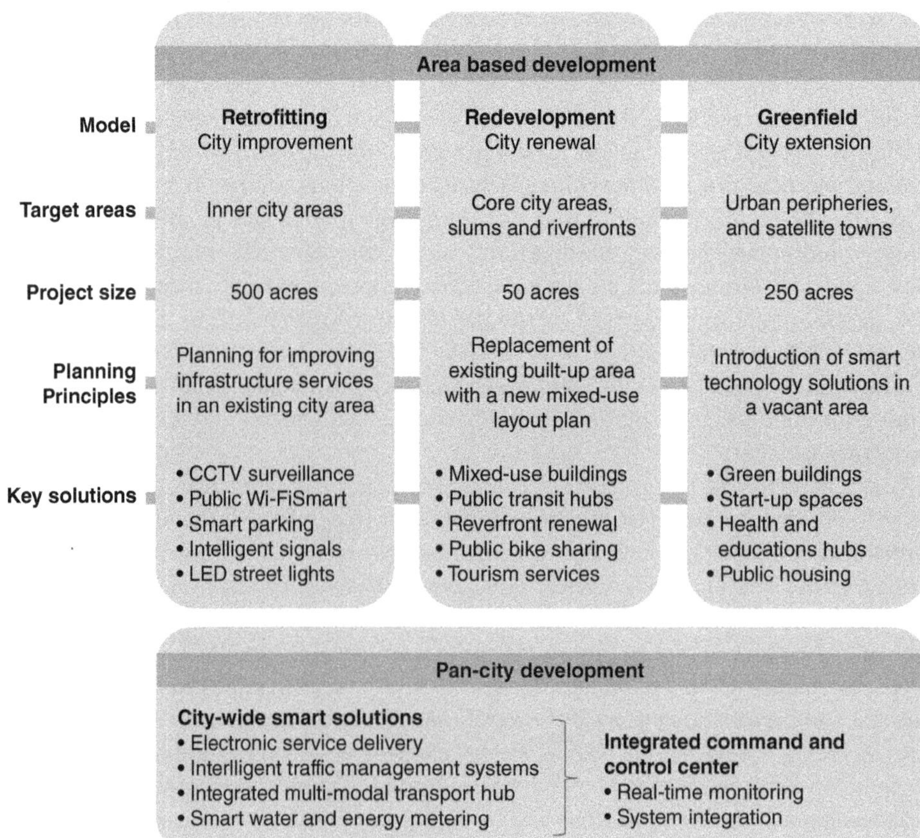

Figure 12.2 Summary of the Indian smart cities' development model.

The smart cities in India have set out an ambitious plan to develop integrated command and control centres (ICCCs) for building synergy between various projects and monitoring the performance of urban services and infrastructure (Praharaj 2020). The ICCC is being developed in 89 of the 100 smart cities as critical pan-city infrastructure. These centralised urban monitoring systems connect, integrate and analyse information streamed by the sensors and digitally instrumented devices plotted across the city, somewhat embracing the model of data-driven networked urbanism (Townsend 2013). The command centre projects are based on the premise that rich seams of data can be used to capture, model and predict urban processes (Lock et al. 2019) and guide the design of future urban policies and interventions. Moreover, they act as a systems integrator to increase interoperability between various projects and solutions developed by different vendors within the ABD and city-wide projects and offer a platform for multi-agency collaboration (Leao et al. 2021) for planning and designing rapid responses to emergencies.

12.4 Sustainability Assessment of Indian Smart Cities

12.4.1 Social Sustainability

The definition of 'social sustainability' may vary across communities (Vallance et al. 2011), but researchers overwhelmingly agree that it is the most-neglected and least-researched component of sustainability (Kandachar 2014). Social sustainability is 'a process for creating sustainable, successful places that promote wellbeing, by understanding what people need from the places they live and work. Social sustainability combines the design of the physical realm with the design of the social world – infrastructure to support social and cultural life, social amenities, systems for citizen engagement, and space for people and places to evolve' (Woodcraft et al. 2011). Socially sustainable communities are equitable, diverse, connected and democratic and provide a good quality of life (Adams 2006). This section examines the smart cities' development process and outcomes in India, measuring to what extent they are socially sustainable. In doing so, we accessed the smart city policy document for each of the 100 cities under the SCM available through https://smartnet.niua.org. A content analysis of these proposals was undertaken to identify the smart city vision and processes employed for the plan preparation by the cities. The proposals were also scanned for identifying the area and population coverage of the ABDs across cities that allow measuring the potential social outcomes.

As outlined in Section 12.2, each designated smart city in India selected a particular area for ABD project development. Hence, the contentious issue in the entire development process was as follows: which area of the city is chosen and how the authorities made the selection? An analysis of 100 smart city proposals in this research suggests that cities primarily relied on voting through online portals, social media and community meetings to prioritise and identify locations for ABD. A review of the smart city proposal of Bhubaneswar – the first-ranked smart city by the GoI – reveals that 26% of the urban population voted in the consultation process, and a substantial number of votes were recorded on the government website and through social media drives. A relatively smaller town in the Himalayan region of India, Agartala could only engage 11.3% of the city population as per its smart city proposal. On average 10–15% of the citizens per city took part in the public participation processes. It is not just that a high majority of people were left out of the consultation process, but an over-reliance on digital means of engagement undermines the value and contribution of the non-digitally inclined population.

The citizen engagement mechanisms deployed in the aspiring smart cities failed to acknowledge the challenges posed by the 'digital divide', i.e. the deep-rooted social and economic inequalities as a result of who has access to communication technology. A quick fact finding from the Census 2011 data shows that a meagre average of 9.8% of households in Indian cities have an internet connection at home. Household internet access in smaller towns under the SCM such as Pasighat, Dahod, Kavaratti, Satna, Rampur, Sagar, Agartala and Dindigul is even below 5%. Only 7 out of 100 smart cities provide more than 20% of households with internet infrastructure. We found the reflection of this digital divide in the areas prioritised for ABD. In the Indian capital, New Delhi, Cannaught Place was selected under the ABD project as it attracted a high number of citizens' votes. This exclusive zone has a significant concentration of skilled workforce and political class who could voice their preferences over the rest of the city.

A recent report by the Housing and Land Rights Network (HLRN 2018) verifies that the participation of low-income communities in the smart cities' consultation process has been inadequate. The global agency surveyed five different locations in Bhubaneswar, where nearly half the population affected by ABD lives in slums. The report finds that none of the disadvantaged settlements was consulted, or the residents were aware of what smart city development meant for the city or them. These findings point to cutting loopholes in the project visioning and prioritisation process adopted by emerging smart cities, challenging their social sustainability. There is an increasing risk of spatial inequality in physical infrastructure development and the potential for the digital divide between communities because of such flawed designs.

This study further investigates the scale and reaches of the ABD projects across the 100 cities to measure the mission's inclusiveness. An analysis of the smart city proposals in Figure 12.3 shows that ABD projects impact less than 5% of the total area per city. This research analysed data from 75 submissions for which information is available. While there is some variation in the area covered by these cities, an overwhelming number of them (58 out of 75) allocated less than 10% of the urban authority area for smart precinct development. This analysis found that 41 out of the 75 cities earmarked less than 5% of the urban expanse for ABD. In Ludhiana, only 0.3% of the city's total area is covered under the SCM, while in Ahmedabad and Jaipur, it is 0.5%. The site that developed as a smart city in New Delhi accounts for 0.6% of the total city area, which is only 0.0015% of the total area of the National Capital Territory of Delhi. The cities of Pune, Indore, Lucknow, Amravati, Guwahati, Ranchi, Nashik and Kota have defined just 1% of the municipal area under the smart city ABD.

Research into the population served by the ABD projects undertaken in this study reveals that only up to one-tenth of the urban population per city is targeted under the proposals. In Pune, a meagre 0.8% of the population will be impacted; in Nashik, 1%; in Ahmedabad, 1.5%; in Bhopal, 1.7%; in Lucknow, 2.5%; in Nagpur, 2.6%; and in Chennai, 3.4%. The scatterplot in Figure 12.3 characterises this trend; in Quadrant B, several smaller towns showcase a higher share of population and city area served by area-based development, whereas

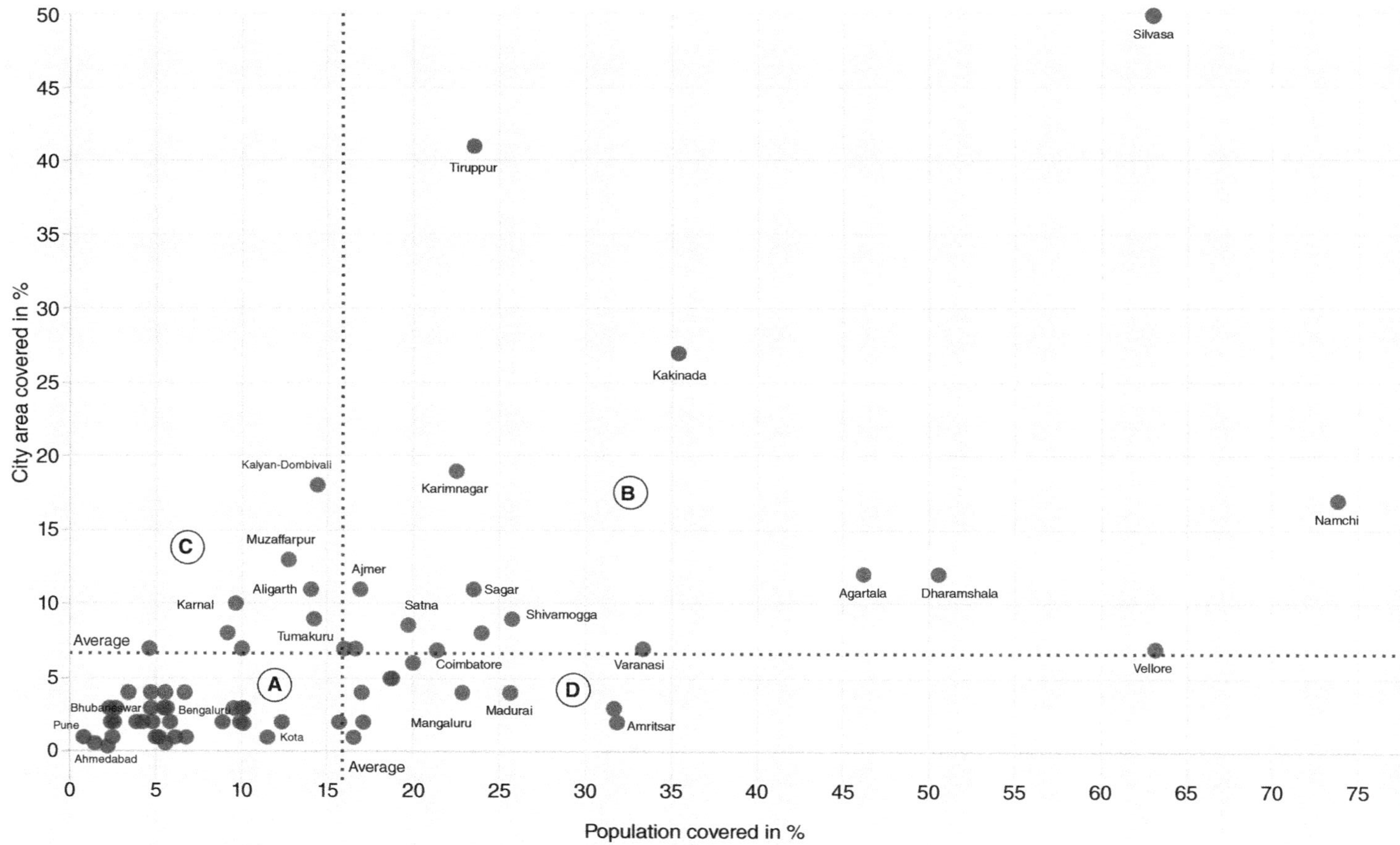

Figure 12.3 City-wise area and population coverage under the ABD projects.

in Quadrant A, as many as 29 cities, mostly large ones, show below-par area and population coverage under the ambit of the smart city urban renewal programme. Cities that have adopted a more inclusionary approach to include a more substantial share of the population under the ABD are Port Blair (77%), Namchi (74%), Pasighat (63%) and Vellore (63%).

Thus, the analysis reveals that 81% of the investments under the SCM allocated for ABD will impact 1 out of 10 populations in the 100 selected cities. In large cities, an average of 3 out of 100 people will be benefited from ABD projects. The skewed investment towards the development of elite enclaves reinforces existing power geometries and social inequalities rather than eroding or reconfiguring them. These pilot urban redevelopments fuel the debate on the possible class inequality effects of policies oriented towards creating smart cities by prioritising certain areas over others, deepening intra-city inequalities against principles of democratic and socially sustainable urban development processes. The findings resonate with what Kitchin (2014) emphasised – smart cities are the vision of particular vested interests, and they serve the interests of those constituencies, both through enacting their sensibilities and politics and by advancing their economic concerns and material desires (Kitchin 2014). Such approaches have long-term consequences on how infrastructure is planned and distributed, how resources are allocated and who can access them within and across cities and regions, severely impacting social sustainability.

12.4.2 Environmental Sustainability

Environmental sustainability refers to a condition of balance that allows human society to satisfy its needs without exceeding or disrupting the capacity of its supporting ecosystems (Adams 2005). Paul Hawken (2007) wrote, 'environmental sustainability is about stabilising the currently disruptive relationship between earth's two most complex systems: human culture and the living world' (Hawken 2007). In the context of smart cities, actions such as the use of environmentally responsible and sustainable energy sources, investment in improving energy efficiency, design for reusability and recyclability, minimising air pollution and providing transportation infrastructure that prioritises low-impact transportation modes are considered (Praharaj et al. 2018a). In urban studies literature, we also find references to cultural services, including access to parks, green recreation spaces and community facilities (library, schools and health services), as critical measures for achieving environmental sustainability (Pancholi et al. 2015). These cultural services build a bridge between ecosystems and human beings while facilitating spiritual enrichment, cognitive development, reflection, recreation, aesthetic experiences and overall well-being. In the last few decades, debates around addressing climate change through building resilience in cities and communities have gained significant momentum which is also key to building smart cities and regions (Wang et al. 2021). This section analyses whether there is an adequate focus on environmental sustainability in the smart cities discourse in India. In achieving this, a series of smart city project-level information were collected from https://smartcities.gov.in/cities-profiles to identify the project names and itemised costs outlined by cities in the smart city proposal. These data were later transformed to present the sector-wise analysis depicting the networks of investments.

The analysis presented in Figure 12.4 shows investment share in various sectors across 100 smart cities and the number of projects undertaken by the 33 cities selected in the first

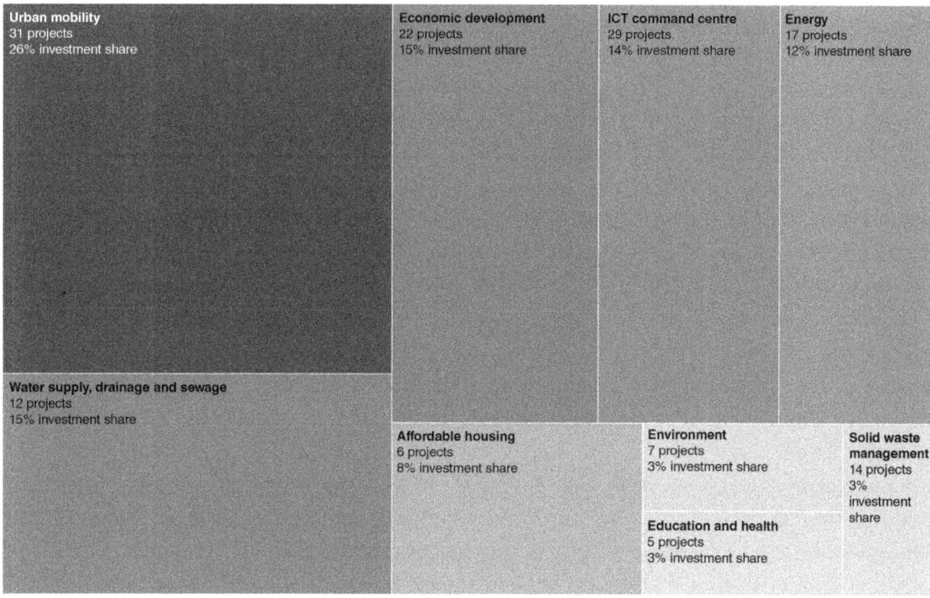

Figure 12.4 Sector-wise share of investments identified in smart city proposals.

and second phases of the SCM. The urban mobility sector received the highest priority, with more than one-fourth of the investment allocated through the programme. The smart city mobility projects commonly found in proposals are intelligent traffic and signalling systems, real-time transit information apps, bus rapid transit systems, smart cards for public transport, transport operations and monitoring centres, public bike-sharing systems and automatic number plate recognition cameras to monitor traffic violations. Economic development, including establishing start-up clusters and innovation spaces within ABD, was issued over 15% of the funds under the SCM. ICT and integrated command and control centre (ICCC) projects attracted nearly 14% of the total spending. As discussed in Section 12.3, 89 of the 100 smart cities are developing these centralised IT systems aimed at building synergy between various projects and monitoring the performance of urban services and infrastructure (Praharaj 2020).

The energy sector was allocated 12.3% of the allocated investments, whereas affordable housing projects received 7.7% of the total funds. Energy sector projects outlined in the smart city proposals include the installation of smart meters, LED streetlights, solar rooftop PV installations in public buildings and municipal solid waste-to-energy plants. A meagre 3.3% of the SCM allocations went to the environment sector, which includes projects such as parks, open spaces, riverfront developments, green buildings, air pollution-controlling measures and other green infrastructure systems. Education and health sector solutions also received a negligible share of investment (just over 3% each). Smart classrooms, telemedicine services, community libraries, welfare and aged-care services, personalised health monitoring technologies and apps, integrated digital patient information management system and remote care services are some of the solutions adopted by Indian cities.

The analysis suggests that Indian smart cities design favours 'hard infrastructure' domains such as transport and utilities, where large-scale technology deployment is likely with a significant presence of vendors in the market having numerous 'plug and play' digital solutions. In contrast, 'soft infrastructure' domains like education, health and environment, where ICT application has a seemingly limited role (Caragliu et al. 2011), were ignored with limited developments proposed for the sector. While a substantial investment allocation for the energy sector under the SCM is a positive trend, the lack of emphasis on social and environmental well-being emerges as a significant barrier to achieving environmental sustainability in aspiring smart Indian cities.

While the smart city plans suffer a narrow spatial focus and developments targeted exclusively at a handful of sectors and ICT solutions, recent studies emphasise the need for critical social and environmental infrastructure for sustainable development in the region. Praharaj and Han (2019a), while building a typology of these 100 smart cities using a series of key performance indicators (KPIs), found that social capital, including education, health facilities and lack of access to green public spaces, severely impact the quality of life in small- and medium-sized towns in India (Praharaj and Han 2019a). Similarly, Rana et al. (2019) asserted that environmental challenges of growing pollution, social challenges of urban safety and lack of community services are significant barriers to smart cities' development in the Indian context (Rana et al. 2019). The 2020 air quality report by IQAir (IQAir 2020) identifies, 10 of the world's 15 most polluted cities are in India. The Lancet estimates that air pollution kills more than one million people in the country every year. Overall, these analyses show that the cities under the SCM somewhat lag in addressing the mammoth environmental and social challenges comprehensively. The attempted design of high-tech urban precincts and new digital solutions through vendor-driven policy and technology transfers in aspiring smart cities tend to concentrate on building 'smart utopias' rather than 'smart realities'. As Datta (2015) explains, attempts to provincialise Western models of smart urbanism in India show no concern for producing socio-political and environmental outcomes (Datta 2015). Instead, such utopian urban planning envisages urbanisation as a business model rather than a model of people's welfare for shaping sustainable cities.

12.4.3 Economic Sustainability

Economic sustainability of cities refers to practices that support long-term economic growth without negatively impacting the community's social, environmental and cultural aspects (Office of Sustainability 2021). The interaction among the environment, society and the economy at all levels from the global to local scale is emphasised in the literature, beyond a typical focus on economic progress. Economic sustainability is often seen as a matter of inter-generational equity. As Anand and Sen (2000) refer, 'sustaining deprivation cannot be our goal, nor should we deny the less privileged today the attention that we bestow on generations in the future (Anand and Sen 2000, p. 2030)'. Researchers over the last decades emphasised the human rights approach to economic sustainability in a world that is marked, on the one hand, by enormous inequities in contemporary living conditions, and on the other, by real threats to the prospects of human life in the future. There is a constant shift from the typical approach of wealth maximisation and the tradition of judging economic success by the overall opulence of society as they failed to come to terms with an adequate understanding of social justice and human development (Anand and Sen 2000). Following the human rights-based approach, this study explores the relationship between the smart cities' paradigm and their impact on distributional equity and sustainable development.

A content analysis of the smart city proposals conducted through this study reveals that 92 of the 100 cities have adopted redevelopment and retrofitting models for area-based development that target urban core and central business districts. Ironically, a large share of populations in the inner-city areas across India live in slums. As per the 2011 Census, nearly 14 million households live in urban slums, with a further 3 million homeless living on city streets in India. The drive for smart cities is triggering evictions of people from slums and informal settlements, as reported by the Human and Land Rights Network (HLRN) in India. There are already reports of slum demolition from Indore, Bhubaneswar, Delhi and Kochi without adequate compensation or alternate accommodation (HLRN 2018). Besides state-sponsored evictions, there is a growing likelihood of market-driven evictions in some of these smart city enclaves due to a sharp increase in rental rates and housing prices. Experience from global cities suggests that high-quality services and concentrated prosperity of workers in the digital economy led to rising rents and increased demand for housing that places disadvantaged citizens, such as slum dwellers and urban poor, at risk. For example, Caprotti (2014), examining smart Chinese cities, found that materialisation of flows of capital and smart technologies in urban centres leads to glittering real estate developments built from scratch by low-paid workers, in which they cannot afford to live themselves, and these geographies of inequality often see low-income households denied access to schooling and critical healthcare (Caprotti 2014). This constitutes the generation of a class of 'new urban poor' on the fringes of flagship smart cities and other spectacular urban developments.

The area-based smart city development in India promises high-quality infrastructure and ICT-driven services, including free Wi-Fi, improved traffic control, intelligent sensors and better utilities with significant contribution from the private sector participation and investment. This research analysed the projects identified through the 100 smart city proposals to find that 167 projects amounting to US$ 1.80 billion were conceived under the public–private partnership (PPP) model. Projects such as multi-level car parking (19), smart power grid (17), solid waste management (12), water supply (10), housing/slum redevelopment (9) and public bike-sharing schemes (8) have been prioritised through the PPPs. While involving private sectors provides significant opportunities to address the public sector's lack of capital and expertise in the execution and operation of large technology-driven projects, a lack of strategy for allocations of risks and sustained revenue generation works as barriers. In the first three years of the SCM, only 11 out of the 167 projects have taken off, amounting to less than 10% of the overall estimated PPP investments. A total of 21 PPP projects were at various stages of the tendering process. Notably, over half of the PPP projects being implemented were in the state of Gujarat, Maharashtra and Delhi – the most affluent regions of India with a well-developed investment climate. On the contrary, cities located within the more impoverished states like Odisha, Uttar Pradesh and Rajasthan could not ground many projects due to a lack of local government capacities and attractiveness to lure private investments. Such trends could potentially further deepen structural economic inequalities between regions, acting as barriers to sustainable development.

A review of the SCM guidelines (Government of India 2015) shows that GoI encouraged cities to charge higher taxes and user fees to provide modern infrastructure – a convenient policy to attract private capital. The HLRN (2018) quotes the Pune city CEO, saying: 'The government is spending through the smart cities vehicle for better infrastructure and facilities in the ABD area. Since the area residents are getting better infrastructure, they won't mind paying increased service charges and taxes (HLRN 2018)'. The high quantum of municipal fees for essential services like water and sanitation could potentially increase the financial burden on the poor, pushing the low-income groups to city peripheries. The principal concern emerging from the smart city ABDs is that they are not innovation districts but gentrified 'consumption districts' (Almirall et al. 2016) designed for upscale services, restaurants, bars and real estate development, which do not generate any real production economies (PC). PCs create significant employment opportunities via ripple (second-order) effects (services needed to support the production economies). In other words, having trendy real estate does not do as much for the economy as a factory or research institution, even if employment numbers are similar. The discussions indicate that smart city policies and developments in India threaten to marginalise particularly those in acute poverty and informal economies, leading to irreversible consequences for cities' economic and social organisation. The programme does not tackle holistically the issues of job creation or skills that have a long-term impact on sustainable urban economies.

12.4.4 Institutional Sustainability

The Handbook in Assessment of Institutional Sustainability defines: 'an institution should be deemed sustainable if it has the strength to survive and develop to fulfil its functions on a permanent basis with decreasing levels of external support (Oslo 2000, p. 7)'. It is a process by which individuals, organisations and social systems increase their capacities and performance in relation to goals, resources and environment. This means that urban authorities and service providers at the local, state and national levels are clear on their own roles, tasks and responsibilities. They can effectively fulfil these roles and are transparent with each other, establishing a multi-stakeholder approach to development (Wash Alliance International 2020). Institutional sustainability as a process goes beyond human resources and organisational development but ensures that urban systems, institutions, policies and procedures at the local level are functional and meet the demand of users of public services (Oslo 2000). In this study, we measure institutional sustainability of the Indian smart cities movement from both the perspectives of institutions (referring to the urban system level and the norms, values and regulations which guide and constrain the behaviour of individuals and authorities) and organisations (the actors or players within the urban governance system). The goal is to evaluate how the governance system is being transformed to absorb digital urban (re)developments and how efficient and sustainable they are emerging to be.

As discussed in Section 12.2.2, smart cities in India adopted an SPV model for implementing smart city projects, which are companies formed by joint partnerships among the state, city government and the private sector (Praharaj et al. 2018b). Our analysis shows that the SCM has infused a new model of urban governance in Indian cities, having the ability to speed up the infrastructure development process. However, this

SPV-driven urban development bypasses the existing governance mechanisms that include dully elected municipal corporations and state-level utility-providing agencies. Specific groups (Chattaraj 2015) observe this as an informed design to drive a broader agenda of sidestepping the democratic process of local self-government by replacing them with more capitalistic business-oriented entities. However, for the last two decades, the GoI continuously encouraged the strengthening of municipal corporations in the true spirit of the 74th Constitutional Amendment Act (CAA), this sudden shift in government policy is set to undermine the entire process of empowerment of the local corporations (Praharaj et al. 2018c). The 74th CAA provisioned to create a process for greater financial autonomy, devolution of power and functions from the state to cities and increased human resource capacities for municipal governments (Praharaj et al. 2017). Rather than simplifying and integrating the governance systems, the SPV has only added a new layer, as the state-level agencies continue to play a key role in overall urban infrastructure development (Praharaj et al. 2018b). Therefore, smart cities' governance in India has fallen short of addressing the systematic capacity-building needs and strengthening of local municipal bodies.

This study also positions the smart cities policies within the broader context of urban planning in India to find that besides the smart cities programme, the Indian government announced 'Atal Mission for Rejuvenation and Urban Transformation (AMRUT)' focussing on infrastructure upgradation across 500 cities in India and 'Housing for All' mission to supply affordable housing for poor in 305 cities. Ministry of Culture simultaneously implements the 'National Heritage City Development and Augmentation Yojana (HRIDAY)' to preserve and revitalise cities with heritage character. These mega schemes are accompanied by the 'Clean Indian Mission' for improving urban sanitation and the 'Digital India' programme to ramp up fiver infrastructure connectivity across large cities. Most of the 100 Smart Cities are entrusted with implementing these schemes simultaneously, raising apprehension about the occurrence of multiple planning documents with different visions emerging from these programmes. A close look at the SCM guideline [5] shows that cities outlined a convergence funding to allocate funds from other missions to smart cities' projects. However, we did not find a framework or outline regarding how these different initiatives will be coordinated for the city's common good.

As cities are beginning to engage in smart city projects, it is becoming apparent that they are looking to implement pilot projects focussing on a narrow scope with high visibility. There is an absence of strategies to integrate smart city projects with existing planning instruments. For example, this study finds at least eight different plans simultaneously being executed in Bhubaneswar city, all aiming at the city's development, either holistically or focussed on a sector. While the state-level agencies deliver the Comprehensive Development Plan and Perspective Plan, the local Municipal Corporation implements the City Development Plan, City Sanitation Plan and Slum Free City Plan. Global aid agency USAID supports the City Disaster Management Plan and Environmental Management Plan. Most recently, the SPVs are charged with smart city plan preparation and development. The smart city proposal is stand-alone, showing little convergence with the existing state and city-level statutory plans. In stark contrast to the Comprehensive Development Plan, which emphasises strengthening existing traditional industries and heritage zones, the Smart City Plan promotes technology and knowledge-based enterprises.

A lack of cohesive approach to institutional development, poor coordination among agencies and continuous experimentation with new plans work as barriers to project implementation and achieving set development goals. Figure 12.5 shows the fund utilisation efficiency and project completion rate of the 20 cities selected in the first phase of the SCM. These cities were supposed to complete all project delivery by January 2021 as per the SCM deadline. However, the data collected through this research (up to 20 April 2021) from the GoI website (https://smartcities.gov.in/dashboard) highlights that nearly half of those cities are struggling even to spend one-fifth of the funding allocated. Guwahati, Ludhiana and Solapur were able to spend less than 10% of the allocated funds. Davangere, Jabalpur, Vishakhapatnam, Jaipur and Pune spent between 10% and 20% of the estimated investment. Chennai was the most successful city in consuming the SCM funds at a 62% fund utilisation rate. Kakinada, Indore and Surat moderately utilised 40–50% of the earmarked investments. The project completion rate was highest (above 80%) in Chennai and Indore. The least number of projects were completed in Ludhiana (8 out of 46), Jaipur (35 out of 143) and Guwahati (3 out of 10). The below-par performance in project execution by the so-called top-ranked smart cities exposes the acute lack of capacity, planning and desired impact of the utopian scheme. These trends establish that the debate around India smart cities in India needs to engage more with the underlying organisational and policy systems

Cities	Fund utilization efficiency (%)	Number of project tender issued	Number of projects completed
Chennai	62	45	37
Kakinada	47	93	51
Indone	46	280	226
Surat	43	87	64
Udaipur	36	128	68
Belagavi	33	217	126
Bhopal	28	96	55
Ahmedabad	24	73	42
Coimbatore	22	72	55
Pune	18	60	19
Jaipur	17	143	35
Visakapattinam	17	66	41
Jabalpur	15	97	42
Davanagere	11	106	36
Solapur	6	46	23
Ludhiana	6	46	8
Guwahati	1	10	3

Figure 12.5 Project implementation status in the 20 cities selected in Phase 1 of the SCM.

to ensure that transformation of governance and social systems is successfully realised for delivering sustainable outcomes.

12.5 Discussion and Conclusions

The debate on cities in the twenty-first century is primarily characterised by the vision of an electronically mediated future. The eventual outcomes in cities are increasingly branded as the production of digital information architecture anchored by global technology corporations and big businesses (Hollands 2008). The emerging information economy is reconstituting the relationships between people and governments and enabling new social constructions and urban patterns. Investment and jobs appear to be migrating to precincts, cities, regions and nations that can quickly put digital infrastructure in place and effectively exploit it (Giffinger et al. 2007). Cities that are basing their policies on technology and digital urban renewal are labelled 'smart cities'. The discourse on smart cities promises an era of innovative urban planning driven by smart technologies that are capable of sensing, analysing and efficiently managing urban dynamics (Praharaj 2019). City planning has thus become a continuous experiment with developing smart cities model that promotes the use of urban spaces as testing grounds for new products, services and solutions.

The projected potential benefits of this technology-enabled transition are so great, and momentum for it is building so rapidly that urban regeneration policies are fast moving from the Global North to the South (Peck et al. 2009). However, these policy transfers are not necessarily adapted to the local conditions and political settings of Global South cities. This chapter sheds light on smart city development in India, where a mega-plan to transform 100 cities across the country into 'smart' ones is underway. The genesis of the SCM is presented along with the interpretation of the smart city development models adopted by Indian cities. A major contribution of this chapter is that it provides a novel framework to measure the smart cities' design and implementation outcomes through four pillars of sustainability (social, environmental, economic and institutional). This study finds that although the smart cities programme is a timely initiative considering the surging urbanisation and multifaceted challenges facing Indian cities, there are flaws and inconsistencies in the overall narrative, planning and implementation. Without a strategic approach to the selection of projects and solutions in place, cities have ventured into the development of a few handpicked technological fixes, and the majority of investments are skewed towards exclusive enclaves. Such an approach cannot negate regional disparities across the vast country (Praharaj and Han 2019a) and the varying level of capacities and needs of the different scales of cities.

The results show that area-based development projects in Indian smart cities follow an exclusionary development approach, targeting an average of less than 5% of the city area and denying nine out of ten people new services and opportunities. The SCM allocates over 80% of the investment for just 10% of the urban populations across the 100 selected cities set to exacerbate intra-urban contrasts. The selection of elite localities through online and social media voting throws light on the new-age smart city that promotes the 'digital divide' (Partridge 2004) rather than enforcing 'digital inclusion'. Instead of developing social infrastructure (education and health) and affordable housing to meet the challenges of urban poverty, slums and basic services, the SCM predominantly invests in building smart city command and control centres and ICT-oriented transport systems. In the promise of better high-quality infrastructure, the ABD model poses threats of evictions of slum dwellers and forced migration of low-income communities out of the business districts due to the likely increase in rents, taxes and amenity fees (HLRN 2018). These findings suggest that cities with a broader portfolio of investments in 'smart' initiatives are not necessarily better or socially sustainable. Instead of attaining a fairer democracy and quality of life, it may turn into dystopian, anti-people environments where urban spaces are used to create business opportunities for corporate profits.

The smart cities in India have allocated a significant share of resources for the energy sector, which can be regarded as a positive approach towards addressing environmental sustainability. However, there is minimal focus on providing better access to environmental services, such as parks, open spaces, riverfront developments, green buildings, air pollution controlling measures and other green infrastructure systems. The analysis also suggests that Indian cities had the least emphasis on education, health and cultural infrastructure as part of the smart city endeavours. There is a lack of recognition towards the fact that smart cities can also invest in social and cultural services that build bridges between ecosystems and human beings – critical to achieving population well-being and environmental sustainability in cities (Pancholi et al. 2015; Yigitcanlar et al. 2019). From an institutional sustainability perspective, the analysis in this chapter reveals that Indian cities have created new corporate-style SPVs for smart city project development, which undermines the long-term constitutional goals of strengthening municipal corporations (Praharaj et al. 2017, 2018b). The smart city pilot projects are focussing on a narrow scope with high visibility, and there is an absence of strategies to integrate smart city projects with existing planning instruments. A lack of coordinated approach to development poses a significant barrier to project implementation as the study finds that even the cities with significant existing municipal governance capacity (e.g. Ahmedabad, Pune and Jaipur) could only utilise less than one-fifth of the earmarked investments at the conclusion of the five years of SCM. Such trends highlight gaps in policy and governance in these aspiring smart cities that need significant work to ensure institutional sustainability.

This study recognises the need for Indian cities to design holistic urban planning approaches that engage and impact the diverse demographics and contested landscapes unique to the region. Mere 'technological fixes' in smart enclaves will not meet the complex urbanisation challenges facing cities of the global south; instead, they will lead to increased spatial inequality and socially polarising outcomes. Labelled as a timely response to India's fast-paced urbanisation (Government of India 2015), the SCM accepts the fatalistic 'inevitability of urbanization'. It does so without addressing its structural causes, imbalances and adverse impacts on the ecosystem. The exclusive, area-based development approach in the smart city paradigm in India, thus, appears to have a somewhat restrictive and fragmented outlook. Instead, they require a human rights-based approach to development (Anand and Sen 2000; HLRN 2018). The smart city pilot projects should engage with the underlying social, environmental, economic and institutional sustainability issues from the sprawling urban fringes to the congested slums and deteriorated older city areas. Smart city plans should not be conceived as stand-alone documents but designed around the city's core long-term visions and master plans. The findings support existing literature that suggests sustainable cities require strategies and methods that incorporate the perspectives and needs of a broad range of stakeholders (Hollands 2008; Praharaj et al. 2018b; Praharaj and Han 2019b). Simply adopting sophisticated ICT infrastructure or showcasing self-promotional business hubs will not lead to sustainable cities. Real smart cities require a more holistic approach to urban planning that involves the devolution of power and strategies that address urban inclusion and the various facets of sustainability. The findings from this chapter and other global case studies build on the knowledge base, supporting a just transition of cities to meet the unprecedented sustainable development challenges in an increasingly digitally moderated society.

References

Adams, L.W. (2005). Urban wildlife ecology and conservation: a brief history of the discipline. *Urban Ecosystem* 8 (2): 139–156. https://doi.org/10.1007/s11252-005-4377-7.

Adams, W.M. (2006). The future of sustainability: re-thinking environment and development in the twenty-first century. *In Report of the IUCN Renowned Thinkers Meeting*, vol. 29, p. 31. IUCN The World Conservation Union.

Almirall, E., Wareham, J., Ratti, C. et al. (2016). Smart cities at the crossroads: new tensions in city transformation. *California Management Review* 59 (1): 141–152. https://doi.org/10.1177/0008125616683949.

Anand, S. and Sen, A. (2000). Human development and economic sustainability. *World Development* 28 (12): 2029–2049. https://doi.org/10.1016/S0305-750X(00)00071-1.

Bharatiya Janata Party (2014). Election manifesto 2014. http://www.bjp.org/images/pdf_2014/full_manifesto_english_07.04.2014.pdf (accessed 30 August 2016).

Bhattacharya, S., Rathi, S., Patro, S. A. et al. (2015). Reconceptualising smart cities: a reference framework for India. Bangalore. https://cstep.in/drupal/sites/default/files/2019-01/CSTEP_RR_Reconceptualising_smart_cities_2015.pdf (accessed 20 October 2020).

Caprotti, F. (2014). Critical research on eco-cities? A walk through the Sino-Singapore Tianjin Eco-City, China. *Cities* 36: 10–17. https://doi.org/10.1016/j.cities.2013.08.005.

Caragliu, A., del Bo, C., and Nijkamp, P. (2011). Smart cities in Europe. *Journal of Urban Technology* 18 (2): 65–82.

Census of India (2011). Provisional population totals urban agglomerations and cities. New Delhi. http://censusindia.gov.in/2011-prov-results/paper2/data_files/India2/1.DataHighlight.pdf (accessed 20 October 2020).

Chattaraj, S. (2015). Are 'smart cities' enough for India? – IGC, International Growth Centre. https://www.theigc.org/blog/are-smart-cities-enough-for-india/ (accessed 29 June 2021).

Datta, A. (2015). A 100 smart cities, a 100 utopias. *Dialogues in Human Geography* 5 (1): 49–53. https://doi.org/10.1177/2043820614565750.

Giffinger, R., Fertner, C., Kramar, H. et al. (2007). *Smart Cities Ranking of European Medium-Sized Cities.* Vienna University of Technology https://doi.org/10.1016/S0264-2751(98)00050-X.

Government of India (2015). Smart cities mission statement and guidelines. New Delhi. http://smartcities.gov.in/writereaddata/smartcityguidelines.pdf (accessed 20 October 2020).

Hawken, P. (2007). *Blessed Unrest: How the Largest Movement in the World Came into Being, and Why No One Saw It Coming*. Penguin.

HLRN (2018). India's smart cities mission: smart for whom? Cities for whom? https://www.hlrn.org.in/documents/Smart_Cities_Report_2018.pdf (accessed 18 April 2021).

Hollands, R.G. (2008). Will the real smart city please stand up? *City* 12 (3): 303–320. https://doi.org/10.1080/13604810802479126.

IQAir (2020). 2020 world air quality report. https://www.iqair.com/world-most-polluted-cities/world-air-quality-report-2020-en.pdf (accessed 29 June 2021).

Kandachar, P. (2014). Chapter 7: materials and social sustainability. In: *Materials Experience* (ed. E. Karana, O. Pedgley, and V. Rognoli), 91–103. Boston: Butterworth-Heinemann https://doi.org/10.1016/B978-0-08-099359-1.00007-2.

Kitchin, R. (2014). The real-time city? Big data and smart urbanism. *GeoJournal* 79 (1): 1–14. https://doi.org/10.1007/s10708-013-9516-8.

Leao, S.Z., van den Nouwelant, R., Shi, V. et al. (2021). A rapid analytics tool to map the effect of rezoning on property values. *Computers, Environment and Urban Systems* 86: 101572. https://doi.org/10.1016/j.compenvurbsys.2020.101572.

Lock, O., Bednarz, T., Leao, S.Z., and Pettit, C. (2019). A review and reframing of participatory urban dashboards. *City, Culture and Society* 20: 100294.

Office of Sustainability (2021). Economic sustainability. University of Mary Washington. https://sustainability.umw.edu/areas-of-sustainability/economic-sustainability/ (accessed 29 June 2021).

Oslo (2000). *Handbook in Assessment of Institutional Sustainability*. Oslo.

Pancholi, S., Yigitcanlar, T., and Guaralda, M. (2015). Place making facilitators of knowledge and innovation spaces: Insights from European best practices. *International Journal of Knowledge-Based Development* 6 (3): 215–240. https://doi.org/10.1504/IJKBD.2015.072823.

Partridge, H.L. (2004). Developing a human perspective to the digital divide in the 'smart city,'. *Australian Library and Information Association Biennial Conference*, 21–24 September 2004.

Peck, J., Theodore, N., and Brenner, N. (2009). Neoliberal urbanism: models, moments, mutations. *SAIS Review of International Affairs* 29 (1): 49–66. https://doi.org/10.1353/sais.0.0028.

Praharaj, S. (2019). A comprehensive analysis of the challenges and opportunities of the smart cities mission in India. http://handle.unsw.edu.au/1959.4/61972 (accessed 29 June 2021).

Praharaj, S. (2020). Development challenges for big data command and control centres for smart cities in India. In: *Data-Driven Multivalence in the Built Environment. S.M.A.R.T. Environments* (ed. N. Biloria). Cham: Springer https://doi.org/10.1007/978-3-030-12180-8_4.

Praharaj, S. and Han, H. (2019a). Building a typology of the 100 smart cities in India. *Smart and Sustainable Built Environment* 8 (5): https://doi.org/10.1108/SASBE-04-2019-0056.

Praharaj, S. and Han, H. (2019b). Cutting through the clutter of smart city definitions: a reading into the smart city perceptions in India. *City, Culture and Society* 18: 100289. https://doi.org/10.1016/j.ccs.2019.05.005.

Praharaj, S., Han, J.H., and Hawken, S. (2017). Innovative civic engagement and digital urban infrastructure: lessons from 100 smart cities mission in India. *Procedia Engineering* 180: 1423–1432. https://doi.org/10.1016/j.proeng.2017.04.305.

Praharaj, S., Han, J.H., and Hawken, S. (2018a). Evolving a locally appropriate indicator system for benchmarking sustainable smart cities in India. *World Sustainability Series*, Springer, 253–274. https://doi.org/10.1007/978-3-319-73293-0_15.

Praharaj, S., Han, J.H., and Hawken, S. (2018b). Towards the right model of smart city governance in India. *International Journal of Sustainable Development and Planning* 13 (2): https://doi.org/10.2495/SDP-V13-N2-171-186.

Praharaj, S., Han, J.H., and Hawken, S. (2018c). Urban innovation through policy integration: critical perspectives from 100 smart cities mission in India. *City, Culture and Society* 12: 35–43. https://doi.org/10.1016/j.ccs.2017.06.004.

Rana, N.P., Luthra, S., Mangla, S.K. et al. (2019). Barriers to the development of smart cities in Indian context. *Information Systems Frontiers* 21 (3): 503–525. https://doi.org/10.1007/s10796-018-9873-4.

Sankhe, S., Vittal, I., Dobbs, R. et al. (2010). *India's Urban Awakening: Building Inclusive Cities, Sustaining Economic Growth*. McKinsey Quarterly.

Townsend, A.M. (2013). *Smart Cities: Big Data, Civic Hackers, and the Quest for a New Utopia*. WW Norton & Company.

UNDP (2020). Human development report 2020. https://report.hdr.undp.org/ (accessed 27 April 2021).

Vallance, S., Perkins, H.C., and Dixon, J.E. (2011). What is social sustainability? A clarification of concepts. *Geoforum* 42 (3): 342–348. https://doi.org/10.1016/j.geoforum.2011.01.002.

Wang, C., Turner, V.K., Wentz, E.A. et al. (2021). Optimization of residential green space for environmental sustainability and property appreciation in metropolitan Phoenix, Arizona. *Science of the Total Environment* 763: 144605. https://doi.org/10.1016/j.scitotenv.2020.144605.

Wash Alliance International (2020). Institutional sustainability. https://wash-alliance.org/our-approach/sustainability/institutional-sustainability/ (accessed 29 June 2021).

Woodcraft, S., Hackett, T., and Caistor-Arendar, L. (2011). *Design for Social Sustainability: A Framework for Creating Thriving New Communities*. Young Foundation.

Yenneti, K., Rahiman, R., Panda, A., and Pignatta, G. (2019). Smart energy management policy in India—a review. *Energies* 12 (17): 3214. https://doi.org/10.3390/en12173214.

Yigitcanlar, T., Kamruzzaman, M., Foth, M. et al. (2019). Can cities become smart without being sustainable? A systematic review of the literature. *Sustainable Cities and Society* 45: 348–365. https://doi.org/10.1016/j.scs.2018.11.033.

13

Energy Transitions and Smart Cities in Russia

Irina Ilina[1] *and Michinaga Kohno*[2]

1 Faculty of Urban and Regional Development, HSE University, Moscow, Russia
2 Michi Creative City Designers Inc., Hokuto, Japan

13.1 Introduction

Cities today are seen at the forefront of addressing climate challenges, while the restructuring of the energy sector has a very significant impact on urban development. Climate plans and strategies of cities should be based on the technologies that ensure the development of urban economies and improve living standards and comfortable living for the population, with the ambitious goal of reducing greenhouse gas emissions.

Russia, as a country with a large share of the economy specialising in the production and export of hydrocarbons and industrial products with a high carbon footprint, faces a number of long-term development risks. To reduce these, as well as to realise the opportunities that can be used, it is necessary to create an effective national strategy for low-carbon development. In this chapter, we consider this agenda in the context of Russian energy and climate policy and emerging strategies for digitalisation of energy transitions and smart cities.

13.2 National Climate Policy

The Russian Federation fully supports activities under SDG 7, which is designed to unite the efforts of various parties in the field of sustainable energy and energy security, energy efficiency and global development of renewable energy. At the same time, the carbon intensity of the economy is twice as high as the carbon intensity for OECD countries in 2018: 0.411 kg CO_2/\$1000 vs. 0.269 kg CO_2/\$1000 (Olivier and Peters 2020). According to the *Strategy for long-term development with low greenhouse gas emissions until 2050*, development in the field of sustainable energy should be comprehensive and balanced and it should take into account the complex and diverse natural and climatic conditions of the country.

Smart Cities, Energy and Climate: Governing Cities for a Low-Carbon Future, First Edition.
Edited by Oleg Golubchikov and Komali Yenneti.

The national climate policy of Russia developed rather slowly. Its development entered an active phase only in the 2010s – with the adoption in 2009 of *the Climate Doctrine of the Russian Federation* (2009), and now there is a gradual detailing of action plans aimed at reducing greenhouse gas emissions and adapting to climate change. The climate doctrine has secured the priority of national interests in the development and implementation of climate policy, but the driver of its current development is the international agenda.

The Russians are generally not prepared to pay more for low-carbon goods and services, even if these funds are used to introduce alternative energy sources or improve energy efficiency. Businesses too, while understanding the problems of climate change, oppose the introduction of additional 'climate' charges. The strengthening of the national climate policy is supported by the overwhelming part of the scientific community (although there are also climate sceptics) and a number of experts who point to a significant potential increase in costs associated both with the need to adapt to climate change and with the further spread of climate initiatives in the world.

Thus, climate policy in Russia can be classified as one of the most controversial and difficult policy areas. Opportunities for mitigating these and other risks, as well as for realising new opportunities that are opening up in a number of technological niches, lie in the field of the formation of national low-carbon strategy, the key elements of which may be:

- Tightening the national target to reduce emissions and linking it with economic dynamics;
- The formation of a national system for reporting, monitoring and economic regulation of greenhouse gas emissions; and
- The development of national registers and institutions for regulating green finance, integration of Russian legislation with widespread international practices of green finance and low-carbon reporting.

In 2019, Russia joined the Paris Agreement, and within the framework of this agreement, it adopted a goal of reducing emissions to 70% of the 1990 level by 2030, subject to the maximum possible account of the absorptive capacity of ecosystems, particularly forests. The Decree of the President of the Russian Federation '*On the reduction of greenhouse gas emissions*' was introduced, followed by the *Order of the Government of the Russian Federation on the development of investment activities and the attraction of extra-budgetary funds in projects of sustainable (including green) development of the Russian Federation* (Government of the RF 2020). These have noticeably intensified the activities of government bodies and businesses in this area (Table 13.1).

In accordance with the recommendation of the Paris Agreement, Russia developed a national action plan for the first stage of adaptation to climate change for the period up to 2022 (Government of the RF 2019), which provides for three blocks of climate change management. The first block is a federal one, which includes:

- Organisational, legal and methodological support, including the development of a standard 'passport' of the climatic safety of the territory;
- A system of target indicators for achieving adaptation goals;

Table 13.1 Russian goals to reduce gas emissions in the framework of international agreements and at the national level.

Period	Objectives of Russia
2008–2012	Commitment to maintaining GHG emissions in the first period of the Kyoto Protocol (2008–2012) at the 1990 level
2013–2020	Ensuring the reduction in greenhouse gas emissions to a level of no more than 75% of the volume of these emissions in 1990 (Decree of the President of the Russian Federation of 30.09.2013 No. 752)
2021–2030	Ensuring a reduction in greenhouse gas emissions by up to 70% compared to the 1990 level taking into account the maximum possible absorbing capacity of forests and other ecosystems and subject to sustainable and balanced socio-economic development in order to implement the Paris Agreement (decree of the President of the Russian Federation of 04.11. 2020 No. 666)

- Methodological recommendations for assessing climate risks and ranking adaptation measures according to their priority;
- Guidelines for the formation of sectorial, regional and corporate plans for adaptation to climate change;
- Draft national action plan for the second stage of adaptation to climate change for the period up to 2025; and
- Information and scientific support.

The second block is industry specific and is aimed at solving the following problems:

- Updating sectorial strategic planning documents; and
- Sectorial plans for adaptation to climate change.

The third block is regional, which includes:

- Correction of strategic planning documents for regions and cities of Russia; and
- Regional and city plans for adaptation to climate change.

Within the framework of solving the problem of ensuring access to energy, it is necessary to consider the possibility of using more environmentally friendly types of fossil fuels, such as natural gas instead of coal, and also pay attention to measures where it is impossible to abandon fossil fuels. This approach to energy development is recorded in the strategic planning documents, including the *Energy Security Doctrine of the Russian Federation,* approved in May 2019.

As a part of measures to prevent climate change, the following has been prepared and is awaiting approval: a draft Federal Law '*On Limiting GHG Emissions*' (Ministry of Economic Development of Russia), which includes a draft *Concept for a system of accounting, registration, release into circulation, transfer and offset of the results of climate projects implemented in the Russian Federation* (Ministry of Economic Development of Russia). An important document, being developed in accordance with the recommendation of the Paris Agreement, is the *draft Strategy for the socio-economic development of the Russian Federation with low greenhouse gas emissions until 2050* (Ministry of Economic Development of Russia 2020).

At the moment, Russia is reforming the system of programme-targeted management in the field of energy conservation and energy efficiency. The purpose of the reform is to achieve targets in the field of energy conservation and energy efficiency by including energy efficiency indicators in sectorial strategic planning documents at all levels. Also, the country is actively working to stimulate the development of renewable energy sources (RES). By the end of 2018, the total installed electric capacity of facilities operating on the basis of renewable energy sources reached 52.6 GW, which amounted to 20.5% of the total installed electric capacity in Russia (256.6 GW) (Figures 13.1 and 13.2).

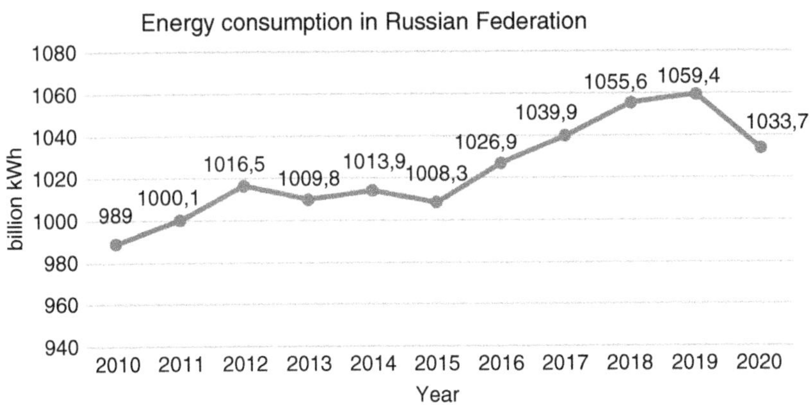

Figure 13.1 Energy consumption in Russian Federation, 2010–2020, billion kWh. *Source:* Rosstat. https://rosstat.gov.ru/enterprise_industrial.

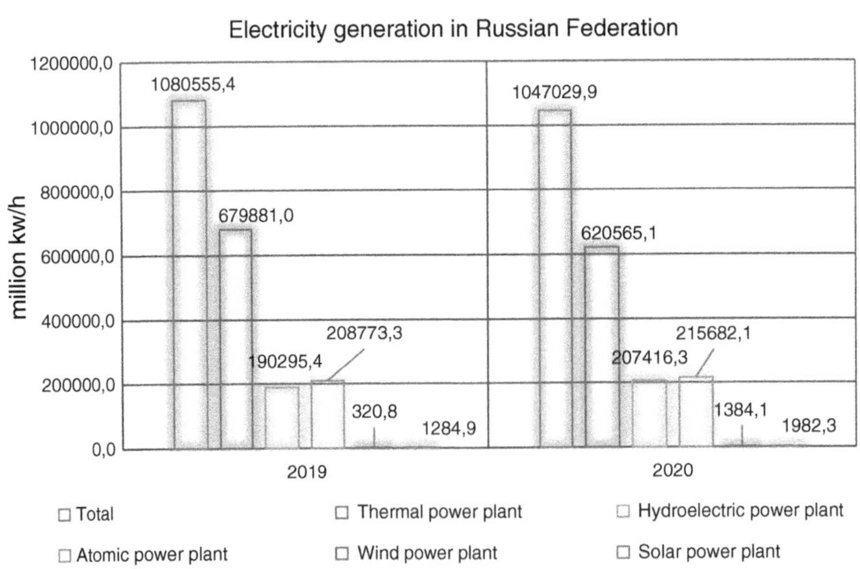

Figure 13.2 Electricity generation in Russian Federation, 2019–2020. *Source:* Data from System operator of the Unified Energy System RF. https://www.so-ups.ru/fileadmin/files/company/reports/disclosure/2020/ups_rep2019.pdf https://www.so-ups.ru/fileadmin/files/company/reports/disclosure/2021/ups_rep2020.pdf

Expert assessments by Bertelsmann Stiftung and the UN Partnership 'Sustainable Development Solutions Network (SDSN)' for Russia in SDG 7 'Affordable and Clean Energy' indicate significant progress in achieving this SDG (91.2 points out of 100) (Sustainable Development Report Dashboards 2019). However, the greatest progress is noted in such SDG indicators as access to electricity and clean fuels. Ensuring universal access to affordable, reliable and modern energy supply (target 7.1) is at a fairly high level since Russia is provided with primary and secondary energy resources, the domestic production of which significantly exceeds consumption.

13.3 Cities in the Climate and Energy Agendas

Climate change and adaptation measures should be taken into account in the mid-term and long-term planning of the development of territories and cities (Table 13.2). The consequences of climate change are different for the regions of the Russian Federation, and within the same region have different effects on population groups, economic sectors and natural objects. Therefore, the development of climate strategies and plans should be based on the assessment of the risks and vulnerabilities of energy, transport, utilities, public health, natural environments, etc., individually justified by taking into account the climatic, economic and social characteristics of cities.

Most of the territories of Russia are located in climatic zones of low temperatures. The average heating period in most parts of the country lasts for half a year, and in the north

Table 13.2 Main trends in the development of the energy sector in Russian cities.

The projected energy sector trends	Forecasted factors		
	Climatic	**Economic**	**Technological**
Unstable dynamics of electricity consumption	Growth of average annual temperatures (up to 2°C), increased demand for cooling of premises in summer, while reducing consumption in winter	Low rates of economic growth	Implementation of energy-efficient technologies
Reducing the need for thermal energy	Increase in temperatures in the winter period of the year and reduction in the heating period	Unprofitable production of thermal energy due to imperfect pricing mechanisms	Implementation of energy-efficient technologies in the construction of residential and public buildings ('0' energy consumption)
Increase in seasonal unevenness of natural gas consumption by city consumers	Total reduction in consumption in winter	Increase in prices for consumed gas	Refusal to use gas stoves and switch to electricity supply

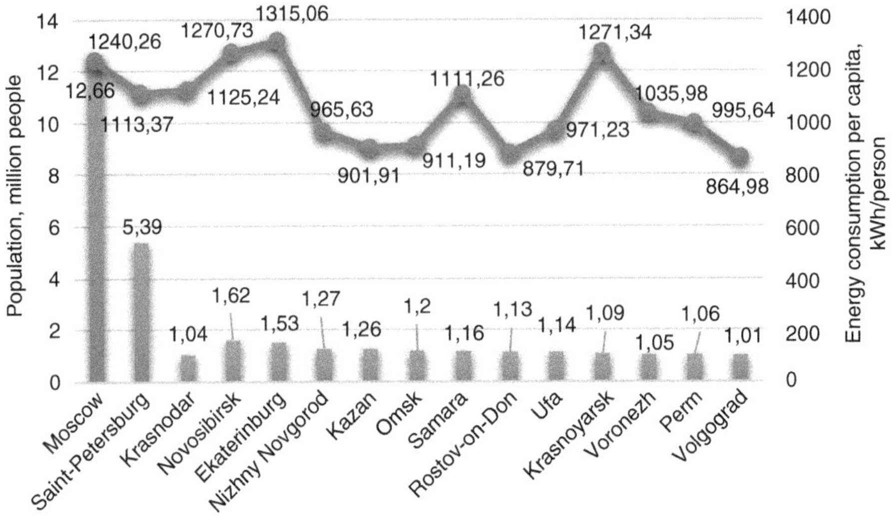

Figure 13.3 Per capita household electricity consumption in million-plus cities of the Russian Federation, 2019–2020, kWh/year. *Source:* Data from Rosstat, Minenergo RF, System operator of the Unified Energy System RF. https://rosstat.gov.ru/storage/mediabank/RZLCLxM7/Region-Goroda-2020.pdf https://www.gks.ru/free_doc/new_site/business/prom/el_potr.htm, https://minenergo.gov.ru/node/20706.

of the European part of Russia, Siberia and the Far East – up to nine months a year. In this regard, energy consumption firmly takes its place in the total consumption of fuel and energy resources (Figure 13.3). In these conditions, the use of fossil fuels is of great importance. Due to the peculiarities of the climate, the volume of consumption of heat energy is twice that of electricity, and in terms of the capacity of centralised heat supply systems, the country is ahead of the rest of the world. At the same time, during the production of electricity at power plants, heat is generated as a by-product in an amount corresponding to the heating load of all buildings in the country. Most of this heat either heats the atmosphere through cooling towers or heats water in specially designed reservoirs (Gashot et al. 2015).

Russian cities consume 2.5–3 times more energy for heating than electricity. For example, Moscow consumes about 110 million MWh of heat and 53 million MWh of electricity. Since heat cannot be delivered far away, the sources are located directly in the city, having a corresponding effect on the urban atmosphere. At the same time, the installed capacity of urban heat sources in Moscow (over 60 thousand MW) is more than four times higher than the capacity of electric sources (about 14 thousand MW).

13.4 The Digitalisation of Energy and Smart Cities

The energy infrastructure of the Russian Federation is being actively transformed through the introduction of digital technologies and platform solutions. Digitalisation is a key driver of technological transformation, and this area is actively supported by the state (Government of the Russian Federation 2014).

The project of the Ministry of Energy of the Russian Federation *'Digital Energy'* is designed for the period 2018–2021 and provides for the creation of conditions for the introduction of digital technologies and platform solutions in the electric power industry, oil and gas complex and the coal industry in Russia.

An important part of the state policy to promote the energy efficiency of the national economy and the introduction of new advanced technologies in the fuel and energy complex is the development of a 'smart city' plays an important part in the state policy to promote the energy efficiency of the national economy and in the introduction of new advanced technologies in the fuel and energy complex. There are 209 active participants in the project of the Ministry of Construction of Russia 'Smart City'. These are cities with a population of over 100 thousand people: all the administrative centres of the regions and also 'pilot' cities, on their own initiative, wanted to join the project. Cities included in the list of 'smart' cities are divided into two groups: (i) major cities of Russia that 'must be smart', and (ii) pilot cities that participate in the project largely in order to receive financial support from the federal government. This approach, which we could call the 'directive-stimulating approach' leads to the implementation of development programmes based on the principle of 'we did what the allocated money was enough for'.

Russian 'smart' cities are mostly focused on the digitalisation of the housing and utilities sector. First of all, the regulatory framework for the project is being developed. The standard has already been developed (basic and additional requirements for smart cities) (Ministry of Construction 2019). At the end of December 2020, the Ministry of Construction of Russia also approved *the concept of the project for digitalisation of the urban economy 'Smart City'*. These documents explain the advantages of smart cities for residents, businesses and government, and help determine the sequence of stages of digitalisation and the direction of the project in order to help cities go through this thorny path.

Based on the UN terminology, the Russian experts defined the 'smart city' as an approach to city development that uses digital tools to improve living standards, quality of services and management efficiency while meeting the needs of present and future generations in all relevant aspects of life. This implies the introduction of highly intelligent systems in the following areas: urban environment, a safe city, digital urban governance, investment, climate and human well-being.

For the first time, the levels of urban economy digitalisation were measured in Russia in 2018 using the urban economy digitalisation index, 'IQ of Cities', which allows determination of the basic level of urban economy digitalisation and the effectiveness of solutions that are being implemented by cities and regions under the 'Smart City' project. In addition to the eight areas of the Smart City Standard, the IQ of Cities index takes into account two additional areas – intelligent systems of social services, economic condition and investment climate – which contains 47 indicators. The IQ index of cities is taken into account to determine the amount of subsidies from the federal budget to support the state and municipal programmes for the formation of a modern urban environment.

In order to be invited into the project of the Ministry of Construction 'Smart City', the city must pass a preliminary selection. First of all, the city needs to create a management body for the smart city project, based on which a competence centre is formed.

In practice, they are created either under the Ministry of Communications or under the Ministry of Construction. In addition, a three-year implementation plan must be approved. Competence centres prepare project descriptions, road maps and implementation plan. Upon successful completion of the selection, a special agreement is made with the city administration. After that, the city becomes a 'systemic' participant in the project: the IQ index is calculated for it, and the city is connected to internal systems of interaction within which the Ministry of Construction of Russia provides methodological and informational support.

Cities are still very different between themselves including with respect to climates, cultural characteristics, levels of digitalisation and digital transformation of departments. This has led to a varied choice of priorities for each city. Of course, there are typical problems, for example, reducing energy consumption and intelligent transport systems needed for every city. However, in Yakutsk, it will be a more pressing problem; in Rostov-on-Don, it will be less of a priority. For the unified development of 'smart' cities in Russia, a 'Smart Cities Solution Bank' ('Smartteka'), a bank of solutions has been created.

The main group of technologies for the power supply of a smart city includes renewable energy sources, electric transport, energy storage systems, smart distribution networks of power supply, smart houses–buildings–enterprises, smart metering systems and aggregation of sources.

The next step towards the development of 'Smart City' is horizontal integration, where data from different sectors can be combined to manage the city better and reduce risks. One of the characteristic features of this 'smart' energy paradigm is the focus on the interests of the consumer, not the energy producer. This can be noted as a fundamental difference from the current situation, where the logic of the development of power systems is dictated by generation, then by networks, but in general by the so-called 'energy lobby', which is difficult to compete with even for the largest industrial consumers, e.g. metallurgical holdings, and even entire industries, not to mention municipalities. Another distinctive feature is the development and even localisation of energy supply systems, and the growth of local types of fuel in the fuel and energy balances. This is seen as a measure to ensure the energy security of cities and settlements in the context of the growing instability of the external energy markets and rising prices for hydrocarbons. Electric power, with its three qualities (versatility, convenience and manageability), is the main priority of generation and consumption.

Manageability in the electricity sector of urban infrastructure is achieved in the most effective way using the smart grid design technology – a 'smart', intelligent (due to the active participation of humans) and energy-informational (interconnected with each other physically and in information management) system.

Another important trend in the development of 'smart' cities in Russia is the use of 'green standards' in new construction, which significantly increases the environmental and economic efficiency of the implementation of the climate programme. On average, new buildings consume 15–20% less energy for heating compared to the existing stock due to improved thermal protection.

'Smart' technologies in the field of housing and communal services allow responding quickly to weather changes and carrying out preventive work timely in order to prevent accidents or minimise their consequences. All this reduces energy waste and saves money. For example in Yekaterinburg, an automated heat supply control system was introduced, which can adapt to weather conditions, ensure a stable temperature regime in flats and industrial facilities and can also distribute the load to power supply sources remotely. Telemetry modules were installed in Tula's schools, according to which the operator remotely monitors the main indicators of the devices. In addition, smart street lights can change their brightness depending on the level of natural light. In furthermore, sensors monitor the water level in rivers and give early warning of impending floods.

Furthermore, a set of climate adaptation measures is applied in Russian 'smart' cities:

1) Orientation of spatial solutions to reduce wind loads and air flows in public urban spaces.
2) Use of the aeration effect (dry fountains and water vapor distributors) and shading of urban spaces: tall shrubs, stationary or mobile sheds, canopies that prevent overheating, drying, dusting of urban spaces during high temperatures and air conditioning of public transport stops.
3) Mobile and stationary protection of the southern, western and south-western facades (internal premises) from overheating (using a loggia, veranda, blinds, shutters, canopies and awnings) and from wind cooling of the northern, north-western and north-eastern facades.
4) Protection against hypothermia and ice formation during low temperatures, placement of stationary and mobile warm objects in public urban spaces (warm stops of public transport, mobile cafes, etc.) and heating of canopies, sidewalks, storm drains from building facades in places of high pedestrian transit.
5) Integration of natural urban systems into the drainage system and utilisation of rainfall; the maximum possible restoration of natural drainage systems in the city.
6) In the nexus of urban planning and urban mobility, the following principles are taken account of:
   - Priority of short structural connections within the city. Short distances increase the efficiency of using urban spaces (reduce the costs of land use and development and increase the number of users per unit of territory). Short links form new models for using resources (e.g. car-sharing and sharing space models).
   - Transport modality and spatial differentiation of links. These reduce the need to use personal transport as urban (daily) transport and lead to increased effects of short-link policies.
   - Accessibility and barrier freeness of the urban environment, including the elimination of large, isolated areas and linear objects that impede the capillary connectivity of urban spaces and natural systems in the city.

In the smart city context, the mobility issue is mainly focused on the change in people's transport behaviour from individual cars to public transportation. Light rail transit (LRT) and bus rapid transit (BRT) are frequently discussed to improve the ridership of public transportation networks. The manufacturers of trams and buses are focusing on the commercialisation of 'fuel cell' vehicles both on rails and roads.

Two other main streams can be found in Russian smart cities. One is the e-Government approach, which can be seen in Moscow, where the digitalisation of city services is citizen-focused. The residents can easily get access to city services such as applications and document submissions, as well as doctor's appointments. The other is the green building approach, which has been promoted by the Federal Ministry of Construction and Utilities, with the energy-efficient houses and apartments being the main focus.

A significant part of the energy supply to houses and apartments is covered by heat supply through district heating and hot water pipeline systems. The source of energy in this heating system is fossil fuel, and it seems difficult to substitute the source for low-emission energy sources. The use of electricity for heating is not efficient in terms of energy transformation coefficient, although electricity generated by nuclear power plants is zero emission. The use of hydrogen can be the answer, but the construction of hydrogen production and distribution infrastructure will require vast investment, while the technologies are still premature.

Local generation of energy is another keyword to characterise the next-generation smart cities. The generation of electricity at large-scale plants, the transmission of electricity by transmission companies through high-voltage grids and the distribution of electricity by retailers represent an old business model of the previous century. This system was believed to be most efficient under the economy of the scale. Under the conditions of using renewable energy sources, the scale merit cannot be expected but the self-contained structure offers more flexibility in operation to the consumers of energy. Large facilities such as schools, hospitals and shopping malls could have their own power generation systems at their hands to balance demand and supply and to control the mixture of energy sources. Russian smart cities should be planned taking these global trends in mobility and local energy grids into account in order to become really smart.

13.5 Conclusions

The transition to smart cities has just begun. However, for now, the smart cities' "pilots" are implementing the minimum list of measures of the standard of the 'Smart City' project.

The main barriers to the introduction of smart city technologies in Russia are as follows:

- *Unclear economics*: there are very few well-developed business models that cover the sphere of smart cities and have proven effective for business. A problem is the payback period. Businesses need market strategies that can generate profits, and the introduction of innovations in a smart city is an investment in long-term prospects;
- *Difficulties with replication*: the creation of one technology in a particular city does not guarantee that such a project can be repeated in another city since climatic, technological and social factors do not allow making an exact analogue;
- *Legacy*: as in any large IT project, when building a smart city, one has to deal with lock in effects and an infrastructure that has not exhausted its capacity. Previously implemented smart infrastructure may be lacking integration into complex systems, or the integration turns out to be too costly.

References

Climate Doctrine of the Russian Federation (2009). Decree of the President of the Russian Federation № 861-rp of 17.12.2009. http://www.kremlin.ru/acts/bank/30311 (accessed July 2023).

Gashot, E.G., Puzakov, V.S., and Stepanov, M.V. (2015). *Reserves and Priorities of Heat and Power Supply of Russian Cities in Modern Conditions*. Moscow.

Government of the RF (2014). The order of the Government of the Russian Federation of March 19, 2014 № 398-r. On a set of measures aimed at abandoning the use of outdated and ineffective technologies, the transition to the principles of the best available technologies and the introduction of modern technologies. https://www.garant.ru/products/ipo/prime/doc/70519522/.

Government of the RF (2019). National action plan for the first stage of adaptation to climate change for the period up to 2022 (Order of the Government of the Russian Federation dated December the 25th, 2019 № 3183-r).

Government of the RF (2020). Order of the Government of the Russian Federation on the development of investment activities and the attraction of extra-budgetary funds in projects of sustainable (including green) development of the Russian Federation, of 18.11.2020 № 3024-r.

Minstroy (2019) The 'Smart city' standard https://minstroyrf.gov.ru/docs/18039/ (accessed July 2023).

Ministry of Economic Development of Russia (2020). Project of the strategy for the socio-economic development of the Russian Federation with low greenhouse gas emissions until 2050. https://economy.gov.ru/material/news/minekonomrazvitiya_rossii_podgotovilo_proekt_strategii_dolgosrochnogo_razvitiya_rossii_s_nizkim_urovnem_vybrosov_parnikovyh_gazov_do_2050_goda_.html (July 2023).

Olivier, J.G.J. and Peters, J.A.H.W. (2020). Trends in global CO_2 and total greenhouse gas emissions: 2019 report. PBL Netherlands Environmental Assessment Agency, The Hague, p. 46. https://www.pbl.nl/sites/default/files/downloads/pbl-2020-trends-in-global-co2-and-total-greenhouse-gas-emissions-2019-report_4068.pdf (July 2023).

Sustainable Development Report Dashboards (2019). SDSN web resource. Transformations to Achieve the Sustainable Development Goals. https://dashboards.sdgindex.org/#/RUS (accessed July 2023).

14

Energy Poverty in Cities: A Behaviourally Informed Perspective

Nives Della Valle

European Commission, Joint Research Centre (JRC), Ispra, Italy

14.1 Introduction

In 2018, nearly 34 million Europeans were unable to afford to keep their houses adequately warm. This is only one dimension of energy poverty, a complex phenomenon depicting insufficient access to essential energy services like adequate heating, cooling and lighting, and energy needed to power household appliances (Thomson and Bouzarovski 2018). These energy poverty conditions prevent citizens from both meeting their most basic human needs and having a full control of their lives. Living in energy poverty has negative consequences not only for physical health and daily well-being, but also for mental health (Ochoa and Graizbord 2016). Therefore, tackling the problem of energy poverty would enable not only to ensure an adequate level of comfort and private budgeting for all citizens, but also to create the right conditions that enhance well-being (Della Valle 2019). Since 2019, the European Union has identified the fight against energy poverty as one of the main policy priorities in Europe. The legislative package "Clean Energy for All Europeans" requires all EU countries to implement actions tackling and monitoring energy poverty, as well as protecting vulnerable energy consumers.[1] With this package, EU countries are also required to reduce energy poverty in their national energy and climate plans and long-term renovation strategies.

The European Commission has further strengthened its commitment to tackle energy poverty with the Recommendation on Energy Poverty[2] issued as part of the Renovation Wave package[3]. More particularly, the recommendation and its supporting staff document provide specific guidance on how to measure, define and tackle energy poverty based on the available EU-level support. The Renovation Wave identifies renovation of the built

1 https://ec.europa.eu/energy/topics/energy-strategy/clean-energy-all-europeans_en.
2 http://data.europa.eu/eli/reco/2020/1563/oj.
3 https://ec.europa.eu/energy/topics/energy-efficiency/energy-efficient-buildings/renovation-wave_en.

Smart Cities, Energy and Climate: Governing Cities for a Low-Carbon Future, First Edition.
Edited by Oleg Golubchikov and Komali Yenneti.

environment as a central lever to address energy poverty, by specifically promoting housing that is not only healthy but also energy efficient (therefore lowering also energy costs).

However, a key challenge remains as to how European Member states will implement tailored actions, as the manifestations of energy poverty are highly heterogeneous across Europe (Dobbins et al. 2019). Initiatives funded by the European Commission, like the European Covenant of Mayors[4] and the European Energy Poverty Advisory Hub[5], are supporting European local and regional authorities in implementing actions against energy poverty through capacity building and knowledge sharing.

These policy initiatives focus on reducing material causes underlying energy poverty, namely poor energy efficiency, low income and high-energy costs (Simcock and Walker 2015). However, these underlying causes are not context neutral. In particular, the physical boundaries in which people are located can produce and reproduce those inequalities (Yenneti et al. 2016) that shape vulnerability to energy poverty (Bouzarovski 2014). In particular, the causes underlying energy poverty are associated with socio-economic and political-legal issues that are intertwined with spatial inequities. Cities are an exemplary material landscape that defines the spatial inequities underlying vulnerability to energy poverty (Bouzarovski and Simcock 2017).

The urban interdependencies between cores and peripheries result in spatially differentiated 'opportunity structures', wherein people living in peripheries have fewer economic opportunities (and thus income) than others (Roberts 2009). These spatially induced local low-income and low-opportunities concentrations make certain areas more vulnerable to energy poverty.

As an example, low-income areas are characterised by less efficient housing, wherein citizens are unable to invest in energy efficiency due to a financial and power gap (Bertoldi et al. 2021; Boardman 2013). At the same time, residential segregation (Massey and Denton 1988) results from market and power relations in the housing supply (Marcuse and Van Kempen 2002). In particular, property owners' allocation strategies and upgrade decisions can exhibit statistical discrimination, with lower-status households having lower opportunities to be offered efficient dwellings for rent (Grossmann 2019). As a result, in large cities where housing is less affordable, the greatest concentrations of groups living in rented or multiple-occupancy houses live in dwellings that are energy inefficient (Cauvain and Bouzarovski 2016).

These spatially induced material and economic inequalities further corroborate inequalities. Citizens living in low-income areas consume more energy to overcome the spatially induced concentration of emissions that reduces the efficiency of solar panels (Scheier and Kittner 2022). They also have higher energy needs to meet spatially induced medical issues (Ormandy and Ezratty 2012). In particular, health outcomes are poorer in low-income areas (Dorling and Dorling 2013), with lower job opportunities and financial pressure resulting in chronic stress (Pickett and Wilkinson 2012), and higher outdoor and indoor air pollution resulting in respiratory diseases and chronic stress (Sovacool 2012).

4 https://www.eumayors.eu/support/energy-poverty.html.
5 https://energy-poverty.ec.europa.eu/discover/epah-atlas_en.

This spatially induced chronic stress and systemic injustice, by impairing citizens' agency, can also negatively affect engagement with energy (Della Valle 2019) and energy citizenship (Della Valle and Czako 2022). Thus, a vicious cycle reinforcing the behavioural mechanisms and the procedural injustice underlying energy poverty may arise.

Interventions that apply insights from behavioural sciences can provide a solution to break this cycle (Della Valle 2019). As an example, nudges have proven to reduce socio-economic disparities, being effective at encouraging choices that are in people's best interests more on vulnerable than on high-income citizens (Mrkva et al. 2021). Yet, a policy approach based solely on nudges might raise concerns about the autonomy of people and their actual civic engagement (Della Valle and Sareen 2020). In particular, citizens can be vulnerable to profiling and discriminatory practices (Ranchordás 2019) and be valued only because they make choices that produce data (Ruppert et al. 2017).

With the world's urban population expected to rise from 55% in 2018 to 68% by 2050 and the increasing use of data-driven nudges in cities (Della Valle et al. 2021), policymakers must be able to consider and tackle the uneven spatially induced behavioural mechanisms underlying energy poverty in cities. This chapter examines the potential and the limitations of behavioural sciences for the analysis and the fight against the spatially induced behavioural mechanisms underlying energy poverty in cities.

14.2 An Additional Lens to Approaching Urban Energy Poverty

In the policy and research circles, energy poverty has been generally associated with the occurrence of three conditions: living in energy-inefficient homes, facing high-energy costs and having a low income (Pye et al. 2015). While it is crucial to support a general approach to understand and tackle energy poverty, adopting a policy lens that only looks at quantifiable conditions might increase the risk for policymakers to fail capturing the subjective and context-specific dimensions underlying energy poverty (Thomson and Bouzarovski 2018). For this reason, energy poverty scholars have recently proposed subjective indicators based on self-assessments of those directly affected by energy poverty (Thomson et al. 2017). Thanks to this more holistic analysis, scholars have identified a fourth driver of energy poverty: behaviour (Kearns et al. 2019). This line of research highlights that the analysis of the material determinants of energy poverty cannot be separated from an analysis of the processes that drive people's decisions related to energy poverty, such as those related to energy efficiency, energy consumption and household budget (Della Valle 2019), and active participation to the energy decision-making process (Della Valle and Sareen 2020).

In this regard, the lens of behavioural economics and psychology (thereafter, behavioural sciences) offer useful evidence-based insights for understanding and improving the processes that drive these types of decisions. This potential has already been widely recognised by many governments and international organisations, such as the European Commission, which have begun to integrate the results of behavioural sciences in the formulation of public policies in various policy areas (Baggio et al. 2021; Van Bavel et al. 2013).

To date, EU Member States have taken several measures to address energy poverty and protect people at risk of falling into energy poverty, including (i) financial interventions;

(ii) consumer protection measures; (iii) energy-efficiency measures; and (iv) information measures (Pye et al. 2015). Although essential for addressing material determinants of energy poverty, these measures rarely succeed in addressing a multifaceted problem in a fully effective way. In addition to the complex nature of the problem, underlying this partial effectiveness is the reference model on which these policy measures are based, namely the model that depicts the actors targeted by policies as fully rational decision-makers (Della Valle 2019).

The assumptions of rational choice theory have been a reference model for public policy-making for decades, since they allow making predictions about the effects of public policies. Nevertheless, extensive empirical evidence has suggested that citizens make their choices in ways that are not entirely rational. In particular, citizens can make rational decisions, but they need cognitive resources to do so, and these are limited (Simon 1955, 1957). Therefore, to overcome these cognitive limitations, they use the so-called heuristics (Tversky and Kahneman 1974). While useful to deal with complex decision contexts, these tools may lead to systematic and predictable errors (Kahneman 2003, p. 1450). The evidence of these errors not only has enabled to describe the observed 'irrational' patterns of behaviour and to more accurately predict the impacts of existing policies, but also offered complementary policy tools that leverage people's actions to solve societal problems, including climate change (Chetty 2015).

One of the major contributions of behavioural economics is that it enabled to acknowledge the fundamental role that context has on citizens' actions. Citizens are rational, but some situational factors expose them to an automatic and often more error-prone decision-making process (Simon 1955). These errors create externalities, i.e. costs to society, and internalities, i.e. self-imposed costs that people are unable to internalise (Loewenstein and Chater 2017). Among the main situational factors that expose citizens to these types of decisions, of particular relevance to the problem of energy poverty, is the condition of resource scarcity. Several experimental studies have shown that when people face a gap between their needs and the resources available to meet them, they have fewer cognitive resources available to make optimal decisions for themselves and society (Mani et al. 2013). These studies point out that people living in scarcity conditions are as capable of making optimal decisions as those living in normal conditions, but the context of resource scarcity in which they live imposes a tax on their cognitive resources (Della Valle 2019).

The urban context can illustrate these cognitive effects more evidently. In particular, spatial segregation imposes a tax on individuals' cognitive resources that will be less available for other decisions requiring mental effort. In some urban areas, the chronic stress induced by the spatially induced lower job opportunities and financial pressure (Dorling and Dorling 2013) can activate a vicious cycle by reinforcing the behavioural mechanisms that drive energy poverty. As an example, it can lead to disengagement with energy (Della Valle 2019) or demanding energy rights (Della Valle and Czako 2022).

When people live in contexts of resource scarcity, they tend to be more prone to use cognitive shortcuts to process information (Simon 1972), like social norms (Bicchieri 2005). These are context-specific injunctions to act or to abstain from acting and are sustained by other people's approval and disapproval (Elster 1989). Thus, depending on the prevalent behaviour in the surrounding context, social norms might also prescribe behaviours that exacerbate suboptimal conditions. As an example, people might fail to engage in certain behaviours because of the fear of being stigmatised, like sharing their experience of unmet energy needs (Della Valle and Sareen 2020).

Behavioural sciences not only enable to detect the behavioural mechanisms underlying energy poverty but also to propose concrete applications to improve the quality of decision-making and reduce disparities.

First, as spatially induced cognitive scarcity makes people more vulnerable to manipulations by malevolent actors, behavioural sciences can inform the design of regulatory instruments to protect consumers from the risk of private actors who aim to exploit people's weaknesses (Loewenstein and Chater 2017). In particular, malevolent actors might implement 'dark patterns' (also called 'sludges' or 'bad nudges') to increase their profits, while decreasing people's welfare (Hansen and Jespersen 2013). As an example, private actors might exploit people's tendency to accept the status quo (Samuelson and Zeckhauser 1988) by designing unethical privacy default options in the context of energy monitoring enabled by smart meters. This might have consequences for citizens' privacy who might unconsciously agree that their data will be used for commercial purposes (Mylrea 2017). In addition, malevolent actors might exploit information overload with a simple 'information visit' to trick residents into agreeing to technological measures, losing their right to tenant protection (Grossmann 2019).

Second, behavioural sciences can inform the design of complementary interventions to tackle some of the roots of spatially induced energy poverty. Above all, behavioural sciences can enrich the policy toolbox through the nudging approach (Thaler and Sunstein 2008). Nudges are cost-effective interventions that alter people's behaviour in a systematic and predictable manner while respecting people's freedom. Practically, in contexts at risk of spatially induced cognitive scarcity, like cities, policymakers can exploit the nudging approach to save people's cognitive resources, being these already challenged by the experience of resource scarcity, and make these resources available for other daily tasks requiring cognitive capacity. This approach has the potential to dilute some of the negative effects of spatial injustice by improving the quality of the decisions, such as those related to savings and energy (Della Valle 2019).

For example, policymakers can exploit the insight that people facing scarcity are more likely to be influenced by a tendency to prefer the status quo (i.e. they tend to prefer to do nothing and stick to the given option (Samuelson and Zeckhauser 1988) and enable them to save cognitive resources for other decisions. Practically, in the context of designing a subsidy programme or regulating energy contracts, policymakers can set as a default the enrolment to a subsidy programme or a more convenient and green energy contract (Della Valle 2019).

Also, to promote better saving and budgeting capacities, policymakers, in collaboration with banks and community groups, can promote the implementation of commitment devices. As an example, they can create 'dedicated accounts' that offer concrete targets with mild commitment (if the committed target is not met, a penalty occurs (Bertrand et al. 2004). They can also exploit reminders as a way to increase people's attention to long-term consequences, such as through messages reminding them to save or to reach a previously committed target (Della Valle 2019).

In addition to improving the quality of daily decision-making related to savings and energy, policymakers can use the nudging approach to promote the effectiveness of energy-efficiency interventions. In particular, the often-observed energy performance gap (i.e. the difference between expected energy savings and actual energy savings (Galvin 2014) suggests that it is better to implement interventions that make domestic engagement with energy 'easy', such as by delegating tasks to technology (Marres 2016). Practically, nudges can be implemented in the form of defaults, such as by making green options the default (like efficient thermostat settings (Brown et al. 2013), or in the form of a social norm, such as by providing information about similar peers' energy behaviour in a real-time display (Caballero and Della Valle 2021).

At the same time, it is not obvious that these interventions end up unlocking the full energy efficiency potential, thus containing the energy performance gap. To be effective, these interventions must carefully account for the context-specific behavioural patterns that precede technological interventions (Della Valle et al. 2018). As an example, when people experience resource scarcity, they tend to prefer engaging in behaviours providing immediate (but smaller) benefits and perceive they have little control over their life outcomes (Della Valle 2019). As a result, they might fail to engage with technologies, like smart meters, and improve their energy bills (Fanghella and Della Valle 2021). People facing conditions of scarcity might also have different energy needs from those assumed by the techno-centric estimates, and in the context of energy efficiency interventions, a 'prebound effect' might arise (Sunikka-Blank and Galvin 2012).

These insights suggest that interventions cannot afford to be designed without detecting first the context-specific energy needs and practices. Eliciting people's experiences and needs with energy is, however, a complex endeavour, especially when people feel that their personal experiences (i.e. making few laundries) and needs (i.e. higher indoor temperatures to cope with medical conditions) can be stigmatised (Hards 2013). In this case, nudges can help people overcome the stigma associated with sharing their experiences and needs. For instance, a nudge can be implemented in the form of messages that embed dynamic norms ('more and more people are sharing their experiences related to energy poverty') (Della Valle and Sareen 2020).

Overall, interventions that delegate tasks to home automation technologies and make decisions 'easy' can contribute to reducing the energy performance gap and saving cognitive resources for other complex daily decisions. At the same time, these interventions might also contribute to further reducing people's control over their domestic life and possibility to actively engage in energy decision-making process (Della Valle and Czako 2022). As the condition of deprivation already negatively affects how people perceive the way they control their life (Sen 2001), it is crucial that behaviour change strategies that aim to tackle energy poverty by promoting energy efficiency and energy conservation do not further suffocate the exercise of one's agency, but rather empower it (Della Valle and Sareen 2020).

These concerns provide a rationale for a composite approach to address the behavioural mechanisms underlying energy poverty. In particular, in contexts of spatially induced cognitive scarcity, nudges should be complemented by other strategies that also boost one's agency.

The nudging approach, relying on the view that human cognitive architecture consists of a dual system (System 1 leads to more automatic and intuitive decisions, whereas System 2 enables more reflective, controlled and analytical ones (Kahneman 2003, 2011), represents only one illustration of a wider range of behaviourally informed policy tools (Loewenstein and Chater 2017). In particular, policymakers can enhance the quality of people's decision-making process not only by harnessing System 1 through the nudging approach but also by engaging and strengthening System 2. The boosting approach aims exactly to accomplish this goal (Grüne-Yanoff and Hertwig 2016). This approach is often traced back to a capabilities-based approach and a socio-psychological understanding of human beings as inherently capable of acquiring greater capabilities and degrees of freedom of action. That is why it is often seen as a key approach to tackle energy poverty in contexts characterised by injustices (Della Valle and Sareen 2020).

As for nudging, the basis for boosting interventions dates back to findings in the behavioural sciences. This approach acknowledges that heuristics are tools enabling to deal with complex decision environments; however, they lead to errors only when the surrounding environments provide misleading information (Gigerenzer et al. 2007; Wegwarth and Gigerenzer 2013). More particularly, even when under cognitive stress, people can make decisions that are beneficial for themselves and their surroundings, if they use rules that are compatible with their socio-material environment (Hertwig 2017).

Practically, boosts do not target behaviour, but empower human agency by targeting core competencies (Hertwig 2017). These can be associated with education-like interventions and increase key skills that are not taught in school. In particular, boosts aim to increase area-specific (e.g. understanding energy information) (Blasch et al. 2022) and general competencies (e.g. statistical literacy) (Kozyreva et al. 2020). They also aim to target human cognition (e.g. decision strategies) and the decision environment (e.g. information representation) (Hertwig 2017).

Therefore, to respond to the urban spatial injustices that make citizens vulnerable to energy poverty, policymakers can aim to train and strengthen citizens' cognitive system and promote the use of simple decision rules in the appropriate context.

As an example, policymakers, in collaboration with community groups, can promote the competence of acting as one's choice architect (or self-nudge). In particular, people can be trained to design daily environments that induce them to make optimal decisions, such as by using social pressure and self-commitment to stick with the plan to save a certain daily amount (Reijula and Hertwig 2020). They can also equip people with experience-based guidelines, or fast-and-frugal decision trees (Hafenbrädl et al. 2016), to provide prescriptive guides for daily decision making.

Interventions that boost core competencies, such as digital, statistical and financial literacy, might not only empower people to navigate through complex situations autonomously, but also to better hold key actors accountable. As an example, inoculation strategies, which are usually used to tackle digital misinformation (Roozenbeek and van der Linden 2019), would help citizens recognise manipulative strategies before they encounter them in real life.

Boosts can also be implemented in the form of information provision in a way that is not cognitive demanding. As an example, presenting statistical information in a transparent numerical format can be an effective strategy to boost people's statistical literacy (Hoffrage et al. 2000).

Finally, training people in democratic deliberation practices (including participatory budgeting) can be an effective strategy both to empower people to hold key actors accountable (such as during the negotiations with landlords) and to promote the active citizen engagement that is required to make cities truly people-centric (Ranchordás 2019).

14.3 Limitations and Ways Forward

There are many criticisms on the appropriateness of the use of insights from behavioural sciences to address complex policy problems (Bhargava and Loewenstein 2015), like energy poverty. In particular, concerns have been raised against their ability to tackle the deeper dimensions of policy problems, such as their economic, social and spatial ones (John 2018). The use of behavioural insights to address policy problems has also been criticised for being technocratic and top-down, as these draw on scientifically gathered behavioural evidence and expertise, and not on the views and knowledge of involved laypeople (Ewert 2020). In addition, the use of behavioural insights has been blamed for potentially contributing to make the relationship between decision-makers and citizens even more asymmetric (Jones et al. 2013), and to limit the actual and active engagement of citizens in the betterment of their surrounding (Button 2018). Finally, low-cost behavioural interventions, like nudges, have been criticised as they can provide an excuse to policymakers not to implement interventions involving more costly investments in infrastructures (Della Valle and Sareen 2020).

In the urban context, where segregated people face systemic injustices in terms of opportunities, health and income, the cost-effectiveness discourse should not serve as an excuse to jeopardise stronger interventions. Complex problems can be solved only with strong regulatory changes (i.e. introduction of minimum energy performance standards, like those advanced in the Revision of the Energy Performance of Buildings Directive)[6]. Foregoing such regulatory changes would otherwise result in unfairly shifting the responsibility to change only to vulnerable citizens.

Yet, behavioural sciences can provide an additional useful lens to understand and tackle some dimensions of complex problems. Complex problems often embed a behavioural component, and behavioural sciences enable to detect whether this behavioural component exists and whether it can be engineered with behavioural levers in a broad policy package (Loewenstein and Chater 2017). In particular, to enhance policy effectiveness, insights from behavioural sciences should be used always in combination with traditional policy tools (such as subsidies, taxes and bans). This would not only enable to address the individual and structural aspects of complex problems, but also to prevent policymakers from having an advantage to act opportunistically in neoliberal times (Ewert 2020).

Remarkably, the boosting approach shed light on the limitations of the technocratic approach to (behaviourally informed) policy-making. In particular, it opened the way for reconsidering the epistemological assumptions underlying the behavioural public policy process, where only experts' knowledge can point in the right direction. By promoting the acknowledgement of citizens as capable actors rather than cognitive misers (John and Stoker 2019), this approach calls for a process where interventions are more effective when they are co-designed not only by public officials and experts, but also by citizens (Evans and Terrey 2016).

---

6 https://ec.europa.eu/commission/presscorner/detail/en/QANDA_21_6686.

Only by making the process of intervention design built around partnership among people, state and communities, the approach to behaviour change as a solution to policy problems can become a common endeavour to be reached through co-production, participation and dialogue with representatives of different groups (Brandsen et al. 2018).

This discourse especially applies to the urban context, where spatially induced uneven patterns in opportunities prevent people to have their voices represented in the decision-making process. Contrariwise, enabling spatially urban segregated citizens to have their say on policy decisions not only will improve the quality of the policy agenda through experiential knowledge (Lieu et al. 2019), but also increase their agency by granting more control over the decisions affecting their lives (Della Valle and Sareen 2020).

14.4 Conclusions

With nearly 34 million Europeans who could not afford to keep their houses adequately warm in 2018, Energy Poverty has become a key policy priority in Europe. However, with its heterogeneous manifestations across and within the Member States, a challenge remains as to how effectively implement countering actions. Energy poverty might even differently manifest within cities due to spatially induced inequities, where many causes come into play, such as spatially reduced people's agency.

With the world's urban population expected to rise from 55% in 2018 to 68% by 2050, and the increasing use of data technology to promote positive behaviour changes in cities, it is essential for policymakers to be able to detect the mechanisms that drive vulnerability to energy poverty in cities.

This chapter provided a short perspective on how behavioural sciences can help detect and address some of the mechanisms that drive energy poverty in cities, namely the behavioural ones. Practically, it illustrated the potential and limitations of two main applications of behavioural sciences to address the spatially induced behavioural mechanisms underlying energy poverty in cities: nudging and boosting. Among these, the boosting approach offered an opportunity to reflect on reducing disparities through people's involvement in the policy process (Evans and Terrey 2016) and civic engagement (Della Valle and Sareen 2020).

The chapter also acknowledged that while the scale and the scope of behaviourally informed interventions could be limited for complex problems like energy poverty, a behavioural perspective is key to capture the behavioural dimensions of complex problems and better inform the policy-making process. That is why the approach to behaviour change should be a part of a broader solution package to tackle energy poverty in cities, co-designed by representatives of different groups (Brandsen et al. 2018).

Disclaimer

The views expressed are purely those of the author and may not in any circumstances be regarded as stating an official position of the European Commission.

References

Baggio, M., Ciriolo, E., Marandola, G., and van Bavel, R. (2021). The evolution of behaviourally informed policy-making in the EU. *Journal of European Public Policy* 28 (5): 658–676.

Bertoldi, P., Boza-Kiss, B., Della Valle, N., and Economidou, M. (2021). The role of one-stop shops in energy renovation-A comparative analysis of OSSs cases in Europe. *Energy and Buildings* 250: 111273.

Bertrand, M., Mullainathan, S., and Shafir, E. (2004). A behavioral-economics view of poverty. *American Economic Review* 94 (2): 419–423.

Bhargava, S. and Loewenstein, G. (2015). Behavioral economics and public policy 102: beyond nudging. *American Economic Review* 105 (5): 396–401.

Bicchieri, C. (2005). *The Grammar of Society: The Nature and Dynamics of Social Norms*. Cambridge University Press.

Blasch, J.E., Filippini, M., Kumar, N., and Martinez-Cruz, A.L. (2022). Boosting the choice of energy-efficient home appliances: the effectiveness of two types of decision support. *Applied Economics* 54 (31): 3598–3620.

Boardman, B. (2013). *Fixing Fuel Poverty: Challenges and Solutions*. Routledge.

Bouzarovski, S. (2014). Energy poverty in the European Union: landscapes of vulnerability. *Wiley Interdisciplinary Reviews: Energy and Environment* 3 (3): 276–289.

Bouzarovski, S. and Simcock, N. (2017). Spatializing energy justice. *Energy Policy* 107: 640–648.

Brandsen, T., Steen, T., and Verschuere, B. (2018). *Co-production and Co-creation: Engaging Citizens in Public Services*. Taylor & Francis.

Brown, Z., Johnstone, N., Hašcic, I. et al. (2013). Testing the effect of defaults on the thermostat settings of OECD employees. *Energy Economics* 39: 128–134.

Button, M.E. (2018). Bounded rationality without bounded democracy: nudges, democratic citizenship, and pathways for building civic capacity. *Perspectives on Politics* 16 (4): 1034–1052.

Caballero, N. and Della Valle, N. (2021). Tackling energy poverty through behavioral change: a pilot study on social comparison interventions in social housing districts. *Frontiers in Sustainable Cities* 2: 66.

Cauvain, J. and Bouzarovski, S. (2016). Energy vulnerability in multiple occupancy housing: a problem that policy forgot. *People, Place and Policy* 10 (1): 88–106.

Chetty, R. (2015). Behavioral economics and public policy: a pragmatic perspective. *American Economic Review* 105 (5): 1–33.

Della Valle, N. (2019). People's decisions matter: understanding and addressing energy poverty with behavioral economics. *Energy and Buildings* 204: 109515.

Della Valle, N. and Czako, V. (2022). Empowering energy citizenship among the energy poor. *Energy Research & Social Science 89*: 102654.

Della Valle, N. and Sareen, S. (2020). Nudging and boosting for equity? Towards a behavioural economics of energy justice. *Energy Research & Social Science* 68: 101589.

Della Valle, N., Bisello, A., and Balest, J. (2018). In search of behavioural and social levers for effective social housing retrofit programs. *Energy and Buildings* 172: 517–524.

Della Valle, N., Gantioler, S., and Tomasi, S. (2021). Can behaviorally informed urban living labs foster the energy transition in cities? *Frontiers in Sustainable Cities* 3: 11.

Dobbins, A., Nerini, F.F., Deane, P., and Pye, S. (2019). Strengthening the EU response to energy poverty. *Nature Energy* 4 (1): 2–5.

Dorling, D. and Dorling, D. (2013). *Unequal Health: The Scandal of Our Times*. Policy Press.

Elster, J. (1989). *The Cement of Society: A Survey of Social Order*. Cambridge University Press.

Evans, M. and Terrey, N. (2016). Co-design with citizens and stakeholders. In: *Evidence-Based Policy Making in the Social Sciences: Methods that Matter* (ed. G. Stoker and M. Evans), 243–262. Bristol University Press.

Ewert, B. (2020). Moving beyond the obsession with nudging individual behaviour: towards a broader understanding of behavioural public policy. *Public Policy and Administration* 35 (3): 337–360.

Fanghella, V. and Della Valle, N. (2021). A behavioral model for in-home displays usage in social housing districts. In: *Smart and Sustainable Planning for Cities and Regions: Results of SSPCR 2019* (ed. A. Bisello, D. Vettorato, H. Haarstad, and J. Borsboom-van Beurden), 511–524. Springer International Publishing AG.

Galvin, R. (2014). Making the 'rebound effect' more useful for performance evaluation of thermal retrofits of existing homes: defining the 'energy savings deficit' and the 'energy performance gap'. *Energy and Buildings* 69: 515–524.

Gigerenzer, G., Gaissmaier, W., Kurz-Milcke, E. et al. (2007). Helping doctors and patients make sense of health statistics. *Psychological Science in the Public Interest* 8 (2): 53–96.

Grossmann, K. (2019). Energy efficiency for whom? a conceptual view on retrofitting, residential segregation and the housing market. *Sociologia Urbana E Rurale* 119: 78–95.

Grüne-Yanoff, T. and Hertwig, R. (2016). Nudge versus boost: how coherent are policy and theory? *Minds and Machines* 26 (1-2): 149–183.

Hafenbrädl, S., Waeger, D., Marewski, J.N., and Gigerenzer, G. (2016). Applied decision making with fast-and-frugal heuristics. *Journal of Applied Research in Memory and Cognition* 5 (2): 215–231.

Hansen, P.G. and Jespersen, A.M. (2013). Nudge and the manipulation of choice: a framework for the responsible use of the nudge approach to behaviour change in public policy. *European Journal of Risk Regulation* 4 (1): 3–28.

Hards, S.K. (2013). Status, stigma and energy practices in the home. *Local Environment* 18 (4): 438–454.

Hertwig, R. (2017). When to consider boosting: some rules for policy makers. *Behavioural Public Policy* 1 (2): 143–161.

Hoffrage, U., Lindsey, S., Hertwig, R., and Gigerenzer, G. (2000). Medicine. Communicating statistical information. *Science* 292 (5518): 853–855.

John, P. (2018). *How Far to Nudge?: Assessing Behavioural Public Policy*. Edward Elgar Publishing.

John, P. and Stoker, G. (2019). Rethinking the role of experts and expertise in behavioural public policy. *Policy & Politics* 47 (2): 209–226.

Jones, R., Pykett, J., and Whitehead, M. (2013). Psychological governance and behaviour change. *Policy & Politics* 41 (2): 159–182.

Kahneman, D. (2003). Maps of bounded rationality: psychology for behavioral economics. *American Economic Review* 93 (5): 1449–1475.

Kahneman, D. (2011). *Thinking, Fast and Slow*. Macmillan.

Kearns, A., Whitley, E., and Curl, A. (2019). Occupant behaviour as a fourth driver of fuel poverty (aka warmth & energy deprivation). *Energy Policy* 129: 1143–1155.

Kozyreva, A., Lewandowsky, S., and Hertwig, R. (2020). Citizens versus the internet: confronting digital challenges with cognitive tools. *Psychological Science in the Public Interest* 21 (3): 103–156.

Lieu, J., Virla, L.D., Abel, R., and Fitzpatrick, C. (2019). "Consensus building in engagement processes" for reducing risks in developing sustainable pathways: indigenous interest as core elements of engagement. In: *Understanding Risks and Uncertainties in Energy and Climate Policy* (ed. H. Doukas, A. Flamos, and J. Lieu), 55. Cham: Springer.

Loewenstein, G. and Chater, N. (2017). Putting nudges in perspective. *Behavioural Public Policy* 1 (1): 26–53.

Mani, A., Mullainathan, S., Shafir, E., and Zhao, J. (2013). Poverty impedes cognitive function. *Science* 341 (6149): 976–980.

Marcuse, P. and Van Kempen, R. (2002). *Of States and Cities: The Partitioning of Urban Space.* Oxford University Press on Demand.

Marres, N. (2016). *Material Participation: Technology, the Environment and Everyday Publics.* Springer.

Massey, D.S. and Denton, N.A. (1988). The dimensions of residential segregation. *Social Forces* 67 (2): 281–315.

Mrkva, K., Posner, N.A., Reeck, C., and Johnson, E.J. (2021). Express: do nudges reduce disparities? choice architecture compensates for low consumer knowledge. *Journal of Marketing* 67–84.

Mylrea, M. (2017). Smart energy-internet-of-things opportunities require smart treatment of legal, privacy and cybersecurity challenges. *The Journal of World Energy Law & Business* 10 (2): 147–158.

Ochoa, R.G. and Graizbord, B. (2016). Privation of energy services in Mexican households: an alternative measure of energy poverty. *Energy Research & Social Science* 18: 36–49.

Ormandy, D. and Ezratty, V. (2012). Health and thermal comfort: from who guidance to housing strategies. *Energy Policy* 49: 116–121.

Pickett, K. and Wilkinson, R. (2012). Income inequality and psychosocial pathways to obesity. Avner Offer, Rachel Pechey, and Stanley Ulijaszek (eds), *Insecurity, Inequality, and Obesity in Affluent Societies*, Proceedings of the British Academy (London, 2012; online edn, British Academy Scholarship Online, 31 Jan. 2013), https://doi.org/10.5871/bacad/9780197264980.003.0010.

Pye, S., Dobbins, A., Baffert, C. et al. (2015). Addressing energy poverty and vulnerable consumers in the energy sector across the EU. *L'Europe en Formation* 4: 64–89.

Ranchordás, S. (2019). Nudging citizens through technology in smart cities. *International Review of Law, Computers Technology* 34 (3): 1–23.

Reijula, S. and Hertwig, R. (2020). Self-nudging and the citizen choice architect. *Behavioural Public Policy* 6 (1): 119–149.

Roberts, K. (2009). Opportunity structures then and now. *Journal of Education and Work* 22 (5): 355–368.

Roozenbeek, J. and van der Linden, S. (2019). Fake news game confers psychological resistance against online misinformation. *Palgrave Communications* 5 (1): 1–10.

Ruppert, E., Isin, E., and Bigo, D. (2017). Data politics. *Big Data & Society* 4 (2): 2053951717717749.

Samuelson, W. and Zeckhauser, R. (1988). Status quo bias in decision making. *Journal of Risk and Uncertainty* 1 (1): 7–59.

Scheier, E. and Kittner, N. (2022). A measurement strategy to address disparities across household energy burdens. *Nature Communications* 13 (1): 288.

Sen, A. (2001). *Development as Freedom*. Oxford Paperbacks.

Simcock, N. and Walker, G. (2015). *Fuel Poverty Policy and Non-Heating Energy Uses*. Lancaster: DEMAND Centre, Lancaster University.

Simon, H.A. (1955). A behavioral model of rational choice. *The Quarterly Journal of Economics* 69 (1): 99–118.

Simon, H.A. (1957). *Models of Man; Social and Rational*. Wiley.

Simon, H.A. (1972). Theories of bounded rationality. *Decision and Organization* 1 (1): 161–176.

Sovacool, B.K. (2012). The political economy of energy poverty: a review of key challenges. *Energy for Sustainable Development* 16 (3): 272–282.

Sunikka-Blank, M. and Galvin, R. (2012). Introducing the prebound effect: the gap between performance and actual energy consumption. *Building Research and Information* 40 (3): 260–273.

Thaler, R.H. and Sunstein, C.R. (2008). *Nudge: Improving Decisions about Health, Wealth, and Happiness*. Yale University Press.

Thomson, H. and Bouzarovski, S. (2018). *Addressing Energy Poverty in the European Union: State of Play and Action*. Manchester: EU Energy Poverty Observatory.

Thomson, H., Bouzarovski, S., and Snell, C. (2017). Rethinking the measurement of energy poverty in Europe: a critical analysis of indicators and data. *Indoor and Built Environment* 26 (7): 879–901.

Tversky, A. and Kahneman, D. (1974). Judgment under uncertainty: heuristics and biases. *Science* 185 (4157): 1124–1131.

Van Bavel, R., Herrmann, B., Esposito, G., and Proestakis, A. (2013). Applying behavioural sciences to EU policy-making. Doc. EUR, 26033, 8.

Wegwarth, O. and Gigerenzer, G. (2013). Trust-your-doctor: a simple heuristic in need of a proper social environment. In: *Simple Heuristics in a Social World* (ed. R. Hertwig and U. Hoffrage), 67–102. Oxford University Press.

Yenneti, K., Day, R., and Golubchikov, O. (2016). Spatial justice and the land politics of renewables: dispossessing vulnerable communities through solar energy mega-projects. *Geoforum* 76: 90–99.

Part III

Technologies and Data for Smart and Low-carbon Urban Futures

15

Smart Energy Future and Smart Cities

Wadim Strielkowski[1,2]

[1] *Department of Agricultural and Resource Economics, University of California, Berkeley, USA*
[2] *Department of Trade and Finance, Faculty of Economics and Management, Czech University of Life Sciences Prague, Prague, Czech Republic*

15.1 Introduction

On 7 May 2021, the US fuel pipeline operator Colonial Pipeline which operates the main gas artery along the East Coast became a target of the ransomware cyberattack carried out by an Eastern European-based criminal cybergang called 'DarkSide' (Nakashima et al. 2021). The attack had significant implications, including the imposition of a state of emergency, temporary gasoline shortages caused by panicked motorists stocking up on fuel for fear of a gasoline shortage caused by the hack and rising gasoline prices.

Three days later, the hackers published a statement on their dark website saying that the attack was 'apolitical' and all they wanted was money, perhaps in order to get rid of the accusations of working on behalf of any hostile government or regime which would make the matters worse (Cox 2021). The attack forced the energy company to shut down, threatening the safety of its pipeline system, followed by a ransom of $5 million that was paid to the hackers a few days later.

The event described above clearly demonstrates the importance of energy security in our digitalised, post-pandemic world. The Colonial Pipeline ransomware attack that for a brief period of time crippled America's infrastructure and triggered gas shortages is just one example of the growing cyber risks threatening our energy systems.

Nowadays, the concept of energy security gains special importance with the growth of cities and large urban agglomerations. The cities of the twenty-first century are not just population hubs but also what is now known as 'smart cities'. They dominate economic growth and innovation and constitute the centres of digitalisation and progress.

This chapter focuses on smart energy systems and energy security in smart cities. It tackles the concept of the smart city, explains the importance of cities nowadays as the centres of innovation, progress and economic growth and discusses the means and tools for ensuring that smart cities operate safe and secure energy infrastructures that are fully protected

Smart Cities, Energy and Climate: Governing Cities for a Low-Carbon Future, First Edition.
Edited by Oleg Golubchikov and Komali Yenneti.

from all kinds of natural and man-made disasters such as the cyberattack on the Colonial Pipeline fuel operator.

15.2 The Concept of the Smart City

In general terms, a smart city can be defined as an innovative city or community that wants to improve the quality of life of its inhabitants by internalising intelligent technologies (Picon 2015; Yang 2020; Silva et al. 2018; Ismagilova et al. 2019). This points to a future in which computer systems are connected to the Internet of Things (IoT) – smart devices and sensors can receive valuable information or data that we can use to manage the city efficiently (Teng et al. 2019; Kumar et al. 2020; Strielkowski 2020). Smart cities can be implemented in a number of ways, including smart buildings, smart infrastructure and smart services. Smart cities use connected technologies and data to improve the efficiency of urban service delivery, improve quality of life and increase fairness and prosperity for residents and businesses (Shen et al. 2018). Smart city is a term for a city that uses information and communication technology to improve its ability to reduce resource consumption, waste and overall costs (Strielkowski et al. 2020). It is an urban space that uses data collected by smart devices such as smart buildings, smart infrastructure and smart services, and then uses the insights from the data to efficiently manage assets, resources and services (Zhang 2020). IoT is typically being used in smart cities to make cities safer, greener and more efficient (Alavi et al. 2018).

Moreover, a smart city is a city that uses sensor-guided data collection to automate and orchestrate a wide range of services, thereby reducing costs and reducing environmental impacts through more sustainable development. IoT devices are used to measure and monitor everything from air quality and water quality to traffic, congestion and air pollution. Translating raw data into alerts, insights and actions requires the right tools, and that is where technology providers and app developers come in. We can reap the benefits of the smart city, not only in terms of economic development but also in terms of the quality of life of its dwellers and citizens. In order to do that, we should work together to collect and contextualise huge amounts of data that can be used to improve the components and systems that run our cities. For example, the ICT framework consists of various machines and sensors that connect to the IoT network and transmit data using radio technology and the cloud. Cloud-based IoT applications can then receive, manage, analyse and process traffic and data in real time. There are many different types of IoT, ranging from consumer IoT and enterprise IoT to manufacturing and industrial IoT (Osman 2019; Qureshi et al. 2020).

15.3 Growing Importance of Cities

With regard to the above, one should understand the growing importance of the world's cities as the centres of economic and social life and the motors of innovations and growth. The United Nations (UN) estimate that 66% of the world's population will be urban by 2050 (United Nations 2014). Moreover, another prediction by UN provides the expected numbers of dwellers in world's major cities (see Figure 15.1).

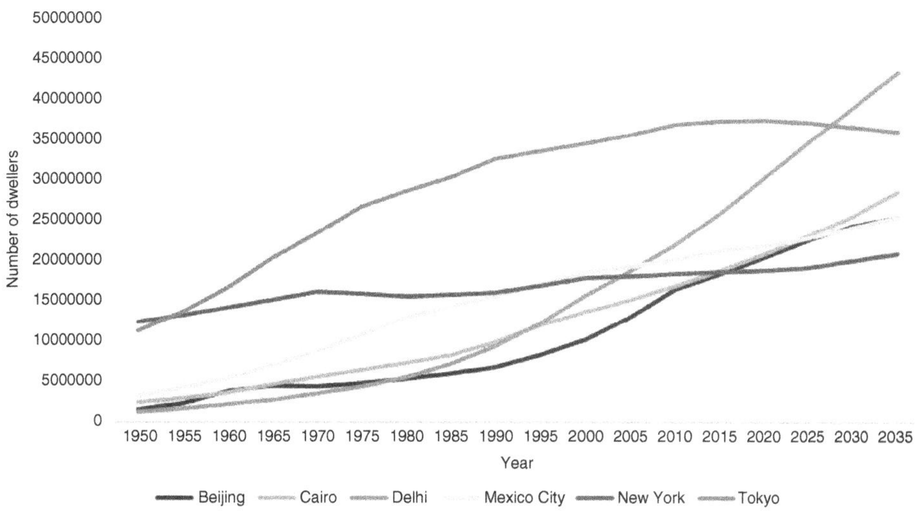

Figure 15.1 Projection of population of the six selected megacities (1950–2035). *Source:* Adapted from United Nations (2018).

Large urban centres and world's major metropolises are capable of sustaining the reliability and security of their energy supply based on the advancements in the development of their energy sectors as well as on enhancements in social efficiency and the quality of life of their dwellers (Omer 2008; Lisin et al. 2018). However, all of the above requires massive investments into novel managerial and information and communication technology solutions that would create the prerequisites for a decent level of quality of life, ensure the focus on the socio-economic direction of the city economy as well as enable the smart cities to play their key functional roles (Newbery et al. 2018; Ahad et al. 2020).

It becomes clear that a technology-based intelligent city is not only a concept but there are also several combinations of technological infrastructure that constitute the concept of smart cities. Furthermore, smart city has social capital and it could be possible and easier to create an intelligent urban concept if there is a mix of education, culture and arts, business and commerce.

A smart city can have value, considering what it does with technology and how much technology it can have. As urbanisation continues to grow globally, with the total population of cities expected to increase by 75% by 2050, there is an increasing demand for smart and sustainable environments that offer citizens a better quality of life. Solution such as 'Living Labs', an innovative and interactive method to improve citizens' sustainable growth and lifestyles, exist for achieving just that (Voytenko et al. 2016).

By using smart technologies like ICT and other technologies, smart cities also optimise territorial, economic as well as environmental resources. The best option they offer is a solution to the problems of urban planning, urban mobility and urban development in the modern world. These services are increasingly made possible by internet-based connected applications with a wide range of sensors, electrification and power generation solutions, including smart lighting, air conditioning, water and wastewater systems, energy management systems and much more. With the urbanisation increasing and the economic and environmental crises that call for new urban governance and citizens, smart cities must find new ways to implement policies and urban projects, as well as new forms of efficient energy management, governance systems as well as urban infrastructure.

15.4 Smart Technologies and Energy

In the next few decades, many new IoT and smart city projects are likely to emerge as available technologies become more widely deployed and targeted. Public lighting or traffic lights are one of the most common applications of IoT in smart cities. The solution includes a robust, multi-node smart pole that provides connectivity and data routing to multiple devices and nodes within the smart pole.

Portable devices, sensors and software collect and analyse user data and send messages to other technologies. Smart homes in smart cities and beyond can be controlled with smart lights, smart thermostats and smart appliances such as air conditioners, refrigerators, air conditioners and air conditioners. New York City is investing in smart sensors that allow drivers to determine which areas of the environment need to be changed to reduce the potential for accidents and congestion. For example, smart traffic lights can determine where drivers most often make sharp turns or stop hard. Portable devices can also be used for public safety to improve first responders' response times in emergencies by providing optimised routes to sites of action and tracking construction workers and firefighters in life-threatening locations. There are also a number of game-changing initiatives such as the digitisation of waste management systems and many more to come.

The more things are interconnected, the more cyber attackers can infiltrate these systems to access sensitive data and disrupt potentially critical systems used for public health, security, education, transportation and other critical infrastructure. There is a great potential for cyberattacks on interconnected systems and sensors, devices and the infrastructure itself.

The IoT is particularly vulnerable, as its use is increasing exponentially. According to the US Department of Homeland Security, the number of IoT devices in the United States alone will increase to 740 million by 2026 and 1.5 billion by 2030. It was found that 98% of IoT traffic is unencrypted, meaning that cybercriminals who successfully bypass the first line of defence can collect personal and confidential information (WEF 2021).

Urban planners seeking to take advantage of smart urban technology must apply the lessons of the past and take into account known weaknesses to ensure that the systems and people we serve remain safe. Whether it is the management of critical water and electricity systems or the handling of sensitive voter registration data, the technology that enables modern life should be protected against cyberattacks.

Infrastructure scheme of the typical smart city involves the connection of individual sectors which is carried out according to the principle of 'a system of a sub-systems', which means achieving compliance with the mechanism of integrated management (see Minoli et al. 2017). A characteristic feature of 'smart' systems is that their degree of intelligence is ensured by a variety of infrastructure options providing a controlled adaptation of the structure to changing external conditions, the active development of systems based on information and communication technologies as well as the presence of a human as a subject and object of functioning and development of the 'socially oriented meta-system' (SoM). This 'triple SoM' (which is a system based on the three pillars – infrastructural, informational and intellectual) represents a framework of the 'smart' metropolis and, energy as a system of not only life support but also the life of the city as a 'living' organism, should develop as a meta-system (see Hefnawy et al. 2016).

With respect to the above, it becomes clear that the concepts of the IoT, in which all devices are interconnected via networks, as well as the Internet of Energy (IoE), in which each subject (consumer) not only selects the information she or he needs but also making a decision about his or her own or collective energy supply, taking into account the interests of other members of society and the entire socio-natural environment, becomes a key requirement for the smart cities (Cabelkova et al. 2020).

In addition, smart grid also becomes part of the overall smart city model, not only uniting physical infrastructure objects (power system, traffic and communications) and high-level internet information into a single intellectual complex but also forming the systems' energy information system as a single human–machine ecosystem (Harrison et al. 2010; Gubbi et al. 2013; Strielkowski 2017; Strielkowski 2021).

This is a single SoM that integrates individual infrastructure subsystems according to the types of energy used as well as the degree of their localisation and centralisation. The main idea, however, is that they manage (independently) the individual objects by several agents and coordinate these actions at the level of common goals, strategic plans and risk assessment of decisions taken.

Decentralised systems are not autonomous, and their interconnection to a common power grid structure requires a sufficiently developed information and control network built on the smart grid with smart method (Alanne and Saari 2006). When it comes to the power scheme of large world megapolises, the problem of using the smart grid taking into account the adjoining areas both with regard to the development of decentralised systems and their integration into the general infrastructure as well as with regard to the increase in the social efficiency of the energy saving become quite relevant.

15.5 Energy Infrastructure for Smart Cities

The growing role of human beings in creating the IoE requires the development of new information technology and intelligent network-centric control algorithms. Network-centric management is a multi-agent management of individual (local) objects while coordinating their operation with the help of the network energy–information infrastructure. At the same time, it is not the control actions themselves that are centralised but only the general idea of management and risk assessment of the decisions made.

The presence of a single target network-centric architecture is a distinctive feature of most projects in smart cities and smart grids in major cities of the world. With regard to the above, it is useful to recall the spectacular large-scale projects of the smart cities that were implemented in Masdar (Abu Dhabi), Stockholm (Sweden), Amsterdam (the Netherlands) and Yokohama (Japan) (Madu et al. 2017). Most of these projects focus on the development of 'green' energy using renewable energy sources (RES), passive (zero-power) homes, the use of incineration, the development of electric lighting systems, electric vehicles and energy saving.

For example, Reiche (2010) describes the process of transforming oil dependency into renewable energy leadership and a transition from a twentieth-century carbon-based economy into a twenty-first-century sustainable economy. He is using an example based on the case study of 'Masdar City', a carbon-neutral, zero-emission town located in Abu Dhabi which is almost solely based on solar photovoltaic (PV) energy. He outlines the major characteristics of Masdar City, analyses the key features behind the project, describes the main actors for its implementation and identifies the main obstacles to creation and development as well as the policy behind Masdar City. The project is still ongoing (due to the constraints imposed by the financial crisis in 2008) but it raises many important concerns and sets up a good example for other smart cities around the world to follow.

The suburb of Stockholm – Hammarby Sjostad – is to become an innovative 'city of the future' with a complete renovation of the former industrial estate to create a 'green city' in the big city (Glazebrook and Newman 2018). The concept of a new urban project focuses on creating a district heating and cooling system with electrical installations, using solar panels on the roofs of 'passive houses' for local electric heating to reduce the number of cars that will be replaced by bicycle transportation and electromobility.

The megaproject 'Smart City of Amsterdam' also deals with the creation of a 'green' energy company (an intelligent energy city). It is estimated that by 2025, up to 30% of the city's energy consumption (only 6% in 2012) will come from environmentally friendly sources of energy. Over 10 million square meters of roof space will be equipped with solar panels and the office buildings of the Zuidas Business Park will be equipped with autonomous fuel and energy elements. The city becomes an independent part of Amsterdam with its own energy complex and a developed intellectual network that integrates the transport, social and energy infrastructure of the region. The 'Old South' district, considered to be one of the dirtiest and poorest districts in Amsterdam, is being rebuilt: the site is creating a new 'city in the city' with innovative business centres, office apartments and luxury apartments. It also requires the conversion of the local power supply system, with the main electricity supply transition, both via external cables and its own power plants (Van Winden 2016).

One of the most ambitious projects for an energy-efficient smart city is the reconstruction of Yokohama's power system in Yokohama, Japan. The project includes various household energy management systems (HEMS), building energy systems (BEMS), municipal energy services (CEMS) and a dispatch control and data acquisition system (SCADA). A peculiarity of the Yokohama Smart Grid model is the widespread use of storage devices as an interconnection between consumers and the urban power grid from autonomous and external sources. The implementation of this system is expected at the expense of funds from electricity companies supplying equipment, including control systems for power plants. The city administration guarantees the companies a nationwide implementation of the smart grid system with positive operating experience in pilot projects.

Innovative development of energy in the cities of the world is on the way to the development of electric transport, fuel cells for autonomous power supply and the transition from the initial technology to the use of autonomic power systems and system batteries. And all these innovations are included in the structure of modern smart grid-powered cities.

Progress in the development of lithium-ion batteries stimulates their widespread use as household and network energy storage. In the world, the production and use of fuel cells

(FC) for autonomous power supply, especially in the fixed version, is actively growing. Their efficiency is 3.5 times higher than that of the internal combustion engine and 2 times higher than that of the best gas turbines. Therefore, the production of fuel cells in the world in five years increased from 7 to 35 thousand units and their capacity from 60 to 170 MW. These fuel cells are supposed to be actively used as backup power sources for the most responsible consumers (in hospitals, business centres, banking structures, etc.). It is possible that they will find applications for the power supply of individual objects in the zones of decentralised power supply.

All innovative solutions lead to the need to revise the infrastructure of the urban energy sector. In the world, this work is accompanied by the active development of smart grid-based systems. The cost of implementing intelligent networks has reached $20 billion, including the growth in the EU countries estimated to be at a rate of about 27–25%, about 16% in the United States, around 12% in China and about 30% in other Asia-Pacific countries. It is assumed that the development of smart grid technologies, among other things, will reduce the needs of energy-deficient regions in fuel and energy resources.

The Russian Federation began the development of a smart city concept in its largest city and its capital Moscow in 2014. The concept of an energy-efficient megacity Smart City called 'New Moscow' leverages global experience in creating smart grid systems as the infrastructure foundation for smart cities. At the same time, special attention will be paid to the territorial hierarchy of the smart city model, taking into account the new scheme for infrastructure development of a megapolis with adjacent areas (Drozdova and Petrov 2018). One of the qualitatively new principles of the organisation of the development of the adjacent areas is the cluster approach, which helps to integrate the social production and the infrastructure integration of the settlements. The development of the cluster principle of territorial production development of the economy also implies the corresponding development of its energy supply systems.

Previously, connecting all systems to the unified power distribution network was a requirement for an extension of the grid meaning additional costs for the construction of these networks, an increase in light load losses and an increase in the accident rate of the power grid complex. In clusters, due to a combination of adequate energy use and adequate energy production, the emphasis is placed on the energy balance of these internal units, eliminating the need for a major expansion of the distribution network complex.

At the same time, there is still a need for inter-cluster integration of these energy centres along limited and high-voltage network routes, with a focus on the future evolution of the load in new geographical nodes. All clusters have their own power scheme that takes into account the characteristics of local consumers, including the use of micro–co-generation plants, while the centralised electricity and heat supply from the power plants is operated in the territory of the EU 'old' city (Farzaneh et al. 2016).

The integration of energy sources and responsible consumers requires the creation of appropriate multi-agent management systems for energy information systems, as well as the development of an integrated (multi-functional) smart grid, linking these centres into a common system.

For various standard units of large cities, these future systems will be universal (in terms of reliability and energy supply requirements) and qualitatively individual, depending on the structure of the objects themselves and the role of the human being as the subject of management. At the level of a single consumer, the intelligent infrastructure can be represented as an integrated power system of a typical smart home.

The innovation of public services in smart cities, for example, by upgrading buildings to optimise energy consumption or improve the efficiency of public transport, can be combined with the use of technology tailored to the needs of companies to reduce operational costs (Simionescu et al. 2020). In today's knowledge-based economy, digital connectivity and solutions are key factors in terms of efficiency and profitability, new and innovative economic activity and ultimately a sustainable competitive advantage.

Generally speaking, some authors report that infrastructure improvements, such as intelligent street lighting, can significantly reduce the energy demand of a modern city. Smart grids have the potential to help distribution systems to better integrate intermittent renewable energies such as wind and solar, and smart city projects are investigating how smart, connected local energy storage systems can support more renewable energy sources on the grid.

15.6 Concluding Remarks

Overall, it becomes obvious that energy management witnessed the emergence of 'smart' technologies: intelligent networks, intelligent water and intelligent energy solutions. Big Data that will be widely applied in smart cities of the future would help to coordinate the extraction and use of wind and solar energy in a mix together with traditional energy sources. In order to work effectively with large data and smart cities, government and public efforts should create infrastructure and prevent crime. As a fertile land for business owners, data-based smart cities are where technology can ultimately increase the quality of life, improve security and protect the environment of large urban hubs.

Smart cities use data and technology to create efficiency, improve sustainability, improve economic development and improve the quality of life of the people who live and work in the city. Electric companies, in collaboration with city officials, technology companies and a number of other institutions, are among the main players to help drive the development of smart cities worldwide.

Today, smart city-safe energy infrastructure is paving the way for the next generation of investments and services for consumers. In order to function properly, smart cities need to establish a functioning two-way flow of information on IoE, peer-to-peer (P2P) intelligent neighbourhoods, micro-grids, electric vehicles, fibres and other interconnected intelligent solutions. All of these involve multiple complexities of delivering products and services in the areas of distributed production, implementation of renewable energy, achieving energy efficiency and infrastructure that still need to be solved.

Smart city employs a combination of data collection, processing and dissemination technologies, as well as network and computer technologies and data security and privacy measures that encourage application innovation to improve the overall quality of life for its citizens and include utilities, healthcare, transport, entertainment and government services. Whether it is improving security, resilience, sustainability, traffic jams, public security or municipal services, every community can have several reasons to invest in building smart cities based on smart systems.

The local governments represent the key actor of the public sector in smart city space. However, not all projects and plans should cover the whole city. In view of the management, scale and other implementation barriers, there is pressing need for strong networks of leaders to drive smart city policies and investments which might become a leading element of economic sustainability.

Few organisations or jurisdictions are affected as quickly or as deeply as smart cities, and the impact of cyberattacks and embedded computing on their infrastructure is felt in many industries and areas. At the same time, the number of potential attack areas that the smart cities need to protect and defend themselves from has also increased enormously. This has opened up the possibility for smart cities to integrate cyber, physics, technology and infrastructure. Energy security is becoming a key element of smart cities and it has to be fostered by the authorities and embedded into the daily life routines of its inhabitants.

References

Ahad, M.A., Paiva, S., Tripathi, G., and Feroz, N. (2020). Enabling technologies and sustainable smart cities. *Sustainable Cities and Society* 61: 102301. https://doi.org/10.1016/j.scs.2020.102301.

Alanne, K. and Saari, A. (2006). Distributed energy generation and sustainable development. *Renewable and Sustainable Energy Reviews* 10 (6): 539–558. https://doi.org/10.1016/j.rser.2004.11.004.

Alavi, A.H., Jiao, P., Buttlar, W.G., and Lajnef, N. (2018). Internet of Things-enabled smart cities: state-of-the-art and future trends. *Measurement* 129: 589–606. https://doi.org/10.1016/j.measurement.2018.07.067.

Araral, E. (2020). Why do cities adopt smart technologies? Contingency theory and evidence from the United States. *Cities* 106: 102873. https://doi.org/10.1016/j.cities.2020.102873.

Bibri, S.E. and Krogstie, J. (2020). The emerging data-driven smart city and its innovative applied solutions for sustainability: the cases of London and Barcelona. *Energy Informatics* 3 (1): 1–42. https://doi.org/10.1186/s42162-020-00108-6.

Brożyna, J., Strielkowski, W., Fomina, A., and Nikitina, N. (2020). Renewable energy and EU 2020 target for energy efficiency in the Czech Republic and Slovakia. *Energies* 13 (4): 965. https://doi.org/10.3390/en13040965.

BV (2018). Smart cities and utilities report. https://www.bv.com/sites/default/files/gated-content/strategic-directions-report/18-SDR-Smart-Cities-Utilities.pdf (accessed 29 April 2021).

Cabelkova, I., Strielkowski, W., Wende, F.D., and Krayneva, R. (2020). Factors influencing the threats for urban energy networks: the inhabitants' point of view. *Energies* 13 (21): 5659. https://doi.org/10.3390/en13215659.

Cox, J. (2021). Pipeline hackers say they're 'apolitical,' will choose targets more carefully next time. Vice (10th of May). https://www.vice.com/en/article/bvzzez/colonial-pipeline-hackers-statement-darkside (accessed on 24 May 2021).

Drozdova, I. and Petrov, A. (2018). World practice and Russian experience of housing and utilities sector digitization. *SHS Web of Conferences* 44: 00031. https://doi.org/10.1051/shsconf/20184400031.

Farzaneh, H., Doll, C.N., and De Oliveira, J.A.P. (2016). An integrated supply-demand model for the optimization of energy flow in the urban system. *Journal of Cleaner Production* 114: 269–285. https://doi.org/10.1016/j.jclepro.2015.05.098.

Glazebrook, G. and Newman, P. (2018). The city of the future. *Urban Planning* 3 (2): 1–20. https://doi.org/10.17645/up.v3i2.1247.

Gubbi, J., Buyya, R., Marusic, S., and Palaniswami, M. (2013). Internet of Things (IoT): a vision, architectural elements, and future directions. *Future Generation Computer Systems* 29 (7): 1645–1660. https://doi.org/10.1016/j.future.2013.01.010.

Harrison, C., Eckman, B., Hamilton, R. et al. (2010). Foundations for smarter cities. *IBM Journal of Research and Development* 54 (4): 1–16. https://doi.org/10.1147/JRD.2010.2048257.

Hefnawy, A., Bouras, A., and Cherifi, C. (2016). IoT for smart city services: lifecycle approach. *Proceedings of the International Conference on Internet of Things and Cloud Computing*, p. 55. https://dl.acm.org/citation.cfm?id=2896440 (accessed 24 May 2021).

Ismagilova, E., Hughes, L., Dwivedi, Y.K., and Raman, K.R. (2019). Smart cities: advances in research-An information systems perspective. *International Journal of Information Management* 47: 88–100. https://doi.org/10.1016/j.ijinfomgt.2019.01.004.

Kumar, H., Singh, M.K., Gupta, M.P., and Madaan, J. (2020). Moving towards smart cities: solutions that lead to the smart city transformation framework. *Technological Forecasting and Social Change* 153: 119281. https://doi.org/10.1016/j.techfore.2018.04.024.

Kylili, A. and Fokaides, P.A. (2015). European smart cities: the role of zero energy buildings. *Sustainable Cities and Society* 15: 86–95. https://doi.org/10.1016/j.scs.2014.12.003.

Lisin, E., Strielkowski, W., Chernova, V., and Fomina, A. (2018). Assessment of the territorial energy security in the context of energy systems integration. *Energies* 11 (12): 3284. https://doi.org/10.3390/en11123284.

Madu, C.N., Kuei, C.H., and Lee, P. (2017). Urban sustainability management: a deep learning perspective. *Sustainable Cities and Society* 30: 1–17. https://doi.org/10.1016/j.scs.2016.12.012.

Minoli, D., Sohraby, K., and Occhiogrosso, B. (2017). IoT considerations, requirements, and architectures for smart buildings-energy optimization and next-generation building management systems. *IEEE Internet of Things Journal* 4 (1): 269–283. https://doi.org/10.1109/JIOT.2017.2647881.

Nakashima, E., Torbati, Y., and Englund, W. (2021). Ransomware attack leads to shutdown of major U.S. pipeline system. *The Washington Post* (9th of May). https://www.washingtonpost.com/business/2021/05/08/cyber-attack-colonial-pipeline (accessed on 22 May 2021).

Newbery, D., Pollitt, M.G., Ritz, R.A., and Strielkowski, W. (2018). Market design for a high-renewables European electricity system. *Renewable and Sustainable Energy Reviews* 91: 695–707. https://doi.org/10.1016/j.rser.2018.04.025.

Omer, A.M. (2008). Energy, environment and sustainable development. *Renewable and Sustainable Energy Reviews* 12 (9): 2265–2300. https://doi.org/10.1016/j.rser.2007.05.001.

Osman, A.M.S. (2019). A novel big data analytics framework for smart cities. *Future Generation Computer Systems* 91: 620–633. https://doi.org/10.1016/j.future.2018.06.046.

Picon, A. (2015). *Smart Cities: A Spatialised Intelligence*. New York: John Wiley & Sons.

Qureshi, K.N., Rana, S.S., Ahmed, A., and Jeon, G. (2020). A novel and secure attacks detection framework for smart cities industrial Internet of Things. *Sustainable Cities and Society* 61: 102343. https://doi.org/10.1016/j.scs.2020.102343.

Rausser, G., Strielkowski, W., and Štreimikienė, D. (2018). Smart meters and household electricity consumption: a case study in Ireland. *Energy & Environment* 29 (1): 131–146. https://doi.org/10.1177/0958305X17741385.

Reiche, D. (2010). Renewable energy policies in the Gulf countries: a case study of the carbon-neutral "Masdar City" in Abu Dhabi. *Energy Policy* 38 (1): 378–382. 10.1016/j.enpol.2009.09.028.

Shen, L., Huang, Z., Wong, S.W. et al. (2018). A holistic evaluation of smart city performance in the context of China. *Journal of Cleaner Production* 200: 667–679. https://doi.org/10.1016/j.jclepro.2018.07.281.

Silva, B.N., Khan, M., and Han, K. (2018). Towards sustainable smart cities: a review of trends, architectures, components, and open challenges in smart cities. *Sustainable Cities and Society* 38: 697–713. https://doi.org/10.1016/j.scs.2018.01.053.

Simionescu, M., Strielkowski, W., and Tvaronavičienė, M. (2020). Renewable energy in final energy consumption and income in the EU-28 countries. *Energies* 13 (9): 2280. https://doi.org/10.3390/en13092280.

Strielkowski, W. (2017). Social and economic implications for the smart grids of the future. *Economics and Sociology* 10 (1): 310–318. https://doi.org/10.14254/2071-789X.2017/10-1/22.

Strielkowski, W. (2020). *Social Impacts of Smart Grids: The Future of Smart Grids and Energy Market Design*. London: Elsevier.

Strielkowski, W. (2021). Energy research and social sciences: thinking outside the box. *E3S Web of Conferences 250: 07001*. https://doi.org/10.1051/e3sconf/202125007001.

Strielkowski, W., Veinbender, T., Tvaronavičienė, M., and Lace, N. (2020). Economic efficiency and energy security of smart cities. *Economic Research-Ekonomska Istraživanja* 33 (1): 788–803. https://doi.org/10.1080/1331677X.2020.1734854.

Teng, H., Liu, Y., Liu, A. et al. (2019). A novel code data dissemination scheme for Internet of Things through mobile vehicle of smart cities. *Future Generation Computer Systems* 94: 351–367. https://doi.org/10.1016/j.future.2018.11.039.

United Nations (2014). World urbanization prospects: the 2014 revision, highlights. Department of Economic and Social Affairs, Population Division. https://population.un.org/wup (accessed 25 May 2021).

United Nations (2018). World urbanization prospects 2018. https://population.un.org/wup (accessed 25 May 2021).

Van Winden, W. (2016). Smart city pilot projects, scaling up or fading out? Experiences from Amsterdam. Regional Studies Association Annual Conference in Austria, Graz, 3rd–6th of April. https://pure.hva.nl/ws/files/811939/RSA_paper_upscaling_RG.pdf (accessed 22 May 2021).

Voytenko, Y., McCormick, K., Evans, J., and Schliwa, G. (2016). Urban living labs for sustainability and low carbon cities in Europe: towards a research agenda. *Journal of Cleaner Production* 123: 45–54. https://doi.org/10.1016/j.jclepro.2015.08.053.

WEF (2021). The Internet of Things and connected devices: making the world smarter. https://reports.weforum.org/digital-transformation/the-internet-of-things-and-connected-devices-making-the-world-smarter (accessed 24 May 2021).

Yang, C. (2020). Historicizing the smart cities: genealogy as a method of critique for smart urbanism. *Telematics and Informatics* 55: 101438. https://doi.org/10.1016/j.tele.2020.101438.

Zhang, C. (2020). Design and application of fog computing and Internet of Things service platform for smart city. *Future Generation Computer Systems* 112: 630–640. https://doi.org/10.1016/j.future.2020.06.016.

16

Governing the Transition Towards Smart Grids Through Organised Industry Events

Suyash Jolly[1,2]

[1] *Business Research Group, Nordland Research Institute, Bodø, Norway*
[2] *Division of History of Science, Technology, and Environment, KTH Royal Institute of Technology, Stockholm, Sweden*

16.1 Introduction

A tremendous buzz over the years has surrounded the multiple potential benefits of smart grids, such as improving the efficiency of the power sector, reducing grid losses and increasing penetration of renewable energy sources in the electricity grids (Ballo 2015; Vesnic et al. 2016). A key challenge is that many promising demonstration and pilot projects and smart grid initiatives often rest on the assumption that active, technically inclined and competent users will adapt to them (Skjølsvold et al. 2015). There are considerable debates around how smart grids are defined, implemented and assimilated into daily user practices. Particularly the role of users is critical, and there is a need for shifting the focus from pushing smart technological solutions to users to the development of energy services and the creation of new service-oriented business models to support the development of smart grids (Tongia 2017; Verbong et al. 2013, 2016).

Smart grids have been defined as '*an electrical grid with automation, communication and IT systems that can monitor power flows from points of generation to points of consumption (even down to the appliances level) and control the power flow or curtail the load to match generation in real-time or near real-time*' (ISGF 2013b, p. 3). However, other studies have also argued for broadening the scope of smart grids and moving beyond narrow technological fixes as the social context is essential for implementing smart grids (Tongia, 2014a). Furthermore, smart grids need to be seen as making electricity grids more responsive to consumer needs and socially inclusive and financially robust in developing economies (Kumar 2019). A key challenge for implementing smart grids in developing economies like India is the capital-intensive investment by financially struggling utilities in India, making the cost–benefit analysis and generating a viable business case for smart grids challenging (Tongia 2016). Furthermore, it is essential to understand the impact of smart technologies

Smart Cities, Energy and Climate: Governing Cities for a Low-Carbon Future, First Edition.
Edited by Oleg Golubchikov and Komali Yenneti.

on vulnerable citizens and ensure they are not excluded from the proposed benefits of smart grids (Kumar 2019).

The chapter focuses on the role of industry events in supporting the development of novel technological fields (Karlsen 2018). It examines the case of India Smart Grid Week (ISGW) 2016 and how it facilitates concrete measures such as creating new expectations, developing new collaborations between firms, setting up new initiatives related to smart grids and providing policy recommendations. ISGW 2016 was a conference and exhibition to discuss relevant policy and regulatory and technological trends related to smart grids and accelerate the development of smart grids by bringing in the necessary global expertise to India. ISGW 2016 was organised as a business-to-business (B2B) event aimed exclusively at smart grid industry experts and stakeholders and not an event involving the public.

This study uses insights from the field-configuring events (FCEs) and the business conferences, and trade shows literature (Henn and Bathelt 2015) to guide the development of new technological fields (Burke and Knight 2011). The literature on FCEs has examined the role of future visions, expectations and promises about new technological fields in particular social settings. FCEs focus on conferences, trade shows, award ceremonies, technology contests, trade shows, professional gatherings, business meetings, etc. to develop new technological fields (Lampel and Meyer 2008; Möllering and Müller-Seitz 2018). This stream of literature has emphasised the role of conferences in developing future expectations about new technological innovations and strengthening the advocacy of new technological fields by facilitating multiple narratives if they do not result in immediate results (Nissilä et al. 2014). The literature emphasises that events can shape the development of new markets even though a single event or isolated events might not always create momentum for the development of new fields despite the field-level uncertainty (Möllering 2010). By building upon these prior insights from the literature, this chapter focuses on answering the following question: *What role do events such as ISGW* 2016 *play in shaping the development of smart grids?*

This chapter is structured as follows. In Section 16.2, I integrate the literature on FCEs with the shows and business conferences. Section 16.3 explains the research methods utilised for the study by highlighting the research design, data collection and data analysis efforts. In Section 16.4, I analyse the role of ISGW 2016 in mainstreaming smart grids in India by discussing the critical narratives around smart grids during the event. In Section 16.5, I focus on examining the field-configuring potential of the event and its usefulness for triggering future developments for smart grid development in India. Finally, Section 16.6 concludes the chapter by highlighting the limitations and providing avenues for future research.

16.2 Theoretical Background

The organisation and management studies have focused on temporarily organized events (i.e. FCEs) for understanding the development of new fields (Schulte Römer 2018). The literature on FCEs has focused on the role of conferences, trade shows, film festivals, award ceremonies and technology contests where diverse participants gather together to discuss and debate the development of a novel field for a particular period (Lampel and Meyer 2008). Furthermore, FCEs stimulate trust building and momentum to develop new organisational fields by bringing different actors together and resolving fundamental conflicts between them (Dobusch and Schüßler 2014; Garud 2008). Organisers of FCEs try to anticipate developments in the future, and secure resources from national governments, international organisations and firms to position the event in the media as well as attract exhibitors and participants (Schüßler et al. 2014, 2015). Prior studies have also indicated that FCEs are not independent entities but events embedded within a stream of ongoing activities within the field (Garud 2008; Schüßler et al. 2015).

FCEs are also avenues where different actors propagate different narratives and counter-narratives to disrupt the institutional status quo and shape new agendas (Zilber 2007, 2011). Hardy and Maguire (2010) highlight the role of discursive spaces where actors discuss, debate and dispute issues vital to them by propagating multiple narratives. The narratives are circulated in reports, pamphlets, advocacy and position papers, press statements, formal and informal talks in plenary sessions and panel discussions. New field-level rules emerge due to conflicting narratives emerging during the event and domination of narratives after the event (Gross and Zilber 2020; Hardy and Maguire 2010; Schüßler et al. 2014).

Trade shows, fairs and business conferences have been discussed within economic geography to focus on exhibitions, industry conventions and other professional meetings for marketing (Henn and Bathelt 2015; Karlsen 2018). These avenues are essential for increasing face-to-face interaction between actors worldwide with access worldwide and help create global knowledge pipelines by bringing together actors who would not normally interact with each other (Andreae et al. 2013; Maskell 2014; Bathelt and Gibson 2015; Schulte Römer 2018). Trade shows are also helpful in discovering the latest technological trends, gathering new market information and developing new contacts with competitors from other countries (Bathelt and Zeng 2014; Rinallo et al. 2017).

Organisers of trade shows play an essential role in designing the event's layout, improving exhibitor knowledge release and finding better ways of communicating the knowledge shared during the event in enriching visitors' and exhibitors' learning experiences. The organisers carefully select the type of participants who are allowed and not invited to attend the event and shape the agenda for the event (Bathelt and Zeng 2014; Schüßler et al. 2015; Rinallo et al. 2017). Firms regularly look for new business partners and gather new knowledge about competitors by regularly attending such events. They often evaluate their activities and achievements by comparing themselves to other firms present during the event and deciding on a future course (Bathelt and Zeng 2014; Bathelt and Gibson 2015).

Follow-up activities after the trade shows include face-to-face interactions for facilitating new business relations, setting up formal collaborations among different firms, suppliers and media coverage of the event to disseminate the knowledge generated at the event (van Tuijl and Dittrich 2015). However, not all the business activities at the trade shows generate actual sales activities despite the learning opportunities at the events. The interactions between the different participants often do not lead to any specific conclusion and financial benefit for the firms (Bathelt et al. 2017). Furthermore, trade shows are less agenda oriented and less rigidly structured than FCEs, as participants attending them have different motivations and are less likely to follow predefined themes. As a result, trade shows sometimes reflect ongoing reproduction and maintenance of the emerging field (Henn and Bathelt 2015).

A vital issue debated in both the stream of literature is that events are often not highly structured and purposefully planned, less dynamic, preserving the status quo, and result in field reproduction rather than field configuration (Schüßler et al. 2015). Furthermore, it is sometimes challenging to determine whether an event can be field configuring or not. The analysis of the field-configuring nature of an event might require a retrospective approach (Jolly and Raven 2016; Schüßler et al. 2015). Even if an event is planned well in advance with a specific agenda and organisational resources dedicated to it, the event might not result in intended results due to the highly uncertain nature of the events. On the other hand, business conferences provide multiple opportunities for face-to-face interaction and facilitate knowledge circulation (Henn and Bathelt 2015). Karlsen (2018) discusses that the impact of events can be suggested ex ante by focussing on the new measures negotiated during the events. In this way, a single event is just among the several events that configure an emerging field. However, it is challenging to estimate the direct impact of an event on the development of an emerging field (Karlsen 2018).

In summary, the existing literature suggests that FCEs provide an arena for collective sensemaking and an opportunity for actors to present their views and resolve ongoing conflicts between them (Zilber 2018). Some events can substantially impact the field compared to other events by providing avenues for disruptive change (Nissilä 2015). However, many other events are more likely to maintain the field rather than reconfigure them or challenge the field-level power structures (Schüßler et al. 2015; Schulte-Römer 2018). On the other hand, trade shows and business conferences are likely to influence the development of new technological fields in a decentralised, unpredictable and incremental manner and often do not result in radical changes (Henn and Bathelt 2015). In this chapter, I take a balanced perspective by integrating insights from the FCEs and the trade shows literature to answer the main research question.

16.3 Research Method

For this study, I utilised a single exploratory case study (Yin 2018) as qualitative case studies are helpful regarding the detailed understanding of organisation fields and analysing field-level events (Zilber 2020). The events are not independent entities and are always a small part of ongoing field processes and offer detailed insights into field-level discursive dynamics. Furthermore, collecting data from a particular field event or field events and complementing it with archival data before and after the event helps understand its broader context (Delgado and Cruz 2014; Zilber 2014, 2015).

FCEs and trade shows provide rich data sources for understanding the development of new industries over time. Analysing events also reduces the cost of fieldwork. It also helps get access to many informants and critical discussions during the development of a novel technological field in a short period (Karlsen 2018).

The study relied on observations during the different seminars, keynote speeches by experts and discussions during and after presentations made by relevant experts in the ISGW 2016. I focused on audio and video recording the main speeches by experts and discussions among experts during panel discussions. I also prepared notes based on the presentations discussed by the experts. I tried to observe discussions between the experts in the exhibitor halls and during informal social events.

Apart from the observations made during the conference, I also collected information from the trade show organised along with the conference by observing and asking questions from the exhibitors and observing the discussion between the exhibitors and visitors. I took pictures and videos of the exhibits in the stands of the different companies and organisations that presented and interacted with their representatives. After the event, I interviewed the representatives from the ISGF to reflect on the occasion and personal experiences in organising the event. I also asked the organisers about the experts, exhibitors, visitors' feedback and the event's potential impact on implementing smart grids. While participant observation is essential for collecting rich data, there is also a limitation. It was sometimes challenging to observe all the informal discussions between the experts while planning the event and private talks between the different experts.

The study also draws on several archival data sources such as policy and regulatory documents by the Ministry of Power, India, ISGF, published research articles and reports on smart grids in India, press releases, promotional materials about the event, homepages of the government smart grid projects, newspaper articles, blogs and newsletters by the ISGF. The event summary reports prepared by the ISGF summarised the discussions in the different sessions in the conference sessions and the recommendations for the policymakers (2015–2021) were helpful to understand the potential field-configuring impact of the event. I also investigated different media sources (e.g. Times of India, Hindu Business Line, Economic Times, Business Standard, DNA India, Energetica and PowerLine) to capture the media reporting of ISGW 2016.

I used content analysis to identify the central thematic debates during the event. The summarised data were analysed to understand the different actors' subjective interpretations (Zilber 2018). While few narratives during the event were optimistic about the immense potential of smart grids, others were critical by highlighting the challenges associated with smart grids. Therefore, I tried to have a balanced analysis by looking at both the narratives and counternarratives surrounding the implementation of smart grids (Gross and Zilber 2020; Zilber 2007, 2011). Furthermore, I focused on considering a contextual understanding of the event by analysing the developments happening before and after the event (Garud 2008) to understand the potential impact of the event regarding concrete work established such as future decisions made and important recommendations (Delgado and Cruz 2014; Zilber 2014).

16.4 India Smart Grid Week 2016

16.4.1 Background About the Event

FCEs are not independent entities but are often embedded within a stream of ongoing activities in the institutional field (Garud 2008). As a result, adequate attention is required to understand developments in the field before and after the event. This section provides a short historical development of smart grid development in India before the ISGW 2016.

The Indian government created the Accelerated Power Development Program (APDP) in 2000 for undertaking distribution reforms, reducing distribution losses and theft and improving the financial performance of the Indian power sector. In 2002, APDP was renamed APDRP (Accelerated Power Development and Reform Program). The programme focussed on enhancing transmission and distribution in the power sector using advanced technologies such as digital meters. It also focussed on using information technology (IT) to improve billing and collection efficiency and reduce electricity theft (Tongia 2003, 2004). The Government of India improved the efficiency of the Indian power sector by introducing power sector reforms and improving the efficiency of the state energy utilities. Still, minor changes have occurred regarding the working and functioning of the state electricity utilities (Talbott 2013). The state energy utilities in India have suffered from problems like inefficiencies in billing, illegal connections, unregistered consumption of electricity, meter tampering and many non-paying users (Tongia 2014b; Gaur and Gupta 2016).

In 2008, the APDRP programme was relaunched as the Restructured Accelerated Power Development and Reforms Programme (R-APDRP) to support the implementation of critical smart grid technologies such as SCADA, GIS, consumer indexing and DT metering. The R-APDRP aimed to reduce aggregate technical and commercial (AT&C) losses, starting applications for energy accounting, automation of the power system using IT applications, strengthening distribution and transmission systems and capacity-building measures (ISGF 2013a; Pargal and Banerjee 2014; Tongia 2015).

The Ministry of Power (MoP) had taken early steps in 2010 by constituting the India Smart Grid Task Force (ISGTF) to develop smart grids. To accelerate the deployment of the Smart Grid initiatives, the Government of India has started various industries such as creating the ISGTF, ISGF (India Smart Grid Forum), National Smart Grid Mission (NSGM) and a series of smart grid demonstration and pilot projects. ISGF is a public-private partnership initiative to accelerate the development and deployment of smart grids. The Government of India established the ISGF to develop and deploy smart grid-related activities (Ministry of Power 2013; Government of India 2017). In 2013, the Ministry of Power, Government of India and ISGF released a roadmap for more significant deployment of smart grids (ISGF 2013b; ISGF and Accenture 2016). In 2015, the Government of India also initiated the NSGM to accelerate smart grids and implement goals laid down in the smart grid roadmap. The NSGM focussed on planning and monitoring smart grid policies and regulations allocating financial resources for smart grid programmes (ISGF and Accenture 2016).

ISGW 2016 was the second edition of the conference and trade exhibition on smart grids organised by ISGF. The first event, ISGW 2015, was organised for the first time in Bengaluru. The first event was a conference and exhibition for bringing utilities, smart grid technology suppliers, regulators, investors and representatives from international agencies, researchers, etc. Around 722 participants attended the event. It resulted in several new initiatives such as drafting the smart grid knowledge paper, a guide for smart cities, a directory of the Indian power sector, etc. The discussions during the event were centred on a range of issues such as capacity constraints faced by Indian utilities, including financial and human resources rules and limited IT capabilities for new smart grid initiatives. The discussions were also focused on taking a holistic perspective on the actual costs and benefits of smart grids instead of overstating the potential benefits of smart grids and developing good business cases by including all relevant stakeholders early on during the projects. Furthermore, clear success criteria were needed to establish pilot projects to ensure they have a clear pathway for commercial deployment and to set clear and realistic project timelines to tackle the delays in demonstration projects (Interview ISGF; ISGW 2016).

16.4.2 Key Narratives During the Event

In this section, I analyse ISGW 2016 as an arena of negotiations addressing the critical challenges of implementing smart grids in India while emphasising the multiple benefits of smart grids. The different experts present during the event acknowledged that with robust regulatory mechanisms, a predictable environment can be created for scaling up smart grids in India but reducing the high costs was still a critical challenge. The critical narratives during the event were (i) ambiguity around the definition of smart grids; and (ii) cost impact on the consumers while implementing smart grids.

16.4.2.1 Ambiguity Around the Definition of Smart Grids

The event started with experts discussing the need to define the context and the ambiguity surrounding their purposes. An expert questioned:

> We still need clarity about the purpose of smart grids. What exactly do we mean by smart grids? Are we going to do everything remotely? Are we going to reduce the price of electricity, are we going to promote clean energy through it, are we going to optimize the energy mix through it, are we going to supply interruption-free electricity, what exactly are we going to do? The common person needs to know what exactly the smart grid is. What is the purpose of the smart grid?

Another expert remarked on the different levels of smartness about smart grids regarding functionalities so that the intended benefits are clear:

> Smart grid is a concept everyone agrees with, but everyone has different ideas about smart grid can be from dumb to genius. There are various levels of smartness one can think about. We should think about what functions do we want smart grids to fulfill and then identify the level of smartness we need [. . ..].

The discussion focused on learning from smart grid project deployment experiences in Europe, Japan, the United Kingdom, Australia and the United States. Lessons were also shared from the ongoing pilot Tata Power project and lessons from Tata Power regarding implementing technologies such as automated demand response. Another expert also discussed challenges with implementing pilot projects to bridge the rural and urban divide and ensuring the benefits of smart grids to all citizens.

> The user can participate in the decision-making process. The mindset regarding the monopolistic business in the power sector has to change. 100% of consumers must be metered. Unless they are metered, they cannot be monitored and controlled. First

> 100% metering has to be done, and after that, various measures have to be taken to reach a stage before doing smart grids and start talking about the concept of smart grids.

Another expert remarked on the difficulties while conducting the pilot projects and ongoing lessons learned from demonstration projects:

> There have been learning lessons regarding technology rollout and the benefits which are likely to accrue due to the smart grid. Like any other case, we first faced difficulties defining smart grids, relevant facts, and implementation issues. Fortunately, we have awarded 8 out of 14 projects, and are just at the initial implementation stage.

Similarly, another expert also remarked on the experiences gained from the different smart grid pilot projects:

> Pilot projects aimed to provide insight into the integration of consumers by deploying smart meters for peak load management through demand-side management, test functionality of rooftop solar, rural integration, and partial load shedding. These experiences have been gained in the last few years and will help plan smart grids.

Despite the negative lessons from the demonstration projects, the experts were hopeful about the potential of India's smart grid pilot projects. They also deliberated on the customer engagement strategies, future regulations required for supporting the demonstration projects and the future implementation challenges.

16.4.2.2 Cost Impact on the Consumers While Implementing Smart Grids

In this theme, the experts suggested that the implementation of smart grids can significantly impact the financial position of state utilities and poor consumers as most state energy utilities have not even implemented the previous R-APDRP programme for improving their operational performance. The critical issue raised was whether poor consumers would pay for advanced smart meters and the high installation costs of the smart meters compared to the conventional electricity meters. For instance, one expert argued citing difficulties associated with the high costs of smart meters for consumers:

> The major barrier to going to the smart grid is the high cost of the smart meter. The majority of the domestic customer is within Rs. 1000 a month. If we go for the smart meter, which costs Rs. 8000, and other additional installation costs, smart meters will be around Rs. 12000. The cost of power will be very high.

On the other hand, other experts believed that we should also look at lost opportunity costs for not implementing smart grids. There was a need to develop a detailed cost–benefit analysis rather than discuss and debate very generic. One expert suggested the need for precise cost–benefit research for smart grids:

> The economics of smart grids is certainly not clear; probably we have yet to spend time working out a clear case of the economics of smart grids so far. In that case,

> gatherings like this are useful for enhancing awareness about these issues. Particularly what is needed as the next step is probably having key stakeholders together and making them understand some of the detailed aspects of the economics of smart grids (cost-benefit analysis) of these interventions.

Another expert also highlighted the need for going beyond short-term issues and looking at the future picture:

> Key issue is a cost-benefit analysis. These are the costs, and these are the benefits. Even if the costs are high, there are still benefits to implementing smart grids, even if we get wrong. Even if smart meters are costly in the short term, they might be a small segment of the population in India, where smart meters are commercially viable. There are enough benefits to recover the costs.

The discussion even focused on the fact that it might not be suitable to compare the cost of a regular electricity meter with an advanced one now without taking future scenarios into account. The findings reveal two opposing discourses propagated by different actors present during the event: a coalition of actors supporting smart grids arguing for developing new business models and conducting accurate cost–benefit analysis. An important issue discussed was the energy utilities' need to use the existing consumer data available to create a comprehensive database that other service providers such as water and gas distribution services can also use to create smart cities (ISGW 2016).

16.5 Discussion: Role of the Event in Facilitating Future Developments

The chapter explores whether ISGW 2016 resulted in broader field-level implications (Möllering and Müller-Seitz 2018) for implementing smart grids in India. The event resulted in discussions around essential issues regarding the rollout of smart meters and the suggestion to use the leasing and service model for the rollout of smart meters. New agreements were developed for new projects between Reliance Infrastructure and Innovari, Inc., CESC Limited, and Silver Spring Networks. ISGF allowed the leading Swedish, Finnish, China, Korean, American and British firms and international companies from other nations to showcase smart grid innovations to Indian firms and interact with the Indian utilities regarding smart grid project experiences. Another significant result of the event was that it resulted in positive expectations about the cost of smart as the Minister of Power emphasised the enormous market potential for 250 million smart meters in India for INR 999 (ISGW 2016; Sengupta 2016).

The event organiser and a representative from the ISGF mentioned that the event resulted in signing of a memorandum of understanding (MOU) between the Indian utilities and global technology suppliers. The organisers collected lessons from the different sessions and submitted policy recommendations to India's Ministry of Power. The event organisers focused on critical themes such as micro-grids and renewable energy, electric vehicle charging infrastructure, AMI rollout and integration of renewables into the grid

(Interview, ISGF). ISGF also published a report on rolling out smart meters with cost-benefit analysis (Pillai et al. 2016).

At the end of the event, the organisers collected feedback from the different participants regarding their crucial learning from the discussions in the plenary sessions during the conference and the exhibition. One critical input from the participants was that conference sessions were planned without giving adequate time for talks between the different sessions, often making it difficult for them to attend the trade exhibition and the conference sessions. The event organisers decided to keep a few plenary sessions and have more parallel sessions on essential themes to participate in the trade exhibition. The organisers also decided not to have extensive significant panel member discussions and instead focus on short plenary sessions followed by a question-and-answer round. The feedback was helpful to address the criticisms by the participants and design and organise the next event in a better manner for the event in 2017 (Interview, ISGF). ISGW 2016 received significant media attention in prominent newspapers and media sources (e.g. Times of India and Economic Times). A few months after the event, the ISGF published a detailed report summarising the critical results, key findings and policy recommendations for mainstreaming smart grids in India (ISGF 2016a,b,c).

After the ISGW 2016, the next conference and exhibition (ISGW 2017) had over 2000 participants from more than 50 countries. The themes in the 2017 event were developed by taking feedback from the ISGW 2016. The key themes discussed were advanced metering infrastructure, energy storage, renewable energy integration and novel themes such as smart city gas distribution and electric mobility (ISGF 2017). ISGW 2018 was the fourth international conference and exhibition on smart grids and was attended by over 2200 participants, with more than 245 global experts from 38 different countries. Several new collaborations were set up during this event, including a memorandum of understanding between ISGF and Think Smart Grids, France. A new partnership was initiated among TPDDL, SAP and Choice, Brazil, to develop a unique solution for preventing electricity theft in India (Interview, ISGF).

In 2019, the event's format was modified, and the fifth edition of the ISGW was renamed India Smart Utility Week (ISUW). Around 1800 participants attended the event, and there were more than 288 experts from 41 different countries. The experts during this event suggested that water and city gas distribution utilities need to focus on the latest technologies to improve operational efficiency and integrate them into smart city projects. ISUW 2019 focused on providing a powerful platform for utilities, regulators, technology providers and government officials to discuss issues around the digital transformation of utilities, cybersecurity, mainstreaming renewable energy, energy storage and enhancing smart mobility (Interview, ISGF).

In 2020, the event was organised as an international conference and exhibition on smart energy and smarter cities to bring Indian electricity, gas and water utilities together with smart grid technology suppliers, regulators, investors and government officials. A total of 1570 participants attended the event, and around 269 experts participated in the different sessions (e.g. thematic sessions, special plenaries, bilateral workshops, parallel events, master classes and technical tours). The event included diverse themes such as E-mobility, digitalisation, advanced cybersecurity and energy storage systems. In 2021, the seventh edition of the event (International Conference and Exhibition on Smart Energy and

Mobility for Smarter Cities) was modified to a virtual event due to COVID-19. It included 3D exhibition booths and different virtual discussion forums. The next event in 2022 has already been planned by building on the lessons learned in the previous events (India Smart Grid 2021).

16.6 Conclusion

This chapter contributes to the literature on FCEs, trade shows and business conferences by suggesting that events provide avenues for negotiation between different actors to support developing new industries over time (Burke and Knight 2011; Karlsen 2018). This study demonstrated that the ISGW 2016 was an essential avenue for understanding emerging discourses around smart grids and the challenges involved in implementing smart grids after the first event in 2015. The study focussed on answering the following question: What role do events like 'ISGW 2016' play in shaping the emerging field for smart grids?

In this study, I try to take a middle ground, i.e. the event did not have significant field-configuring potential but resulted in concrete results, as highlighted in Section 5. The specific impact includes setting up new collaborations and providing the participants with benefits such as new business opportunities and the latest know-how on smart grid deployment from experiences worldwide (see Henn and Bathelt 2015; Karlsen 2018). I also showcase the field-configuring potential of the event by tracking down the critical debates and discussions in subsequent ISGW events from 2017 to 2021 and demonstrating how the event has evolved and changed over time. The chapter extends prior studies (Karlsen 2018) by including developments in successive events while understanding the potential of an event.

The chapter has a few limitations. First, I focused on two specific critical narratives in the ISGW 2016 around the implementation of smart grids. The other different narratives are related to developing smart grid standards, challenges related to power theft, integrating renewable energy and micro-grids, dynamic pricing based on peak demand and peak supply, developing electric vehicle infrastructure, smart and cyber cities, advanced analytics and the future of utilities (ISGW 2016).

Second, limited insights were provided about the citizens, consumers and civil society (Rommetveit et al. 2021) and their engagement with the critical debates during the event. There were limited avenues for getting public input and building trust among civil society, citizens, the smart grid industry and the government.

Future studies can focus on quantifying the event's impact and examining its effectiveness using different quantitative metrics (e.g. participation satisfaction, business and trade linkages developed, actual sales by firms present during the event and media attention gained). The impact can be assessed by linking the different metrics with the initial aims and objectives of the event (Tafesse and Skallerud 2017). The relationship between pre-event and post-event activities and the interaction with the activities during the event requires further exploration (Mair and Hehenberger 2014).

Governance for future events also needs to reduce the environmental impact by setting up several self-governing local events in different geographical locations coordinated by a central entity. Such measures will be helpful to avoid excessive GHG emissions and improve the inclusivity of the events (Etzion et al. 2022).

References

Andreae, M., Hsu, J.Y., and Norcliffe, G. (2013). Performing the trade show: the case of the Taipei international cycle show. *Geoforum* 49: 193–201.

Ballo, I.F. (2015). Imagining energy futures: sociotechnical imaginaries of the future smart grid in Norway. *Energy Research and Social Science* 9: 9–20.

Bathelt, H. and Gibson, R. (2015). Learning in organized anarchies: the nature of technological search processes at trade fairs. *Regional Studies* 49 (6): 985–1002.

Bathelt, H. and Zeng, G. (2014). The development of trade fair ecologies in China: case studies from Chengdu and Shanghai. *Environment and Planning A* 46 (3): 511–530.

Bathelt, H., Li, P., and Zhu, Y.W. (2017). Geographies of temporary markets: an anatomy of the Canton Fair. *European Planning Studies* 25 (9): 1497–1515.

Burke, G. and Knight, L. (2011). Institutional work at field-configuring events: shaping industry change within sustainable transitions. *Paper Presented at 2nd International Conference on Sustainability Transitions* (13–15 June 2011) Lund, Sweden (accessed 14 January 2021).

Delgado, N.A. and Cruz, L.B. (2014). Multi-event ethnography: doing research in pluralistic settings. *Journal of Organization Ethnography* 3 (1): 43–58.

Dobusch, L. and Schüßler, E. (2014). Copyright reform and business model innovation: regulatory propaganda at German music industry conferences. *Technological Forecasting and Social Change* 83: 24–39.

Etzion, D., Gehman, J., and Davis, G. (2022). Reimagining academic conferences: towards a federated model of conferencing. *Management Learning* 53 (2): 350–362. https://doi.org/10.1177/13505076211019529.

Garud, R. (2008). Conferences as venues for the configuration of emerging organizational fields: the case of cochlear implants. *Journal of Management Studies* 45 (6): 1061–1088.

Gaur, V. and Gupta, E. (2016). The determinants of electricity theft: an empirical analysis of Indian states. *Energy Policy* 93: 127–136.

Government of India. (2017). India country report. Research, development, demonstration and deployment of smart grids in India. Retrieved from: https://dst.gov.in/sites/default/files/India%20Country%20Report%20on%20Smart%20Grids.pdf (accessed 18 January 2021).

Gross, T. and Zilber, T.B. (2020). Power dynamics in field-level events: a narrative approach. *Organization Studies* 41 (10): 1369–1390.

Hardy, C. and Maguire, S. (2010). Discourse, field-configuring events, and change in organizations and institutional fields: narratives of DDT and the Stockholm convention. *Academy of Management Journal* 53: 1365–1392.

Henn, S. and Bathelt, H. (2015). Knowledge generation and field reproduction in temporary clusters and the role of business conferences. *Geoforum* 58: 104–113.

Hess, D.J. (2021). Undone science and smart cities: civil society perspectives on risk and emerging technologies. In: *Knowledge and Civil Society (Knowledge and Space, Vol 17)* (ed. J. Glückler, H.-D. Meyer, and L. Suarsana). Cham: Springer International. https://www.springer.com/gp/book/9783030711467.

India Smart Grid Forum (ISGF) (2013a). Post-R-APDRP strategy. *A White Paper by ISGF*. Retrieved from: http://www.indiasmartgrid.org/reports/Post%20R-APDRP%20Strategy%20%20-%20A%20White%20Paper%20by%20ISGF%20-%20Final.pdf (accessed 13 February 2021).

India Smart Grid Forum (ISGF) (2013b). Smart grid vision and roadmap for India. Retrieved from: https://www.nsgm.gov.in/sites/default/files/India-Smart-Grid-Vision-and-Roadmap-August-2013.pdf (accessed 9 February 2021).

India Smart Grid Forum (ISGF) (2016a). India smart grid week (ISGW) 2016. Retrieved from: http://www.isgw.in/about-isgw-2016/ (accessed 8 February 2021).

India Smart Grid Forum (ISGF) (2016b). India smart grid forum. Retrieved from: http://www.indiasmartgrid.org/document/ISGF%20FLYER.pdf (accessed 5 February 2021).

India Smart Grid Forum (ISGF) (2016c). India smart grid week 2016. *Post Show Report*. Retrieved from: http://www.indiasmartgrid.org/reports/Event%20Report%20ISGW%202016.pdf (accessed 11 February 2021).

India Smart Grid Forum (ISGF) (2017). India smart grid week 2017. Retrieved from: http://www.indiasmartgrid.org/document/India%20Smart%20Grid%20Week%20(ISGW)%20Brochure%202017.pdf (accessed 4 February 2021).

India Smart Grid Forum (ISGF) (2021). India, smart grid knowledge portal. Retrieved from: https://www.indiasmartgrid.org/ (accessed 3 January 2021).

ISGF & Accenture (2016). Recommendations for updating India smart grid roadmap: 2016. Retrieved from: https://indiasmartgrid.org/reports/India%20Smart%20Grid%20Roadmap_2016.pdf (accessed 10 February 2021).

ISGW. (2016). India smart grid week 2016. Retrieved from https://www.smart-energy.com/wp-content/uploads/2016/03/ISGW-2016-Brochure.pdf (accessed 8 January 2021).

Jolly, S. and Raven, R. (2016). Field-configuring events shaping sustainability transitions the case of solar PV in India. *Technological Forecasting and Social Change* 103: 324–333.

Karlsen, A. (2018). Framing industrialization of the offshore wind value chain – a discourse approach to an event. *Geoforum* 88: 148–156.

Kumar, A. (2019). Beyond technical smartness: rethinking the development and implementation of sociotechnical smart grids in India. *Energy Research & Social Science* 49: 158–168.

Lampel, J. and Meyer, A.D. (2008). Field-configuring events as structuring mechanisms: how conferences, ceremonies, and trade shows constitute new technologies, industries, and markets. *Journal of Management Studies* 45 (6): 1025–1035.

Mair, J. and Hehenberger, L. (2014). Front-stage and backstage convening: the transition from opposition to mutualistic coexistence in organizational philanthropy. *Academy of Management Journal* 57 (4): 1174–1200.

Maskell, P. (2014). Accessing remote knowledge-the roles of trade fairs, pipelines, crowdsourcing, and listening posts. *Journal of Economic Geography* 14 (5): 883–902.

Ministry of Power. The Government of India (2013). Smart grid vision and roadmap for India. http://www.nsgm.gov.in//upload/files/India-Smart-Grid-Vision-and-Roadmap.pdf (accessed 15 February 2021).

Möllering, G. (2010). Collective market-making efforts at an engineering conference. *MPIfG Discussion Paper, No. 10/2*, Max Planck Institute for the Study of Societies, Cologne (accessed 15 February 2021).

Möllering, G. and Müller-Seitz, G. (2018). Direction, not the destination: institutional work practices in the face of field-level uncertainty. *European Management Journal* 36 (1): 28–37.

Nissilä, H. (2015). Conferences as sequential arenas for creating new sustainable fields. *Industry and Innovation* 22: 209–228.

Nissilä, H., Lempiälä, T., and Lovio, R. (2014). Constructing expectations for solar technology over multiple field-configuring events: a narrative perspective. *Science and Technology Studies* 27: 54–75.

Pargal, S. and Banerjee, S.G. (2014). More power to India: the challenge of electricity distribution directions in development--energy and mining. World Bank, Washington, DC. Retrieved from: https://openknowledge.worldbank.org/bitstream/handle/10986/18726/889060PUB0978100Box385252B00PUBLIC0.pdf?sequence=1&isAllowed=y (accessed 27 January 2021).

Pillai, R.K., Bhatnagar, R., and Thukral, H. (2016). AMI rollout strategy and cost-benefit analysis for India. *ISGF White Paper*. India Smart Grid Forum. Retrieved from: https://www.caba.org/wp-content/uploads/2020/04/IS-2017-220.pdf (accessed 3 February 2021)

Rinallo, D., Bathelt, H., and Golfetto, F. (2017). Economic geography and industrial marketing views on trade shows: collective marketing and knowledge circulation. *Industrial Marketing Management* 61: 93–103.

Rommetveit, K., Ballo, I.F., and Sareen, S. (2021). Extracting users: regimes of engagement in Norwegian smart electricity transition. *Science, Technology & Human Values* 1–28: 01622439211052867. https://doi.org/10.1177/01622439211052867.

Schüßler, E., Ruling, C.C., and Wittneben, B.B.F. (2014). On melting summits: the limitations of field-configuring events as catalysts of change in transnational climate policy. *Academy of Management Journal* 57: 140–171.

Schüßler, E., Grabher, G., and Müller-Seitz, G. (2015). Field-configuring events: arenas for innovation and learning. *Industry & Innovation* 22 (3): 165–172.

Schulte-Römer, N. (2018). Extraordinary LED installations: events for user–innovation interaction. *Industry & Innovation* 25 (7): 699–727.

Sengupta, D. (2016). Smart grid market to touch Rs 50,000 crore in five years. Retrieved from: http://energy.economictimes.indiatimes.com/news/power/smart-grid-market-to-touch-rs-50000-crore-in-five-years/51103001 (accessed 13 January 2021).

Skjølsvold, T.M., Ryghaug, M., and Berker, T. (2015). A traveler's guide to smart grids and the social sciences. *Energy Research & Social Science* 9: 1–8.

Tafesse, W. and Skallerud, K. (2017). A systematic review of the trade show marketing literature: 1980–2014. *Industrial Marketing Management* 63: 18–30.

Talbott, W. (2013). Lighting the way: unlocking performance gains in electricity distribution and retailing in India. *Discussion Paper 2013-46*. Cambridge, MA: Harvard Environmental Economics Program. Retrieved from: https://heep.hks.harvard.edu/files/heep/files/dp46_talbott.pdf (accessed 9 February 2021).

Tongia, R. (2003). The political economy of the Indian power sector reforms. *PSED Working Paper Number 4*, Stanford, CA (accessed 13 February 2021).

Tongia, R. (2004). What it can and cannot do for the power sector and distribution in India: link to reforms, incentives, and management: program on energy and sustainable development. *Working Paper #19*. Retrieved from: https://fsi-live.s3.us-west-1.amazonaws.com/s3fs-public/IT_India.pdf (accessed 11 February 2021).

Tongia, R. (2014a). Smart grids in India: separating hype from hope. Retrieved from: https://www.brookings.edu/blog/planetpolicy/2014/05/29/smart-grids-in-india-separating-hype-from-hope/ (accessed 15 Febraury 2021).

Tongia, R. (2014b). Growing complexity and transformations of the power sector: India as an example of developing regions using Enterprise Architecture and Smart Grids. *A Systemic Perspective on Managing Complexity with Enterprise Architecture*. IGI Global, pp. 454–501. https://doi.org/10.4018/978-1-4666-4518-9.ch014 (accessed 16 February 2021).

Tongia, R. (2015). Making India's electricity utilities viable enterprises: economics, institutions, & the social contract. *Brookings India IMPACT Series*, Brookings Institution India Center (accessed 18 Febraury 2021).

Tongia, R. (2016). A developing country's perspective of the smart grid. Retrieved from: http://smartgrid.ieee.org/newsletters/march-2016/a-developing-country-s-perspective-of-the-smart-grid (accessed 20 February 2021).

Tongia, R. (2017). Making a smart energy grid work for India. Retrieved from: https://www.brookings.edu/opinions/making-a-smart-energy-grid-work-for-india/ (accessed 22 January 2021).

van Tuijl, E. and Dittrich, K. (2015). Events as spaces for upgrading ERIM report series research in management ERS-2014-013-LIS. Erasmus Research Institute of Management (ERIM). Available at: https://repub.eur.nl/pub/76070/ERS-2014-013-LIS.pdf (accessed 14 February 2021).

Verbong, G.P.J., Beemsterboer, S., and Sengers, F. (2013). Smart grids and smart users Involving users in developing a low carbon electricity economy. *Energy Policy* 52: 117–125.

Verbong, G.P.J., Verkade, N., Verhees, B. et al. (2016). Smart business for smart users? A social science agenda for developing smart grids. In: *Smart Grids from a Global Perspective: Bridging Old and New Energy Systems* (ed. A. Beaulieu, J. de Wilde, and J.M.A. Scherpen), 27–42. (Power Systems). Springer https://doi.org/10.1007/978-3-319-28077-6_3.

Vesnic, A.L., Breitegger, M., and Pereira, A.G. (2016). What smart grids tell about innovation narratives in the European Union: Hopes, imaginaries, and policy. *Energy Research & Social Science* 12: 16–26.

Yin, R.K. (2018). *Case Study Research and Applications: Design and Methods*. London: Sage.

Zilber, T.B. (2007). Stories and the discursive dynamics of institutional entrepreneurship: the case of Israeli high-tech after the bubble. *Organization Studies* 28 (7): 1035–1054.

Zilber, T.B. (2011). Institutional multiplicity in practice: a tale of two high-tech conferences in Israel. *Organization Science* 22 (6): 1539–1559.

Zilber, T.B. (2014). Beyond a single organization: challenges and opportunities in doing field-level ethnography. *Journal of Organizational Ethnography* 3 (1): 96–111.

Zilber, T.B. (2015). Studying organizational fields through ethnography (in press). In: *The Handbook of Qualitative Organizational Research* (ed. K. Elsbach and R. Kramer). New York: Taylor & Francis/Routledge.

Zilber, T.B. (2018). Know thy place: location and imagined communities in institutional field dynamics. In: *Knowledge and Institutions. Knowledge and Space*, vol. 13 (ed. J. Glückler, R. Suddaby, and R. Lenz). Cham: Springer https://doi.org/10.1007/978-3-319-75328-7_9.

Zilber, T.B. (2020). The methodology/theory interface: ethnography and the microfoundations of institutions. *Organization Theory* 1 (2): 263178772091943.

17

Emission Reduction and Renewables Integration Through Distributed Ledger Technology

Hamid M. Pouran[1], *Komali Yenneti*[2], *Mariana Padilha Campos Lopes*[3], *Louis Gyoh*[1], *and Yong Sheng*[4]

[1] *Faculty of Science and Engineering, University of Wolverhampton, Wolverhampton, UK*
[2] *School of Architecture and Built Environment, University of Wolverhampton, Wolverhampton, UK*
[3] *Federal University of Rio de Janeiro, Rio de Janeiro, RJ, Brazil*
[4] *Faculty of Science and Engineering, University of Hull, Hull, UK*

17.1 Introduction

Fifty years ago, the first long-distance electronic message was sent from the University of California, Los Angeles (UCLA), to Stanford Research Institute (SRI International). This modest start grew into what we now know as the internet (Leiner et al. 2009), with billions of devices connected to this global network (Nordrum 2016). Internet has created an unprecedented capacity for overcoming institutional barriers and changed every aspect of our lives (Atzori et al. 2017). Relying on this network now we see another milestone emerging with a capacity to subvert established standards and present a blueprint for a new structure – distributed ledger technology (DLT), which is also known as Blockchain (Wright and De Filippi 2015; Zheng et al. 2017). This chapter aims to provide a perspective on the nature of DLT and how this platform could catalyse the reduction in greenhouse gas (GHG) emissions and the integration of renewable energies. In this work, DLT and Blockchain are used interchangeably.

17.2 What is the DLT or Blockchain Platform?

Regardless of the fate of cryptocurrencies (the technology behind it), the DLT/Blockchain is here to stay (Chowdhury et al. 2019). The DLT platform was initially invented for creating cryptocurrencies, but recently has been considered a general purpose technology that could be developed for novel applications (Dinh et al. 2018; Kamran et al. 2020). Figure 17.1 is a schematic illustration of the differences between blockchain technology and other methods.

Smart Cities, Energy and Climate: Governing Cities for a Low-Carbon Future, First Edition.
Edited by Oleg Golubchikov and Komali Yenneti.

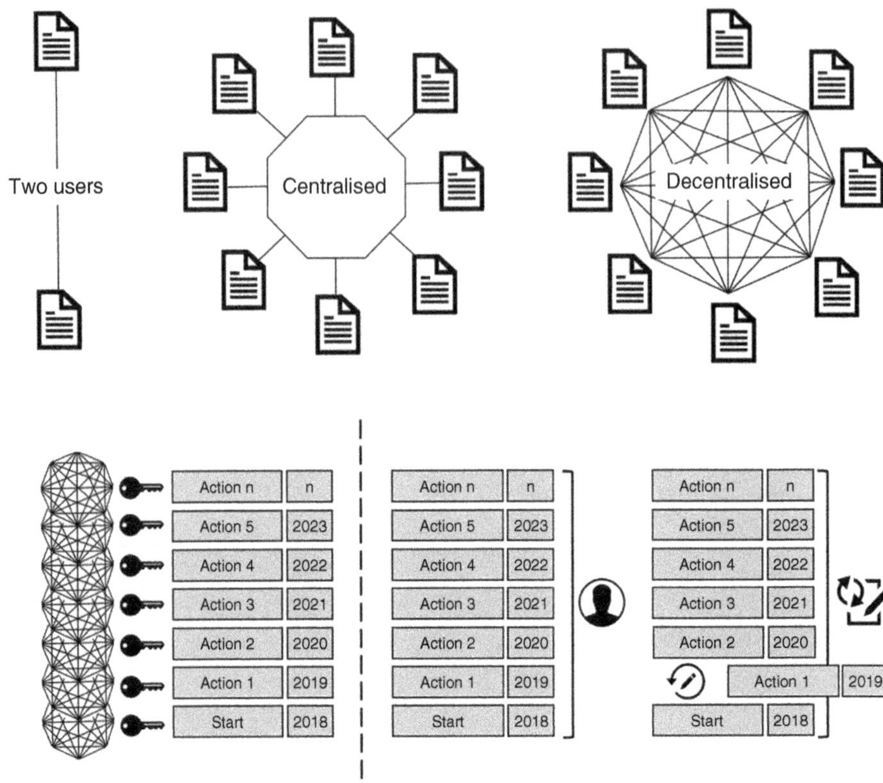

Figure 17.1 Two users, multi-users centralised and decentralised networks (top) and the comparison of decentralised ledger system with common and existing methods (bottom).

A Blockchain platform keeps the records and distributes them through a decentralised network (Aitzhan and Svetinovic 2018; Dinh et al. 2018). In such digital system, blocks of information are timestamped and chained (linked) to each other. Copies of these blocks are kept and monitored by verified ledgers (Wright and De Filippi 2015). It is worth noting that the network only allows new data to be chained to the existing blocks of information. Modifying previous data requires changes to all the transactions that happened after that block. If at some point there is an attempt to rewrite the history of operations from a specific data point/block, this would be immediately recognised by other members who can reject such a change (Karame and Capkun 2018; Khan and Salah 2018).

The DLT timestamps the blocks of data and keeps them secure and readily available, but this process does not guarantee the accuracy of the protected information. If the input is incorrect, the Blockchain platform would represent false information, which will be extremely challenging to rectify due to the nature of decentralised networks (Davidson et al. 2016; de Vos 2017). This is one of the current shortcomings of the DLT platforms.

Figure 17.1 shows two users, multi-users centralised and decentralised networks. In a decentralised network, its integrity is maintained and constantly monitored based on the protocols. Every action is recorded and cannot be tampered by an individual, while for a centralised network, a person or a group of people who centrally manage the network can interfere. Figure 17.1 also illustrates the reliability of a decentralised ledger system compared to common and existing methods. In a DLT, any changes to the existing data require rewriting the entire history, which would be immediately identified by other network participants. As a basic analogy, we could use Encyclopedia Britannica, where a central entity determines the end product, and Wikipedia, a transparent system to which all members can actively contribute to and all changes are recorded to compare the centralised and decentralised systems.

Blockchain is a complex and evolving system for managing digital transactions and information, through a peer-to-peer cryptographic technology (Christidis and Devetsikiotis 2016). The participants in this network, who contribute to its integrity, receive monetary incentives, which makes it work based on financial rewards (Dinh et al. 2017). Blockchain could be used for any subject that completely or partly can be digitised, from supply chain to GHG emissions (Casado-Vara et al. 2018). As an operating system, it provides transparent and reliable chains of information where each transaction can be traced back and validated immediately (Nikolakis et al. 2018; Ølnes et al. 2017). Within this context, it could play a key role in reinforcing climate action and supporting countries towards becoming net-zero and climate-neutral economies by 2050.

Two crucial aspects of addressing climate change are mitigating GHG emissions and using clean energies (Cherry et al. 2018; Mi et al. 2019). Some countries are not transparent about their emissions. A prime example is China which produces over 25% of global GHG but does not provide reliable data on its emissions (Li et al. 2018; Miller et al. 2019; Shan et al. 2018). This lack of transparency could be potentially minimised by using a Blockchain system to verify the emissions in real time. With respect to clean energies, we currently face a new phenomenon in electricity consumption. After decades of consuming electricity from the power grid, now many small entities, houses and companies have become both consumers and producers or as it is often referred to prosumers (Kubli et al. 2018; Mengelkamp et al. 2018). This shift is taking place because of the availability of renewable energies, e.g. deployment of solar panels. A DLT platform can potentially facilitate the integration of renewable energies particularly within micro-grids by making transactions secure, verifiable and the entire grid more connected (Aitzhan and Svetinovic 2018; Siano et al. 2019).

The development of Blockchain technology to address climate change requires overcoming the existing barriers. Table 17.1 lists some of the main challenges and opportunities

Table 17.1 Challenges and opportunities of Blockchain for climate change and energy.

Opportunities (Features that Blockchain technology provides as a network of trust)	**Challenges (Existing challenges for implementing Blockchain technology/DLT network)**
Instant verification and transparency	Intensive energy consumption
Real-time information	Absences of regulations and legal complexities
Cost reduction	Secure (private) data storage
Automation	Infrastructure requirements
Simplifying the approval process	Adopting and social concerns
Decentralised and immutable	System integration
Increasing accountability	Credibility
Protecting the stakeholders	Inability to scale

that DLT platform presents. The Blockchain platform was initially designed for cryptocurrencies, and therefore to be used for other purposes it needs to overcome major physical and technological barriers (Nofer et al. 2017; Reyna et al. 2018; Cao et al. 2019; Salah et al. 2019).

17.3 How DLT Could Catalyse Mitigation of GHG Emissions and Integration of Renewables

17.3.1 Creating a Verifiable Global Network of Emission Reduction Accounts

Blockchain technology is still in early stages and its current applications, apart from cryptocurrencies, are on a pilot scale (Dinh et al. 2018). When developed, this platform could reinforce global efforts to mitigate environmental challenges including climate change (Pecl et al. 2017). At the Paris Conference (COP21) (Stern 2016; Tobin et al. 2018), countries across the world were asked to submit their intended nationally determined contributions (INDCs), meaning what they intend to do after 2020 to reduce their emissions (UNFCCC n.d.; Tobin et al. 2018). A continuous debate on COP21 agreement is verifying the INDCs as well as its Article 6, which elaborates on creating an international carbon market that would also affect carbon pricing approaches (Rose et al. 2018). Implementing such a concept on a global scale requires efficient verification and transparency of the claims. For instance, China, the number one polluter in the world, does not provide officially published data on its emissions (Liu et al. 2015). Consequently, there are drastic differences in the estimated numbers (Shan et al. 2018). Like China, there are many countries that have a young fleet of fossil fuel-based power plants and growing economies with expected upward-trend emissions in the coming years. (Pouran 2018). With such a crucial role for fossil fuels, validating the accuracy of the world's progress as well as these countries' claims on their commitment to COP21 is not feasible without having a reliable monitoring network.

Considering the nature of the DLT platform (Casado-Vara et al. 2018; Dinh et al. 2018; Politou et al. 2019; Siano et al. 2019), implementing Blockchain-based carbon management system can create a digital flow of information, alongside the physical flow of fossil fuels from the source to the end users. It would provide a viable solution, by increasing transparency and simplifying the approval process. Figure 17.2. is a conceptual presentation of coal flow, as a fossil fuel, from coal mines to its consumers. As seen in every step of all financial transactions, ions (Treleaven et al. 2017) and carbon footprints are registered in a decentralised network and updated immediately, which helps to identify any discrepancies and bring reliability to the emission reduction and carbon market.

Transparency and verifiability are crucial for the Paris Agreement and its Article 6, which elaborates on the concept of Internationally Transferred Mitigation Outcomes (ITMOs) (Mehling et al. 2018). This concept allows developed countries to finance lower-cost emissions in developing economies as part of their commitments and is a tool for transferring the value of emission reduction. It also supports creating a network that connects the carbon accounts of the COP21 countries (UNFCCC n.d.). Using a DLT platform would be imperative for achieving this goal. It could trace every transaction and its corresponding carbon

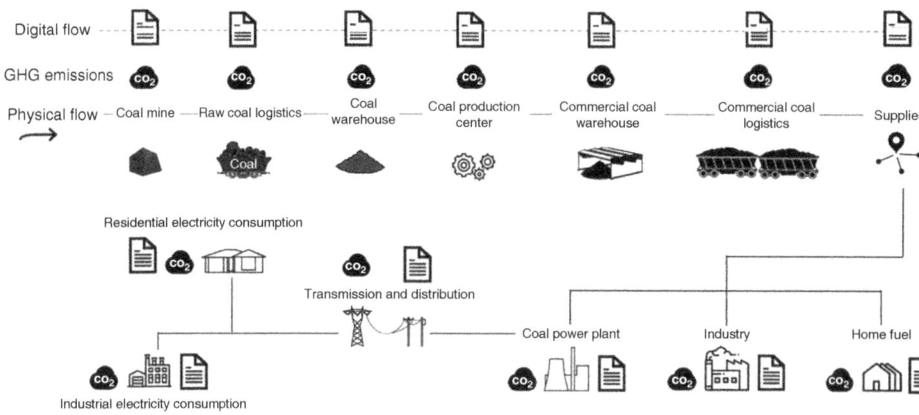

Figure 17.2 Schematic representation of a simplified Blockchain-based carbon accounting system for coal.

footprint and make the carbon credit market more liquid and inclusive. A digital flow of information, immutable from being altered by third parties, is immediately created for each transaction of the physical flow of carbon and safely stored across a distributed network.

In 2019, the United Kingdom became the first major economy that passed the laws to end its contribution to global warming by 2050, aiming to bring its GHG emissions to net zero by the next three decades (the previous target was at least 80% reduction compared to 1990 levels). Since then, an increasing number of countries have expressed their commitments to achieving net-zero emissions over the coming decades, which includes EU countries' 2050 long-term strategy. This strategy provides a road map for the transition to a climate-neutral society as an urgent challenge (*2050 long-term strategy|Climate Action*, no date) and involves restructuring in all aspects of the economy from industry, mobility and building to agriculture. Among all these sectors and required changes, the most critical one, as highlighted in the recent International Energy Agency (IEA) report, Net-Zero by 2050 (*Net Zero by 2050 – Analysis – IEA*, no date), is the energy sector. The report (*Net Zero by 2050 – Analysis – IEA*, no date) suggests that even the current pledges towards reducing GHG are not enough to create the chance of limiting the global temperature rise to 1.5°C and bring global energy-related carbon dioxide emissions to net zero by 2050. These reductions need profound structural changes in the energy system and will require the intervention of governments for such transition.

Energy transition for this purpose, which in essence needs complete transformation of the global energy system, cannot be achieved without constant, real-time and verifiable monitoring of the implemented changes over the coming decades. The only existing technology that could address such a task is the DLT/Blockchain concept, which still is in its early stages. Cities, as the centres of populations, the backbone of the economy and where most of the energy is consumed globally, are more vulnerable to the impacts of climate change and often suffer from failing and ageing infrastructures that limit their resilience. Socio-economic services provided in urban areas are heavily dependent on energy infrastructures, particularly electricity, and therefore deploying the DLT systems not only can reduce energy consumption and GHG emissions but also could make the entire electricity infrastructures more resilient and reliable in such environments (Mengelkamp et al. 2018).

17.3.2 Reinforcing Integration of Renewables Through Supporting Prosumers and Micro-Grids

Smart cities have become an increasingly important concept in urban development as they rely on digital infrastructures that improve communication platforms and support sustainable urban growth (Ahvenniemi et al. 2017). Such connectivity integrates physical and social capital, ensuring efficient utilisation of available resources (Biswas and Muthukkumarasamy 2016). However, the interconnectivity that exists between different stakeholders, which includes smart devices, is prone to corruption and interference by third parties (De La Torre Parra et al. 2019; Siano et al. 2019). Integrating Blockchain technology in smart city platforms could minimise the risk of fraud and insecurity of these systems (Lin and Ouyang 2014; Lu et al. 2016). Using secure networks would support the expansion of smart grids which improve the reliability of power systems and lowers the peak demand (Al-Turjman and Abujubbeh 2019; Siano et al. 2019).

However, the benefits of deploying DLT are far beyond supporting flexible demand response and lowering the carbon footprints (Ahl et al. 2019; Al-Turjman and Abujubbeh 2019; Kubli et al. 2018). It catalyses development of distributed energy resources (Li et al. 2017; Mengelkamp et al. 2018), which is an ecosystem where prosumers (e.g. rooftop solar energy) can sell their surplus to consumers without the need of an intermediary trusted party. In such a micro-grid, peer-to-peer transactions occur immediately and locally (Sousa et al. 2019). Such generation, storage and trade system, microgrid that allows peer to peer transactions, is an efficient way of using clean energies in a community. This is increasingly important as electricity storage capacities are improving and becoming more versatile, for instance, an electric vehicle can be charged by solar power in daytime and become a power bank at night (Kong and Karagiannidis 2016; Zhou et al. 2016). Figure 17.3 conceptualises peer-to-peer electricity trade within a local energy community.

The DLT could transform the electricity sector by bridging consumption and generation. Moreover, it would have profound impacts on the market (Mannaro et al. 2017; Sousa et al. 2019). When the households can trade energy directly with each other and eliminate the intermediary entities, they have flexibility and incentive to offer more attractive prices for the consumers within the micro-grid community (Mengelkamp et al. 2018). When people are assured that every transaction can be traced and verified, and the immutable digital information flow accompanies every micro-generation and consumption of electricity, decentralised smart grid energy trading will start to expand and consequently accelerate the integration of renewable energies (Andoni et al. 2019; Mengelkamp et al. 2018)

Transactions to reduce energy consumption in demand response programmes can apply the Blockchain structure to offer transparency, robustness and data security in the process (Sypatayev et al. 2020). The blockchain is the basis for the negawatt[1] market, which involves

1 Negawatts can be defined as the electric power saved or demand reduction (kW) in one hour, estimated by comparing the baseline electrical load and the measured electrical consumption after implementing selected control measures (Jing et al. 2019).

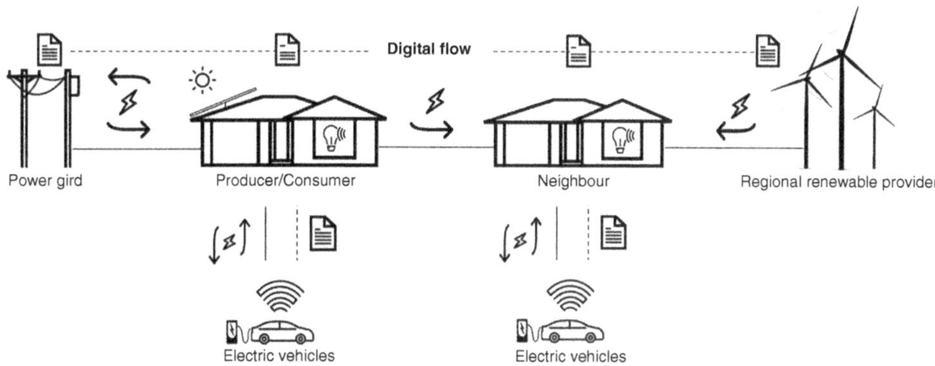

Figure 17.3 Electricity trade among the prosumers, consumers, power grid and renewable provider in a smart micro-grid – all the transactions are recorded and can be immediately verified through a DLT platform.

aggregators of response to demand and transactions by building owners/residents or the P2P negawatt trading network (Jing et al. 2019), thus promoting the reduction in energy consumption that is directly related to GHG emissions and other environmental impacts.

We can see one of the first examples of DLT-based micro-grid energy systems in New York City (Mengelkamp et al. 2018). This city has an ageing power grid and is also prone to natural disasters. For example, Hurricane Sandy which hit New York in 2012 caused blackout in lower Manhattan, left hundreds of thousands of New Yorkers without power for days (Dobson 2016) and resulted in billions of dollars of damages. The city has tried to improve the reliability of its power grid by developing smart systems and deploying sensors that can report the issues in real-time; but like other power grids, it is susceptible to the power outage due to external damages (Mylrea and Gourisetti 2017). However, a decentralised energy distribution platform could mitigate this problem.

Brooklyn micro-grid is a DLT-based micro-grid energy system in New York that demonstrates the Blockchain potential in localising energy markets and incentivising integration of renewable energy resources (Mengelkamp et al. 2018). This micro-grid is the first project that catalyses electricity transactions through a DLT-based platform. It could provide major lessons in deploying more advanced systems and make energy sector more technology dependent. Nevertheless, there are many socio-economic aspects that still need to be evaluated and analysed (Brisbois 2019).

The other challenge that the Blockchain platform can address is adding small- and micro-scale renewable energy resources (e.g. rooftop solar in off-grid regions), especially where the society faces economic challenges (e.g. rural areas in developing countries) (Creutzig et al. 2017; Sandwell et al. 2016). A Blockchain-based energy market in these regions would mitigate the investment risks and protect both residents and external investors as every micro-scale of energy generation and transaction could be automated and verified, and this provided energy security would accelerate sustainable economic growth of these societies.

DLT is a general purpose technology and its applications are still being explored. It is on pilot scale and faces significant barriers, which need to be addressed before its commercial viability could be demonstrated. Given these challenges are addressed, e.g. developing new algorithms and cloud-based systems that drastically reduce Blockchain technology energy demand, it could transform the energy sector and facilitate achieving Paris Agreement goals, in particular through reinforcing emission reductions and renewables integrations.

17.4 Concluding Remarks

The DLT platform is based on reliability of decentralised networks, which cuts the presence of the central ledger and eliminates unnecessary costs and unlawful interferences. Yet, despite all the benefits that Blockchain technology could offer, it is in early stages of evolving. Its applications are still unknown and surrounded by some degrees of uncertainty, which makes standardisation of this platform crucial. Apart from legal complexities, it suffers from major technological shortcomings, which need to be addressed to see DLT expanded beyond pilot scales. When Blockchain technology overcomes these challenges, it could play a crucial role in climate action by creating a framework that connects carbon accounts globally, making the carbon market reliable and reshaping the electricity sector by facilitating smart micro-grids which intrinsically add more renewables.

References

Ahl, A., Yarime, M., Tanaka, K., and Sagawa, D. (2019). Review of Blockchain-based distributed energy: implications for institutional development. *Renewable and Sustainable Energy Reviews* 107: 200–211. https://doi.org/10.1016/j.rser.2019.03.002.

Ahvenniemi, H., Huovila, A., Pinto-Seppä, I., and Airaksinen, M. (2017). What are the differences between sustainable and smart cities? *Cities* 60: 234–245. https://doi.org/10.1016/j.cities.2016.09.009.

Aitzhan, N.Z. and Svetinovic, D. (2018). Security and privacy in decentralized energy trading through multi-signatures, Blockchain and anonymous messaging streams. *IEEE Transactions on Dependable and Secure Computing* 15 (5): 840–852. https://doi.org/10.1109/TDSC.2016.2616861.

Al-Turjman, F. and Abujubbeh, M. (2019). IoT-enabled smart grid via SM: an overview. *Future Generation Computer Systems* 96: 579–590. https://doi.org/10.1016/j.future.2019.02.012.

Andoni, M., Robu, V., Flynn, D. et al. (2019). Blockchain technology in the energy sector: a systematic review of challenges and opportunities. *Renewable and Sustainable Energy Reviews* 100: 143–174. https://doi.org/10.1016/j.rser.2018.10.014.

Atzori, L., Iera, A., and Morabito, G. (2017). Understanding the Internet of Things: definition, potentials, and societal role of a fast evolving paradigm. *Ad Hoc Networks* 56: 122–140. https://doi.org/10.1016/j.adhoc.2016.12.004.

Biswas, K. and Muthukkumarasamy, V. (2016). Securing smart cities using Blockchain technology. *2016 IEEE 18th International Conference on High Performance Computing and Communications; IEEE 14th International Conference on Smart City; IEEE 2nd International*

Conference on Data Science and Systems (HPCC/Smart City/DSS). IEEE, pp. 1392–1393. 10.1109/HPCC-SmartCity-DSS.2016.0198.

Brisbois, M.C. (2019). Powershifts: a framework for assessing the growing impact of decentralized ownership of energy transitions on political decision-making. *Energy Research and Social Science* 50: 151–161. https://doi.org/10.1016/j.erss.2018.12.003.

Cao, B., Li, Y., Zhang, L. et al. (2019). When Internet of Things meets Blockchain: challenges in distributed consensus. *IEEE Network* 33 (6): 133–139. https://doi.org/10.1109/MNET.2019.1900002.

Casado-Vara, R., Prieto, J., De la Prieta, F., and Corchado, J.M. (2018). How Blockchain improves the supply chain: case study alimentary supply chain. *Procedia Computer Science* 134: 393–398. https://doi.org/10.1016/j.procs.2018.07.193.

Cherry, C., Scott, K., Barrett, J., and Pidgeon, N. (2018). Public acceptance of resource-efficiency strategies to mitigate climate change. *Nature Climate Change* 8 (11): 1007–1012. https://doi.org/10.1038/s41558-018-0298-3.

Chowdhury, M.J.M., Ferdous, M.S., Biswas, K. et al. (2019). A comparative analysis of distributed ledger technology platforms. *IEEE Access* 7: 167930–167943. https://doi.org/10.1109/ACCESS.2019.2953729.

Christidis, K. and Devetsikiotis, M. (2016). Blockchains and smart contracts for the Internet of Things. *IEEE Access* 4: 2292–2303. https://doi.org/10.1109/ACCESS.2016.2566339.

Creutzig, F., Agoston, P., Goldschmidt, J.C. et al. (2017). The underestimated potential of solar energy to mitigate climate change. *Nature Energy* 2: 17140. https://doi.org/10.1038/nenergy.2017.140.

Davidson, S., De Filippi, P., and Potts, J. (2016). Disrupting governance: the new institutional economics of distributed ledger technology. *SSRN Electronic Journal* https://doi.org/10.2139/ssrn.2811995.

De La Torre Parra, G., Rad, P., and Choo, K.K.R. (2019). Implementation of deep packet inspection in smart grids and industrial Internet of Things: challenges and opportunities. *Journal of Network and Computer Applications* 135: 32–46. https://doi.org/10.1016/j.jnca.2019.02.022.

Dinh, T.T.A., Wang, J., Chen, G. et al. (2017). Blockbench: a framework for analyzing private Blockchains. *Proceedings of the 2017 ACM International Conference on Management of Data – SIGMOD '17*. New York, USA: ACM Press, pp. 1085–1100. https://doi.org/10.1145/3035918.3064033.

Dinh, T.T.A., Liu, R., Zhang, M. et al. (2018). Untangling Blockchain: a data processing view of Blockchain systems. *IEEE Transactions on Knowledge and Data Engineering* 30 (7): 1366–1385. https://doi.org/10.1109/TKDE.2017.2781227.

Dobson, I. (2016). Electricity grid: when the lights go out. *Nature Energy* 1 (5): 1–2. https://doi.org/10.1038/nenergy.2016.59.

Jing, Z., Pipattanasomporn, M., and Rahman, S. (2019). Blockchain-based negawatt trading platform: conceptual architecture and case studies. *IEEE PES GTD Grand International Conference and Exposition Asia*, GTD Asia 2019, pp. 68–73. https://doi.org/10.1109/GTDAsia.2019.8715890.

Kamran, M., Khan, H.U., Nisar, W. et al. (2020). Blockchain and Internet of Things: a bibliometric study. *Computers and Electrical Engineering* 81: 106525. https://doi.org/10.1016/j.compeleceng.2019.106525.

Karame, G. and Capkun, S. (2018). Blockchain security and privacy. *IEEE Security and Privacy* 16 (4): 11–12. https://doi.org/10.1109/MSP.2018.3111241.

Khan, M.A. and Salah, K. (2018). IoT security: review, Blockchain solutions, and open challenges. *Future Generation Computer Systems* 82: 395–411. https://doi.org/10.1016/j.future.2017.11.022.

Kong, P.Y. and Karagiannidis, G.K. (2016). Charging schemes for plug-in hybrid electric vehicles in smart grid: a survey. *IEEE Access* 4: 6846–6875. https://doi.org/10.1109/ACCESS.2016.2614689.

Kubli, M., Loock, M., and Wüstenhagen, R. (2018). The flexible prosumer: measuring the willingness to co-create distributed flexibility. *Energy Policy* 114: 540–548. https://doi.org/10.1016/j.enpol.2017.12.044.

Leiner, B.M., Cerf, V.G., Clark, D.D. et al. (2009). A brief history of the internet. *ACM SIGCOMM Computer Communication Review* 39 (5): 22. https://doi.org/10.1145/1629607.1629613.

Li, Z., Kang, J., Yu, R. et al. (2017). Consortium Blockchain for secure energy trading in industrial Internet of Things. *IEEE Transactions on Industrial Informatics* 14: 1–1. https://doi.org/10.1109/TII.2017.2786307.

Li, S., Bush, R.T., Santos, I.R. et al. (2018). Large greenhouse gases emissions from China's lakes and reservoirs. *Water Research* 147: 13–24. https://doi.org/10.1016/j.watres.2018.09.053.

Lin, B. and Ouyang, X. (2014). Energy demand in China: comparison of characteristics between the US and China in rapid urbanization stage. *Energy Conversion and Management* 79: 128–139. https://doi.org/10.1016/J.ENCONMAN.2013.12.016.

Liu, Z., Guan, D., Wei, W. et al. (2015). Reduced carbon emission estimates from fossil fuel combustion and cement production in China. *Nature* 524 (7565): 335–338. https://doi.org/10.1038/nature14677.

Lu, X., McElroy, M.B., Peng, W. et al. (2016). Challenges faced by China compared with the US in developing wind power. *Nature Energy* 1 (6): 16061. https://doi.org/10.1038/nenergy.2016.61.

Mannaro, K., Pinna, A., and Marchesi, M. (2017). Crypto-trading: Blockchain-oriented energy market *2017 AEIT International Annual Conference: Infrastructures for Energy and ICT: Opportunities for Fostering Innovation, AEIT 2017*. Institute of Electrical and Electronics Engineers Inc., pp. 1–5. https://doi.org/10.23919/AEIT.2017.8240547.

Mehling, M.A., Metcalf, G.E., and Stavins, R.N. (2018). Linking climate policies to advance global mitigation. *Science* 359: 997–998. https://doi.org/10.1126/science.aar5988.

Mengelkamp, E., Gärttner, J., Rock, K. et al. (2018). Designing microgrid energy markets: a case study: the Brooklyn Microgrid. *Applied Energy* 210: 870–880. https://doi.org/10.1016/J.APENERGY.2017.06.054.

Mi, Z., Guan, D., Liu, Z. et al. (2019). Cities: the core of climate change mitigation. *Journal of Cleaner Production* 207: 582–589. https://doi.org/10.1016/j.jclepro.2018.10.034.

Miller, S.M., Michalak, A.M., Detmers, R.G. et al. (2019). China's coal mine methane regulations have not curbed growing emissions. *Nature Communications* 10 (1): 303. https://doi.org/10.1038/s41467-018-07891-7.

Mylrea, M. and Gourisetti, S.N.G. (2017). Blockchain for smart grid resilience: exchanging distributed energy at speed, scale and security. *Proceedings – 2017 Resilience Week, RWS 2017*. Institute of Electrical and Electronics Engineers Inc., pp. 18–23. https://doi.org/10.1109/RWEEK.2017.8088642.

Nikolakis, W., John, L., and Krishnan, H. (2018). How Blockchain can shape sustainable global value chains: an evidence, verifiability, and enforceability (EVE) framework. *Sustainability* 10 (11): 3926. https://doi.org/10.3390/su10113926.

Nofer, M., Gomber, P., Hinz, O. et al. (2017). Blockchain. *Business & Information Systems Engineering* 59 (3): 183–187. https://doi.org/10.1007/s12599-017-0467-3.

Nordrum, A. (2016). The internet of fewer things. *IEEE Spectrum* 53 (10): 12–13. https://doi.org/10.1109/MSPEC.2016.7572524.

Ølnes, S., Ubacht, J., and Janssen, M. (2017). Blockchain in government: benefits and implications of distributed ledger technology for information sharing. *Government Information Quarterly* 34 (3): 355–364. https://doi.org/10.1016/J.GIQ.2017.09.007.

Pecl, G.T., Araújo, M.B., Bell, J.D. et al. (2017). Biodiversity redistribution under climate change: impacts on ecosystems and human well-being. *Science* 355: eaai9214. https://doi.org/10.1126/science.aai9214.

Politou, E., Casino, F., Alepis, E. et al. (2019). Blockchain mutability: challenges and proposed solutions. *IEEE Transactions on Emerging Topics in Computing* 9 (4): 1972–1986. https://doi.org/10.1109/TETC.2019.2949510.

Pouran, H.M. (2018). From collapsed coal mines to floating solar farms, why China's new power stations matter. *Energy Policy* 123: 414–420. https://doi.org/10.1016/j.enpol.2018.09.010.

Reyna, A., Martín, C., Chen, J. et al. (2018). On Blockchain and its integration with IoT. Challenges and opportunities. *Future Generation Computer Systems* 88: 173–190. https://doi.org/10.1016/J.FUTURE.2018.05.046.

Rose, A., Wei, D., Miller, N. et al. (2018). Policy brief—achieving paris climate agreement pledges: alternative designs for linking emissions trading systems. *Review of Environmental Economics and Policy* 12 (1): 170–182. https://doi.org/10.1093/reep/rex029.

Salah, K., Rehman, M.H.U., Nizamuddin, N., and Al-Fuqaha, A. (2019). Blockchain for AI: review and open research challenges. *IEEE Access* 7: 10127–10149. https://doi.org/10.1109/ACCESS.2018.2890507.

Sandwell, P., Chan, N.L.A., Foster, S. et al. (2016). Off-grid solar photovoltaic systems for rural electrification and emissions mitigation in India. *Solar Energy Materials and Solar Cells* 156: 147–156. https://doi.org/10.1016/j.solmat.2016.04.030.

Shan, Y., Guan, D., Zheng, H. et al. (2018). China CO2 emission accounts 1997–2015. *Scientific Data* 5: 170201. https://doi.org/10.1038/sdata.2017.201.

Siano, P., De Marco, G., Rolán, A., and Loia, V. (2019). A survey and evaluation of the potentials of distributed ledger technology for peer-to-peer transactive energy exchanges in local energy markets. *IEEE Systems Journal* 13 (3): 3454–3466. https://doi.org/10.1109/JSYST.2019.2903172.

Sousa, T., Soares, T., Pinson, P. et al. (2019). Peer-to-peer and community-based markets: a comprehensive review. *Renewable and Sustainable Energy Reviews* 104: 367–378. https://doi.org/10.1016/j.rser.2019.01.036.

Stern, N. (2016). Economics: current climate models are grossly misleading. *Nature* 530 (7591): 407–409. https://doi.org/10.1038/530407a.

Sypatayev, D., Kumar Nunna, H.S.V.S., and Shintemirov, A. (2020). A novel peer-to-peer negawatt trading transactive energy system for prosumers. *Proceeding of 2020 IEEE 14th International Conference on Compatibility, Power Electronics and Power Engineering*

CPE-POWERENG 2020, pp. 181–186. https://doi.org/10.1109/CPE-POWERENG48600.2020.9161457.

Tobin, P., Schmidt, N.M., Tosun, J., and Burns, C. (2018). 'Mapping states' Paris climate pledges: analysing targets and groups at COP 21. *Global Environmental Change* 48: 11–21. https://doi.org/10.1016/J.GLOENVCHA.2017.11.002.

Treleaven, P., Brown, R.G., and Yang, D. (2017). Blockchain technology in finance. *Computer* 50 (9): 15–17. https://doi.org/10.1109/MC.2017.3571047.

UNFCCC (n.d.). https://unfccc.int/process-and-meetings/the-paris-agreement/what-is-the-paris-agreement (accessed 29 June 2018).

de Vos, T. (2017). The implications of unlocking decentralised flexibility: a comparative analysis of the intermediary aggregator vs. Blockchain technology by considering a system integration perspective. https://repository.tudelft.nl/islandora/object/uuid:b0f5adc5-194f-43b8-99d4-8c5f9854bb79?collection=education (accessed 11 December 2018).

Wright, A. and De Filippi, P. (2015). Decentralized Blockchain technology and the rise of lex cryptographia. *SSRN Electronic Journal* https://doi.org/10.2139/ssrn.2580664.

Zheng, Z., Xie, S., Dai, H. et al. (2017). An overview of Blockchain technology: architecture, consensus, and future trends. *Proceedings – 2017 IEEE 6th International Congress on Big Data, BigData Congress 2017*. Institute of Electrical and Electronics Engineers Inc., pp. 557–564. https://doi.org/10.1109/BigDataCongress.2017.85.

Zhou, B., Li, W., Chan, K.W. et al. (2016). Smart home energy management systems: concept, configurations, and scheduling strategies. *Renewable and Sustainable Energy Reviews* 61: 30–40. https://doi.org/10.1016/j.rser.2016.03.047.

18

Just Comfort: District Heating and Cooling as a Sustainable Energy Solution

Keith Baker and Ronald Mould

Built Environment Asset Management (BEAM) Centre, Glasgow Caledonian University, Glasgow, Scotland, UK

Glossary of Abbreviations

DHC	District Heating and Cooling system(s)
DHS	District Heating System(s)
DSO	Distribution Service Operator
EPC	Energy Performance Certificate
HSA	Heat Supply Act
PV	(Solar) Photovoltaic

18.1 Introduction – A Brief History of District Heating

Although there has been a drive towards individual heating solutions in the past century in the United Kingdom, there is a long history of district heating systems (DHS), with evidence of their use being found in archaeological and historical records (Bean et al. 2010; Bostock 1855, Book ix, Chapter 79, pp. 54; Marshal 1931, pp. 81–206; Guo 2002; Rogers 2011; Steingisser 2009). Indeed, the system installed in Chaudes-Aigues in Cantal, France, began operating in 1322 and continues to do so today (Gilbert and Jaudin 1999).

Moving on nearly 500 years, in the early 1800s, a Scottish engineer, Robertson Buchanan, described how a number of neighbouring buildings could be supplied from a single boiler plant (Buchanan 1807; Buchanan 1810, p. 194). A hundred years later, in 1906, when Thomas Edison built his first electricity plant in Philadelphia, his financial modelling predicted that the only way to make the plant profitable was to sell the generated heat to customers rather than waste this important revenue stream (Granovetter and McGuire 1998). Edison supplied the waste heat as steam to the nearby Thomas Jefferson Hospital. This district heating system is still in operation today.

Smart Cities, Energy and Climate: Governing Cities for a Low-Carbon Future, First Edition.
Edited by Oleg Golubchikov and Komali Yenneti.

Throughout the twentieth century this was not a widely replicated model in Western countries as new power plants were built farther from urban districts. The use of waste heat for district heating became economically less viable as the distance between the generating plant and urban areas increased. Furthermore, the focus on housing in Western countries began to be shifted to provide individual, self-contained units. Le Corbusier's designs for multi-storey living in modern cities did not include provision for DHS. Other notable architects from the early part of the century (Frank Lloyd Wright, Walter Gopius, Ludwig Mies van der Rohe and Charles Rennie Macintosh) did not include district heating in their designs as they focused on individual properties (Sharp 2002).

In the United Kingdom, following World War II, house-building programmes were focused on addressing the shortage of housing. The emphasis was then on quick modern building using new materials and techniques and this led to prefabricated housing being built as quickly as possible (Tarn 2004). Under these pressures, the additional time on the design, installation and commissioning of a DHS was not deliverable within the building programmes. However, other countries, notably Denmark (see: *Lessons from Denmark's district heating revolution*), persisted with developing new DHS.

Today the proportion of households supplied by district heating varies throughout Europe. This is outlined in Table 18.1. DHS is more common in Eastern bloc countries

Table 18.1 Prevalence of DHS in Europe, as of 2012/2013.

Country	Per cent heat demand from DHS	Notes
Austria	21	Extensive DHS in Vienna and other urban areas
Czechia	40	Largest DHS is in Prague
Denmark	60	
Eire	<1	Uncommon, some retrofits in urban areas
England[a]	0.01	
Finland	50	
Germany	14	
Greece	<1	Mostly limited to Western Macedonia
Hungary	15	
Iceland	95	Systems were built to exploit the easily available geothermal resource
Italy	3	Most common in northern urban areas. Turin is the largest, supplying 550,000 homes
Norway	2	Cheap hydro-generated electricity means that electric heating is favoured
Poland	40	
Romania	7	Limited but the largest in Bucharest supplies 72% of heat demand
Spain	Insufficient data	Limited amounts of retrofit
Sweden	60	

Source: Adapted from Andrews et al. (2012) except [a]DECC (2013).

which have legacy DHS from Soviet-era plans, e.g. Russia, Hungary, Poland and Bulgaria. These are mostly urban developments and DHS development is viewed in part as a reflection of the change in how our societies deliver and depend on energy systems (Rutter and Keirstead 2012; Webb 2012). It has been estimated that there is a 33 GW potential of DHS development in urban areas in the United Kingdom (Andrews et al. 2012).

In contrast to the drive towards individual home solutions in the United Kingdom, the Soviet Union emphasised the importance of co-generation in its first five-year energy programme in the 1920s. Extensive DHS were installed in urban areas in Russia and many other Soviet states. When the Soviet Union collapsed and the component countries moved to market-based economies, a significant portion of their urban population was supplied by DHS (Raizada 2012).

Following the collapse of the Soviet Union and in the free-market environment which then prevailed, investment declined as did the quality of the service provided to customers. Customers of these systems now find themselves with no alternative but to take the heat delivered and pay the costs levied. This disenfranchisement through the inability to control the amount of energy they use in their homes and the resulting bills has been called a new form of fuel poverty (Tirado-Herrero et al. 2011).

The other group of countries that have favoured DHS are the Scandinavians. Within the Scandinavian bloc, Norway is an exception in relation to DHS. Norway has low electric costs and as a result domestic heat pumps feeding wet distribution systems are favoured lately (Boait and Rylatt 2010; Karytsas and Theodoropoulou 2014). Throughout the Nordic states, the thermal qualities of houses tend to be high and therefore their relative heating costs are low (Andrews et al. 2012).

Iceland has the highest portion of its domestic stock supplied by DHS. This was developed to exploit the abundant near-surface geothermal resources. Denmark has one of the most developed DHS sectors. In response to the oil crisis in the early 1970s, Denmark looked to increase the market penetration of DHS. Iceland, Denmark and Norway illustrate that the development or lack of DHS within any country is driven by many specific factors.

Danish DHS practitioners have proposed that DHS may be a fundamental aspect of energy saving throughout Europe (Andrews et al. 2012). This is based on a strategic view of energy and heat throughout Europe where DHS act as a tool to maximise flexibility of the system by acting as an energy store for increased renewable generation (Andrews et al. 2012).

18.2 Defining District Heating and Cooling by Generation

Over time, district heating and cooling systems (DHC) have become classified by their generation (i.e. their age) and their operational relation to each other, which makes for not only useful but also potentially confusing shorthand terminology. Therefore, to enable understanding, we provide the following summary.

First-generation systems date back to the 1880s to the 1930s and burn coal to feed steam at very high temperatures and high pressures, which is fed to buildings through concrete ducts. These systems were popular in the United States and parts of Europe. As might be expected, these systems were very inefficient and prone to breaking down (Lund et al. 2014).

Second-generation systems began appearing in the 1930s and lasted through the 1970s. The design switched from using steam to using water piped in pipes placed within concrete ducts to buildings at high temperatures (>100°C). In existing systems, this change was made incrementally as part of upgrading first-generation systems. Heat was generated by burning fossil fuels (not only coal but also gas and oil), and improvements in efficiency were gained by using combined heat and power (CHP) plants to generate electricity as well as heat. These systems are typified by those found in the former Soviet bloc and parts of Eastern Europe (e.g. Poland) where many remain operational (Lund et al. 2014).

Third-generation systems began life in the 1970s and were a response to the oil crises of that decade. Also known as 'Scandinavian district heating systems' because of where many of their components were produced, they became popular worldwide and are probably what most people think of as a DHS. The design of third-generation systems made the leap from concrete ducts and high-temperature water/steam to insulated piping buried underground and distributing water at lower temperatures (<100°C), and began to unlock one of the big advantages of DHS by incorporating non-fossil fuel supplies, predominantly biomass and waste but also some solar and geothermal (Lund et al. 2014).

Fourth-generation systems were pioneered by Denmark (and are referred by us and others as 'Danish model' systems). They use water heated to lower temperatures (<70°C), along with incorporating CHP, to increase system efficiencies and reduce grid losses. Fourth-generation systems have emerged as a response to climate change and are typified by including a range of technologies, critically using inter-seasonal thermal stores and large-scale solar thermal arrays, and often make use of biomass but can be powered one hundred per cent renewably (David et al. 2018; Lund et al. 2014; Marderos et al. 2018).

Fifth-generation systems, also known as cold district heating, distribute water at ambient temperatures (10–25°C) and use heat pumps to enable buildings to 'share' both heating and cooling. The system design reduces the need for pipes to be insulated and enables the recovery of waste heat from buildings and the use of water bodies (rivers, lakes, seas, etc.) to provide both heating and cooling (Buffa et al. 2019; Pellegrini and Bianchini 2018).

18.3 Technologies and Fuel Supplies

Historically, much of the energy used by DHS has come from fossil fuels, and the remaining capacity of these systems means that, proportionally, this strong bias still exists today. Globally, fuel use is still dominated by fossil gas (43.2%) and coal and coal products (43%). Coal remains in high use in countries where second-generation systems remain in use, such as China (90%), Poland (73%) and the former Soviet bloc, whereas the use of fossil gas remains high in the United States (73%) and Japan (55%), and Germany stands as an example of an even split (45%/42% gas vs. coal). Denmark's older systems also show a similar split (22%/24% gas vs. coal) but with the largest source of supply coming from biomass (32%) as a result of its mix of older systems running on biomass waste and newer systems incorporating biomass in conjunction with other fuels. Biofuels and waste also constitute the next largest fuel globally (6.5%), followed by oil and oil products (4.3%), 'other sources' (e.g. heat pumps) at 2.5%, nuclear and geothermal both at 0.2% and solar at close to 0% (IRENA 2020).

However, the development of fourth- and fifth-generation systems has the potential to change this mix dramatically. One of the most significant advantages of district heating is the ability to plug in a wide range of energy technologies, including conventional fossil fuel combustion, heat recovery technologies, biomass and biogas, energy from combustible waste, anaerobic digestion, heat pumps, geothermal and solar thermal. Furthermore, with sufficient strategic planning, heat networks can also expand organically to utilise local sources of energy (waste heat, water bodies, existing and new sources of biomass, etc.) and to connect to sufficient sources of demand from existing and new buildings. Here again, the Danish experience demonstrates the importance of a Heat Supply Act for enabling this.

Another critical factor for ensuring that new heat networks are successful and existing ones can expand effectively is the availability of robust and accurate data on supply and demand. Multi-technology heat networks require significant up-front investment and longer lead-in times than for standalone grid-connected renewable solutions, and this investment (in costs and time) plays to the professional instincts of engineers to design networks to meet maximum levels of aggregate peak demand, resulting in many district heating networks being greatly oversized. This makes sense if you are an engineer worried about being held responsible for complaints from householders when a heat network does not meet demand under the worst-case scenario of all householders on a network boosting their heating to the maximum settings and running showers and baths when outside temperatures drop deep into negative numbers. However, in reality, such scenarios rarely, if ever, happen, and so oversizing invariably works to drive up capital and operational costs to developers and customers, extending the time needed to recoup investments and reducing the marginal profits from selling heat long beyond the point at which the costs break even (Millar et al. 2020).

A further complication arises when demand is based on modelled data, which is prone to both over- and under-predicting demand. Here, in the UK buildings, energy demand is commonly calculated by the building energy models used to produce energy performance certificates (EPCs) rather than using actual (measured) data, and the continued belief in their validity is, we would argue, the single biggest failing of government policies for decarbonising the built environment and tackling fuel poverty. A full discussion of the failings of EPCs could easily form a new chapter and so we point readers to a number of useful sources (Alembic Research et al. 2019; Baker and Mould 2018; Jenkins et al. 2017).

Naturally, demand for heat (and cooling) varies over both daily and yearly cycles but the most successful heat networks overcome this by integrating inter-seasonal thermal stores and/or share heating and cooling demands, and these should be viewed as critical components to all new heat networks (Guelpa and Verda 2019). And although smoothing daily demand curves is less important for a DHS delivering large amounts of heat to communities of buildings compared to small and building-integrated systems, domestic heat batteries such as those developed by Sunamp Ltd can be used to further optimise system performance, and have the additional benefit of having significantly lower embodied energy costs than electric batteries (Sunamp 2021).

Moving on, designing successful heat networks requires a holistic approach that considers not only whole fuel supply chains but which also accounts for their direct and co-benefits to communities. It follows that, of all energy technologies, the development of sustainable DHS is by far the most intricately linked with the changes we will need to make to the way we use and manage the global commons to secure our social, economic and environmental future.

If DHS are to realise their full potential, then wherever possible, new DHS and heat network projects should seek to maximise the use of waste heat, minimise heat losses using heat recovery and energy storage technologies and capitalise on the availability of local renewable heat sources such as geothermal hotspots (Scottish Government 2019), flooded mine workings (Hytiris et al. 2015) and major water bodies (Yu 2018). However, these are site-specific resources and it is likely that many new projects will need to source significant proportions of their heat from, in order of preference: small- and large-scale solar thermal; combined heat and power (CHP), anaerobic digestion (AD); thermal hydrolysis plants fuelled by sustainable, locally sourced biomass and biofuels; and, where the only other option is landfilling, energy from waste (EfW) plants.

Across the globe, there are examples of land use patterns where the benefits to the resident populations are diminished and the value that land generates is not directly beneficial to the residents. Through changing land use, local benefits can be increased while associated carbon footprints are lessened. DHS and renewable generation can have an important role to play in these changes. Here in Scotland, in a policy paper that proposes an alternative vision for Scotland's grouse moors, the 'think and do' tank Common Weal has calculated the annual gross value added (GVA) per hectare for different land uses (Common Weal et al. 2018). Of these, solar PV farms offer the highest GVA of any energy technology at £10,952 per hectare, and biomass offers a GVA of £2,596 (the paper does not consider large-scale solar thermal but we would expect a similarly high GVA to solar PV). However, some forms of biomass and biofuels, particularly the redevelopment of native woodlands for sustainable forestry products, provide additional value as land for recreation, tourism and enhancing biodiversity (Pridmore et al. 2017).

While, wherever possible, DHS and heat networks should maximise the use of renewable energy technologies and the recovery of waste heat, it is likely, as per the Danish experience, that a substantial proportion of projects will require new fuel supplies. This is another issue that the governments have not tackled, whether that is through a lack of understanding or a lack of a strategic vision, often due to the 'siloisation' of energy policy. For example, the Scottish Government has so far failed to grasp this matter in any degree of detail to the point that the latest (2016) Land Use Strategy actually backtracks on the previous (2011) version by failing to mention biomass or biofuels specifically at all (Scottish Government 2016). We are strongly of the view that the impact of policy silos and the lack of a strategic approach to developing new fuel chains will only serve to hinder the development of successful DHS and heat networks in Scotland, and in other nations that have yet to take on board the lessons from the Danish experience. This is a critical failing of policy with potential implications for securing energy security and enabling greater environmental protection and the sustainable development of renewable and low-carbon energy.

The development of bioenergy in its many forms has been widely criticised, and while entering into that debate is beyond the scope of this chapter, it is probably best viewed as, at worst, a necessary evil (Ravilious 2020). It is important to note that using wood as a fuel for biomass risks significantly tarnishing national reputations if gaps emerge between demand and supply, resulting in countries becoming increasingly reliant on imported supplies, with the associated costs of greenhouse gas emissions from shipping. As well as the additional emissions, relying on imports leaves countries at the mercy of schemes which are supposed to certify the sustainability of timber products; however, the most well-established of these, the Forest Stewardship Council's (FSC) label, has long been subject to serious allegations of greenwashing and providing a cover for trafficking in illegal timber (Conniff 2018).

If we are to minimise carbon associated with the transport of biofuels for DHS, it is absolutely imperative that countries meet demand from within their own borders, where they can safely dictate and police their own requirements for sourcing sustainable supplies. However, as we have noted previously in this paper, doing so may open up substantial opportunities, particularly in deprived rural and island areas. One example of progress in Scotland is on Orkney, where the Agronomy Institute has planted 150,000 willow trees as a pilot towards developing a new local source of sustainable energy that has the potential to deliver a wide range of co-benefits to the islands, which continues to suffer from the highest levels of fuel poverty in Scotland. However, planting mono-crops such as willow, although native to the United Kingdom, does not get us all the way to a truly sustainable solution, which would see the restoration and management of Scotland's woodlands and forests in ways that encourage biodiversity, enhance the country's natural beauty, provide new sources of other native timber products and new opportunities for recreation and tourism. Similar progress could, and should, be made in other countries where biofuels are likely to form either a transitional or long-term part of decarbonising energy supplies.

Finally, we should again stress that biomass is only one component of a range of technological solutions to equitably decarbonising heat that must be harnessed in a coordinated and collective manner. In this case, and again following the Danish experience, the use of large-scale solar thermal and heat recovery and storage technologies should be employed to substantially reduce the demand for biomass supplies as part of multi-technology DHS projects.

18.4 District Cooling

Historically, the development of DHS and the discourse around fuel/energy poverty has been driven by work in countries where energy demand for cooling has been minor or negligible. Exceptions to this include the United States, the United Arab Emirates and Japan (IRENA 2020). However, as global temperatures continue to increase with a lack of adequate mitigations to lessen those increases, it will become increasingly important that not only do homes have the provision for heating during colder months but also cooling during the hottest times (Giannakopoulos et al. 2009; Petri and Caldeira 2015). From a domestic stance, the impact of the need for cooling is comparable to the need for heating since it can be reduced to a cost associated with maintaining a target temperature and the associated comfort and health benefits. It is important that we note that heat stress can exacerbate existing conditions and lead to excess avoidable deaths (Hajat et al. 2013; Kovats and Hajat 2007), just as hypothermia does (Neno 2005; Noe et al. 2012). Although research has focussed on the needs and cost of cooling in areas where cooling dominates the domestic heat loads (Giannakopoulos et al. 2009; Oh et al. 2016), there is an increasing body of research that affirms the increasing demand for cooling loads in the United Kingdom and other temperate countries (Aebischer et al. 2007; Ahmed et al. 2009; Fong and Alwan 2013; Hawkey et al. 2013; Noe et al. 2012).

DHS is merely a distribution system that carries heat from a source to a demand and therefore can act both as a delivery system and as a recipient of heat (Rezaie and Rosen 2012). The ability to offer cooling capacity integrated with heating capacity is a characteristic of fourth (and above)-generation DHS (Lund et al. 2014; Werner 2017). In fourth- and fifth-generation DHS, it is typical to include an element of thermal storage to operate over short-term and seasonal variations in demands. Therefore, by utilising heat stores, DHS/DHC can offer the ability to cool and recover waste heat from buildings at return it at times when heating is needed. In doing this, systems can deliver efficiencies, emissions savings and affordable heat and cooling better than any other solution. However, in order to be able to do this requires them to reach a scale with a diversity of load and integrated infrastructure for diverse heat generation and storage.

One of our challenges in achieving the large-scale deployment of fourth- and fifth-generation DHS/DHC is how do we account for, design and manage the waste heat collected from cooling (Gong and Werner 2015; Li and Svendsen 2012). One mechanism we can use is the thermodynamic term exergy (Dincer and Rosen 2013). In relation to a DHS, exergy is the energy available in order to return a system to equilibrium. In simple terms, a building has a target temperature. In order to reach that temperature, it absorbs heat. If the building is above its target temperature, then it sheds heat. It is the heat the building sheds that is captured by the term exergy, but as yet exergy figures are rarely factored into policy-making, often making DHS/DHC appear an unfavourable option compared to other low-carbon and renewable technologies (Dincer 2002; Li and Svendsen 2012). That heat may be transferred directly from one building to another, or it may be captured and returned from and to an individual building over short or long periods of time in response to daily and seasonal changes in heating demand. Short-term displacement refers to heat shedding during the day to be used in the evening, or through the working week to be used at the weekend. Long-term displacement refers to seasonal displacement where heat shed during the summer is deployed during the winter. Either way, optimal system efficiency can only be achieved through using thermal stores and/or heat and cooling sharing with other sources of supply and demand. A successful example of such a project is the Mijnwater (mine water) project in the Heerlen municipality in the Netherlands, which uses flooded former coal mines to balance a fifth-generation DHC to service local buildings (Brummer 2019; Verhoeven et al. 2014).

However, now the urgency of tackling climate change has become a major driver for changing the ways we heat and cool buildings, it has also become far too tempting for policymakers and politicians to look at successful fourth- and fifth-generation projects in other countries and simply assume that they can be copied at home without understanding the legislation and other background conditions behind their successes (Baker and Mould 2019; Mould 2018). In the case of fifth-generation systems, those lessons are still being learned, but they are rooted in the experience of the development of fourth-generation systems and the critical roles of strategic planning and the incorporation of heat stores. Therefore, the next section summarises how Denmark has gone from an early adopter of modern DHS to the world leader in fourth-generation systems, and the lessons that must be learned from its district heating revolution.

18.5 Lessons from Denmark's District Heating Revolution

Aside from Iceland, which uses heat from its substantial geothermal resources as a source for its heat networks, Denmark leads the world in developing sustainable DHS and has set an example for others to follow. The country now has six large central district heating areas, with a total heat production of approximately 60 petajoules (PJ) per year. There are also around 400 smaller decentralised district heating areas supplying an additional 75 PJ per year. In 2013, the total production of district heating in Denmark amounted to 134 PJ, 72.8% of which was produced in co-generation with electricity (CHP), saving around 30% of fuel compared with separate generation of heat and power. The Danish Energy Agency notes that the inclusion of heat storage in all district heating areas has been central to its many successes (Danish Energy Agency n.d.).

Another critical factor in Denmark's DHS revolution has been the adoption of a succession of Heat Supply Acts (HSAs), beginning as far back as 1979, when local authorities were first required to zone areas for the development of new networks. The HSAs are built around four key principles:

- Local authorities/municipalities are responsible for the approval of new heat supply projects;
- Local authorities/municipalities have to make sure that project with the highest socio-economic benefits is chosen;
- Production of heat must, if possible, be produced as combined heat and power; and
- The collective heat supply price must offer consumer prices based on 'true costs', meaning that the heat price cannot be higher or lower than the actual heat production costs (Danish Energy Agency n.d.).

Here in Scotland, this approach has since been adopted in a limited fashion by the Scottish Government, but critically, and contrary to recommendations, it has paid scant attention to assessing the full socio-economic benefits of DHS projects and shied away from regulating to require connection and co-location of waste heat sources with existing and new housing (Baker et al. 2012).

The Danish Government is already planning for adapting to the 2–3% rise in temperatures predicted for Copenhagen by 2050, which is already reaching 35°C in the summer. Two cooling networks utilising a mix of seawater and abstraction cooling from district heating have been developed and are expected to save around 14,000 tonnes of carbon dioxide per year (Danish Energy Agency n.d.).

With its established DHS market, Denmark certainly has much from which other countries can learn. However, simply lifting Denmark's ethos, legislation and regulation now and forcing them into different fiscal and politico-social environments is likely to create friction. Indeed, we have to ask ourselves if Denmark had applied its current legislation in the early- to mid-twentieth century would their DHS sector have expanded as it did? If other countries are to learn from Denmark, we should be looking at the history of DHS in Denmark and identifying the tipping points in the development of the DHS sector.

Table 18.2 summarises the history of the development of DHS and heat networks in Denmark. The Danish sector is over a hundred years old and is generally considered to have started in 1903 in Frederiksberg, a municipality lying to the west of Copenhagen.

Table 18.2 Key dates in the development of district heat systems in Denmark.

Decade	Driver	Key developments	Market status
1900–1920	Industrial expansion	Denmark's first district heating power plant opened in Frederiksberg in 1903. Initially, the heat fed a hospital, children's home and a poor house. The Frederiksberg system was seen as success and other Danish municipalities developed small-scale DHS, often using diesel generators	Expanding market
1920–1950	Maintaining existing demand	As electricity generation became centralised into larger plants, the demand to build municipal-based CHP diminished, however, there was an established market for heat and that demand continued to be met	Stagnant market. About 4% of heat demand from DHS
1960	Technology push	The next step that increased the development of DHS came in the 1960s with the technological advances in affordable insulated pipework and new waste-to-energy plants being built	Expanding market
1970	Socio-political and financial push	The fuel crisis hits in 1972/1973. The costs of fuel increase and fossil fuels are less readily available. Danish municipalities react to fuel shortages by introducing regulation on energy efficiency including a preference for DHS use. Zoning legislation is introduced in 1979 under Denmark's first Heat Supply Act, which incentivised the development of DHS for heat and gas networks for cooking	Expanding market Around 30% of domestic heat demand is met from DHS
1980	Environmental and financial push	Expanding networks connect to each other creating multi-nodal DHS. Established market with multiple ownership models for plant. In 1989, Dronninglund becomes the first DHS system in the country to install natural gas-driven engines for combined heat and power production	Expanding market
2000	Environmental push	In 2005, Dronninglund begins the replacement of natural gas with renewable energy (e.g. biomass and large-scale solar thermal)	Transitioning market
2010 onwards	Environmental and financial push	Significant expansion of fourth-generation DHS utilising large-scale solar thermal and inter-seasonal heat storage	Above 60% of heat demand from DHS

Source: Danish Energy Agency (n.d.), Mould (2018), Plan Energi and Niras (2015).

Frederiksberg had developed as a municipality through industrialisation, and in the late 1800s there was growing concern for the amount of waste building up. In 1896, an incinerator for municipal waste had been built in Hamburg, and after waste was sent to Germany and tested for its combustibility in the Hamburg incinerator, the first waste-to-energy plant in Denmark was built in Frederiksberg, producing both electricity and heat.

In addition to the time frame for the development and implementation of key legislation, we must also consider the ownership of DHS assets by municipalities in Denmark, which is rooted in the earliest history of DHS there. This historical context and an emphasis on public ownership for the public good has enabled Danish municipalities to invest in DHS assets with long payback periods of 20 or more years, resulting in high levels of investor confidence as the drivers for change have moved from industrial expansion to tackling climate change. This is probably the most significant contextual difference between Denmark and most of the other countries now pushing to develop district heating and cooling networks but, with climate change now driving more radical solutions across the policy landscape, we are not convinced that it is an insurmountable one.

Learning from the Danish Energy Agency, we point to a number of key lessons for regulation that should enable other countries to develop successful DHS/DHC regardless of differences in levels of public ownership of the building stock. An effective district heating regulator must:

- Be fully resourced financially and with appropriately skilled staff.
- Have strong powers to investigate charges levied against householders.
- Have strong powers to enforce cost caps or adjustments.
- Have strong powers to ensure vulnerable householders are supported by their network operators.
- Have strong powers to enforce operators to provide support to vulnerable householders or contract those services and charge the operators.
- Have oversight of new developments and ensure they meet national standards.
- Work with landlords to ensure that they have strategic long-term management plans for the networks.

Finally, returning to technology, most recently, Denmark has pioneered fourth-generation DHS, using large-scale solar thermal arrays, inter-seasonal heat storage and one or more of a range of technologies – heat recovery systems, biomass and heat pumps – to meet demand for heat. Significantly, the most successful projects are now achieving over 50% of heat demand from solar thermal. This multi-technology approach has been successfully implemented by a growing number and size of district heating networks across Europe. Denmark leads the world and hosts nine of the largest solar thermal plants in Europe, including Dronninglund (26MW_{th}) (PlanEnergi and Niras 2015), Marstal on the island of Aeroe (23MW_{th}) (GSTE 2014) and 13MW_{th} installations at Grasten and Braedstrup. Dronninglund and Marstal use gravel-lined pits to provide inter-seasonal thermal storage, whilst Braestrup uses boreholes (Stadler 2014). Significant investments in district heating using this same model have also been made by Norway, and Austria (Baker and Mould 2019; IET 2012; Mauthner et al. 2014; Wien Energie 2019) is now making significant investments in solar thermal. In 2011, the estimated capacity was a mere 13MWh, however, in 2012, the development of a large installation to supply the Lillestrøm's DHS added 4GWh to capacity.

Of particular note is that generation costs for this multi-technology approach are highly competitive, currently around 30 to 50 €/MWh (Solar District Heating 2018), making it a direct competitor to heat pumps, individual biomass boilers and other common low- and zero-carbon forms of heating, and a game changer for the economics of DHS. Costs for connection to systems based on this model are also low and, counter-intuitively, are not significantly higher in rural areas, even before accounting for co-benefits such as job creation and tackling fuel poverty. Connection costs for Danish model systems deployed in Vienna, Austria, are as low as £5,000 per household for normal-density urban housing (Pöyry et al. 2009), rising only to £7,000 per household for very low-density suburban to semi-rural housing around Copenhagen, Denmark. In comparison, costs of connection to more conventional district heating in urban Sheffield have amounted to £6,100 to £7,200 per household (Pöyry et al. 2009). This makes the approach highly attractive both for investors and as a highly effective option for tackling energy poverty.

18.6 District Heating and Cooling as a Solution to Energy Poverty

In general terms, fuel poverty is the condition under which a household has to spend disproportionately more of its income to maintain an adequate level of warmth, and energy poverty is the condition under which a household has to spend disproportionately more of its income to meet all its essential energy needs. By far the most common definitions of these determine whether a household is fuel/energy poor by using a threshold of 10% of net household income (Baker et al. 2018; Boardman 1991). The terms fuel poverty and energy poverty are sometimes used interchangeably and the latter is sometimes used (we would argue, erroneously) to describe households, particularly in the developing world, who lack access to adequate energy services.

This is a deliberate over-simplification we make here as, in the context of discussing how best to develop district heating and cooling, the exact definitions make very little difference. For example, we often argue that a standard definition of 'fuel poverty' should include both heating and cooling costs. A household in a colder climatic region which is fuel poor by virtue of being unable to afford heating costs but can afford the minimal cooling costs associated with its local climate is still in a form of poverty, and as is a household in a hot climate where energy demand for cooling dominates and which can afford minimal heating costs but cannot afford cooling costs. Keeping to this over-simplification, fuel/energy poverty has three core components – income, energy efficiency and vulnerability – all of which are complex variables which can be disaggregated. However, for the purposes of this discussion, it is only the emergent property (fuel/energy poverty, however, defined) that we are interested in.

Designing successful DHC can beneficially impact all three of those complex variables but it is the technical aspects of the design of these systems, including factors such as strategic planning and the consideration of fuel supply chains, which are critical to defining 'success'. Again, to put it simply, a successful DHC is one which maximises the number of connected buildings, minimises unit energy costs to customers (while at least doing better than breaking even on investors' returns), minimises or eliminates greenhouse gas emissions and creates co-benefits for the community it serves (which include not only alleviating fuel/energy poverty but also creating jobs and tackling socio-economic deprivation, etc.) (Pridmore et al. 2017).

Staying with income and emissions for a moment, two ways that district heating can push down unit costs of heat is through high specification and scale. In order for this to be consistently achieved, regulators need to issue clear performance specifications for district heating networks. Otherwise, we will continue to count aspirational carbon savings rather than actual savings. What we need is for carbon savings, which were claimed at design stage for district heating, to be verified through annual reporting of actual fuel use. This reporting should inform:

- Whether our models and how we use them are reliable.
- Whether the network has been built properly.
- Whether the network can be used by the users in the way that design engineers envisaged.
- Whether it delivers the energy savings and reductions in costs expected.

Such regulation and reporting need to be cognisant of the fact that each component of a DHS/DHC system offers different strengths, weaknesses, opportunities and costs for tackling fuel/energy poverty. At the extreme, some countries in the former Soviet bloc have historically used heat networks to provide free, unmetered and unregulated, heat to buildings (Mould 2018) and there is a strong argument that waste heat should be taxed where producers do not recover and recycle it to local buildings at minimal or no cost to those customers (Baker 2017).

Fundamentally, successful heat and cooling networks capitalise on the existence of geographical nodes where capacity to recover waste heat and/or generate it renewably is greatest. So, if we were to take any town or city and plan a new DHS/DHC network, we would first identify the most significant supply node and build the network out from it, which is exactly what happened in Copenhagen over one hundred years ago. We would then map other supply nodes, add demand nodes and introduce legislation to encourage or, if necessary, compel the owners of these to connect to our network. This practical and legislative progress would be staged over a time frame that ensures that the upfront capital costs are not so high, and the returns on investment not so low, as to limit the expansion of our nascent network and ultimately risk killing it off by the owners of those nodes simply refusing to connect. This gets us to Denmark in 1979, the passing of the first Heat Supply Act and the origins of fourth-generation DHS.

However, we are now on the verge of a climate change catastrophe, so if DHS/DHC is to be part of our energy future, we need a realistic plan for taking Denmark's revolution from 1979 onwards and accelerating it into a politically and scientifically acceptable time frame. So, if we were to take a country or region in which we had free reign to formulate a DHS/DHC revolution, we would start by mapping those supply and demand nodes, and include nodes where new and/or expanded sources of supply (e.g. from sustainable biomass or renewably generated waste heat) could be realistically and rapidly established. We would then subject each node to an environmental, social and economic assessment, the results of which would be used to prioritise the development of our networks, a determination which would include accounting for where economies of scale could be accelerated by networks merging as they expand over time. Doing this today is not difficult thanks to Geographical Information Systems, provided we select the right criteria for those assessments and conduct them using real, robust data.

This leads us to other common failings of policies intended to spur the development of DHS/DHC, i.e. the use of assumptions. We have already pointed to the need to consider exergy and the fundamental problem of the United Kingdom's use of modelled data to produce energy performance certificates and related estimates of demand, the latter of which, as well as resulting in oversizing, also lead to the conclusion that the greatest gains from using DHS/DHC to reduce emissions and tackle fuel poverty lie in installing new networks in highly urbanised areas (Mould 2018). This latter problem is further compounded by the assumption that installing networks in lower-density semi-urban and rural areas is significantly more costly, yet the evidence from Denmark shows that the additional infrastructure costs are not significant (see: *Lessons from Denmark's district heating revolution*). So (here in Scotland at least), once we factor in the benefits and savings from tackling energy poverty and other problems that are endemic in socio-economically deprived rural and island areas, the balance should tip strongly in favour of prioritising the development of new DHS/DHC towards lower-density semi-urban and rural areas. Therefore, it comes as no surprise to us that many of the world-leading examples of fourth-generation DHS, such as Dronninglund, Marstal and Braestrup, are located in such areas.

And there is one final, and no less critical, component of achieving our revolution, which is ensuring we have the skills to deliver it. Developing and managing heat networks requires skilled engineers, scientists and other professionals working together and over long periods of time. This is yet another area where matching supply and demand is essential and strategic planning needs to take account of it. While it is well beyond the scope of this chapter to address the successes and failures of any given educational system in meeting this demand, we would point to the numbers of such skilled staff employed in the fossil fuel industry, and industry which has to be eliminated if we are to tackle climate change (Baker and Wood 2019), as one source of existing skilled labour who will need to be transitioned from being part of the problem to part of the solution.

Similarly, while the solutions for tying all this together will necessarily vary according to different national contexts, we have endeavoured to highlight the cornerstones on which any revolution must be built. However, here in Scotland, we have addressed this task in full by setting out our proposals for the establishment of a Scottish Energy Development Agency (Baker et al. 2019), which were based on taking the development of successful DHS/DHC as a highly complex task and working backwards to ensure they capture the development of other low-carbon and renewable energy systems. We note this here as something we hope may be adopted in other comparable countries and national contexts.

That work is rooted in a question we have alluded to throughout this chapter – what exactly defines a 'successful' DHC? Therefore, before concluding this chapter, we return to that question by presenting a set of criteria which we first published in 2019. These are based on the principles of Strategic Environmental Assessment adopted by the European Union (Baker and Mould 2019) and are intended, broadly, to be internationally applicable.

18.7 Defining Successful District Heating and Cooling Schemes

In order to fully realise the potential of DHS/DHC networks, it is essential to define what we mean by 'successful'. It follows that the significant investment decisions that need to be made must be based on comprehensive and holistic assessment criteria to ensure the benefits of new projects best meet the needs of national and local policy agendas and of local communities. This has led us down a line of research that produced the following set of criteria, which were initially developed as part of a collaboration with the Danish engineering and design consultancy Ramboll, which has been responsible for developing many of the successful DHS projects covered in this chapter. These were formulated for use here in Scotland with all the relevant data needs and sources checked for their availability to the Scottish Government, however, subject to checks on data availability, we see no reason why they should not be usable in other national contexts.

National and Local Policy Objectives

- Supporting national, regional and local development plans
  - Development should be consistent with planning and zoning (where this is in use) for new and existing housing, industry, business, commerce, conservation, etc.
- Supporting and consistent with regeneration and economic development objectives
  - Where biomass and biofuels will be used as a fuel supply, development should include the establishment and management of fuel supply chains that generate additional benefits in terms of new employment, new economic opportunities from complementary agricultural and forestry products, creating new recreation and tourism opportunities and enhancing biodiversity.
- Empowering local communities
  - Development should be in partnership with community and third-sector groups, local authorities and (where appropriate) local housing associations, which should have financial stakes in the networks and be represented at all levels of planning, management and operation.

Environmental Objectives

- Reducing greenhouse gas emissions
  - Developments should minimise the production of greenhouse gas emissions. Larger developments should seek to maximise the contribution of solar thermal and other renewables, as well as heat storage and recovery technologies, to minimise the contribution of (low-carbon) biomass and biofuels. Smaller developments and those with access to geothermal sources or suitable water bodies should aim to eliminate the need for low-carbon fuel sources.
- Air quality
  - Developments should be consistent with meeting local and national air quality objectives and zoning, particularly for NO_x and PM_{10} emissions.
- Land use
  - Developments should be consistent with current and future distributions of housing, public buildings, industry and commercial buildings in order to maximise both the demand on systems and the heat (including waste heat) available to them, and this should be backed by a Danish-style Heat Supply Act. Where biomass, biogas and alternative fuel sources will be used, the development of these supplies should maximise the potential of existing local sources and/or include the development of new local supply chains.
- Industry
  - Where appropriate, developments should seek to maximise the recovery of waste heat from existing industrial sources. Where new industry is planned, this should

incorporate the development of heat recovery systems for connection to existing or new heat networks. Planning permission for new industries should also include this as an assessment criterion.

- Visual impact
  - Developments should not contravene planning and environmental restrictions as regards visual impact.
- Water quality
  - Where water bodies are to be used as a source for heating or cooling, this should not be to the detriment of the local aquatic environment, and developments should aim to minimise the impact under all relevant indicators as determined by a national Environment Agency or equivalent body.
- Noise pollution
  - Developments should minimise noise pollution during both construction and operation, meeting all appropriate local restrictions and conditions.
- Waste
  - As well as recovering waste heat, developments may seek to utilise local waste supplies as a heat source, particularly the use of agricultural and industrial waste to feed anaerobic bio-digestors. In each case, the assessment of potential developments should consider the relative environmental impacts of any alternative options for managing these waste streams.

Social Objectives

- Fuel/energy poverty
  - Developments should prioritise the provision of low-cost, renewable and low-carbon heating (and cooling) to fuel-poor and otherwise vulnerable householders (if and as these are defined in any given national context). The assessment of this should include the analysis of real (as opposed to modelled) data on household energy consumption and costs, as well as real data and information on the technical benefits to be gained from connection to a DHS. It should also include an assessment of the co-benefits to the local community and economy of using the project to address the needs of local fuel-poor and otherwise vulnerable householders.
- Health
  - Developments should consider any potential health impacts to householders and local communities, including any emissions from energy centres and impacts on internal air quality and comfort arising from converting properties to using DHS and connecting them to heat networks.
  - Developments should seek to maximise the provision of heat to householders whose physical and mental health and well-being will benefit most from them. Where groups of such householders can be identified, these should be prioritised for the assessment of the benefits that connection to a DHS may offer.

Economic Objectives

- Whole life cycle costs
  - Developments should seek to minimise their whole life costs in terms of greenhouse gas emissions and cost-effectiveness.

- Economic growth
  - Developments should seek to maximise the creation of skilled employment opportunities through creating new jobs and new demand for skilled labour and to maximise the retention of skilled labour in areas suffering from depopulation.
- Regeneration of deprived areas
  - The prioritisation of support for new developments should emphasise the development of projects in deprived communities, particularly in rural and island areas. Developments should seek to maximise the potential new inward investment, the development of local fuel supply chains, job creation, tourism and recreation opportunities and the enhancement of local biodiversity.

Technical Objectives

- Electricity grid impacts
  - Developments should be consistent with, and support, the needs of the distribution service operators (DSOs), or equivalent energy companies, as regards the management, upgrading, reinforcement and expansion of the national electricity grid, as well as with the needs of those developing and managing any local micro-grids.
- Gas grid impacts
  - Developments should be prioritised towards communities off the gas grid and should seek to maximise the potential of networks for future expansion and the incorporation of new heat sources.
- Retrofitting existing buildings
  - Developments should be prioritised towards those where there is the greatest potential to retrofit energy centres and heat recovery technologies into existing buildings and to retrofit heat networks to supply existing buildings and homes, particularly in deprived areas.
- Technical feasibility
  - The technical aspects of developing DHS/DHC and heat networks require highly specialist expertise, and so the feasibility of all developments should be assessed by appropriately qualified independent experts.

Since first publishing these criteria, we have sought to engage further with other experts and policy specialists, and to date they have stood up to scrutiny. By reproducing them here, we are hoping to expand that scrutiny by reaching out to new audiences, and we would be grateful to hear from any readers of this book who share our view that heating and cooling networks should be part of the transition to a zero-carbon, zero energy poverty, future.

18.8 Concluding Remarks

The development of district heating and cooling has a long history behind it and, we argue, a long future ahead of it. While there is no silver bullet 'smart' technology for tackling climate change and eliminating energy poverty, the development of fourth- and fifth-generation DHS/DHC has shown how, by combining a mix of proven technologies, old and new and, crucially, putting infrastructure at the heart of our responses to these challenges, we can successfully implement smart, sustainable energy systems and futureproof our built environments and the communities they serve.

The technologies we need to deliver a just transition to a zero-carbon, zero energy poverty future are already here and are proven at scale, the need to implement them is more urgent than ever and public opinion is behind us. However, developing successful DHS/DHC is a complex task requiring strategic planning and long-term investment and policy signals, and our experiences here in Scotland lead us to conclude that the last remaining challenge is embedding smart thinking among policymakers and politicians. With more and more fifth-generation systems coming online, we could be – indeed we should be – on the verge of a new revolution in district heating and cooling and surely, in the face of climate change and increasing social inequality, bringing about that revolution would be a truly smart achievement.

References

Aebischer, B., Henderson, G., Jakob, M. et al. (2007). Impact of climate change on thermal comfort, heating and cooling energy demand in Europe. *Proceedings of the European Council for an Energy Efficient Economy 2007 Summer Study*. European Council for an Energy Efficient Economy.

Ahmed, A., Ip, K., Miller, A. et al. (2009). Thermal performance of earth-air heat exchanger for reducing cooling energy demand of office buildings in the United Kingdom. *Proceedings of the Eleventh International IBPSA Conference*. Glasgow, Scotland (27–30 July 2009).

Alembic Research, Energy Action Scotland and Waterford, P. (2019). *A Review of Domestic and Non-Domestic Energy Performance Certificates in Scotland*. Scottish Government.

Andrews, D., Riekkola, A.K., Tzimas, E. et al. (2012). Background report on EU-27 district heating and cooling potentials, barriers, best practice and measures of promotion. European Commission - Joint Research Centre, 25289 EN.

Baker, K.J. (2017). Renewable heat: the perfect storm? In: *A Critical Review of Scottish Renewable and Low Carbon Energy Policy* (ed. G. Wood and K.J. Baker), 81–101. Cham, Switzerland: Palgrave Macmillan.

Baker, K.J. and Mould, R. (2018). *Energy Performance Certificates - An Alternative Approach*. Glasgow, Scotland: Common Weal.

Baker, K.J. and Mould, R. (2019). *Just Warmth: Developing Equitable and Sustainable District Heating Systems in Scotland*. Glasgow, Scotland: Common Weal.

Baker, K.J. and Wood, G. (2019). Our time is up. In: *The Palgrave Handbook of Managing Fossil Fuels and Energy Transitions* (ed. G. Wood and K.J. Baker), 611–615. Cham, Switzerland: Palgrave Macmillan.

Baker, K.J., Emmanuel, R., and Phillipson, M. (2012). Support for RPP2 - housing futures. Report for the Scottish Government.

Baker, K.J., Mould, R., and Restrick, S. (2018). Rethink fuel poverty as a complex problem. *Nature Energy* 3: 610–612. https://doi.org/10.1038/s41560-018-0204-2.

Baker, K.J., Morgan, G., Mould, R. et al. (2019). *Powering Our Ambitions: The Role of Scotland's Publicly Owned Energy Company and the Case for a Scottish Energy Development Agency*. Glasgow, Scotland: Common Weal.

Bean, R., Olesen, B.W., and Kwang, W.K. (2010). A history of radiant heating & cooling systems. *ASHRAE Journal* 1: 40–42.

Boait, P.J. and Rylatt, R.M. (2010). A method for fully automatic operation of domestic heating. *Energy and Buildings* 42 (1): 11–16.

Boardman, B. (1991). *Fuel Poverty: From Cold Homes to Affordable Warmth*. London, UK: Belhaven Press.

Bostock, J. (1855). *The Natural History. Pliny the Elder*. London, UK: Taylor and Francis.

Brummer, N. (2019). *Mijnwater Heerlen*. Construction21 International.

Buchanan, R. (1807). *On the Warming of Mills and Other Buildings*. Glasgow, UK: J. & A. Duncan.

Buchanan, R. (1810). *Practical and Descriptive Essays on the Economy of Fuel and Management of Heat*. London, UK: J. Weale.

Buffa, S., Cozzini, M., D'Antoni, M. et al. (2019). 5th generation district heating and cooling systems: a review of existing cases in Europe. *Renewable and Sustainable Energy Reviews* 104: 504–522. https://doi.org/10.1016/j.rser.2018.12.059.

Common Weal, Lateral North, and Revive (2018). *Back to Life: Visions of Alternative Futures for Scotland's Grouse Moors*. Glasgow, Scotland: Common Weal.

Conniff, R. (2018). *Greenwashed Timber: How Sustainable Forest Certification Has Failed. Yale Environment 360*. New Haven, CT, USA: Yale University.

Danish Energy Agency (n.d.). *District Heating – Danish Experiences*. Danish Energy Agency.

David, A., Vad Mathiesen, B., Averfalk, H. et al. (2018). Heat roadmap Europe: large-scale electric heat pumps in district heating systems. *Energies* 10 (4): 578. https://doi.org/10.3390/en10040578.

DECC (2013). Summary evidence on district heating networks in the UK. Department of Energy and Climate Change, UK.

Dincer, I. (2002). The role of exergy in energy policy making. *Energy Policy* 30 (2): 137–149. https://doi.org/10.1016/S0301-4215(01)00079-9.

Dincer, I. and Rosen, M.A. (2013). *Exergy: Energy, Environment and Sustainable Development*. Oxford, UK: Elsevier.

Fong, J. and Alwan, Z. (2013). Modelling to predict future energy performance of solar thermal cooling systems for building applications in the north east of England. *Applied Thermal Engineering* 57 (1–2): 81–89. https://doi.org/10.1016/j.applthermaleng.2013.03.004.

Giannakopoulos, C., Le Sager, P., Bindi, M. et al. (2009). Climatic changes and associated impacts in the Mediterranean resulting from a 2°C global warming. *Global and Planetary Change* 68 (3): 209–224. https://doi.org/10.1016/j.gloplacha.2009.06.001.

Gilbert, J.P. and Jaudin, F. (1999). *Using Geothermal Waters in France: The District Heating System of Chaudes-Aigues from the Middle Ages*. Sacramento, CA: Geothermal Resources Council.

Gong, M. and Werner, S. (2015). Exergy analysis of network temperature levels in Swedish and Danish district heating systems. *Renewable Energy* 84: 106–113. https://doi.org/10.1016/j.renene.2015.06.001.

Granovetter, M. and McGuire, P. (1998). The making of an industry: electricity in the United States. *The Sociological Review* 46 (1): 147–173.

GSTE (2014). *Denmark: 23 MWth Cover 55 % of Heat Demand of 1,500 Households*. Milan, Italy: Global Solar Thermal Energy Council, United Nations Environment Programme.

Guelpa, E. and Verda, V. (2019). Thermal energy storage in district heating and cooling systems: a review. *Applied Energy* 252: 113474. https://doi.org/10.1016/j.apenergy.2019.113474.

Guo, Q. (2002). The Chinese domestic architectural heating system [Kang]: origins, applications and techniques. *Architectural History* 45: 32–48.

Hajat, S., Vardoulakis, S., Heaviside, C. et al. (2013). Climate change effects on human health: projections of temperature-related mortality for the UK during the 2020s, 2050s and 2080s. *Journal of Epidemiology & Community Health* 68 (7): 641–648. https://doi.org/10.1136/jech-2013-202449.

Hawkey, D., Webb, J., and Winskel, M. (2013). Organisation and governance of urban energy systems: district heating and cooling in the UK. *Journal of Cleaner Production* 50: 22–31. https://doi.org/10.1016/j.jclepro.2012.11.018.

Hytiris, N., Emmanuel, R., Aaen, B. et al. (2015). Heat recovery from mineworkings: opportunities in the Glasgow area. *Environmental Geotechnics* 4 (6): https://doi.org/10.1680/envgeo.15.00007.

IET (2012). Solar Energy Awakening in Norway. Institute for Energy Technology, Kjeller, Norway.

IRENA (2020). Heating and cooling: key statistics. International Renewable Energy Agency. https://www.irena.org/heatingcooling/Key-Statistics (accessed 10 July 2023).

Jenkins, D., Simpson, S., and Peacock, A. (2017). Investigating the consistency and quality of EPC ratings and assessments. *Energy* 138: 480–489. https://doi.org/10.1016/j.energy.2017.07.105.

Karytsas, S. and Theodoropoulou, H. (2014). Public awareness and willingness to adopt ground source heat pumps for domestic heating and cooling. *Renewable and Sustainable Energy Reviews* 34: 49–57.

Kovats, R.S. and Hajat, S. (2007). Heat stress and public health: a critical review. *Annual Review of Public Health* 29: 41–55. https://doi.org/10.1146/annurev.publhealth.29.020907.090843.

Li, H. and Svendsen, S. (2012). Energy and exergy analysis of low temperature district heating network. *Energy* 45 (1): 237–246. https://doi.org/10.1016/j.energy.2012.03.056.

Lund, H., Werner, S., Wiltshire, R. et al. (2014). 4th generation district heating (4GDH): integrating smart thermal grids into future sustainable energy systems. *Energy* 68: 1–11. https://doi.org/10.1016/j.energy.2014.02.089.

Marderos, A.S., Jadwiszczak, P., Axcell, B.P. et al. (2018). Heat pump placement, connection and operational modes in European district heating. *Energy and Buildings* 166: 122–144. https://doi.org/10.1016/j.enbuild.2018.02.006.

Marshal, J. (ed.) (1931). *Mohenjo-Daro and the Indus Civilization*. Delhi, India: Asian Educational Services.

Mauthner, F., Weiss, W., and Spörk-Dür, M. (2014). Solar heat worldwide-markets and contribution to the energy supply 2011. Technical Report, AEE-Institute for Sustainable Technologies, Gleisdorf, Austria.

Millar, M.-A., Elrick, B., Jones, G. et al. (2020). Roadblocks to low temperature district heating. *Energies* 13 (22): 5893. https://doi.org/10.3390/en13225893.

Mould, R. (2018). An investigation of the district heating – fuel poverty nexus. PhD Thesis, School of Computing, Engineering and the Built Environment, Glasgow Caledonian University, Scotland.

Neno, R. (2005). Hypothermia: assessment treatment and prevention. *Nursing Older People* 17 (7): 24–29. https://doi.org/10.7748/nop2005.10.17.7.24.c2389.

Noe, R.S., Jin, J.O., and Wolkin, A.F. (2012). Exposure to natural cold and heat: hypothermia and hyperthermia medicare claims, United States, 2004–2005. *American Journal of Public Health* 102 (4): 11–18. https://doi.org/10.2105/AJPH.2011.300557.

Oh, S.J., Ng, K.C., Thu, K. et al. (2016). Forecasting long-term electricity demand for cooling of Singapore's buildings incorporating an innovative air-conditioning technology. *Energy and Buildings* 127: 183–193. https://doi.org/10.1016/j.enbuild.2016.05.073.

Pellegrini, M. and Bianchini, A. (2018). The innovative concept of cold district heating networks: a literature review. *Energies* 11 (1): 236. https://doi.org/10.3390/en11010236.

Petri, Y., and Caldeira, K. (2015). Impacts of global warming on residential heating and cooling degree-days in the United States. *Scientific Reports* 5: 12427. https://doi.org/10.1038/srep12427

PlanEnergi and Niras (2015). *Dronninglund Solar tHermal Plant*. PlanEnergi and Niras.

Pöyry, Faber Maunsell and AECOM (2009). The potential and costs of district heating networks. Report for the Department of Energy and Climate Change, UK.

Pridmore, A., Smith, A., Baker, K.J. et al. (2017). Evidence review of the potential wider impacts of climate change mitigation options: built environment sector. Report for the Scottish Government.

Raizada, M. (2012). Renewables and district heating: Eastern Europe keeps it warm. *Renewable Energy World* 15 (5).

Ravilious, K. (2020). *Biomass Energy: Green or Dirty? Physics World*. Bristol, UK: IOP Publishing.

Rezaie, B. and Rosen, M.A. (2012). District heating and cooling: review of technology and potential enhancements. *Applied Energy* 93: 2–10. https://doi.org/10.1016/j.apenergy.2011.04.020.

Rogers, J.S. (2011). Architecture and complex hearth features at the Amaknak Bridge Site, Eastern Aleutian Islands, Alaska. *Artic Anthropology* 48 (2): 92–112.

Rutter, P. and Keirstead, J. (2012). A brief history and the possible future of urban energy systems. *Energy Policy* 50: 72-80. https://doi.org/10.1016/j.enpol.2012.03.072

Scottish Government (2016). *Getting the Best from Our Land: A Land Use Strategy for Scotland, 2016 – 2021*. Scottish Government.

Scottish Government (2019). *Renewable and Low Carbon Energy: Geothermal*. Scottish Government.

Sharp, D. (2002). *Twentieth Century Architecture: A Visual History*. Victoria, Australia: The Images Publishing Group Pty Ltd.

Solar District Heating (2018). *Decarbonising District Heating with Solar Thermal Energy*. Stuttgart, Germany: Solar District Heating.

Stadler, C. (2014). *Large Scale Solar Thermal Power Plants for District Heating*. Skørping, Denmark: ARCON Solar.

Steingisser, M. (2009). Human impacts on geyser basins. *Yellowstone Science* 17 (1): 8.

Sunamp (2021). *Compact Thermal Storage for Heating and Hot Water*. Sunamp Ltd.

Tarn, R.J. (2004). Housing in Liverpool and Glasgow. The growth of civic responsibility. *Town Planning Review* 39 (4): 319–334.

Tirado-Herrero, S., Űrge-Vorsatz, D., Arena, D. et al. (2011). Co-benefits quantified: employment, energy security and fuel poverty implications of the large-scale, deep retrofitting of the Hungarian building stock. *Proceedings of the European Council for an Energy Efficient Economy 2011 Summer Study* (18–20 November 2013), Berlin: European Council for an Energy Efficient Economy.

Verhoeven, R., Willems, E., and Harcouët-Menou, V. (2014). Minewater 2.0 project in Heerlen the Netherlands: transformation of a geothermal mine water pilot project into a full scale hybrid sustainable energy infrastructure for heating and cooling. *Proceedings of the 8th International Renewable Energy Storage Conference and Exhibition, IRES 2013.*

Webb, J. (2012). Climate change and society: the chimera of behaviour change technologies. *Sociology* 46 (1): 109–125.

Werner, S. (2017). International review of district heating and cooling. *Energy* 137 (15): 617–631. https://doi.org/10.1016/j.energy.2017.04.045.

Wien Energie (2019). *How It Works*. Austria: Wien Energie GmbH.

Yu, S. (2018). Introduction of water source heat pump system. In: *Handbook of Energy Systems in Green Buildings* (ed. R. Wang and X. Zhai). Berlin, Heidelberg: Springer https://doi.org/10.1007/978-3-662-49088-4_4-1.

19

The Role of Energy-Efficient Buildings in the Post-Carbon Future

Gloria Pignatta[1,2] and Shayan Naderi[1]*

[1] *School of Built Environment, Faculty of Art, Design and Architecture, University of New South Wales (UNSW), Sydney, Australia*
[2] *City Futures Research Centre (CFRC), University of New South Wales (UNSW), Sydney, Australia*

19.1 Introduction

The role of energy-efficient buildings in a post-carbon future is crucial in the transition to a more sustainable and resilient urban environment. According to the report published by International Energy Efficiency (IEA), buildings are responsible for about 36% of total energy use in the world (IEA 2023). Considering carbon emission of upstream processes for power generation, the building sector accounts for about 28% of energy-related carbon emissions, reaching its highest absolute value of 10 Gt CO_2 in 2019 (IEA 2023). From 2013 to 2016, the emissions from the building sector levelled off because of the measures taken for reducing the carbon footprint of the power generation sector. Since 2016, the demand increased considerably for energy services, including the electricity demand for air conditioning (AC) systems, especially cooling systems, and other electric appliances (IEA 2023). These trends suggest that addressing the energy efficiency of buildings is essential in reducing carbon emissions and achieving a more sustainable and low-carbon future. By implementing energy-efficiency measures and setting ambitious targets, cities can lead the way in addressing the building sector's significant contributions to climate change.

One of the reasons behind the growing energy demand in buildings is the need for cooling (IEA 2018). Global warming increases cooling degree days and the need for cooling in the building sector to provide indoor thermal comfort. Also, the trend of AC ownership is increasing (Imenes and Mills 2004), and the number of households with an installed AC unit increases every year. Figure 19.1 depicts the share of households with at least one installed AC unit for different countries. The total number of installed AC units was around 1.6 billion in 2016, and IEA predicts that around 5.6 billion AC units will be installed in buildings by 2050 (IEA 2018; Jensen et al. 2017), contributing to approximately 37% of global electricity demand.

* Corresponding Author (g.pignatta@unsw.edu.au)

Smart Cities, Energy and Climate: Governing Cities for a Low-Carbon Future, First Edition.
Edited by Oleg Golubchikov and Komali Yenneti.

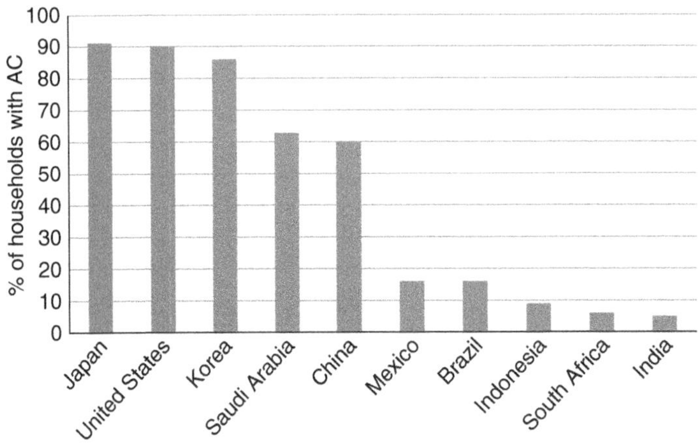

Figure 19.1 Share of households with air conditioning (AC) units in different countries. *Source:* Adapted from IEA (2018).

The AC cooling demand is known as the most temperature-sensitive residential load component (Wang et al. 2022). They are extensively operated during hot days, creating a peak demand in the electricity sector that coincides with the extreme ambient temperatures. During extremely hot days, the AC cooling demand can contribute to around 50% of the electricity network peak demand (Heslop et al. 2019), mainly during the afternoon or early evening when households return home and switch on their AC units.

To meet the cooling demand in 2050, around 3,183 billion USD in investment is required to upgrade electricity generation, transmission and distribution infrastructure (IEA 2018). The main cause of the required investment in the electricity sector is the increased peak demand. It is estimated that installing a 2 kW AC unit costs about 1,500 AUD for an Australian household, when it adds to the peak demand, costs around 7,000 AUD to upgrade the electricity network infrastructures (Department of Resources Energy and Tourism 2012).

In addition to the required investment to upgrade the electricity network infrastructure, the high peak demand increases the levelised cost of electricity. The extra capacity installed to meet the peak demand comes into operation and generates electricity when the demand is high, with a probability of exceedance of less than 10%. The limited operation means the extra capacity makes a profit only during a limited time. Therefore, the price paid for one unit of electricity to such generation units must be high enough to meet the financial needs of their business. Moreover, unit commitment follows a merit order algorithm in the electricity market. The units with lower marginal costs are operated first, and then units with higher marginal costs are operated as the demand increases. Therefore, during peak hours, units with higher marginal costs are operated increasing the levelised cost of electricity (Wood et al. 2013). Consumers finally pay these costs through higher electricity bills.

To encourage end-use consumers to shift their consumption to times with lower demand, Time-of-Use (TOU) pricing has been proposed (Imani et al. 2018). Based on TOU pricing, the price per kWh varies from time to time during the day or week. Lower prices are applied during off-peak hours, and the price is maximum during peak hours. Sometimes, there is a transition between the off-peak and peak hours known as shoulder, with an intermediate electricity price. People can reduce their electricity bills by shifting their consumption to off-peak hours (Tabares-Velasco et al. 2019). In this context, demand response (DR) programmes (Ibrahim et al. 2022; Naderi et al. 2022a; Naderi et al. 2023) are introduced to support customers in reducing their electricity usage during peak periods, typically through financial incentives or rebates, while improving grid reliability and reducing the need for expensive power generation.

In addition to the introduction of TOU pricing and dynamic pricing, a type of pricing structure that charges customers based on the real-time cost of electricity, peak demand shifting can be done in different ways, including:

- Shifting plug loads through behavioural change (Roberts 2020). This involves changing consumer behaviour to reduce peak electricity demand by encouraging customers to use energy-efficient appliances and lighting and promoting conservation practices.
- Using smart grid technologies, such as advanced metering infrastructure and distribution automation to monitor and control electricity usage in real-time and help manage peak demand.
- Investing in building retrofit and energy-efficiency measures, such as building envelope improvement (e.g. insulation, air sealing and window and door upgrades to reduce building heat loss and gain, reducing the need for heating and cooling), lighting upgrades (e.g. replacing incandescent and fluorescent lighting with LED lighting to reduce lighting energy consumption by up to 75%), heating, ventilation and air conditioning (HVAC) system upgrades (e.g. replacing outdated HVAC systems with more efficient equipment to reduce energy consumption and improve indoor air quality), building controls and automation (e.g. installing building automation systems (BAS) to control HVAC and lighting systems, reduce energy consumption and improve the overall building energy performance) and water efficiency (e.g. installing low-flow fixtures to reduce the need for water pumping and the associated energy consumption), among other solutions. It is worth noting that investing in building energy-efficiency measures can also bring many benefits such as reducing energy costs, improving indoor air quality and thermal comfort, increasing the value of the building and reducing the environmental impact.
- Implementing energy storage using electrochemical battery energy storage (BES) (Robyns et al. 2019) or building thermal mass (BTM) (James 2003), and active thermal storage (Meng et al. 2021).
- Implementing distributed generation, where the use of small-scale renewable energy systems, such as rooftop solar panels, micro wind turbines or solar–wind hybrid systems, can help reducing peak demand by generating electricity closer to where it is needed.

Energy storage systems are used to store energy during off-peak periods and release it during peak periods, helping to reduce the need for additional power generation during peak times. Electrochemical BES is a technology with a discharge time ranging from minutes to hours (Chatzivasileiadi et al. 2013). They are available with different battery chemistries, power and energy densities. However, the cost associated with installing BES results in a long payback period for investments and remains a challenge for integration into buildings (Roberts et al. 2019). BTM energy storage does not necessarily require extra investment (Naderi et al. 2022b) and can be implemented using building envelopes, furniture and the existing AC system (Olsthoorn et al. 2017; Sun et al. 2013). However, the utilisation of BTM for energy storage is limited by occupants' thermal comfort, meaning that the indoor temperature must be kept within the thermal comfort range when the building is occupied (Sun et al. 2013). Some works combine active energy storage and BTM storage because their discharge time and ability to retain thermal energy are different (Nelson et al. 2019), and when they are combined, they cover an acceptable time scale. BTM storage can be utilised for short demand response events, but active thermal storage can shift the load for more extended periods, i.e. a water tank retains cold water for longer periods compared to the time that BTM remains cool. Therefore, active energy storage, such as cold-water energy storage, offers more flexibility (Ershad et al. 2020).

Utilising local renewable energy sources to meet building demand has been proposed in the literature. It is projected in IEA Energy Technology Perspectives 2016 (IEA 2016) that the capacity of rooftop photovoltaics (PV) installation in the world will be more than 9 TW by 2050. The highest uptake of solar is reported in Australia, with about 2.7 million homes equipped with rooftop solar PV as of 2 March 2021. Among the homes with rooftop solar PV, about 31,000 systems are integrated with BES (Clean Energy Regulator 2021). Such buildings are called 'prosumer' since they produce and consume electricity simultaneously (Parag and Sovacool 2016).

The falling cost of PV technology enables the prosumers to reduce their on-site generation costs even below the cost of electricity provided by the grid (Luthander et al. 2015). In such situations, the priority is the self-consumption (SC) of locally generated electricity. Consuming the on-site generation means less export to the grid with a cheap feed-in tariff (FIT) (Luthander et al. 2015) and eventually less import from the grid because of the portion of the demand met by on-site generation (Nyholm et al. 2016). In some countries, the FIT is zero and the prosumer is not paid for the export to the grid (Luthander et al. 2015). PV SC increases the profitability of PV systems and reduces the payback period of rooftop PV installations (Chiaroni et al. 2014). However, due to a mismatch between the solar generation profile and building load profile, not all the generated electricity can be locally utilised, limiting the maximum achievable SC in the absence of local energy storage (Nyholm et al. 2016).

The following sections introduce solutions for building retrofit and discuss BTM and BES methods with more details for peak demand shifting and matching building demand with solar PV generation (Sections 19.2–19.4, respectively). Then, the utilisation of solar energy in buildings is discussed in Section 19.5. As the last point, the concepts of nearly zero-energy buildings (nZEB) and positive energy districts (PEDs) are introduced in Section 19.6 and discussed as strategies for sustainable urban development. Furthermore, the importance of cost-effective solutions to increase flexibility in PEDs and support the transition towards sustainable smart energy communities is highlighted.

19.2 Building Retrofitting

As buildings account for a significant portion of global energy consumption and greenhouse gas emissions, building retrofitting has become a crucial aspect in reducing their environmental impact and achieving a low-carbon future. Building retrofitting refers to the process of upgrading and advancing existing buildings to improve their energy efficiency.

Making low-carbon and energy-efficient design decisions at the building retrofitting stage, as well as at the design stages for new buildings, can significantly reduce the environmental impacts of a building over its life cycle (Basbagill et al. 2013). Multiple criteria decision-making methods can serve this aim by identifying and prioritizing the possible retrofitting solutions (Ranjbar et al. 2023).

There are two main approaches to building retrofitting: passive and active retrofitting. The Sections 19.2.1 and 19.2.2, provide a general explanation of passive and active building retrofitting strategies and their methods that are commonly used to improve the energy efficiency of buildings. Last Section 19.2.3 briefly reports the main factors affecting the economics of retrofitting existing buildings.

19.2.1 Passive Retrofitting

Passive retrofitting involves the modification of a building's envelope, such as insulation and replacements, to reduce the amount of heat loss or gain. For example, adding insulation to walls, roofs and floors can prevent heat loss in winter and heat gain in summer, reducing the need for heating and cooling, respectively. According to Schiavoni et al. (2016), the energy consumption of a building is strongly dependent on the characteristics of its envelope, and thermal insulation is one of the best strategies to reduce energy consumption and improve indoor thermal comfort. Similarly, replacing existing building components, such as single-paned windows with double-paned windows or existing doors and roofings with high-performing alternatives, can reduce heat loss and heat gain, further improving energy efficiency. Other examples of passive retrofitting include installing shading devices and upgrading HVAC systems. Additionally, to prevent the infiltration of unconditioned air, advanced air sealing can be done; sealing all gaps and cracks in the building envelope to prevent drafts and improve the overall energy efficiency of the building. Strategies for passive building renovation, which are also commonly employed in new buildings, include:

- *Building orientation* (Tushar et al. 2021): Orienting a building, or part of it, to optimise solar gain in the winter and minimise it in the summer to reduce the need for heating and cooling.
- *Daylighting* (Wong 2017): Incorporating natural light into a building through skylights, clerestory windows and light tubes can reduce the need for artificial lighting and improve the overall comfort of the space.
- *Thermally massive materials* (Verbeke and Audenaert 2018): Using materials like concrete, brick or stone that have a high thermal mass can help regulate the temperature inside a building by absorbing and releasing heat.
- *Green roofs and walls* (Manso et al. 2021): Installing vegetated surfaces in building envelope components can help to insulate a building, reduce heat gain and improve air quality.
- *Cool roofs and walls* (Kolokotroni and Kolokotsa 2013): These cool strategies implemented in the external layer of the building envelope are made of reflective materials, such as cool tiles, cool membranes, cool plasters, natural white gravel coverings (Pisello et al. 2014; Castaldo et al. 2015), etc., that can reflect more of the sun's heat and absorb less heat than a standard roof or wall. This can help to reduce the amount of heat absorbed by the building, lower indoor temperatures and decrease the need for AC,

especially when cool materials are used both on roofs and external walls (Perino and Serra 2015; Pisello et al. 2015). These materials can also help to reduce urban heat island effects (UHI) and improve energy efficiency (Pisello et al. 2016a).

- *Passive solar design* (Albayyaa et al. 2019): Incorporating passive solar design principles such as thermal mass, direct gain and indirect gain can help to reduce the need for heating and cooling by harnessing the energy of the sun.

The design of a building, including retrofitting, plays a crucial role in determining its environmental impact throughout its life cycle. Research has shown that building design decisions account for nearly 70% of a building's total environmental impact (Rebitzer 2002). To identify effective building retrofitting strategies, various methodologies have been applied in literature, such as surveys and inspections (Pignatta et al. 2017), energy monitoring (Pisello et al. 2016b), integrated experimental and numerical methods (Pisello et al. 2016c), energy audits and software simulations to test customised energy conservation measures in existing buildings and evaluate the resulting energy savings (Owolabi et al. 2022).

19.2.2 Active Retrofitting

Active retrofitting involves the integration of renewable energy systems into existing buildings. For example, installing photovoltaic (PV) panels on the roof of a building can generate electricity from sunlight, reducing the need to purchase electricity from the grid. Similarly, installing battery systems can store excess electricity generated by PV panels for later use, further improving energy efficiency. Other examples of active retrofitting include replacing heating and cooling systems with more efficient ones, such as air source heat pumps. More details about active retrofitting strategies are provided in Sections 19.3–19.5.

19.2.3 Economics of Building Retrofitting

Retrofitting existing buildings can be expensive, and the economics of retrofitting depends on several factors, such as the cost of the retrofit, the expected energy savings and the payback period. For example, if the cost of retrofitting a building is $100,000 and the expected energy savings are $10,000 per year, the payback period would be 10 years. This, however, does not consider the cost of capital and inflation. Studies available in the literature discuss various economic analysis methods that can be used to evaluate the cost-effectiveness of retrofit solutions in buildings (Webb 2017). However, retrofitting can also bring other benefits, such as improved indoor air quality, increased property value and reduced environmental impact, that should be taken into account when assessing the economics of retrofitting.

19.3 Thermal Mass Energy Storage

Different materials (Taylor and Miner 2014) and parts of a building can be considered BTM, and be used for passive energy storage (Bastien and Athienitis 2018). Concrete slabs, floors, envelopes, walls and furniture are common types of BTM. Based on their physical properties (Ulgen 2002), quantity and placement inside a building, they have different thermal dynamics and offer different energy storage capacities (Le Dréau and Heiselberg 2016). Based on the quantity of BTM, the building can be either a heavy-weight or light-weight building (James 2003). Since BTM heat capacity correlates with its energy storage potential, some studies considered adding thermal mass to light-weight buildings by installing extra furniture or phase change materials (PCMs) to building walls (Chen et al. 2020). PCMs can store excess heat during the day and release it at night, or store coolness during the night and release it during the day, by undergoing a phase transition between solid and liquid states at a specific temperature. This makes them useful for temperature regulation in buildings. PCMs can be organic or inorganic, and can be incorporated into various building materials, such as insulation, wallboards and paint, to improve energy efficiency.

BTM was not utilised for shifting peak demand in buildings in traditional AC control strategies. Traditional methods for controlling AC systems were based on set-up and setback strategies. In such controllers, when the building is unoccupied, the thermostat setpoint increases during hot seasons and reduces during cold seasons. This setpoint adjustment switches the AC off during unoccupied hours. At the onset of occupancy, the controller changes the temperature setpoint to switch the AC on and make the indoor environment thermally comfortable.

The time that the AC must be working and its required power directly relate to building thermal inertia. The more inertia, the more energy is required to bring the indoor temperature within the thermal comfort range (James 2003). When implementing such control strategies, the thermal mass is considered a barrier to energy saving and contributes to the morning peak of the commercial sector in cold climates like Canada and Germany (Olsthoorn et al. 2017). It is also the same for the afternoon peak of the residential sector (Arababadi 2016; Wang et al. 2020) and the morning peak of commercial buildings (Vishwanath et al. 2019a; Vishwanath et al. 2019b) in hot climates, like some climates of the United States, such as Phoenix (Wijesuriya et al. 2018), and some areas of Australia (Heslop et al. 2019).

However, BTM can be used as a thermal battery to shift the AC demand by adopting a different control strategy. In cooling seasons, pre-cooling can be implemented to reduce the indoor air temperature and BTM temperature before peak hours when the electricity demand is low. At peak onset, the temperature setpoint increases, which can eventually lead to the turning off of the AC or a reduction in its power. Therefore, the peak AC demand is totally or partially eliminated. The reduced peak AC demand means less consumption during times with expensive electricity tariffs.

However, since the storage efficiency of BTM is not 100% (Tang and Wang 2021), the amount of cost savings from pre-cooling depends on the amount of energy used to pre-cool the building. There is always an energy penalty, and the amount of stored thermal energy before peak onset is higher than the discharged energy during peak hours. The difference between peak and off-peak electricity tariff rates should be high enough to justify consuming more energy to pre-cool the building and make pre-cooling profitable (Nelson et al. 2019).

There is not a simple and linear relationship between building technical parameters and potential cost savings. Naderi et al. evaluated the pre-cooling cost savings of a range of Australian buildings located in Sydney, with different construction weights (i.e. thermal

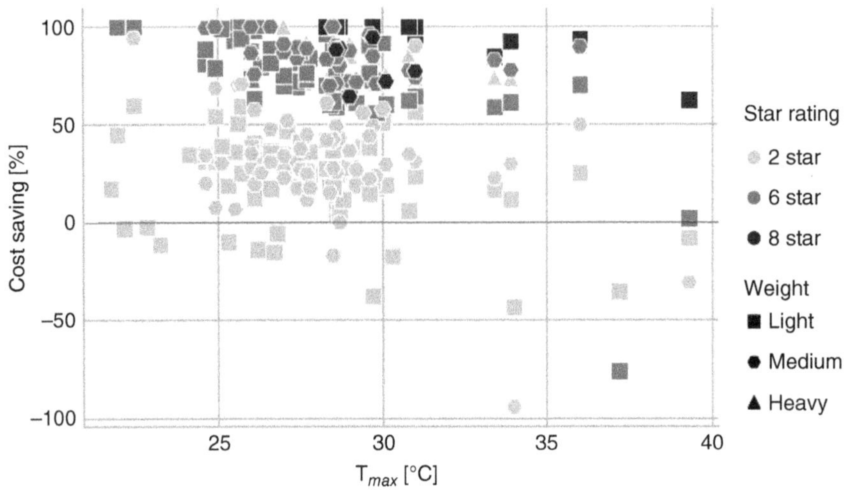

Figure 19.2 Pre-cooling cost savings in a range of Australian buildings. *Source:* Naderi et al. (2021)/MDPI/Licensed under CC BY 4.0.

mass quantity) and cooling demand, as shown in Figure 19.2. Star rating is the scheme to rate Australian buildings based on their annual heating and cooling demand, with zero-star buildings having the highest demand and ten-star buildings having the least demand in a climate (NatHERS National Administer 2022). The figure shows the percentage of daily cost savings for all buildings during the summer. With the same star rating, buildings with heavier construction weights can achieve higher daily cost savings. Moreover, higher cost savings are reported for buildings with lower annual heating and cooling demand (i.e. higher star ratings). It might be due to the higher levels of insulation and the ability of buildings to retain coolness after pre-cooling.

Although studies suggest that incorporating larger amounts of BTM can enhance the potential for pre-cooling, there is a limit to how much demand-shifting potential can be achieved through additional BTM (Chen et al. 2020). Moreover, it was observed that the storage efficiency drops in buildings with higher quantities of BTM (Tang and Wang 2021).

Thermal mass activation is the term used to describe the injection/removal of thermal energy into/from BTM to increase/reduce its temperature. Thermal mass activation can be done in four different ways (Olsthoorn et al. 2017):

- *Surface activation*: This is an indirect method for activating BTM. This is called indirect since the controller can change the indoor air temperature. By conditioning the zone air temperature, the BTM in contact with zone air is also conditioned. It is the common method for pre-heating and pre-cooling in most studies (Turner et al. 2014; Vishwanath et al. 2019a). This approach is advantageous since it does not require installing extra equipment, and there is no extra capital cost associated with surface activation. It should also be noted that the pre-heating and pre-cooling potentials in this method are limited by the thermal comfort range. During occupied hours, the indoor air temperature cannot go beyond the thermal comfort range, which reduces load shifting potentials. In this regard, partially occupied buildings offer higher levels of flexibility using surface activation because the temperature can be maintained outside the thermal comfort range to make a significant difference between the indoor air temperature and thermal mass temperature.

- *Forced air activation*: Using hollow core slab or ventilated slab systems, the conditioned air is used to cool down or heat the BTM during summer or winter, respectively (Park and Krarti 2015). In addition to being a radiant exchange system, the conditioned air is discharged to the control zone, thus providing better thermal comfort. Although the literature on ventilated slabs mostly focuses on thermal comfort (Olsthoorn et al. 2017), some studies show that this technology can reduce a proportion of AC energy use (Patania et al. 2010).
- *Hydronic activation*: The BTM is heated or cooled down using a working fluid in hydronic activation. This configuration enables using a lower temperature difference between the radiant surface and indoor temperature for providing thermal comfort because the surface area of the thermal mass is large. The low temperature difference can be achieved by low-lift chillers or renewable sources like photovoltaic thermal systems (Hassan and Abdelaziz 2020).
- *Electrical activation*: BTM is activated using long thermal resistances in the form of wires in electrical activation. According to the literature, this kind of activation is used for heating applications rather than load shifting and energy storage (Olsthoorn et al. 2017). However, studies show that in cold climates, up to 80% of the heating load can be shifted to nighttime using electrically heated floors (Thieblemont et al. 2016). Electrically heated floors might not have accurate control of floor surface temperature and zone temperature (Lin et al. 2005), so care must be taken not to jeopardise thermal comfort.

19.4 Battery Energy Storage

An alternative option for energy storage is converting electrical energy to electrochemical energy when there is a surplus of renewable generation or cheap electricity and again reconverting it to electrical energy when required. This technology, known as the electrochemical BES, is available in various sizes and power densities, delivering power from microwatts to megawatts (Chatzivasileiadi et al. 2013). However, they have some shortcomings. To name a few, sensitivity to the operating temperature, significant degradation and ageing due to cycling and environmental problems in life cycle assessment (LCA) analysis. Despite the attractive technical applications of BES systems in buildings, their high capital cost is still a barrier to widespread utilisation (Best et al. 2021; Roberts et al. 2019; Xu et al. 2012). Investigation of different battery chemistries and their applications in the building sector does not fall within the scope of this chapter and can be found in the literature (Chatzivasileiadi et al. 2013).

BES packages installed on buildings usually come with a gateway for smart monitoring, control and management. Due to the advantages of Li-ion batteries over other battery chemistries and their reasonable cost, this type of BES is mainly used in buildings. Low rate of self-discharge, high energy density and consequently longer charge retention are some advantages of Li-ion batteries.

Table 19.1 Percentage of average daily electricity consumption of Australian houses versus the capacity of two commercially available BES packages.

	Summer (%)		Autumn (%)		Winter (%)		Spring (%)	
State\season	**Tesla**	**LG**	**Tesla**	**LG**	**Tesla**	**LG**	**Tesla**	**LG**
Queensland	56	66	66	78	61	72	69	82
Victoria	57	68	61	72	38	45	60	71
South Australia	68	81	84	99.5	67	79	90	100

House energy consumption is based on the CSIRO energy use dashboard.
Source: Adapted from The Commonwealth Scientific and Industrial Research Organisation (2017).

BES packages are compact and can be installed indoors or they can be installed outdoors to save space. They are also expandable, and their capacity can be increased by adding new modules. Examples of such packages are Tesla Powerwall (Tesla 2022) and LG Home Battery (LG 2022), with maximum usable capacities of 13.5 and 16 kWh, respectively. Their continuous power outputs are 5 and 7 kW, respectively. Table 19.1 shows the percentage of daily electricity consumption of some Australian houses that these two BES packages can cover. The average daily electricity consumption of houses is extracted from a dashboard generated based on a survey conducted by the Commonwealth Scientific and Industrial Research Organisation (CSIRO) (The Commonwealth Scientific and Industrial Research Organisation 2017). The results highlight that except for winters in Victoria, the investigated BES packages cover more than 50% of the average daily electricity consumption of the houses. It means that installing BES packages can give houses a high level of flexibility for shifting the load and reducing electricity bills through storage during times with cheap tariff rates or surplus renewable generation, and then discharging when tariff rates are high.

A study conducted in Australia found that the main barriers to BES uptake in buildings are capital cost and policy. Policy in this study refers to high FIT that delays the BES revolution (Best et al. 2021). In a review of the effect of battery energy storage on SC of residential consumers, it is observed that SC can be improved by around 13–24% (Jäger-Waldau et al. 2020). Although SC is improved by installing BES (Bertsch et al. 2017), there is an optimum point for the size of BES, after which the increase in SC is negligible (Lund 2018).

19.5 Solar Energy Utilisation

Different types of renewable energy resources can be utilised for electricity generation. Wind, solar, geothermal and hydropower are some candidates. However, when the goal is to integrate them with buildings, several factors can directly affect their efficacy. Buildings are usually built in dense urban areas, where there is no empty land to install large-scale energy systems. If there is space to build such facilities, their corresponding noise, pollution and hazards might sacrifice human comfort.

Solar energy is abundant in most locations of the world and can be harvested, without related noises and loss of comfort. It can be used to supply a considerable part of energy use in a building, including hot water, electricity and space heating and cooling (Kasaeian and Sarrafha 2021). Technologies that harvest solar energy can be easily integrated into buildings or mounted on their roof. Photovoltaic (PV) and photovoltaic thermal (PV/T) systems are among the frequently used technologies to achieve net-zero energy buildings (NZEBs).

19.5.1 Building Integrated PV (BIPV) Systems

BIPV systems are the result of the integration of PV technology into the material of a building or replacing the conventional materials and components with PV systems. They also can be integrated into the architectural design of the building. This integration can be done in different parts of the building, such as walls, roofs and facades. When integrating PV into the roof, for example, by replacing tiles, or slate, one choice is to keep the architecture of the building intact and maintain architectural continuity. If so, the newly replaced materials cannot be distinguished from their conventional counterparts (Peng et al. 2011). On the other hand, opaque or transparent BIPV systems can be designed to improve the appearance of the building. BIPV systems are usually made transparent or semi-transparent, which allows solar radiation to pass, while they can control solar heat gain or loss (Reddy et al. 2020).

Using glass-on-glass semi-transparent BIPV systems reduces the need for artificial lighting since the sunlight enters the indoors. Transparent BIPV systems have a lower g-value than glass resulting in lower solar heat gain, which is favourable in cooling-dominated climate zones due to reducing AC cooling demand. Replacing traditional windows with BIPV systems is rather straightforward than retrofitting other components such as roofs or walls (Reddy et al. 2020). Currently, tile, glazing, module products and foils are the available BIPV products available in the market (Jelle 2016; Shukla et al. 2016). For roof applications, tiles and foils are used, while glazing has been frequently used for vertical installations such as walls, windows and facades (Reddy et al. 2020).

Unlike BIPV system, building-attached photovoltaics (BAPV) systems only generate electricity, without any other contribution to the building construction materials. They are not installed as a construction material and do not eliminate the need for conventional construction. Instead, they are installed or attached to the existing structure, such as roofs.

In addition to playing the role of construction materials and adding to the beauty of the building, BIPV systems can eliminate a part of the initial cost and labour. They require less investment and labour compared to non-integrated systems, like BAPV systems. In other words, in BAPV systems, not only the conventional construction material is purchased and installed but also a separate solar system must be added. BIPV systems have the functionality of these two layers simultaneously. It is reported that around 20% of BIPV/T systems are integrated into the facade, and around 80% are integrated with the roof (Shukla et al. 2017).

Special care must be taken to select an appropriate BIPV system for a specific building type and functionality. Various systems are available in the market, but not all of them lead to a cost-effective integration. The first step is to evaluate the location of the building, its architecture and surrounding buildings. The orientation of panels in buildings located in the southern hemisphere (i.e. Australia) is true north, while in the northern hemisphere, panels should face true south to receive direct solar irradiation. The surrounding buildings might also cause shading effects in dense urban environments and limit BIPV integration.

For instance, if a building's walls are fully or partially shaded, the only possible integration is rooftop PV installation. In complex architectures, it is recommended to integrate thin-film flexible modules (Alim et al. 2019).

After system selection, a feasibility study should be conducted to reveal the economic aspects of the project. All the economic terms must be considered, including capital costs, costs associated with transportation and installation and maintenance costs. The economic inputs and energetic performance of the system are then analysed to investigate the payback period, energy savings and environmental performance of the system. In such analyses, some inputs are specific to the location of the building. For instance, electricity tariff and FIT have major effects on the economic performance of the system throughout its lifetime. In some locations with cheap electricity prices and zero or low FITs, it might not be economically feasible to invest in BIPV systems, despite their potential environmental benefits.

19.5.2 Building Integrated PV Thermal (BIPV/T) Systems

The range of terrestrial solar radiation wavelength is something between 0.25 and 2.5 μm (Solanki 2015) which cannot be fully utilised by solar cells to generate electricity. A proportion of the radiations that include photons carrying energy close to the band gap of the PV cell is converted to electricity. The remaining part, which is around 70% of solar radiation (Yang and Athienitis 2016), generates heat by dissipating its energy and increasing the panel's temperature. It has been reported that when panels work at higher temperatures, their efficiency is significantly lower. It means that the proportion of radiation that is not converted to electricity negatively impacts the operation of PV panels (Skoplaki and Palyvos 2009). It has been reported that for temperatures above 25°C, the power output decreases by an average of 0.5% per degree Celsius (Sathe and Dhoble 2017).

This problem can be dealt with by collecting the heat from the PV panels using different fluids, including air, water, nanofluids, refrigerants and PCMs (Kasaeian and Sarrafha 2021). BIPV/T systems are designed to harvest thermal energy from PV modules and produce thermal energy in addition to electrical energy. They are innovative and practical solutions to achieve NZEBs with worldwide commercialisation potential (Shukla et al. 2016). The collected heat from a BIPV/T system can be used for air heating and ventilation, water heating and space heating (Yu et al. 2021). A schematic of a flat-plate PVT collector is shown in Figure 19.3. The coolant enters the collector and exits at a higher temperature.

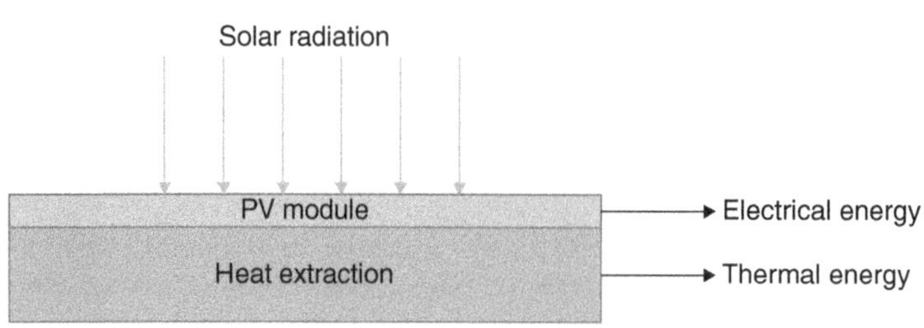

Figure 19.3 Schematic of a flat-plate PVT collector. *Source:* Adapted from Joshi and Dhoble (2018).

From a technological point of view, BIPV/T systems can be classified based on the receiver, which can be either a flat plate or a concentrating device. Also, the medium used to collect the useful heat can be different. Usually, commercially available working fluids like air, water or a refrigerant are used in BIPV/T applications (Babu and Ponnambalam 2017; Maghrabie et al. 2021). There is also another classification that divides BIPV/T systems into façade-based and roof-based systems (Yu et al. 2021).

The performance of BIPV/T systems is evaluated from four main viewpoints:

1) *Electrical performance*: The ratio of produced electricity to the incident solar radiation (Diwania et al. 2020; Kim and Kim 2012).
2) Thermal *performance*: The ratio of utilised heat to the incident solar radiation (Bergene and Løvvik 1995).
3) *Heat gain/losses*: This is the effect of BIPV/T on the heat gain/losses through the roof or facade, and consequently, the cooling/heating loads of the building.
4) *Total efficiency*: This is the overall efficiency of the BIPV/T system. Some researchers convert electricity to heat and then add it to the thermal efficiency because electricity is considered a higher-grade form of energy (Chow et al. 2006; Huang et al. 2001). On the other hand, others directly add electrical efficiency to thermal efficiency and consider an equal value for them (Bergene and Løvvik 1995).

There are some barriers to the widespread utilisation of BIPV/T systems. One of the main barriers is the cost-effectiveness of the systems. The current cost of BIPV/T technologies is still high and people blame investment costs as the reason why they do not install such environmentally friendly systems (Agathokleous and Kalogirou 2020). Moreover, there is a lack of data from the operation and commissioning of BIPV/T systems, operation and maintenance experts and a global standard for the design and operation of such systems (Maghrabie et al. 2021).

19.6 Energy Flexibility in Buildings and Districts

Energy flexibility in buildings and districts is essential for achieving sustainable smart energy communities in the post-carbon future. The ability to shift and control energy demand in buildings is crucial for transitioning from energy-efficient individual buildings (i.e. nZEBs) to PEDs, which present significant advantages (Ullah et al. 2021), including the economy of scale. Energy targets set at the building and district scale can be achieved cost-effectively through integrated design, construction and monitoring (Mavrigiannaki et al. 2021a). Implementing net zero energy concepts at the settlement or district scale presents numerous benefits and barriers that need to be taken into account when transitioning from individual to group buildings (Mavrigiannaki et al. 2021b; Meir et al. 2020).

The concept of nZEB refers to a building that has a very high energy performance and where the nearly zero or very low amount of energy required to run the building is covered to a very significant extent by energy from renewable sources mainly produced on-site or nearby. When a building generates as much energy as it consumes on an annual basis, typically through the use of renewable energy sources such as solar panels or wind turbines, it is defined as a net-zero energy building (NZEB). The goal of an NZEB is to have a balance between the energy used by the building and the energy produced by the on-site renewable energy systems. This means that NZEB goes one step further than nZEB, indeed NZEB generates as much energy as it consumes, while nZEB generates nearly as much energy as it consumes. PED refers to a district, neighbourhood or settlement in which the overall energy consumption of the buildings is lower than the renewable energy produced on-site or in the immediate vicinity. PEDs are designed to be self-sufficient in terms of energy, reduce dependence on fossil fuels and promote the use of renewable energy sources.

NZEBs and PEDs have the potential to significantly reduce energy consumption and promote the use of renewable energy sources by acting as strategies for sustainable urban development (Kolokotsa et al. 2022). Compared to traditional districts, PEDs often have a lower energy demand for heating and cooling due to the high energy efficiency of new buildings and implement advanced control systems that can maximise the utilisation of available energy flexibility. These systems can include demand–response programmes, energy storage solutions and smart grid integration. It is important to explore cost-effective solutions for increasing energy flexibility in buildings and PEDs, such as energy-efficient retrofits, renewable energy integration and building automation. By utilising energy flexibility in buildings and districts, we can pave the way towards a sustainable, low-carbon future for our cities.

19.7 Conclusions

In conclusion, this chapter has outlined the crucial role that energy-efficient buildings play in the transition to a post-carbon future. Strategies such as building retrofits, passive design and the integration of renewable energy generation and storage are essential for achieving net-zero energy buildings (NZEBs) and positive energy districts (PEDs), which have been presented as key strategies for sustainable urban development. This chapter has highlighted the potential of these strategies to shift and control energy demand in buildings, which can have a significant impact on energy consumption. It is well recognised that making low-carbon and energy-efficient design decisions has great potential to reduce the negative environmental impacts of the building sector. Therefore, cities around the world should consider implementing similar design solutions in both new and existing buildings.

Instead of solely focusing on individual buildings, attention is now being directed towards the energy performance of building complexes, in recognition of the important role that cities play in combating climate change and reducing carbon emissions in the energy sector. This shift in perspective allows for the benefits of economies of scale to be leveraged, resulting in a reduction in the additional costs associated with improved performance.

One of the key takeaways from this chapter is the importance of cost-effective solutions to increase energy flexibility and achieve energy targets in buildings and districts to support the transition towards sustainable smart energy communities. This highlights the need for continued research in this area to identify, explore and develop such solutions. The incorporation of 'smart' and advanced control systems and cost-effective solutions can also help to maximise the utilisation of available energy flexibility. The chapter has also emphasised the importance of continuous monitoring for measuring and verifying the expected performance of energy-efficient buildings.

The transition to a post-carbon future requires a concerted effort from all stakeholders, including architects, engineers, policymakers and building owners and occupants, to work towards the common goal of creating energy-efficient and sustainable built environments.

In terms of policy recommendations, government and industry collaboration is needed to support the implementation of NZEB and PEDs. This can include providing incentives for developers and building owners to implement energy-efficient measures, as well as setting regulations and standards for energy performance in buildings. Furthermore, education and awareness campaigns should be implemented to raise public awareness about the benefits of energy-efficient buildings and the role they play in transitioning to a sustainable future.

References

Agathokleous, R.A. and Kalogirou, S.A. (2020). Status, barriers and perspectives of building integrated photovoltaic systems. *Energy* 191: 116471.

Albayyaa, H., Hagare, D., and Saha, S. (2019). Energy conservation in residential buildings by incorporating passive solar and energy efficiency design strategies and higher thermal mass. *Energy and Buildings* 182: 205–213.

Alim, M.A., Tao, Z., Hassan, M.K. et al. (2019). Is it time to embrace building integrated photovoltaics? A review with particular focus on Australia. *Solar Energy* 188: 1118–1133.

Arababadi, R. (2016). *Operational and Technological Peak Load Shifting Strategies for Residential Buildings*. PhD thesis. Arizona State University.

Babu, C. and Ponnambalam, P. (2017). The role of thermoelectric generators in the hybrid PV/T systems: a review. *Energy Conversion and Management* 151: 368–385.

Basbagill, J., Flager, F., Lepech, M., and Fischer, M. (2013). Application of life-cycle assessment to early stage building design for reduced embodied environmental impacts. *Building and Environment* 60: 81–92.

Bastien, D. and Athienitis, A.K. (2018). Passive thermal energy storage, part 1: design concepts and metrics. *Renewable Energy* 115: 1319–1327.

Bergene, T. and Løvvik, O.M. (1995). Model calculations on a flat-plate solar heat collector with integrated solar cells. *Solar Energy* 55: 453–462.

Bertsch, V., Geldermann, J., and Lühn, T. (2017). What drives the profitability of household PV investments, self-consumption and self-sufficiency? *Applied Energy* 204: 1–15.

Best, R., Li, H., Trück, S., and Truong, C. (2021). Actual uptake of home batteries: the key roles of capital and policy. *Energy Policy* 151: 112186.

Chatzivasileiadi, A., Ampatzi, E., and Knight, I. (2013). Characteristics of electrical energy storage technologies and their applications in buildings. *Renewable and Sustainable Energy Reviews* 25: 814–830.

Castaldo, V.L., Coccia, V., Cotana, F. et al. (2015). Thermal-energy analysis of natural “cool” stone aggregates as passive cooling and global warming mitigation technique. *Urban Climate* 14: 301–314. https://doi.org/10.1016/j.uclim.2015.05.006.

Chen, Y., Xu, P., Chen, Z. et al. (2020). Experimental investigation of demand response potential of buildings: combined passive thermal mass and active storage. *Applied Energy* 280: 115956.

Chiaroni, D., Chiesa, V., Colasanti, L. et al. (2014). Evaluating solar energy profitability: a focus on the role of self-consumption. *Energy Conversion and Management* 88: 317–331.

Chow, T.T., Ji, J., and He, W. (2006). Photovoltaic-thermal collector system for domestic application. *Journal of Solar Energy Engineering* 129: 205–209.

Clean Energy Regulator (2021). Postcode data for small-scale installations. http://www.cleanenergyregulator.gov.au/RET/Forms-and-resources/Postcode-data-for-small-scale-installations (accessed 12 July 2023).

Department of Resources Energy and Tourism (2012). Energy white paper 2012 Australia's energy transformation. https://www.aip.com.au/sites/default/files/download-files/2017-09/Energy_%20White_Paper_2012-.pdf (12 July 2023).

Diwania, S., Agrawal, S., Siddiqui, A.S., and Singh, S. (2020). Photovoltaic–thermal (PV/T) technology: a comprehensive review on applications and its advancement. *International Journal of Energy and Environmental Engineering* 11: 33–54.

Ershad, A.M., Pietzcker, R., Ueckerdt, F., and Luderer, G. (2020). Managing power demand from air conditioning benefits solar PV in India scenarios for 2040. *Energies* 13: 2223.

Hassan, M.A. and Abdelaziz, O. (2020). Best practices and recent advances in hydronic radiant cooling systems – Part II: simulation, control, and integration. *Energy and Buildings* 224: 110263.

Heslop, S., Roberts, M., Yildiz, B. et al. (2019). Temporal patterns of residential air conditioning consumption in Australia's eastern capital cities. *Asia Pacific Solar Research Conference*, Canberra (3–5 December 2019). https://apvi.org.au/solar-research-conference/wp-content/uploads/2020/02/Heslop-S-A-characterisation-of-air-conditioning-consumption-in-Australia%E2%80%99s-eastern-capitals.pdf.

Huang, B.J., Lin, T.H., Hung, W.C., and Sun, F.S. (2001). Performance evaluation of solar photovoltaic/thermal systems. *Solar Energy* 70: 443–448.

Ibrahim, C., Mougharbel, I., Kanaan, H.Y. et al. (2022). A review on the deployment of demand response programs with multiple aspects coexistence over smart grid platform. *Renewable and Sustainable Energy Reviews* 162: 112446.

IEA (2016). *Energy Technology Perspectives 2016*. Paris.

IEA (2018). The future of cooling. https://www.iea.org/reports/the-future-of-cooling (accessed 12 July 2023).

IEA (2023). Tracking buildings. https://www.iea.org/energy-system/buildings#tracking (accessed 12 July 2023).

Imani, M.H., Ghadi, M.J., Ghavidel, S., and Li, L. (2018). Demand response modeling in microgrid operation: a review and application for incentive-based and time-based programs. *Renewable and Sustainable Energy Reviews* 94: 486–499.

Imenes, A.G. and Mills, D.R. (2004). Spectral beam splitting technology for increased conversion efficiency in solar concentrating systems: a review. *Solar Energy Materials and Solar Cells* 84: 19–69.

Jäger-Waldau, A., Adinolfi, G., Batlle, A. et al. (2020). Self-consumption of electricity produced with photovoltaic systems in apartment buildings - Update of the situation in various IEA PVPS countries. *2020 47th IEEE Photovoltaic Specialists Conference (PVSC)* (pp. 0938–0950). IEEE.

James, E.B. (2003). Load control using building thermal mass. *ASME Journal of Solar Engineering* 125: 292–301.

Jelle, B.P. (2016). Building integrated photovoltaics: a concise description of the current state of the art and possible research. *Pathways* 9: 21.

Jensen, S.Ø., Marszal-Pomianowska, A., Lollini, R. et al. (2017). IEA EBC annex 67 energy flexible buildings. *Energy and Buildings* 155: 25–34.

Joshi, S.S. and Dhoble, A.S. (2018). Photovoltaic -thermal systems (PVT): technology review and future trends. *Renewable and Sustainable Energy Reviews* 92: 848–882.

Kasaeian, A. and Sarrafha, H. (2021). Solar energy systems: an approach to zero energy buildings. In: *Advances in Clean Energy Technologies* (ed. A.K. Azad), 89–170. Academic Press.

Kim, J.-H. and Kim, J.-T. (2012). Comparison of electrical and thermal performances of glazed and unglazed PVT collectors. *International Journal of Photoenergy* 2012: 957847.

Kolokotroni, M. and Kolokotsa, D. (2013). Energy and environmental aspects of cool materials. *Advances in the Development of Cool Materials for the Built Environment* 231–272.

Kolokotsa, D., Pignatta, G., and Ulpiani, G. (2022). Nearly zero-energy and positive-energy buildings: status and trends. In: *Technologies for Integrated Energy Systems and Networks* (ed. G. Graditi and M. Di Somma), 239–273. Wiley Online Library https://doi.org/10.1002/9783527833634.ch10.

Le Dréau, J. and Heiselberg, P. (2016). Energy flexibility of residential buildings using short term heat storage in the thermal mass. *Energy* 111: 991–1002.

LG (2022). Home battery. https://www.lgessbattery.com/us/home-battery/intro.lg (accessed February 2022).

Lin, K., Zhang, Y., Xu, X. et al. (2005). Experimental study of under-floor electric heating system with shape-stabilized PCM plates. *Energy and Buildings* 37: 215–220.

Lund, P.D. (2018). Capacity matching of storage to PV in a global frame with different loads profiles. *Journal of Energy Storage* 18: 218–228.

Luthander, R., Widén, J., Nilsson, D., and Palm, J. (2015). Photovoltaic self-consumption in buildings: a review. *Applied Energy* 142: 80–94.

Maghrabie, H.M., Elsaid, K., Sayed, E.T. et al. (2021). Building-integrated photovoltaic/thermal (BIPVT) systems: applications and challenges. *Sustainable Energy Technologies and Assessments* 45: 101151.

Manso, M., Teotónio, I., Silva, C.M., and Cruz, C.O. (2021). Green roof and green wall benefits and costs: a review of the quantitative evidence. *Renewable and Sustainable Energy Reviews* 135: 110111.

Mavrigiannaki, A., Gobakis, K., Kolokotsa, D. et al. (2021a). Zero energy concept at neighborhood level: a case study analysis. *Solar Energy Advances* 1: 100002.

Mavrigiannaki, A., Pignatta, G., Assimakopoulos, M. et al. (2021b). Examining the benefits and barriers for the implementation of net zero energy settlements. *Energy and Buildings* 230: 110564. https://doi.org/10.1016/j.enbuild.2020.110564.

Meng, Q., Ren, X., Wang, W. et al. (2021). Reduction in on-off operations of an air source heat pump with active thermal storage and demand response: an experimental case study. *Journal of Energy Storage* 36: 102401.

Meir, I.A., Isaac, S., Kolokotsa, D. et al. (2020). Towards Zero Energy Settlements – A brief note on commissioning and POE within the EU ZeroPlus Settlements. *IOP Conference Series: Earth and Environmental Science* 410(1): 012038. https://doi.org/10.1088/1755-1315/410/1/012038.

Naderi, S., Heslop, S., Chen, D. et al. (2021). Cost-saving through pre-cooling: a case study of Sydney. *Environmental Sciences Proceedings* 12: 2.

Naderi, S., Pignatta, G., Heslop, S. et al. (2022a). Demand response via pre-cooling and solar pre-cooling: a review. *Energy and Buildings* 272: 112340.

Naderi, S., Heslop, S., Chen, D. et al. (2022b). Consumer cost savings, improved thermal comfort, and reduced peak air conditioning demand through pre-cooling in Australian housing. *Energy and Buildings* 271: 112172.

Naderi, S., Heslop, S., Chen, D. et al. (2023). Clustering based analysis of residential duck curve mitigation through solar pre-cooling: a case study of Australian housing stock. *Renewable Energy* 216: 119064.

Nathers National Administer (2022). Nationwide house energy rating scheme (NatHERS) – software accreditation protocol. https://www.nathers.gov.au/ (accessed September 2023).

Nelson, J., Johnson, N.G., Chinimilli, P.T., and Zhang, W. (2019). Residential cooling using separated and coupled precooling and thermal energy storage strategies. *Applied Energy* 252: 113414.

Nyholm, E., Goop, J., Odenberger, M., and Johnsson, F. (2016). Solar photovoltaic-battery systems in Swedish households – self-consumption and self-sufficiency. *Applied Energy* 183: 148–159.

Olsthoorn, D., Haghighat, F., Moreau, A., and Lacroix, G. (2017). Abilities and limitations of thermal mass activation for thermal comfort, peak shifting and shaving: a review. *Building and Environment* 118: 113–127.

Owolabi, A.B., Suh, D., and Pignatta, G. (2022). Investigating the energy use in an Australian building: a case study of a west-facing apartment in Sydney. *Ain Shams Engineering Journal* 14 (8): 102040.

Parag, Y. and Sovacool, B.K. (2016). Electricity market design for the prosumer era. *Nature Energy* 1: 16032.

Park, B. and Krarti, M. (2015). Development of a simulation analysis environment for ventilated slab systems. *Applied Thermal Engineering* 87: 66–78.

Patania, F., Gagliano, A., Nocera, F. et al. (2010). Thermofluid-dynamic analysis of ventilated facades. *Energy and Buildings* 42: 1148–1155.

Peng, C., Huang, Y., and Wu, Z. (2011). Building-integrated photovoltaics (BIPV) in architectural design in China. *Energy and Buildings* 43: 3592–3598.

Perino, M. and Serra, V. (2015). Switching from static to adaptable and dynamic building envelopes: a paradigm shift for the energy efficiency in buildings. *Journal of Facade Design and Engineering* 3 (2): 143–163.

Pignatta, G., Chatzinikola, C., Artopoulos, G. et al. (2017). Analysis of the indoor thermal quality in low income Cypriot households during winter. *Energy and Buildings* 152: 766–775.

Pisello, A., Pignatta, G., Castaldo, V., and Cotana, F. (2014). Experimental analysis of natural gravel covering as cool roofing and cool pavement. *Sustainability* 6(8): 4706–4722. https://doi.org/10.3390/su6084706.

Pisello, A.L., Castaldo, V.L., Piselli, C. et al. (2015). Combined thermal effect of cool roof and cool facade on a prototype building. *Energy Procedia* 78: 1556–1561. https://doi.org/10.1016/j.egypro.2015.11.205.

Pisello, A.L., Castaldo, V.L., Pignatta, G., and Cotana, F. (2015). Integrated numerical and experimental methodology for thermal-energy analysis and optimization of heritage museum buildings. *Building Services Engineering Research and Technology* 37(3): 334–354. https://doi.org/10.1177/0143624415609910.

Pisello, A.L., Castaldo, V.L., Pignatta, G. et al. (2016a). Experimental in-lab and in-field analysis of waterproof membranes for cool roof application and urban heat island mitigation. *Energy and Buildings* 114: 180–190.

Pisello, A.L., Pignatta, G., Piselli, C. et al. (2016b). Investigating the dynamic thermal behavior of building envelope in summer conditions by means of in-field continuous monitoring. *American Journal of Engineering and Applied Sciences* 9: 505–519.

Ranjbar, N., Balali, A., Valipour, A. et al. (2023). Identification and prioritization of energy consumption optimization strategies in the building industry using the hybrid swara-bim model. *Journal of Green Building* 18(1): 37–69. https://doi.org/10.3992/jgb.18.1.37.

Rebitzer, G. (2002). Integrating life cycle costing and life cycle assessment for managing costs and environmental impacts in supply chains. In: *Cost Management in Supply Chains*, 127–146. Physica-Verlag HD.

Reddy, P., Gupta, M.V.N.S., Nundy, S. et al. (2020). Status of BIPV and BAPV system for less energy-hungry building in India—A review. *Applied Sciences* 10: 2337.

Roberts, M. (2020). UNSW CEES (Collaboration on Energy and Environmental Markets) and Solar Analytics. *CRC-P Integrated Smart Home Energy Management Technologies project report*. https://www.ceem.unsw.edu.au/sites/default/files/documents/SmartHEMS%20User%20Needs%20-%20Final%20Report.pdf.

Roberts, M.B., Bruce, A., and Macgill, I. (2019). Impact of shared battery energy storage systems on photovoltaic self-consumption and electricity bills in apartment buildings. *Applied Energy* 245: 78–95.

Robyns, B., Davigny, A., Barry, H. et al. (2019). *Electrical Energy Storage for Buildings in Smart Grids*. Wiley.

Sathe, T.M. and Dhoble, A.S. (2017). A review on recent advancements in photovoltaic thermal techniques. *Renewable and Sustainable Energy Reviews* 76: 645–672.

Schiavoni, S., Bianchi, F., and Asdrubali, F. (2016). Insulation materials for the building sector: a review and comparative analysis. *Renewable and Sustainable Energy Reviews* 62: 988–1011.

Shukla, A.K., Sudhakar, K., and Baredar, P. (2016). A comprehensive review on design of building integrated photovoltaic system. *Energy and Buildings* 128: 99–110.

Shukla, A.K., Sudhakar, K., and Baredar, P. (2017). Recent advancement in BIPV product technologies: a review. *Energy and Buildings* 140: 188–195.

Skoplaki, E. and Palyvos, J.A. (2009). On the temperature dependence of photovoltaic module electrical performance: a review of efficiency/power correlations. *Solar Energy* 83: 614–624.

Solanki, C.S. (2015). *Solar Photovoltaics: Fundamentals, Technologies and Applications*. Phi Learning Pvt. Ltd.

Sun, Y., Wang, S., Xiao, F., and Gao, D. (2013). Peak load shifting control using different cold thermal energy storage facilities in commercial buildings: a review. *Energy Conversion and Management* 71: 101–114.

Tabares-Velasco, P.C., Speake, A., Harris, M. et al. (2019). A modeling framework for optimization-based control of a residential building thermostat for time-of-use pricing. *Applied Energy* 242: 1346–1357.

Tang, H. and Wang, S. (2021). Energy flexibility quantification of grid-responsive buildings: energy flexibility index and assessment of their effectiveness for applications. *Energy* 221: 119756.

Taylor, R.A. and Miner, M. (2014). A metric for characterizing the effectiveness of thermal mass in building materials. *Applied Energy* 128: 156–163.

Tesla (2022). Powerwall. https://www.tesla.com/en_eu/powerwall (accessed February 2022).

Tushar, Q., Bhuiyan, M.A., Zhang, G., and Maqsood, T. (2021). An integrated approach of BIM-enabled LCA and energy simulation: the optimized solution towards sustainable development. *Journal of Cleaner Production* 289: 125622.

The Commonwealth Scientific and Industrial Research Organisation (2017). Energy use of 209 Australian households. https://ahd.csiro.au/other-data/typical-house-energy-use/ (accessed February 2022).

Thieblemont, H., Haghighat, F., and Moreau, A. (2016). Thermal energy storage for building load management: application to electrically heated floor. *Applied Sciences* 6: 194.

Turner, W.J., Roux, J., and Walker, I.S. (2014). Reducing residential peak electricity demand with mechanical pre-cooling of building thermal mass. Lawrence Berkeley National Laboratory. LBNL Report #: LBNL-180912. Retrieved from: https://escholarship.org/uc/item/4x28r51q.

Ulgen, K. (2002). Experimental and theoretical investigation of effects of wall's thermophysical properties on time lag and decrement factor. *Energy and Buildings* 34: 273–278.

Ullah, K.R., Pignatta, G., Prodanovic, V. et al. (2021). Technological advancements towards the net-zero energy communities:a review of 23 case studies around the globe. *Solar Energy* 244: 1107–1126.

Verbeke, S. and Audenaert, A. (2018). Thermal inertia in buildings: a review of impacts across climate and building use. *Renewable and Sustainable Energy Reviews* 82: 2300–2318.

Vishwanath, A., Chandan, V., and Saurav, K. (2019a). An IoT-based data driven precooling solution for electricity cost savings in commercial buildings. *IEEE Internet of Things Journal* 6: 7337–7347.

Vishwanath, A., Hong, Y.-H., and Blake, C. (2019b). Experimental evaluation of a data driven cooling optimization framework for HVAC control in commercial buildings. *Proceedings of the Tenth ACM International Conference on Future Energy Systems*. Phoenix, AZ: Association for Computing Machinery. 25–28 June 2019.

Wang, J., Tang, C.Y., and Song, L. (2020). Design and analysis of optimal pre-cooling in residential buildings. *Energy and Buildings* 216: 109951.

Wang, C., Wang, B., Cui, M., and Wei, F. (2022). Cooling seasonal performance of inverter air conditioner using model prediction control for demand response. *Energy and Buildings* 256: 111708.

Webb, A.L. (2017). Energy retrofits in historic and traditional buildings: a review of problems and methods. *Renewable and Sustainable Energy Reviews* 77: 748–759.

Wijesuriya, S., Brandt, M., and Tabares-Velasco, P.C. (2018). Parametric analysis of a residential building with phase change material (PCM)-enhanced drywall, precooling, and variable electric rates in a hot and dry climate. *Applied Energy* 222: 497–514.

Wong, L. (2017). A review of daylighting design and implementation in buildings. *Renewable and Sustainable Energy Reviews* 74: 959–968.

Wood, A.J., Wollenberg, B.F., and Sheblé, G.B. (2013). *Power Generation, Operation, and Control*. Wiley.

Xu, Z., Guan, X., Jia, Q. et al. (2012). Performance analysis and comparison on energy storage devices for smart building energy management. *IEEE Transactions on Smart Grid* 3: 2136–2147.

Yang, T. and Athienitis, A.K. (2016). A review of research and developments of building-integrated photovoltaic/thermal (BIPV/T) systems. *Renewable and Sustainable Energy Reviews* 66: 886–912.

Yu, G., Yang, H., Yan, Z., and Kyeredey Ansah, M. (2021). A review of designs and performance of façade-based building integrated photovoltaic-thermal (BIPVT) systems. *Applied Thermal Engineering* 182: 116081.

20

Using Bottom-Up Digital Technologies in Technical Decision-Making for Designing a Low-Carbon Built Environment

Clarice Bleil de Souza[1], *Camilla Pezzica*[1], *and Jakob Hahn*[2]

[1] *Welsh School of Architecture, Cardiff University, Cardiff, Wales, UK*
[2] *Research Institute CENERGIE – Center for Energy Efficient Buildings and Districts, Munich University of Applied Sciences, Munich, Germany*

20.1 The Role of Bottom-Up Data in Technical Decision-Making

Technology-driven performance, energy-efficient, smart and net-zero buildings, low-carbon urbanism and sustainability rely on the interconnectivity between a network of sensors and actuators to inform decisions that balance several types of demands with quality of life (O'Dwyer et al. 2019; Thornbush and Golubchikov 2021). On one hand, they open opportunities for whole-system optimisation, for instance, integration of fluctuating renewable technologies, interconnection with local/public energy grids and demand shifts in energy use at peak times to achieve savings (Mahapatra et al. 2017). On the other hand, they require practitioners to better understand how citizens and building occupants behave in different contexts, how they interact with their cities and buildings, besides how they can better participate in decision-making towards producing sustainable solutions for all.

Therefore, much effort is currently spent on improving the understanding of citizen–city and human–building interactions to better inform urban and building design as well as cities and buildings' sustainable operation and control. Several initiatives are put in place to try and capture these interactions via the use of information communication technologies (ICT), integrated via the Internet of Things (IoT), to inform performance-based planning and sustainable design and operation (Kandt and Batty 2021; Pentland 2013; Zheng et al. 2020) while at the same time proclaiming to empower citizens and occupants to better interact with their cities and buildings.

This recognition of the growing impact of people on cities and building performance together with a strong push from performance-based legislation calls for changes in technicians'[1] assumptions about urban and building modelling. Citizens and occupants are

1 Technicians, in this case, include the body of technical stakeholders involved on urban and building design, operation and control (e.g. architects, engineers, project managers, facilities managers, etc.).

Smart Cities, Energy and Climate: Governing Cities for a Low-Carbon Future, First Edition.
Edited by Oleg Golubchikov and Komali Yenneti.

growingly recognised as central decision-making agents, with their complex and dynamic traits and bounded rationality, with their individual, collective and cultural identities, interplaying with a variety of social and spatial contexts and ethical and power relations (D'Oca et al. 2017; Gaetani et al. 2016; Ouf et al. 2018).

For instance, the implementation of 'intelligent metering systems that shall assist the active participation of consumers in the electricity supply market' ('Directive 2009/72/EC' 2009) requires management and flexibility in integrating intermittent renewable energy systems to the grid, including the need to have distributed trading mechanisms in place to ensure fairness and understanding the role of each different stakeholder in the process. Technically speaking, they require optimising energy supply and demand in interconnection with energy trading (Gregoratti and Matamoros 2015; Wang et al. 2014; Xu and Zhang 2016). That is, they require models that attempt to optimise supply, demand and costs based on market-driven interactions with monetary incentives to participants (Karnouskos 2011). To this end, they have to not only inform these participants of their contributions to energy-efficiency operation but also collect data from these participants to inform systems' design and operation ('Directive (EU) 2018/844, EPBD 2018' 2018). Technicians are challenged to work with two-way system models which feed and are fed by people's interaction with systems and its consequent changes in demand, while at the same time manage supply fluctuations that span from the unpredictability of renewables up to market variations.

This two-way context for design and operation pushes practitioners to combine traditional rule-based approaches to decision-making, in which physics-based models are used to inform different types of planning and design interventions, with hybrid approaches in which expert knowledge is used to make sense of bottom-up empirical data post-processed through statistical and machine learning models (O'Dwyer et al. 2019) inserted within 'intelligent' feedback loops. Technicians now should deal with a series of uncertain conditions related to data availability and quality (which go from conversion of data from soft sensors to predefined data types up to adequate aggregation, granularity and trustworthiness), systems' management and complexities and unpredictability of occupant/citizen's behaviour in a context which requests for flexibility. They are then forming several types of alliances to discuss how to improve the quality of information used as well as to develop comprehensive hybrid models which capture, for instance, the human–building interaction from within a whole energy system perspective (e.g. IEA-EBC Annex 66 (2013-2018) 'Simulation and Definition of Occupant Behaviour in Buildings' (Yan and Hong 2018) and IEA-ECB Annex 79 (2018-2023) 'The Behaviour-Centric Building Design and Operation' (O'Brien et al. 2020)).

This chapter showcases examples of how technicians use bottom-up data to reconcile performance demands with citizens' demands, when building and regenerating neighbourhoods and cities. The chapter starts with providing an overview of different possibilities available for bottom-up data gathering and the role these can play in hybrid approaches to technical decision-making. It then presents two case studies in Germany which illustrate how the use of data collected 'on behalf' of the people, through smart meters and indoor climate quality sensors, provided data to design a real-time energy management system while at the same time educating building occupants on their energy use patterns. It expands current discussions happening within the IEA Annex 79 related to empowering occupants when they become citizen scientists, by extrapolating on a case study from Italy which explored how co-production of data using digital technologies could aid post-disaster reconstruction in a resilient and sustainable way. The first two case studies explain the rationale and post-processing techniques to gather, transfer and re-use contextual data in technical decision-making. The second case study explores the difficulties in translating data collected by citizen scientists into useful technical information, exploring these difficulties initially in the context of resilience to sustainability and highlighting how this can impact the transferring of these practices to energy transitions. The chapter finishes with a synthesis of what has happened in these case studies, highlighting the difficulties involved in using bottom-up data in technical decision-making with regards to representativeness and data quality, essential components in producing sustainable solutions for all.

20.2 Possibilities for Bottom-Up Data Gathering

In a data-driven society (Pentland 2013) where computing power is in rapid expansion, the number of urban sensing technologies is growing steadily. Consequently, diverse types of data can be captured from the space, the air and at the street level and below the ground as well as on and from individual people.

All these types of data capture technologies jointly contribute to the understanding and management of the built environment, including how humans interact with it (Hutchby 2001). Specifically, they can support technical decision-making by providing a wide range of spatial, temporal and attribute data inputs to planning, architectural and engineering analytic tools such as Geographic Information Systems (GIS), Building Information Models (BIM), Building Energy Model (BEM) as well as digital simulations (from configurational analysis at an urban scale down to environmental and energy performance analysis at the building scale) as well as smart/intelligent controls. They enable technicians to work with hybrid approaches, integrating expert knowledge and empirical data with statistics, machine learning and physics-based models, to produce evidence to make and/or validate technical decisions (Bleil de Souza 2019), apart from also supporting control-feedback loops and lessons learning.

Types of data capture - include, but are not limited to, a range of techniques, from Interferometric Synthetic Aperture Radar (InSAR) and optical remote sensing technologies mounted on board a satellite, an aircraft or a drone, to Light Detection and Ranging (LiDAR), long- and close-range photogrammetry, integrated mobile mapping (e.g. collecting laser scanning and location data in parallel) and Indoor Positioning Systems (IPS). They also include water and energy meters to record electricity and gas for space heating and Domestic Hot Water (DHW) by energy and volume, as well as ambient sensing technologies and indoor climate quality sensors widely available to collect data such as temperature, relative humidity, lighting levels, carbon dioxide (CO_2), volatile organic components (VOCs). More recently, the wide deployment of portable/wearable devices producing decentralised user-generated content (UGC) as well as the recording of occupancy via motion detectors (passive infrared and ultrasound), WiFi device counts, CO_2 sensors and depth and smart/intelligent cameras, with filtering algorithms which indicate presence and number of people without disclosing their identity, provide large volumes of soft-data as well as promising data for occupant-centric controls. Examples of UGC are (geotagged) photos, videos, GPS or social media data.

Metering and indoor climate quality sensors - The use of metering and indoor climate quality sensor data is emerging in hybrid approaches for decision-making in building design and operation. These are now starting to be integrated into building performance simulation models and/or used to feed design as well as operation and maintenance decisions. However, their direct use in practice is still highly controversial due to issues related to poor representativeness and validity, potentially related to low occupants' uptake or buy-in as well as issues related to data privacy and anonymity, with regulations already in place (e.g. EU General Data Protection Regulation – GDPR) to respond to these to some extent but challenges still remaining in relation to 'infringing on people's rights or misusing their trust' (O'Dwyer et al. 2019). These issues push technicians to either generalise from small datasets or work with synthetic population data (Andrews et al., 2016), in attempts to compensate for data scarcity and have some 'real world' information to act upon. It also explains why technicians apply safety margins and safety coefficients to their decisions and resort to using available normative 'legal minima' when designing, abiding by several building, health and safety regulations (e.g. UK Approved Document L: conservation of fuel and power, Approved Document F Ventilation, to cite a few) and/or 'culturally accepted targets imposed by the market' (Cass, 2017) through a series of standards and guidelines (e.g. ANSI/ASHRAE/USGBC/IES, ISO, EN/BS, DIN/VDI). In addition, controversies also arise in relation to different techniques used in data aggregation (integration or statistics methods) and disaggregation (interpolation methods) as these are dependent on the levels of resolution involved in data gathering and data usage, temporally (hourly, daily, monthly, or annual reads) and spatially (e.g. whole building, per floor, per room).

User-Generated Content - UGC is emerging in demographic, mobility-based, urban morphology, place semantics studies and sentiment analysis as an important complementary approach in urban analytics to understand sociocultural contexts and the public perception of urban spaces (Gao et al. 2021; Paldino et al. 2016). In this case, people themselves are the sensors and can collect data with either a specific scientific aim in mind as in citizen-science approaches (Goodchild 2007) or through user contributed data, enhancing crowdsourcing data streams (Howe 2006), created while simply carrying out activities for their individual purposes. UGC application possibilities are context-specific and depend on the relationship between agents' goals and capabilities and the characteristics of the (data) objects (Gibson 2015). For instance, street-level photos are particularly interesting for planning because they portray streetscapes through a human perspective and have been used, for example, to research physical accessibility of urban space (Hara and Froehlich 2015). If of good quality and in large numbers (to be representative), they can provide in-depth possibilities to capture 'sense of place' and heavily promote user engagement in which virtuous cycles of feedback loops are developed, so technician have up-to-date information to act whenever needed. Via open sourcing and crowdsourcing, UGC could feed technology companies with the amount and variety of data inputs needed to innovate and develop better planning tools (UN Environment 2018). UGC can also contribute to urban resilience building by enhancing disaster response efforts and directly contributing to accelerate the achievement of Sustainable Development Goals

(SDGs). However, citizens may need to be actively encouraged to participate to these initiatives potentially demanding some form of partaking and/or compensation if long-term engagement is pursued (O'Dwyer et al. 2019) or large volumes of data gathering are required, and many times properly instructed or trained to produce useful data to be useful by technicians.

The use of these two groups of sensing techniques in technical decision-making (metering together with indoor climate quality sensors and UGC data capturing) are explored in detail through a set of case studies respectively in Sections 20.3 and 20.4. Two case studies in Germany explore how data gathered by sensors on behalf of the people can be re-used in a similar context to develop control algorithms to manage nearly net-zero energy districts. They are followed by a case study in Italy which illustrates what happens when technicians want to work with data gathered by citizen-scientists to assess and plan resilient and sustainable historic buildings' reconstruction after disaster, extrapolating on how this type of data gathering can be used to aid in energy transitions. Despite incomparable in terms of data being used and technical objectives to be achieved, these case studies illustrate the role of two different types of bottom-up data gathering in hybrid models, highlighting how technicians' factor in uncertainties (mainly related to representativeness and data quality) and lessons learned throughout their decision-making processes.

20.3 Data Gathered by Sensors on Behalf of the People

A key issue technicians face when using bottom-up data in decision-making is to have available and reliable data to act upon. When designing a building, this requires not only occupants' consent but also full client's and technicians' buy-in, a situation quite rare in practice that is permeated by liabilities. Therefore, it is not uncommon to see innovative projects pushing forward the net-zero energy agendas happening in partnership between the private and the public sectors. These initiatives promote the use of hybrid models and open important precedents for future cases in terms of lessons learned and data transfer.

Since data quality, reliability and validity are important aspects to be considered in hybrid models, consortiums are formed with academic institutions through publicly funded research. This ensures scientific monitoring protocols are in place to gather high-quality data through sufficient lengths of time with reasonable amount of spatial resolution (e.g. indoor air quality and electricity consumption at apartment level), so information on occupant behaviour is available at sufficient levels of detail to be scrutinised and correlated with demographic data and climate data and can be extracted for future use in a transferable way.

This section describes and discusses how a heavily monitored residential building district was used as a base case to extract lessons and transfer data to a mixed-mode building complex which is supposed to be a flagship for grid-supportive energy efficiency in South Germany. The two sequential practice-based research projects funded by the German Federal Ministry for Economic Affairs and Energy (BMWi) – the +EINS (0327431R) and the +EQ-Net I/II (03ET1299A, 03EN1046B) – were managed by the Research Institute CENERGIE – Centre for Energy Efficient Buildings and District at Munich University of Applied Sciences – which assisted in the design and monitoring of +EINS, using it as a precedence to design and monitor +EQ-Net.

Aiming to develop a breath of scenarios to inform the design and implementation of control algorithms for the overall energy management system (EMS) of the +EQ-Net energy project, so it could provide efficient and agile responses to different occupant demands and maximise the integration of fluctuating renewable energies, the team had to build on lessons learned from the +EINS project. To this end, similar outdoor climate conditions, similar building standards and socio-economic background were considered sufficient to transfer energy data from the first project to the second one.

20.3.1 Finding Reliable Data: The +EINS Project

The Ludmilla Residential Park in Landshut (Stockinger et al. 2015) in South-East Germany was commissioned to be a heavily monitored low-energy district. Designed to be a near-zero energy complex for its (mostly) occupant–owners, the project aimed to balance in situ generation with grid supply on an annual basis. Advised by the CENERGIE research group, the design team combined code compliance with building energy modelling throughout the decision-making process to ensure legal requirements were met. At the same time, a more efficient and alternative energy system was proposed to achieve nearly net-zero project aims, reducing energy and environmental costs to its occupants.

In addition, a comfort and energy visualisation online platform (called 'o-Vive') was installed in all of the apartment units and dwellings to inform occupants on their energy use and indoor environmental quality. It displays comfort trends as well as energy consumption of each unit in relation to other units within the district normalised by floor area on a daily, monthly and annual basis either in total or disaggregated for space heating and DHW. A traffic light system is in place for occupants to micro-manage their everyday choices with respect to their economic and environmental impacts (Figure 20.1), in attempt to educate occupants about their energy use while at the same time potentially persuading them to reduce consumption.

Case study data – Containing 55 apartments in 8 multi-family houses and 13 single and semi-detached dwellings, the project received funding from 2010 to 2014 to equip all units (apartment sizes of 45–110 m^2) with various sensors which included thermal energy for space heating, DHW and electricity for ventilation with heat recovery and plug loads. In addition, indoor environmental quality data (IEQ), mainly CO_2, humidity and air temperatures, were collected in 17 apartments (99 rooms) to monitor comfort and occupant satisfaction. The residential park is serviced by a district heating network with combined heat and power generation (CHP), a large thermal storage for decoupling of demand and generation, photovoltaic systems (PV), plus gas-fired boilers for backup thermal peak load. The single and semi-detached dwellings (between the apartment buildings) are equipped with micro-heat pumps combined with ground heat collectors and all units have underfloor heating distribution systems and controlled domestic ventilation heat recovery systems.

Case study results – This four-year project (with long-term monitoring still ongoing) provided useful and insightful information for similar projects to be developed in this region as well as the set of lessons summarised below (Hahn et al. 2020; Stockinger et al. 2015; Zink and Hahn 2018):

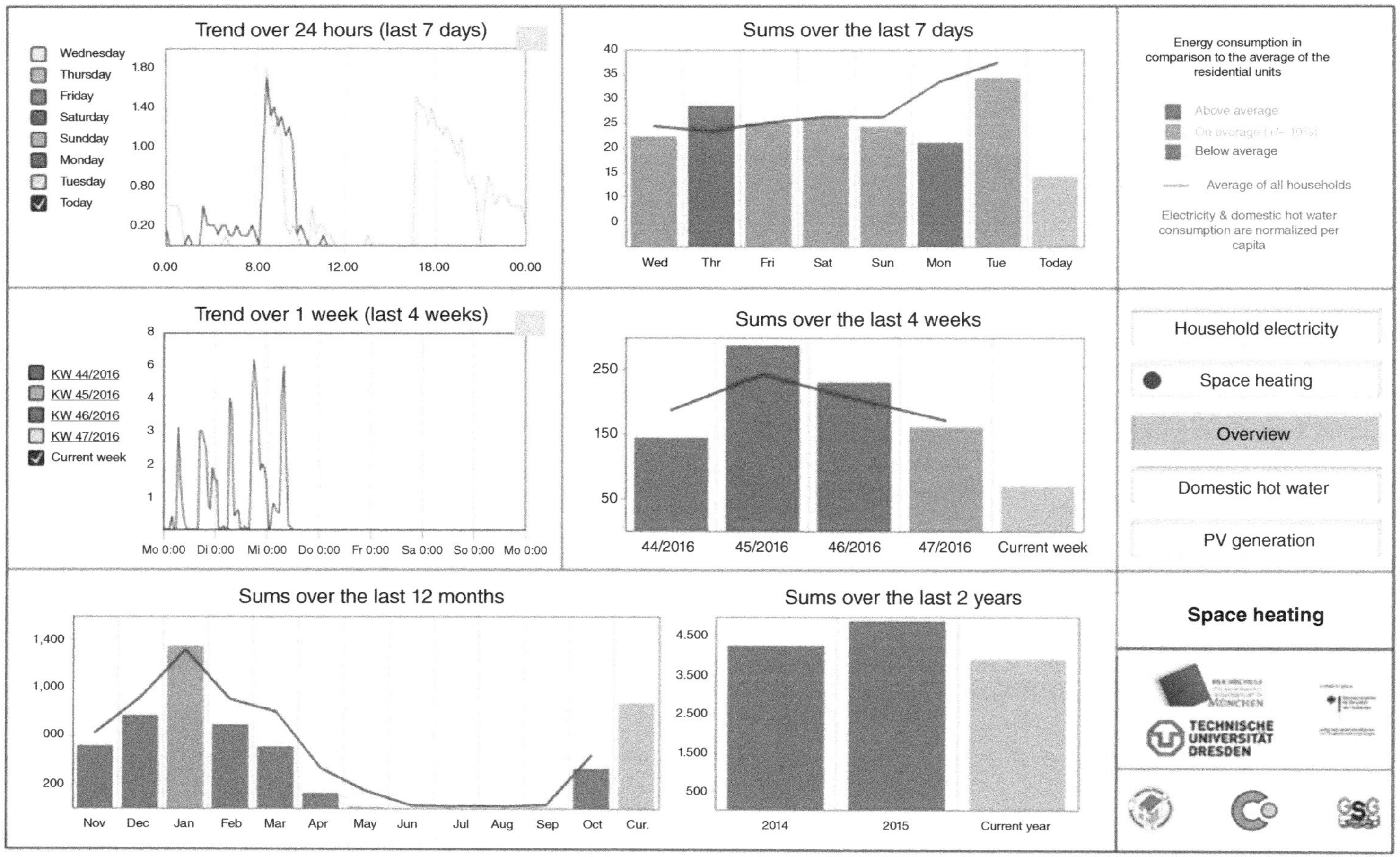

Figure 20.1 Snapshot of the online 'o-Vive' platform installed in the individual units of the Ludmilla Residential Park (labels are translated from German).

- Available renewable energy, co-generation technologies and a local heating network when efficiently combined offer large potential for operation optimisation without comfort limitations for the users in running time, operative temperatures and volume flows;
- Occupant behaviour has large influence on heating energy use and varies a lot among the different units; in this case, up to five times in magnitude in a single block of flats;
- Space heating energy consumption was on average higher than calculated by energy codes (DIN V 18599 and DIN V 4108-6). This can be explained by the fact that occupant-controlled thermostats enabled users to set it for higher or lower temperatures;
- The comfort and energy visualisation platform ('o-Vive') seems to have affected the energy consumption of the units as most of the users showed a decrease in consumption due to a higher awareness of their impact;
- Actual energy consumption for DHW as well as simultaneous peak loads were lower than assumed by the energy codes (DIN V 18599 and DIN V 4108-6), with again consumption variation between units up to five times in magnitude per unit and six times per capita in at least one of the blocks of flats;
- The energy consumption from household appliances per capita varied by a factor of 3 among different units. This variation was consistent across the four years' monitored period showing clear energy consumption fingerprints per unit exclusively due to occupant behaviour.

20.3.2 Transferring Data from One Context to Another: The +EQ-Net Project

The pulsG mixed-use district in Geretsried, also South-East Germany, was commissioned to be a grid-neutral district exploring synergies among residential, commercial and retail use (Hahn et al. 2021a,b; Lauterbach et al. 2021). It was designed to achieve electric grid-neutral interaction through a high degree of self-sufficiency taking advantage of different usages for load shifting and waste heat recovery from cooling processes (supermarket) as well as exhaust air (ventilation).

Its flexible combination of renewable generation and different usage demands together with energy storage systems requires a hybrid approach to technical decision-making, so that system sizing and controls are balanced towards achieving the project aims. Thus, building energy performance needs to be modelled together with the energy management system (EMS) and its respective controls considering several scenarios of weather and occupancy, for which real data can provide enough variations in magnitude and time to gauge robustness and design for flexibility.

Case study data – The complex has 66 flats with a mixture of 7 offices and doctor clinics plus a full-range supermarket and received funding from 2015 to 2021 to develop the energy concept/system, follow the design and construction phase and fully monitor space heating, electricity and DHW comprehensively (e.g. including piping flows and storage) during its operation phase. The complex is serviced by a district heating network with two compressor heat pumps and ground heat collectors, a heat pump for exhaust air heat recovery, two CHP units, thermal and electric storages and PV systems. Units have hybrid DHW (centrally serviced thermally with local electrical boosts) and underfloor heating systems, with space cooling and process cooling installed in commercial units and supermarket, respectively.

Data transferred – The CENERGIE research group used the +EINS data as a precedent, extracting from it a breath of aggregate occupancy energy profiles and figures for energy consumption, normalising them by floor area or per capita to support scenario generation for the +EQ-Net case. For instance, an average per capita profile together with its corresponding amount of DHW thermal energy consumption was transferred from the +EINS to the +EQ-Net. An assumed number of people was then estimated for the +EQ-Net to generate an annual profile to be used in the simulation of its energy and control systems. This would ensure that differences among different DWH consumption in the +EINS would be averaged out when transferred to the +EQ-Net, preserving context (i.e. similar social economic status, climate conditions, building standards, etc.). The same procedure was also done for the occupant-related electric energy consumption, more specifically lighting and appliances within the building units. The space heating energy consumption was not transferable, despite thermal insulation standards being similar. Building parameters such as compacity index, exposed façade area and cladding were all of different meanings and there were not enough comparable parameters to enable data transfer. However, space heating energy consumption normalised by floor area was still used to benchmark the +EQ-Net code calculations and simulations.

Case study results – The system design proposed by the CENERGIE was not adopted in its fullness in practice. Final decisions were made between the design team and the client/owner and resulted in a combination of the CENERGIE proposal with code compliance, capital and operational costs and their associated risks (Figure 20.2). Short-term monitoring indicated that general assumptions for the demands were coherent. That is, demands

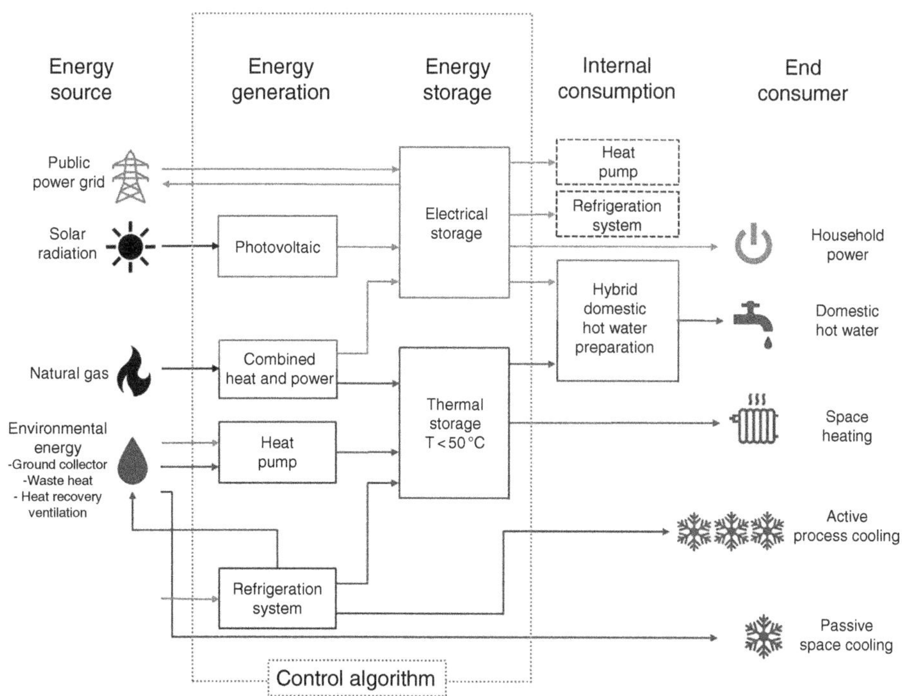

Figure 20.2 The implemented pulsG energy system within the +EQ-Net project in South-East Germany. *Source:* Adapted from Lauterbach et al. (2021).

were met and now the system can be better optimised for operation towards achieving the project's initial goals.

A further important outcome of the experiment is its possibility to enable the validation of initial research assumptions, i.e. use the monitoring data to scale up the implemented project to its original design proposal and assess variations. Through this, researchers can, for instance, assess deviations between the smaller electric storage and solar energy systems built and the original design, scaling these deviations up using simulations to verify if original goals could ever be achieved. This general approach to data transfer proves to be valid and as a result, the project was considered a flagship for using a collection of currently available energy system technologies operating through a robust intelligent energy management system working towards promoting energy transitions. Its success led to the funding of a sister project, the +EQ-Net II, which will monitor the energy consumption of the +EQ-Net for a longer period adding robustness to the overall proposal.

This iconic example shows that, despite not generalisable, high-quality data collected 'on behalf' of the people is essential to develop different scenarios of occupancy to feed energy management systems with a breath of grounded options to act. That is, to feedback control loops in real time and to generalise data for the development of more occupant-centric and energy-efficient controls. It enables technicians to experiment with real-world conditions on the assemblage and control of currently available technologies still not used to their full potential touching an important part of energy transitions; technical limitations in the deployment of current technologies due to uncertainties related to their interactions in the real world, something which requires extensive experimentation to be better understood.

20.4 Data Gathered by Citizen Scientists Using Digital Technologies

Advances in urban analytics in the context of smart cities support the faster generation of new hypotheses from the mining of big urban sensing data. However, they ought to be informed by theory and validated based on contextual understanding from small, low-frequency data to establish casual relationships and guide policymaking and urban planning (Kandt and Batty 2021). Citizens and public authorities can, and perhaps should, collaborate to achieve a successful management of urban environments (Quintili and Tempestini 2017) exploiting open-source data and software to support a better administration and consideration of social, economic and environmental factors. The scope for collecting and using this type of data can span from cities to buildings especially considering the new ISO standards series 55000 (BSI 2014), which see buildings as services in which customer satisfaction is an essential key performance indicator to be reconciled with clients' objectives and emission targets.

However, the use of citizen science initiatives in buildings is rare, if not completely absent, as post-occupancy evaluation data have a very clear top-down purpose: better inform clients, facilities managers and potentially designers on how to better design and run buildings to improve their energy efficiency without compromising occupants' satisfaction, rather than to empower occupants to participate in these decisions.

Therefore, this section presents a case in which civic hacking (bottom-up) data are used to support technical decision-making – mainly historic buildings' reconstruction – after the 2016–2017 Central Italy earthquakes. Through an example of how the co-production of data using digital technologies could aid post-disaster reconstruction in a resilient and sustainable way, it explores the difficulties in translating data collected by citizen scientists into useful technical information, drawing parallels with potential use of civic hacking in the context of energy transitions.

This was an experimental project, part of a collaboration between the DESTEC and the DICI at the University of Pisa and Cardiff University (Pezzica et al. 2018, 2019). The work focussed on the challenges of translating UGC from civic hacking initiatives into technical information useful for 3D reconnaissance operations to aid in identifying structural damage within BIM workflows, potentially acknowledging priorities in reconstruction to be defined by the community involved. This presumes that civic hacking affords the fulfilling of goals from different actors (Table 20.1), and highlights opportunities arising for technicians to enrich their data pool in contexts in which extra data are welcomed to speed up decision-making.

20.4.1 The Civic Hacking Infrastructure Terremoto Centro Italia

The civic hacking infrastructure *Terremoto Centro Italia* (TCI 2016b), built after the 2016 Central Italy earthquakes, aimed to facilitate activists' horizontal collaboration, and thus focussed on an incremental approach and decentralised solutions to aggregate

Table 20.1 Affordances, actors and goals of photographic mapping after a disaster.

Affordances	Actors	Goals
Complementary data source when other types of data are unavailable, too poor or delayed	Surveyor	Improve outputs of 3D reconnaissance operations
	Engineer/conservation architect	Perform remote visual inspections Access additional documentation resources
	Urban planner/designer	Automatic harvesting of points of interest Gather information relevant to placemaking Understand mobility routes Inform urban design
Storytelling, 3D documentation, open-sourcing, collaborative mapping, free choice of subject to be represented, wider coverage and update frequency	Citizen scientist/civic activist	Strengthen civic activism and community participation Include local views in stories (e.g. community priorities and points of interest) Improve communication with the external public

crowdsourced data and re-distribute them in form of open information (Tempestini 2016). Decisions about its objectives, rules and actions were made by the activists following a democratic voting principle and independently from the public administration by exploiting bottom-up collaborative data management processes (Quintili and Tempestini 2017). The idea was to create value and enhance humanitarian decision-making processes by bringing together and making accessible all relevant content from a variety of sources, including UGC, rather than replacing official information (Battaglia 2016).

The TCI project was operationally supported by an international non-governmental organization (NGO) as part of a monitoring, information and collaboration programme called SISMICO, whose mid-term goal was to establish a digitally advanced and effective civic activism model enhancing urban resilience and enabling BBB goals (SISMICO 2017). After the 2016–2017 Central Italy earthquakes, SISMICO supported the organisation of a two-day 'Monitoring and Civic Action School' focussed on the use of digital tools for producing UGC, which was delivered in collaboration with two disaster hit municipalities, different national and local associations, research and academic institutes and *Terremoto Centro Italia* (Sismico 2017). During the event, a contest was launched, and participants in the school and beyond were invited to collaboratively map the disaster-affected territory using Mapillary (TCI 2016a), an open platform for collecting geo-referenced street-level images. The initiative aimed to ignite a self-sustaining collaborative photographic mapping process, enabling the 3D documentation of photographed subjects, the capture and sharing of up-to-date data potentially useful to enhance collaborative vector mapping, freedom in subject choices and the reconstructing of individual narratives and an increase in civic and participatory activities by the community.

Data collected by citizen scientists – During the Monitoring and Civic Action School, 716 images were collected in Arquata del Tronto along 18 km of streets surveyed by 39 people, which reached 1081 images and 22 km after 183 days from the launch of the challenge. Generic instructions were given to participants with regards to camera handling and orientation, content (respecting privacy and avoiding photographing military areas) and data collection process, instructing participants to plan ahead, use geotags to label records at specific times and places, move when the sun is out, go when fewer cars and people are around, make sure windows are clean when photographing from inside cars, walk in groups, etc.

20.4.2 Converting Citizen Scientist Data into Useful Technical Data

Mapillary data coverage from 2018 (i.e. 2 years after the first quake) was compared with Google coverage to assess overlaps in relation to space and time. Google coverage proved to be pre-disaster (latest update from 2011), whereas the Mapillary initiative was, in contrast, a fit-for-purpose post-disaster initiative with community mapping density being influenced by pedestrian accessibility and concentrated in the highly affected urban areas, meaning they could yield interesting information for reconstruction.

A low-cost photogrammetric processing pipeline based on structure from motion and multi-view stereo (SfM-MVS) algorithms (Westoby et al. 2012) was used to attempt the 3D reconstruction of three urban fragments in two municipalities, namely Norcia and Arquata del Tronto, from UGC uploaded to Mapillary by citizen scientists. The process encompassed four steps: (i) data preparation (image grouping/filtering, selection and pre-processing); (ii) data processing (orientation and bundle adjustment); (iii) sparse (SfM) and dense (MVS) 3D point cloud reconstruction; and (iv) model scaling, mesh generation and texture mapping. The images were selected based on subject coverage (i.e. buildings and urban fragments), balancing requirements for good exposition and sharpness (enabling the automated extraction of good feature points) against the need to guarantee a minimum 20% overlap between images to prevent inaccuracies in the SfM triangulation.

Quality for technical use – Although images collected by the public were semantically rich, UGC data presented several challenges to be useful in technical decision-making. Poor quality and insufficient overlapping, occlusions and distancing issues prevented achieving acceptable 3D photogrammetric reconstructions. Thus, the ideal collaborative workflow presented in Figure 20.3, which is based on CRISP-DM (Wirth and Hipp 2000), did not materialise as digital images generated by the activists, although undoubtedly representing an important testimony of sense of place, did not yield sufficient valuable information for decision-makers.

In addition, the initiative did not attract enough participants to be considered inclusive and representative to infer community reconstruction priorities and was not sustainable to enable the exploitation of feedback loops to augment and inform technical decision-making in the long term; thus, reducing the scope for the exploitation of this type of UGC in recovery planning.

Results of this study make it clear that when UGC is to be used for technical purposes, including BIM, this goal needs to be explicit and compatible with participants' capacities and backgrounds and supported by additional user-centred guidelines. This requires careful consideration about how the collection is performed and managed in practice, as it influences the effectiveness, inclusivity and sustainability of the process (Soden 2017). That is, the process effectiveness in generating data that are technically fit-for-purpose with enough community buy-in, so it is inclusive and representative with enough follow-on activities to make it sustainable towards building a transformative partnership long-term. Trust should be built between institutions and civic actors (including knowledge holders, process facilitators and knowledge users) to ensure the success of the approach since expert facilitators can indirectly control the process of knowledge representation and transfer (Chambers 2006) and the community involved clearly needs to see benefits in engaging with the initiative so it can contribute with valuable open data that is representative and

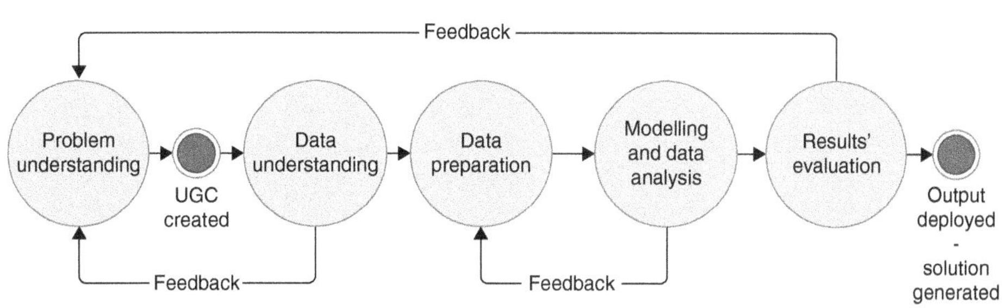

Figure 20.3 Collaborative data analytics workflow.

inclusive. Similar challenges apply if transferring this type of initiative to buildings in the context of promoting energy transitions, which heavily depends on appropriate user engagement, also requiring trust and many times some form of compensation.

Although the value of a civic hacking initiative can potentially be seen as beneficial to empower occupants (particularly in speculative buildings) to better inform building operations towards achieving true occupant satisfaction, this is unlikely to fit with the aims of facilities managers. This is because core asset management principles are primarily focussed on bringing benefits to the organisation managing the asset, i.e. the building owner. Specifically, they comprise improving financial performance, managing risks (e.g. providing good service with reduced liabilities to lower insurance costs), meeting customer expectations, promoting ethical businesses practices through corporate social responsibility, demonstrating compliance and improving sustainability and operation of the managing organisation through increasing efficiencies and effectiveness (BSI 2014). In this context, occupants are just a part of asset management as 'buildings are, and will remain, financial assets that figure in forms of market exchange, the operation of which often revolves around shared conventions and agreed forms of standardized description, measurement and provision' (Cass and Shove 2018).

20.5 Challenges in Using Bottom-Up Data in Technical Decision-Making

The case studies shown in Sections 20.3 and 20.4 illustrate hands-on examples of how bottom-up data can be useful to technicians to generate sustainability-oriented proposals, from narrowing the scope of decisions, so investments could be directed to more fit-for-purpose solutions reducing overall costs, up to augmenting technical information when time and money are limited as well as when further insights are needed. Bottom-up data gathering fed into technical decision-making provided more informed contextual boundaries for technicians to act and was never generalised beyond its context or claimed to be broadly representative. It was understood within its contextual boundaries and exploited acknowledging its limited transferability or use.

This means technicians were not after bespoke occupant- or citizen-centric solutions. They always had to combine data reflecting multiple realities and perspectives into a unitary viewpoint so technical decisions could be made to the benefit of all rather than a weighted combination of different interests objectively justified to detriment of collaboration, cooperative work and social agreement. Thus, despite factoring in the context and understanding the importance of user buy-in, technicians still need to generalise, within context boundaries, data gathered about occupants or from citizen scientists into useful information for decision-making. This context-based or context-dependent generalisation involved, in the case of this chapter, the following careful considerations:

- *Expert judgement on data transfer to new design*: which in the German case study comprised disaggregating overall energy use considering the different parameters affecting it, to judge what could be standardised (e.g. per capita or floor area), and therefore transferred to the second case study, and what could not.

- *Data aggregation and disaggregation appropriate to context*: which in the German case study involved abstracting occupancy energy profiles with enough breath (minima and maxima) in consumption patterns so per capita figures, used to inform scenario generations for the EMS on the second case study, could safely average-out user variations.
- *Strong guidance from technicians on how data are to be gathered by citizen scientists*: which in the Italian case study involved developing a set of more detailed guidelines for data collection so the quality of data is sufficient to be post-processed as useful information to decision-making.
- *Balancing semantic richness with representativeness*: which in the Italian case study involved abandoning the use of data to identify community priorities, as the number of citizen scientists involved in the initiative was small and non-representative, focussing only on the selection of good-quality data useful to decision-making.

Context-based generalisations represent a 'sharable point of view in which the many factors influencing the environment can coexist coherently, so that [people] can work together . . . because [they] share a single holistic view of the unitary goal of life' (Alexander 2014). Technical decision-makers are fundamental enablers of this environment since they design affordances for it to happen. Therefore, the strategic role of bottom-up data collection in a smart city context needs to be carefully thought out so it can represent community-based aspirations and needs and can be 'standardised' accordingly.

Acknowledgements

This chapter has been inspired by valuable discussions with colleagues and works within the IEA EBC Annex 79. Dr. Bleil de Souza acknowledges financial support from the UK Engineering and Physical Science Research Council to participate in the Annex Activities (grant number EP/S03126X/1). Mr. Hahn acknowledges financial support from the German Ministry of Economic Affairs and Energy (BMWi) within the framework of the research project 'Plus-EQ-Net' (2015–2021; grant number 03ET1299A). Section 20.4 of this chapter is based on Dr. Pezzica's PhD dissertation (Pezzica 2021), funded by the Department of Energy, Systems Territory and Construction Engineering (DESTEC), University of Pisa.

References

Alexander, C. (2014). *The Nature of Order: An Essay on the Art of Building and the Nature of Universe*, 3rde. Berkeley, CA: Centre for Environmental Structure.

Andrews, C.J., Allacci, M.S., Senick, J. et al. (2016). Using synthetic population data for prospective modeling of occupant behavior during design. *Energy and Buildings* 126: 415–423. https://doi.org/10.1016/J.ENBUILD.2016.05.049.

Battaglia, R. (2016). Che cosa stanno facendo gli attivisti digitali per il terremoto - wired, wired. https://www.wired.it/attualita/ambiente/2016/08/29/terremoto-centro-italia-terremoto-atttivisti/?refresh_ce= (accessed 6 June 2019).

Bleil de Souza, C. (2019). Risks and needs of digital regenerative design. In: *Regenerative Design in Digital Practice: A Handbook for the Built Environment* (ed. E. Naboni and L. Havinga). Bolzano, Italy: EURAC.

BSI (2014). EN ISO 55000. Asset management - overview, principles and terminology. https://shop.bsigroup.com/products/asset-management-overview-principles-and-terminology/standard (accessed 20 December 2021).

Cass, N. and Shove, E. (2018). Standards? whose standards? *Architectural Science Review* 61 (5): 272–279. https://doi.org/10.1080/00038628.2018.1502158.

Chambers, R. (2006). Participatory mapping and geographic information systems: whose map? who is empowered and who disempowered? who gains and who loses? *The Electronic Journal of Information Systems in Developing Countries* 25 (1): 1–11. https://doi.org/10.1002/j.1681-4835.2006.tb00163.x.

D'Oca, S., Chen, C.F., Hong, T., and Belafi, Z. (2017). Synthesizing building physics with social psychology: an interdisciplinary framework for context and occupant behavior in office buildings. *Energy Research & Social Science* 34: 240–251. https://doi.org/10.1016/J.ERSS.2017.08.002.

Directive (EU) 2018/844, EPBD 2018 (2018). Directive (EU) 2018/844 of the European Parliament and of the Council of 30 May 2018 amending Directive 2010/31/EU on the energy performance of buildings and Directive 2012/27/EU on energy efficiency. The European Parliament and the Council of The European Union. https://eur-lex.europa.eu/legal-content/EN/TXT/?uri=CELEX%3A32018L0844 (accessed 28 July 2021).

Directive 2009/72/EC (2009). Directive 2009/72/EC of the European Parliament and of the Council of 13 July 2009 concerning common rules for the internal market in electricity and repealing Directive 2003/54/EC. The European Parliament and the Council of the European Union. https://eur-lex.europa.eu/legal-content/EN/TXT/?uri=CELEX%3A32009L0072&qid=1627458759379 (accessed 28 July 2021).

Gaetani, I., Hoes, P.J., and Hensen, J.L.M. (2016). Occupant behavior in building energy simulation: towards a fit-for-purpose modeling strategy. *Energy and Buildings* 121: 188–204. https://doi.org/10.1016/J.ENBUILD.2016.03.038.

Gao, S., Liu, Y., Kang, Y., and Zhang, F. (2021). User-generated content: a promising data source for urban informatics. In: *Urban Book Series* (ed. W. Shi et al.), 503–522. Springer Science and Business Media Deutschland GmbH https://doi.org/10.1007/978-981-15-8983-6_28.

Gibson, J.J. (2015). *The Ecological Approach to Visual Perception*, Classice. Psychology Press.

Goodchild, M.F. (2007). Citizens as sensors: the world of volunteered geography. *GeoJournal* 69: 211–221. https://doi.org/10.1007/s10708-007-9111-y.

Gregoratti, D. and Matamoros, J. (2015). Distributed energy trading: the multiple-microgrid case. *IEEE Transactions on Industrial Electronics* 62 (4): 2551–2559. https://doi.org/10.1109/TIE.2014.2352592.

Hahn, J., Schumacher, P., Lang, W., and Jensch, W. (2020). Performance gap and occupant behavior – review and analysis of high-efficiency residential buildings in Germany. *ECOS 2020 – Proceedings of the 33rd International Conference on Efficiency, Cost, Optimization, Simulation and Environmental Impact of Energy Systems*, pp. 2023–2035.

Hahn, J., Lauterbach, S., and Jensch, W. (2021a). Netzdienliche Quartiere – Von der Konzeption bis zum Betrieb. Hochschule München (ed), *Book of Abstracts, INUAS Conference 2021: "Urbane Transformationen - Ressourcen: Material, Zeit, Raum, Energie"*, Munich (03–05 March 2021), Germany, pp. 50–50.

Hahn, J., Rust, C., and Jensch, W. (2021b). Moving towards renewable and grid-supportive district energy systems – experiences gathered from design to operation of a case study in

Germany. *Book of Abstracts of the 2nd International Conference on Evolving Cities (ICEC 2021), University of Southampton* (22–24 September 2021), Great Britain, pp. 22–24.
Hara, K. and Froehlich, J.E. (2015). Characterizing and visualizing physical world accessibility at scale using crowdsourcing, computer vision, and machine learning. *ACM SIGACCESS Accessibility and Computing. Association for Computing Machinery (ACM)*, 113, pp. 13–21. https://doi.org/10.1145/2850440.2850442.
Howe, J. (2006). The rise of crowdsourcing. *WIRED*. https://www.wired.com/2006/06/crowds/ (accessed 9 September 2020).
Hutchby, I. (2001). Technologies, texts and affordances. *Sociology* 35 (2): 441–456. https://doi.org/10.1177/s0038038501000219.
Kandt, J. and Batty, M. (2021). Smart cities, big data and urban policy: towards urban analytics for the long run. *Cities* 109: 102992. https://doi.org/10.1016/j.cities.2020.102992.
Karnouskos, S. (2011). Demand side management via prosumer interactions in a smart city energy marketplace. *IEEE PES Innovative Smart Grid Technologies Conference Europe*. IEEE, pp. 1–7. https://doi.org/10.1109/ISGTEurope.2011.6162818.
Lauterbach, S., Rust, C., Hahn, J., and Jensch, W. (2021). District energy systems: first insights from a high-efficiency building case study in Germany. *Proceedings of ECOS 2021 – 34rd International Conference on Efficiency, Cost, Optimization, Simulation and Environmental Impact of Energy Systems*, Taormina, Italy (27 June–2 July 2021), pp. 1444–1453.
Mahapatra, C., Moharana, A.K., and Leung, V.C.M. (2017). Energy management in smart cities based on internet of things: peak demand reduction and energy savings. *Sensors* 17 (12): https://doi.org/10.3390/s17122812.
O'Brien, W., Wagner, A., Schweiker, M. et al. (2020). Introducing IEA EBC annex 79: key challenges and opportunities in the field of occupant-centric building design and operation. *Building and Environment* 178: 106738. https://doi.org/10.1016/J.BUILDENV.2020.106738.
O'Dwyer, E., Pan, I., Acha, S., and Shah, N. (2019). Smart energy systems for sustainable smart cities: current developments, trends and future directions. *Applied Energy* 237: 581–597. https://doi.org/10.1016/J.APENERGY.2019.01.024.
Ouf, M.M., O'Brien, W., and Gunay, H.B. (2018). Improving occupant-related features in building performance simulation tools. *Building Simulation* 11 (4): 803–817. https://doi.org/10.1007/S12273-018-0443-Y.
Paldino, S., Kondor, D., Bojic, I. et al. (2016). Uncovering urban temporal patterns from geo-tagged photography. *PLoS One* 11 (12): e0165753. https://doi.org/10.1371/journal.pone.0165753.
Pentland, A.S. (2013). The data-driven society. *Scientific American* 309 (4): 78–83. https://www.jstor.org/stable/26018109?seq=1 (Accessed 27 May 2021).
Pezzica, C. (2021). Temporary Housing and Transitional Urbanism: Planning for Sustainable Post-Disaster Recovery under Uncertainty, University of Pisa.
Pezzica, C., Cutini, V., and Bleil De Souza, C. (2018). Opportunities for the use of collaborative 3D mapping in post disaster situations. In: *Environmental and Territorial Modelling for Planning and Design, Smart City, Urban Planning for a Sustainable Future* (ed. A. Leone and G. Carmela), 475–482. Naples: FedOAPress http://orca.cf.ac.uk/117907/ (accessed 12 June 2019).
Pezzica, C., Piemonte, A., Bleil De Souza, C., and Cutini, V. (2019). Photogrammetry as a participatory recovery tool after disasters: a grounded framework for future guidelines. *ISPRS - International Archives of the Photogrammetry, Remote Sensing and Spatial*

Information Sciences XLII-2/W15(2/W15): 921–928. https://doi.org/10.5194/isprs-archives-XLII-2-W15-921-2019.

Quintili, C. and Tempestini, M. (2017). Singolarità e comunità: ICT in fase di emergenza. In: *Building Back Better: idee e percorsi per la costruzione di comunità resilienti*, 1ste (ed. F. Esposito et al.), 71–78. Carocci: Roma.

Sismico (2017). Monitoring report. http://www.garzantilinguistica.it/ricerca/?q=beneficiario (accessed 6 June 2019).

Soden, R. (2017). Citizens' participation and crowdsourcing. https://www.unisdr.org/we/inform/publications/57417 (accessed 29 May 2019).

Stockinger, Jensch; Volker, Werner and Grunewald, J. (2015). Final report: Forschungsvorhaben +Eins - Plusenergiesiedlung Ludmilla-Wohnpark Landshut. Munich, Germany.

TCI (2016a). Racconta e Visita i luoghi del terremoto grazie a Mapillary, Terremoto Centro Italia website. https://terremotocentroitalia.info/2016-10-10-usare-mapillary-per-raccontare/ (accessed 31 May 2019).

TCI (2016b). Terremoto Centro Italia. https://terremotocentroitalia.info/ (accessed 28 May 2021).

Tempestini, M. (2016). TerremotoCentroItalia e la strategia dell' AGOPENDI. https://medium.com/terremotocentroitalia/terremotocentroitalia-e-la-strategia-dell-agopendi-410c590b1d49 (accessed 28 May 2021).

Thornbush, M. and Golubchikov, O. (2021). Smart energy cities: the evolution of the city-energy-sustainability nexus. *Environment and Development* 39: 100626. https://doi.org/10.1016/J.ENVDEV.2021.100626.

UN Environment (2018). Citizen science: crowdsourcing data to support disaster response. https://www.unenvironment.org/news-and-stories/story/citizen-science-crowdsourcing-data-support-disaster-response (accessed 31 May 2019).

Wang, Y., Saad, W., Han, Z. et al. (2014). A game-theoretic approach to energy trading in the smart grid. *IEEE Transactions on Smart Grid* 5 (3): 1439–1450. https://doi.org/10.1109/TSG.2013.2284664.

Westoby, M.J., Brasington, J., Glasser, N.F. et al. (2012). "Structure-from-Motion" photogrammetry: a low-cost, effective tool for geoscience applications. *Geomorphology* 179: 300–314. https://doi.org/10.1016/j.geomorph.2012.08.021.

Wirth, R. and Hipp, J. (2000). CRISP-DM: towards a standard process model for data mining. *Proceedings of the 4th International Conference on the Practical Applications of Knowledge Discovery and Data Mining*. Manchester, UK, 11–13 April 2000, pp. 29–39.

Xu, J. and Zhang, R. (2016). Cooperative energy trading in CoMP systems powered by smart grids. *IEEE Transactions on Vehicular Technology* 65 (4): 2142–2153. https://doi.org/10.1109/TVT.2015.2423317.

Yan, D. and Hong, T. (2018) Definition and simulation of occupant behavior in buildings Annex 66 final report operating agents of annex 66. www.iea-ebc.org (accessed 20 December 2021).

Zheng, C., Yuan, J., Zhu, L. et al. (2020). From digital to sustainable: a scientometric review of smart city literature between 1990 and 2019. *Journal of Cleaner Production* 258 (120689): https://doi.org/10.1016/j.jclepro.2020.120689.

Zink, T. and Hahn, J. (2018). Monitoring in plus-energy residential buildings: assessment of thermal comfort and indoor air quality. In: *ARC 2018: Proceedings of the Applied Research Conference* (ed. J. Mottok, M. Reichenberger, and W. Bogner), 275–281. Deggendorf, Germany: Deggendorf Institute of Technology, Pro BUSINESS Berlin.

21

Street Lighting as a Dimension of Smart Energy Cities

Mary Thornbush[1] *and Oleg Golubchikov*[2]

[1] *Faculty of Environmental Studies, York University, Toronto, ON, Canada*
[2] *School of Geography and Planning, Cardiff University, Cardiff, Wales, UK*

21.1 Introduction

Gil-Garcia et al. (2015) have conceptualised the smart city as a sociotechnical phenomenon that is characterised by 'urban innovation, digital city, intelligent city, innovative city, creative city, learning city, human city, and sustainable city and from a variety of disciplines using these and other characterizations' (p. 65). Their review has revealed 10 core components based on the academic literature, underlain by four dimensions: (i) government, (ii) society, (iii) physical environment, and (iv) technology and data. They have espoused that technology is an enabler of progress in these building blocks of smart cities, operating to enhance and interconnect the components of the smart city.

This chapter further addresses technology and its role in the emergence of smart cities. The aim is to examine literature from which to build an understanding of the relationships between smart cities and energy efficiency. More specifically, our focus is on a case study of smart street lighting and light pollution reduced through the application of sensors monitoring energy use. Cases from street lighting publications convey the different systems that are currently in place or being experimented with. Making street lights 'smart' can make considerable contributions to reduced energy consumption.

21.2 Hardwiring the City

Since at least 2011, the term 'smart energy city' (SEC) has arisen in reference to the energy-relevant component of smart cities. This development stems from the understanding that energy transcends all other dimensions of smart cities and is at the root of the technological advancement necessary for their emergence. To realise SECs, intelligent and integrated systems are required for the generation, distribution and use of energy (Doukas et al. 2016).

Smart Cities, Energy and Climate: Governing Cities for a Low-Carbon Future, First Edition.
Edited by Oleg Golubchikov and Komali Yenneti.

In addition to smart electricity grids and smart district heating and cooling grids, generation systems such as renewable energy production should be incorporated (Thornbush and Golubchikov 2019). Furthermore, as outlined by Fregonara et al. (2013), a multidisciplinary approach is necessary for reducing energy demand in all urban systems including by introducing solutions at an early stage of their design (see also Thornbush and Golubchikov 2021).

Lighting provides a vital opportunity for energy savings through the adoption of smart technologies. The European Commission (2004) has, for instance, conveyed that major electricity savings can be achieved through the use of energy-efficient lighting systems. These systems can also improve lighting quality while reducing maintenance costs and are generally cost-effective investments. In recent years, there has been recognition of 'light pollution' as an environmental problem stemming from the inappropriate use of lighting systems (cf. Korkmazer et al. 2019), in the wrong placement, timing, or quantity. Others have identified excessive night-time artificial lighting as both an environmental and social problem (Zielinska-Dabkowska 2014).

Smart applications can help overcome the waste associated with outdated, untargeted street lighting. The selection of cases presented below focuses on schemes that are designed to reduce energy consumption and improve the quality of street lighting. Moreover, sensors installed on lighting poles are multifunctional in that they can collect different kinds of information, such as about vehicular traffic, air quality, weather information, acoustics, and the presence and activities of people. This means that they can be used to monitor public safety as well as providing other services based on surveillance of the local environment.

21.3 Street Lighting Efficiency

Street lighting is one of the systems possible for reducing energy consumption and, thus, CO_2 emissions, through energy-efficiency management. For example:

- Rijeka city in the Republic of Croatia, where 400 W mercury bulbs have been substituted with 250 W high-pressure sodium (HPS) bulbs, achieved an energy usage reduction of 3.7%, with 13% electrical efficiency improvement and a reduction between 2003 and 2008 of 104 tonnes of CO_2/year (Radulovic et al. 2011);
- there was a 26% CO_2 emission reduction with light-emitting diode light (LED) installation in Bandung city, Indonesia (Yusup et al. 2020);
- based on a retrofit case study in Montepescali, Italy, Valetti et al. (2021) estimated at least 70% reduction in CO_2.

Mohammed (2020) highlights particularly the advantages of high-grade LED lamps. LED bulbs have five times longer lifetime than HPS lamps, and are seven times more energy efficient than incandescent lamps and twice more than fluorescent lamps (p. 667). Their lifetime is up to 100,000 hours, which is also important in terms of reducing maintenance costs. Their energy efficiency can be attributed to a greater lumen output per unit of energy consumed (e.g. 80 W HPS lamp replaced with a 30 W LED lamp equivalent). They can function at a low temperature and are better suited, therefore, to cold climates as well as areas susceptible to vibrations, as they can resist shocks. Their directionality allows them to be controlled more easily so that less light is wasted and there is less light pollution. The LED lamp is able to reproduce more natural light (compared to HPS lamps that produce yellow light or to green light produced by fluorescent street lights). They do not emit UV or IR radiation.

21.4 Smart Lighting

A study by Beccali et al. (2017) – also see Bonomolo et al. (2014) – has reported the use of efficient lamps installed in the city of Bagheria (Italy). This was framed from the perspective of 'smart cities' and represented new lighting infrastructures to improve the urban environment while supporting its rehabilitation – particularly in old towns. LED lamps were controlled through advanced ICT-based control systems (with dimmers). Besides helping to ensure reduced energy consumption, these street-lighting systems also promoted street and pedestrian safety as part of an integrated design. The way the system works is by static voltage controls through a voltage reduction of 67% during the hours between 12 and 5 am. This has resulted in an energy savings of 15%, equivalent to 3,800 operating hours per year (p. 290), although a dynamic dimming system could have afforded a further 3,000 operating hours per year (cf. Miki et al. 2006).

Similar approaches were considered by Cellucci et al. (2015) in a seaside town in Italy. The mere replacement of existing HPS lamps with LED lamps improved light quality and proved to be economical because of the durability offered by LED lamps as well as reduced maintenance costs – even where there were already energy-efficient HPS lamps in place. Such a retrofit resulted in a 33% reduced street lighting energy consumption.

The new lighting system allowed for a controlled luminous flux level in addition to on–off switch mode. A remote control based on the emplacement of sensors on each pole to manage the flux level was expensive, so a new generation ballast was employed instead to dim the power at set times. Smart lighting encompassed the use of remote control that enables controlling traffic and crosswalk lights.

21.5 System of Street Lights Controller

This system was designed in order to reduce energy consumption in Malaysia (with a case study at the International Islamic University of Malaysia) (Abdullah et al. 2019a). The system relies on LED lamps (switched from HPS lamps or metal halide), controlled sensors and object movement. Switching from halide to LED lamps was found to be insufficient to achieve significant energy savings (see also findings by Beccali et al. 2015), although LED provide savings over HPS lamps and have a longer lifetime (e.g. Abdullah et al. 2019b). Instead, energy savings resulted in this study by controlling lamp brightness, with lamp dimming depending on the speed of moving objects detected, like pedestrians (low mode detection speed, e.g. 0–5 km/h), cyclists (medium mode, e.g. 7–15 km/h), and cars (high mode, e.g. >16 km/h), such that higher speeds detected produce greater light intensities.

This differentiates the use of energy-efficient technology from that of its smart application due to 'occupancy' affecting lamps turning on/off or dimming (reducing the light intensity, Cacciatore et al. 2017). The dimming method results in the most reductions in energy consumption (e.g. based on dimming to 60% light intensity when users are absent, see Cacciatore et al. 2017). In particular, the light-dependent resister (LDR) adjusts the lighting intensity in different situations, with lights turning on when <80% of sunlight is detected by the LDR sensors; infrared (IR) sensors, on the other hand, are based on the speed of movement. This study (Abdullah et al. 2019b) conveys an electrical savings of 40–45% per month (30 days, using 9W LED power) based on the use of motion detection devices.

21.6 Connected Security Lighting System

The connected security lighting system (CSLS) serves a dual function of illuminating public spaces as well as allowing for visual recognition of approaching pedestrians (Cho et al. 2019). It is based on the premise that 'security lighting' reduces the fear of crime as it deters potential offenders (for discussion concerning public street lighting and crime see Vogiatzaki et al. 2020). A beacon installed in the CSLS detects the presence of a pedestrian using a registered smartphone App enabled by Bluetooth, which triggers IR sensors to augment the brightness intensity of nearby lights at the same time as to illuminate all other (unregistered) pedestrians in the vicinity.

Users apply to the local government for a secure serial number to access the system. Therefore, it has the advantage (over conventional lighting security systems) that the user activates a brightening of the illumination on demand.

In this way, it reduces wasting electrical energy and the problem of light pollution (refer to Table 2 in Cho et al. 2019, p. 5). Pedestrian routes are recorded on the web server of the beacon service when data are transmitted between the smartphone App and CSLS. If a crime occurs, the time and location where it occurred are recorded in the system. It is particularly suited, for instance, for use in narrow alleys.

21.7 The 'Smart Pole'

According to Gagliardi et al. (2020), a 'smart pole' is '. . . a street light lamp mainly composed by a local control device, a video camera and a computational unit for video processing, weather sensors, communication devices and highly efficient LED lamps' (p. 1499). Smart poles have capabilities of data exchange and remote management. For instance, during the day, they can extract real-time data of traffic using video processing that is shared through the web; during night-time dimming, LED lamps monitor vehicles and pedestrians, with light intensity kept to a minimum in vacant areas. However, where a presence is detected, this increases LED light intensity based on the number and kind of vehicles detected as well as weather conditions. Therefore, energy savings are achieved during periods when there is no traffic during the night. The smartness of this system is based on its ability to: detect vehicle/pedestrian presence in areas that are being monitored; adjust lamp brightness to vehicles/pedestrians and weather conditions; maintain light brightness levels according to government regulations; and share information by alerting the remote management and control station.

In some circumstances, it is possible for only smart 'master' poles to be equipped with motion detection sensors, weather stations, and dimming sensors, leaving smart 'slave' poles to have only dimming sensors in order to reduce infrastructural and operational costs. In this way, local communication is deployed between the master and slave poles (i.e. among wireless devices based on an 802.15.4 IEEE ZigBee wireless protocol, which can be integrated into IoT applications via Zigbee gateways), while there is still the possibility of remote communication via the remote server. Communications are kept secure with the use of a ZigBee protocol stack MAC layer that provides basic security services, allowing for an access control list and an encryption algorithm to protect transmitted data.

There are also different functionalities of devices: the 'coordinator (gateway)' device forms the local network and routes traffic to other devices – and every network has only one coordinator through which the system is managed; the 'router' routes traffic to other devices; and the 'end device' cannot route traffic to other devices. The gateway can use either 'unicast' routing to communicate with a single smart pole or 'multicast' routing to communicate with a group of smart poles or, otherwise, communicate with all smart poles that are connected to the network by way of 'broadcast' routing.

21.8 Dynamic Lighting Control System

This smart approach is based on lighting classes defined by parameters, such as speed limit, traffic volume and composition, junction density, parking, and ambient luminosity (Wojnicki and Kotulski 2018). This means that precise photometric calculations are necessary to identify the necessary lighting system, including type, power level, position on the pole, overhang, and so on. Dynamic control, therefore, selects the class based on sensor data identifying environmental parameters (e.g. average traffic intensity, ambient light, weather conditions, etc.). Specific power settings are allocated to meet a given lighting class on a specific street.

The main costs involved are determined by the sensor system cost, the frequency of observations required to sense the environment, and any changes that occur. Then comes the communication cost, involving maintenance and transmission fees. Data processing and storage become more expensive as more data are handled and kept.

A test case is available for the city of Krakow (Poland), where 324 remotely controlled fixtures were emplaced in 107 loops measuring traffic intensity (Wojnicki and Kotulski 2018). Light class switching, for instance, could involve dimming luminaires, resulting in energy savings. Assuming that lights are turned on at sunset and off at sunrise, an energy savings of 27% has been predicted based on the test case. The preference in this experiment was LED lights (not older technologies, such as HPS) in order to maximise energy efficiency and savings.

21.9 Conclusions

This chapter has provided a brief overview of technology-based solutions to achieve energy efficiency in street lighting. Cases for street lighting, including lamp posts and luminaires, have been deployed to illustrate contemporary developments demonstrating initiatives to support 'actually existing smart cities' as they emerge and develop. The increasing presence of these technologies demonstrates how the digitalization of energy rapidly transforms the very nature of seemingly 'mundane' urban services. Street lighting can be one part of the integrated system for a smarter city, and potentially also for a more people-smarter city. According to Beccali et al. (2017): 'the city can be "smart" because a user can interact with the city and with the place where [s/he] lives. Thus, life in the city becomes better and easier, and users become "planners" of the city, in this case, through the control of... lighting...' (p. 287).

References

Abdullah, A., Yusoff, S.H., Zaini, S.A. et al. (2019a). Energy efficient smart street light for smart city using sensors and controller. *Bulletin of Electrical Engineering and Informatics* 8 (2): 558–568. https://doi.org/10.11591/eei.v8i2.1527.

Abdullah, A., Pambudi, R.L., Purnama, W. et al. (2019b). Redesigning street lighting system using LED and HPS luminaires for better energy saving application. *Journal of Engineering Science and Technology* 14 (4): 2140–2151.

Beccali, M., Bonomolo, M., Ciulla, G. et al. (2015). Improvement of energy efficiency and quality of street lighting in South Italy as an action of sustainable energy action plans. The case study of Comiso (RG). *Energy* 92 (P3): 394–408. https://doi.org/10.1016/j.energy.2015.05.003.

Beccali, M., Bonomolo, M., Galatioto, A., and Pulvirenti, E. (2017). Smart lighting in a historic context: a case study. *Management of Environmental Quality* 28 (2): 282–298. https://doi.org/10.1108/MEQ-06-2015-0109.

Bonomolo, M., Beccali, M., Galatioto, A., Pulvirenti, E. (2014). A smart lighting network design for urban rehabilitation and environmental sustainability: a case study of Bagheria. *Proceedings of the 9th Conference on Sustainable Development of Energy, Water and Environment Systems*. Zagreb, Croatia, (20–27 September 2014). Zagreb: Sunčana Matijašević, Marko Ban, pp. 1–13. http://hdl.handle.net/10447/98223.

Cacciatore, G., Fiandrino, C., Kliazovich, D., et al. (2017). Cost analysis of smart lighting solutions for smart cities. *Proceedings of the 2017 IEEE International Conference on Communications*, Paris, France (21–25 May 2017). IEEE (31 July 2017), pp. 1–6. https://dx.doi.org/10.1109/ICC.2017.7996886.

Cellucci, L., Burattini, C., Drakou, D. et al. (2015). Urban lighting project for a small town: comparing citizens and authority benefits. *Sustainability* 7: 14230–14244. https://doi.org/10.3390/su71014230.

Cho, Y., Jeong, H., Choi, A., and Sung, M. (2019). Design of a connected security lighting system for pedestrian safety in smart cities. *Sustainability* 11: 1308. (11 pp.). https://doi.org/10.3390/su11051308.

Deambrogio, E., Allegretti, S., Turi, P. et al. (2017). Increase sustainability in buildings through public procurements: the PROLITE project for lighting retrofit in schools. *Energy Procedia* 111: 328–337. https://doi.org/10.1016/j.egypro.2017.03.194.

Dominković, D.F., Bačeković, I., Sveinbjörnsson, D. et al. (2017). On the way towards smart energy supply in cities: the impact of interconnecting geographically distributed district heating grids on the energy system. *Energy* 137 (15): 941–960. https://doi.org/10.1016/j.energy.2017.02.162.

Doukas, H., Marinakis, V., Spiliotis, E., Psarras, J. (2016). OPTIMUS decision support tools: transforming multidisciplinary data to energy management action plans. *7th International Conference on Information, Intelligence, Systems and Applications (IISA)*, Chalkidiki, Greece (13–15 July 2016). IEEE Xplore (19 December 2016), pp. 1–5. https://doi.org/10.1109/IISA.2016.7785359.

European Commission. EURELECTRIC (2004). *Electricity for More Efficiency: Electric Technologies and Their Energy Savings Potential; European Commission*, vol. 121, 118 pp. Brussels, Belgium: http://www.uie.org/sites/default/files/REPORT_FINAL_July_2004.pdf.

Fregonara, E., Curto, R., Grosso, M. et al. (2013). Environmental technology, materials science, architectural design, and real estate market evaluation: a multidisciplinary approach for energy-efficient buildings. *Journal of Urban Technology* 20 (4): 57–80. https://doi.org/10.1080/10630732.2013.855512.

Gagliardi, G., Lupia, M., Cario, G. et al. (2020). Advanced adaptive street lighting systems for smart cities. *Smart Cities* 3: 1495–1512. https://doi.org/10.3390/smartcities3040071.

Gil-Garcia, J.R., Pardo, T.A., and Nam, T. (2015). What makes a city smart? Identifying core components and proposing an integrative and comprehensive conceptualization. *Information Polity* 20: 61–87. https://doi.org/10.3233/IP-150354.

Korkmazer, E., Bektas, Y., Aykanat, M. et al. (2019). Study of light pollution in urban lighting in Nisantasi example. *Journal of Contemporary Urban Affairs* 3 (2): 8–15. https://doi.org/10.25034/ijcua.2018.4696.

Miki, M., Asayama, E., and Hiroyasu, T. (2006). Intelligent lighting system using visible-light communication technology. *Proceedings of IEEE Conference on Cybernetics and Intelligent Systems*, Bangkok, Thailand (7–9 June 2006). IEEE (04 December 2006), pp. 1–6. https://dx.doi.org/10.1109/ICCIS.2006.252257.

Mohammed, J.K. (2020). Reduction cost and energy consumption for LED smart lighting street technology in Iraq. *Indonesian Journal of Electrical Engineering and Computer Science* 20 (2): 662–669. https://doi.org/10.11591/ijeecs.v20.i2.pp662-669.

Mrazovac, B., Todorovic, B.M., Kukloj, D., and Samardzija, D.A. (2014). Solution for enabling intelligent street lighting in smart cities. *Proceedings of the VDE-Kongress 2014, Smart Cities – Intelligente Lösungen für das Leben in der Zukunft*, Frankfurt, Germany (October 2014), 5 pp.

Paris, S. and Bagnato, V.P. (2012). Bari, a public Mediterranean city: new projects to valorise public heritage. *TECHNE* 3: 206–215. https://doi.org/10.13128/Techne-10845.

Pasolini, G., Toppan, P., Zabini, F. et al. (2019). Deployment and evolution of heterogeneous smart public lighting systems. *Applied Sciences* 9: 3281. (25 pp.). https://doi.org/10.3390/app9163281.

Petritolia, E., Leccese, F.L., Pizzuti, S., and Pieroni, F. (2019). Smart lighting as basic building block of smart city: an energy performance comparative case study. *Measurement* 136: 466–477. https://doi.org/10.1016/j.measurement.2018.12.095.

Radulovic, D., Skok, S., and Kirincic, V. (2011). Energy efficiency public lighting management in the cities. *Energy* 36 (4): 1908–1915. https://doi.org/10.1016/j.energy.2010.10.016.

Skandali, C. and Lambiri, Y.S. (2018, 2018). Optimization of urban street lighting conditions focusing on energy saving, safety and users. *Journal of Contemporary Urban Affairs* 2 (3): 112–121. https://doi.org/10.25034/ijcua.2018.4726.

Thornbush, M.J. and Golubchikov, O. (2019). *Sustainable Urbanism in Digital Transitions: From Low Carbon to Smart Sustainable Cities*. Springer (65 pp). https://doi.org/10.1007/978-3-030-25947-1.

Thornbush, M. and Golubchikov, G. (2021). Smart energy cities: the evolution of the city-energy-sustainability nexus. *Environment and Development* 100626, 7 pp. https://doi.org/10.1016/j.envdev.2021.100626.

Valetti, L., Floris, F., and Pellegrino, A. (2021). renovation of public lighting systems in cultural landscapes: lighting and energy performance and their impact on nightscapes. *Energies* 14: 509 (25 pp). https://doi.org/10.3390/en14020509.

Vogiatzaki, M., Zerefos, S., and Hoque Tania, M. (2020). Enhancing city sustainability through smart technologies: a framework for automatic pre-emptive action to promote safety and security using lighting and ICT-based surveillance. *Sustainability* 12: 6142. (20 pp.). https://doi.org/10.3390/su12156142.

Wojnicki, I. and Kotulski, L. (2018). Empirical study of how traffic intensity detector parameters influence dynamic street lighting energy consumption: a case study in Krakow, Poland. *Sustainability* 10: 1221. (16 pp). https://doi.org/10.3390/su10041221.

Yusup, G.S.B., Melda, R.J., Maman, I., and Li, J. (2020). Sustainable low carbon urban lighting analysis: a case study of Bandung city. *Journal of Engineering Science* 7 (1): H9–H18. 10.21272/jes.2020.7(1).h2.

Zielinska-Dabkowska, K.M. (2014). Urban city lights. Light pollution as one of the effects of incorrectly designed external illumination. How can a successful lighting masterplan diminish its impact? *Proceedings of SLLP 2014, 1st International Conference on Sustainable Lighting and Light Pollution*, Seoul, South Korea (November 2014), pp. 59–67. https://doi.org/10.13140/RG.2.1.3495.0484.

Index

Please note entries in **bold** refer to tables.

15 minute neighbourhoods, Paris 170–171
20-minute cities
  accessibility 177–180
  citizens 175–176, 184–185
  at city scale 172–173, 177–180
  core concepts 167, 169–176
  Edinburgh 167–193
  Melbourne 173–174
  neighbourhoods 173–175, 180–184
  Portland 172–173
  scales 171–172
+EINS project 352–356
+EQ-NET project 354–356

a

ABD *see* area-based development
Abu Dhabi, Masdar City 270
AC *see* air conditioning
Accelerated Power Development and Reform Program (APDRP) 282
accessibility, 20-minute cities 177–180
active buildings 93–110
  behavioural factors 96–98, 102–103
  customer education 104–105
  decarbonisation 93, 95–96
  demand-side response 97–98, 102–104
  infrastructure constraints 102–105
  key concepts 99–100
active homes
  definition 99
  initial insights 100–105
  smart meters 97–98, 100–105
  see also active buildings
active retrofitting 332
actor constellations, and transferability 80, 83, 86–87
adequacy, of network capacity 41–42
Aeolian architecture 132
Africa
  diaspora-led investment 62–63
  external investment 61–63
  integrated approaches 63–65
  smart energy cities 53–74
African Energy Chamber 58
African Smart Town Networks project 55
AGI *see* artificial general intelligence
AI *see* artificial intelligence
air conditioning (AC) 328
airflow, passive architecture 132
air quality, India 226
Akon 62
Amsterdam 270
APDRP *see* Accelerated Power Development and Reform Program
applications, wind farms 45, 47
apportionment, carbon dioxide budgets 153–154
architecture
  Aeolian 132
  biophilic design 129
  greenhouse gas emissions 129–130, 131

Smart Cities, Energy and Climate: Governing Cities for a Low-Carbon Future, First Edition.
Edited by Oleg Golubchikov and Komali Yenneti.

architecture (*cont'd*)
passive 131–132
roof shapes 132
timber buildings 129–130
area-based development (ABD) 219–221, 227–228
artificial general intelligence (AGI) 114
artificial intelligence (AI)
autonomous cities 111–122
autonomous vehicle-induced urbanism 114–117
and big data 112–113
characteristics 112
sustainability 113–114
urban 112–113
asset life extension, wind farms 46–47
atmospheric science, urban climate 127–128
Australia, Melbourne 20 minute neighbourhoods 173–174
Austria, Ökopark Hartberg pilot project 81–84
autonomous cars
definition 112
induced urbanism 114–117
sharing 116–117
autonomous cities
bicycle urbanism 117–119
big data 112–113
car sharing 116–117
city brains 112
mobility 111–122
sensors 113
sustainability 113–114
vehicle-induced urbanism 114–117
autonomous urbanism, definition 111
autonomous vehicle-induced urbanism 114–117

b

BAPV *see* building-attached photovoltaic
barriers, Africa 57
battery energy storage (BES) 329, 335–336
behavioural economics, energy poverty 251–256
behavioural factors
bicycle urbanism 118
smart homes 96–98, 102–103
BEIS *see* Department for Business, Energy and Industrial Strategy
Berlin, Germany
2030 scenario 157–159
2040 scenario 159–160
2050 scenario 156–157
carbon dioxide budget 148, 152–154
climate neutrality pathway 147–166
emissions apportionment 153–154
energy sources 154–155, 162
restrictions-based approaches 155–160
status quo 154–155
strategy recommendations 160–162
BES *see* battery energy storage
bicycle urbanism 117–119
big data 112–113
biofuels, district heating systems 308
biomass, district heating systems 308, 310–311
biophilic building design 129
BIPV *see* building-integrated photovoltaic
BIPV/T *see* building-integrated PV thermal
blockchain technology 293–304
challenges and opportunities 294–295
core concepts 293–296
emissions reductions 296–298
integrity 294–295
micro-grids 298–300
prosumers 298–300
renewable energy generation 298–300
transparency 296–298
verifiability 296–298
boosting interventions, energy poverty 255
Bosco Verticale 129
bottom-up digital technologies 347–364
+EINS project 352–356
+EQ-NET project 354–356
challenges 360–361
data collection methods 349–351
user-generated content 349–351, 356–360
Bristol, UK, smart local energy systems 23–24
Brooklyn micro-grid 299
building-attached photovoltaic (BAPV) systems 337
building-integrated photovoltaic (BIPV) systems 337–338
building-integrated PV thermal (BIPV/T) systems 338–339
building-level innovations 19

building materials
greenhouse gas emissions 129–130, 131
passive 131
buildings
active 93–110
air conditioning 328
battery energy storage 329, 335–336
biophilic design 129
connection costs 40–41
and decarbonisation 93, 95–96, 327–346
design 129–132
energy flexibility 339–340
energy storage systems 329–330, 332–336
greenhouse gas emissions 129–130, 131
Manchester 39–42
passive architecture 131–132
phase change materials 131
resilience 130
retrofitting 330–332
roof shapes 132
solar energy 330, 336–339
thermal mass energy storage 329–330, 332–335
timber-based 129–130
building thermal mass 329–330, 332–335

C

Cagliari, Italy
sustainable urban mobility plan 195–214
objectives **200**
strategies and actions **201–207**
capacity
and adequacy 41–42
African contexts 57–58
and connection costs 41, 48
carbon dioxide (CO_2) budgets
apportionment 153–154
Berlin 148, 152–154
framework standards 149–151
theory and practice 151–152
car sharing 115–116
Central Africa, Rwanda 55
CFD *see* computational fluid dynamics
CHP *see* combined heat and power
chrono-urbanism 170–171
cities
importance of 266–268
as testbeds 18
citizen-led initiatives 18
citizen scientists 356–360
city brains 112, 221
city improvement 219
city planning, green and blue spaces 132
city renewal 219–220
city selection, India 217–218
civic hacking infrastructure, *Terremoto Centro Italia* project 357–358
clean cooking 60, 64
climate
modelling 135–136
resilience 130–131
sea levels 133
urban 125–146
atmospheric science 127–128
building design 129–132
energy efficiency 128–132
green and blue spaces 132
greenhouse gas emissions 128
modelling 135–136
nationwide infrastructure planning 133–135
climate adaptation measures, Russia 245
climate change, sub-national actors 17–18
Climate Emergency Declarations 18
climate governance architectures, Berlin 160–162
climate neutrality
Berlin 147–166
carbon dioxide budget 148, 152–154
emissions apportionment 153–154
energy sources 154–155
restrictions-based approaches 155–160
status quo 154–155
strategy recommendations 160–162
framework standards 149–151
Paris Agreement 17–18, 148–154
Russia 245
sub-national actors 17–18
theory and practice 151–152
climate policy
20-minute cities 167–193
Berlin 147–166
Russia 237–241
CO_2 *see* carbon dioxide
coal, district heating systems 308
co-evolution, energy system change 20–21
co-location, 20 minute neighbourhoods 180–181

Colonial Pipeline hack 265
combined heat and power (CHP) systems 308
community engagement, 20 minute citizens 176, 184–185
commuting 116
competency measures, energy poverty 255–256
computational fluid dynamics (CFD) 135–136
coniferous trees 132
connected security lighting systems (CSLS) 368
connection costs 40–41, 48
constraints, replication and upscaling 76–78
consumer archetypes, and time of use tariffs 97–98
consumer education
  energy poverty 255
  smart meters 104–105
contestations
  external investment in Africa 62
  Manchester 38–49
contingencies
  connection costs 40–41, 48
  grid investments 42–43
  Manchester 38–39
  service networks 40–42
  urban–rural relationships 43–44
controller systems, street lighting 367–368
cooking fuels 64
  Sierra Leone 60
cooling, passive solar 131
costs
  building retrofits 332
  of connection 40–41, 48
  of electricity units 328
  smart grids for consumers 284–285
COVID-19, and the 20 minute idea 167–168, 170–171
CSLS *see* connected security lighting systems

d

data collection
  +EINS project 352–354
  bottom-up decision-making 347–364
  challenges 360–361
  ethical considerations 64, 98
  sensors 349–351
  user-generated content 349–351, 356–360
data-driven governance 186–187
data transfer, +EQ-NET project 354–356
decarbonisation
  active buildings 93, 95–96
  district heat and cooling 310–311
  energy-efficient buildings 327–346
  and investment planning 43
  smart energy systems 95–96
decentralisation
  20 minute neighbourhoods 174–175
  distributed ledger technology 293–304
  energy security 268–269
decision making
  bottom-up digital technologies 347–364
    +EINS project 352–356
    +EQ-NET project 354–356
    challenges 360–361
    sensors 349–351
    user-generated content 349–351, 356–360
defossilation 126
demand shifting 97–98, 102–104, 328–329
demand-side response (DSR) 97–98, 102–104
Democratic Republic of Congo (DRC), urbanisation 20
demographics
  Sierra Leone 60
  West Africa 54, 59, 60, 64
Denmark, district heating 307, 313–316
densification, 20-minute cities 172–173
Department for Business, Energy and Industrial Strategy (BEIS) 26–27
DERs *see* distributed energy Resources
design, buildings 129–132
DHS *see* district heating systems
diaspora, African investment 62–63
diffusion
  actor constellations 80, 83, 86–87
  economic factors 80, 82–83, 86
  embedding 76
  fundamental constraints 76–77
  institutional factors 80, 83–84, 87
  mission-related factors 80, 82–83, 86
  path dependent constraints 76
  patterns of 78
  spatial-structural aspects 80, 81–82, 84–85

technological dimensions 80, 82, 85–86
types of 79
see also replicability; upscaling
digital inclusion, and smart meter use 104
digitalisation
distributed ledger technology 293–304
pedestrian infrastructure planning 186
Russia 245–246
Sierra Leone 61
direct load control 102–103
Directorate of Science, Technology and Innovation (DSTI), Sierra Leone 61
distributed energy Resources (DERs) 21
distributed ledger technology (DLT) 293–304
challenges and opportunities 294–295
core concepts 293–296
emissions reduction 296–298
integrity 294–295
micro-grids 298–300
prosumers 298–300
renewable energy generation 298–300
transparency 296–298
verifiability 296–298
distribution network operators (DNOs)
connection costs 40–41, 48
investment models 42–43
district cooling 311–312, 318–321
see also district heating systems
district heating systems (DHS) 305–326
Denmark 307, 313–316
economic objectives 320–321
energy poverty 316–318
environmental objectives 319–320
fifth generation systems 308, 311–312
first generation systems 307
fourth generation systems 308, 311–312
fuels 308–311
history of 305–307
Iceland 307
local policy 319
national policy 319
Russia 246
Scandinavian countries 307
second generation systems 308
social objectives 320
Soviet Union 307
success metrics 318–321
technical objectives 321
technologies 308–311
third generation systems 308
United Kingdom 305–306
DLT *see* distributed ledger technology
DNOs *see* distribution network operators
DRC *see* Democratic Republic of Congo
DSR *see* demand-side response
DSTI *see* Directorate of Science, Technology and Innovation
dynamic lighting control systems 369
dynamic pricing 329

e

econometric modelling, distribution network operators 42–43
Economic Community of Western African States (ECOWAS) 59–60
renewable energy policy (EREP) 63
economic factors
building retrofits 332
and transferability 80, 82–83, 86
economic objectives, district heating and cooling 320–321
economic sustainability, Smart Cities Mission, India 226–228
ECOWAS *see* Economic Community of Western African States
Edinburgh, UK, 20-minute idea 167–193
EDTF *see* Energy Data Taskforce
e-Government, Russia 245–246
electrical activation 335
Electricity (Connection Charges) Regulations, The 41
electric vehicles (EV), smart charging 101, 103
electrification
and decarbonisation 95–96
and network capacity 41–43
Sierra Leone 60–61
West Africa 59
elite provisioning 222–224, 227
embedding
innovation 22, 25–26
replication and upscaling 76
emissions reductions
distributed ledger technology 296–298
see also greenhouse gas emissions

energy access
Sierra Leone 60–61
West Africa 56–61, 60–61
energy consumption, Russia 240
Energy Data Taskforce (EDTF) 26
energy efficiency
street lighting 366–367
urban planning 128–132
energy-efficient buildings
air conditioning 328
battery energy storage 329, 335–336
decarbonisation 327–346
energy flexibility 339–340
energy storage systems 329–330, 332–336
retrofitting 330–332
solar energy 330, 336–339
thermal mass energy storage 329–330, 332–335
energy flexibility, buildings 339–340
energy generation
Europe 21
Russia 240–242, 246
United Kingdom 21
verifiability and transparency 296–300
see also renewable energy generation
energy justice, equity; fairness and smart energy systems 22–23, 28
Energy for Londoners programme 23
energy policy
urban planning 128–132
wind farms 44–47
energy poverty 249–261
behavioural economics 251–256
boosting interventions 255
district heating and cooling 316–318
nudging 253–255
opportunity structures 250
spatially induced cognitive scarcity 253
ways forward 256–257
energy security 265, 268–269
energy service platforms 102–103
energy services
African contexts 56–58
connection costs 40–41, 48
decarbonisation 93, 95–96
distributed ledger technology 293–304
district heat and cooling 305–326
equity 54–55, 57–58
check if should be equity **** 1000
Manchester 38–49
as obligatory passage points 36–37
path dependency 37, 42
pre-paid electricity meters 57
replication and upscaling 75–91
Russia 240–242
Smart Cities Mission, India 225
socio-technical analysis 35–52
sustainable development goals 55–56
time of use tariffs 97–98, 100–102
urban planning 133
verifiability and transparency 296–298
energy sources
Berlin 154–155, 162
district heating systems 308–311
Russia 242
energy storage systems 329–330, 332–336
batteries 329, 335–336
thermal mass 329–330, 332–335
Energy Superhub Oxford (ESO) 24
energy system change, momentum 20–21
Energy Systems Catapult (ESC) 22
environmental objectives, district heating and cooling 319–320
environmental sustainability, Smart Cities Mission, India 224–226
equity
20 minute neighbourhoods 174–176
African contexts 54–55, 57–58, 62–63, 64
energy poverty 249–261
Smart Cities Mission, India 221–224, 227
time of use tariffs 97–98, 100–102
urban mobility 116–119
EREP *see* Economic Community of Western African States, renewable energy policy
ESC *see* Energy Systems Catapult
ESO *see* Energy Superhub Oxford
ethics, of data collection 64
Europe
Amsterdam 270
Berlin climate neutrality 147–166
Cagliari sustainable urban mobility plan 195–214
Edinburgh 20-minute idea 167–193
energy poverty 249–261
Paris 15 minute neighbourhoods 170–171
pre-industrial urban architecture 169–170
renewable energy generation 21
Stockholm 270

European Strategic Energy Technologies (SET) plan 75
EV *see* electric vehicles
exclusion
  20 minute neighbourhoods 174–176
  Smart Cities Mission, India 221–224, 227
expansion, smart grids 79
external investment, West Africa 61–63
extreme commuting 116
extreme precipitation 130
extreme weather events, urban planning 130–131

f

fairness
  20 minute neighbourhoods 174–176
  African contexts 64
  Smart Cities Mission, India 221–224, 227
  and smart energy systems 22–23, 28
  time of use tariffs 97–98, 100–102
  see also equity
FC *see* fuel cells
FCC *see* Freetown City Council
feed-in tariffs (FIT) 330
fifth generation district heating systems 308
financial access
  M-Pesa 55
  pre-paid electricity meters 57
  sub-Saharan Africa 54
financing, Smart Cities Mission, India 218–219, 227–229
first generation district heating systems 308
FIT *see* feed-in tariffs
flexibility, and decarbonisation 95
forced air activation 335
fourth generation district heating systems 308
France, Paris 15 minute neighbourhoods 170–171
Freetown City Council (FCC) 54
fresh air corridors 133
fuel cells (FC) 270–271
functional separation, cities 133

g

gardening, urban 132
Geretsreid, Germany 354–356
German Advisory Council on the Environment (SRU) 152–153
Germany
  Berlin climate neutrality 147–166
  Geretsreid pulsG mixed-use district 354–356
  Ludmilla Residential Park, Landshut 352–354
Ghana, Hope City 55
GHG *see* greenhouse gas
GLA *see* Greater London Authority
global climate change, sub-national actors 17–18
Global South, West Africa 53–74
governance
  20 minute cities and neighbourhoods 174–184
  Berlin climate neutrality 160–162
  Cagliari sustainable urban mobility plan 195–214
  data-driven 186
  Russia 243–244
  Smart Cities Mission, India 218, 228–231
  smart local energy systems 26–28
  United Kingdom 26–28
Greater London Authority (GLA) 23
greenfield development 220–221
greenhouse gas (GHG) emissions
  active buildings 93, 95–96
  alternate building materials 129–130, 131
  Berlin 148, 152–154
  budgets 148, 152–154
  distributed ledger technology 296–298
  Paris Agreement 149–151
  Russia 237–241
  standards 149–151
  UK policy 93–94
  urban 128
grid technologies
  interoperability 77
  replication and upscaling 75–91
  scalability 77–78
  socio-technical aspects 78–79
  transferability 79–88
  see also smart grids
gross value added (GVA) 42–43
GVA *see* gross value added

h

Hammarby Sjostad, Stockholm 270
Hartberg, Austria, Ökopark pilot project 81–84

heating
  passive solar 131
  phase change materials 131
homes, active 99
  see also active buildings
Hope City, Ghana 55
horizontal scaling 79, 244
hot spots, connection costs 41
hydronic activation 335
Hyllie, Malmö, smart grid pilot project 81, 84–88

i

ICCCs *see* integrated command and control centres
Iceland, district heating 307
imaginaries, active buildings 93–110
implementation, Smart Cities Mission, India 218
inclusive development, Africa 54–55, 62, 64
India
  Accelerated Power Development and Reform Program 282
  air quality 226
  poverty 227
  smart cities mission 215–235
    economic sustainability 226–228
    environmental sustainability 224–226
    equity 221–224, 227
    financing 218–219, 227–229
    institutional sustainability 228–231
    integrated command and control centres 221
    model components 219–221
    overview 217–219
    selection process 217–218
    social sustainability 221–224
    taxation 228
India Smart Grid Week 2016 (ISGW) 281–287
  background 281–283
  key narratives 283–284
  memorandum of understanding 285–286
  outcomes 285–287
Industrial Strategy (2017), UK 21–22
industry events 277–291
  India Smart Grid Week 2016 281–287
  research methods 280–281
  theoretical background 278–280
inequality, Smart Cities Mission, India 221–224, 227
infrastructure
  20 minute cities 177–180
  20 minute neighbourhoods 174
  connection costs 40–41, 48
  distributed ledger technology 293–304
  and electrification 41–43
  energy security 268–269
  investment decisioning 42–43, 49
  policy issues 44–49
  smart homes 102–105
  street lighting 365–372
  and urbanisation 20, 269–272
  see also smart grids
infrastructure planning 133–135, 224–226
  bottom-up digital technologies 347–364
innovation
  diffusion of 78–88
  embedding 22, 25–26
  replication and upscaling 75–91
institutional factors, and transferability 80, 83–84, 87
institutional sustainability, Smart Cities Mission, India 228–231
integrated command and control centres (ICCCs) 221
integrity, distributed ledger technology 294–295
Intergovernmental Panel on Climate Change (IPCC) 150
Internationally Transferred Mitigation Outcomes (ITMOs) 296–297
Internet of Energy (IoE) 269
internet of things (IoT) 113, 268–269
  bottom-up decision-making 347–364
interoperability 77
intricate mingling 167
investment
  network capacity 41–43
  West Africa 59–61
invisible grid 37
IoE *see* Internet of Energy
IoT *see* internet of things
IPCC *see* Intergovernmental Panel on Climate Change
IQ of Cities index 243–244
ISGW *see* India Smart Grid Week 2016
isoprene emitting trees 132

Italy
  Cagliari sustainable urban mobility plan 195–214
  Terremoto Centro Italia project 357–358
ITMOs *see* Internationally Transferred Mitigation Outcomes

j

Japan, Yokahama 270
justice
  see also energy justice; equity; fairness
  energy poverty 249–261
  and smart energy systems 22–23, 28, 54

k

Kenya, pre-paid electricity meters 57
key services 179
Kigali, Rwanda, Smart City Masterplan 55

l

Landshut, Germany 352–354
LEO *see* Local Energy Oxfordshire
life extension, wind farms 46–47
Living Well in Low Carbon Homes research project 93–110
  initial insights 100–105
Local Energy Oxfordshire (LEO) 24
local energy systems
  Bristol 23–24
  competing logics 25–26
  governance 26–28
  London 23
  Oxford 24
  and privatisation 22–23, 25–26
  scalability 22–23, 25–26
  United Kingdom 21–28
local policy
  district heating and cooling 319
  versus national policy, UK 22–28
London, UK, smart local energy systems 23–24
long-term scenarios, Berlin climate neutrality 156–157
low-carbon technologies
  active buildings 93–110
  bottom-up 347–364
  district heating and cooling 305–326
  equity 28, 64, 98, 100–102, 174–176, 221–224, 227
  industry events 227–291
  passive architecture 131–132
  passive retrofitting 331–332
  replication and upscaling 75–91
  storage 329–330, 332–336
  see also buildings; decarbonization; energy...; renewable energy generation; smart...
Ludmilla Residential Park, Landshut 352–354

m

Malmö, Sweden, Hyllie pilot project 81, 84–88
Manchester Digital Strategy 39
Manchester, UK 38–49
  description of 38–39
  grid investment 42–43
  hot areas 47–49
  regional wind farms 44–47
  rural relationships 43–44
  smart buildings 39–42
marginal cost pricing 328
market-based systems, and fairness 22–23, 28
Masdar City 270
MCC *see* Metropolitan City of Cagliari
MED *see* Modernising Energy Data
medium-term scenarios, Berlin climate neutrality 157–160
Melbourne, Australia, 20 minute neighbourhoods 173–174
Metropolitan City of Cagliari (MCC)
  sustainable urban mobility plan 195–214
    objectives **200**
    strategies and actions **201–207**
micro-grids, distributed ledger technology 298–300
mission, and transferability 80, 82–83, 86
mobile phone penetration, Sierra Leone 61
mobility 111–122
  20-minute idea 169–176
    cities 172–173, 177–180
    citizens 175–176, 184–185
    Edinburgh 176–185
    neighbourhoods 173–175, 180–184
  autonomous vehicle-induced urbanism 114–117
  bicycle urbanism 117–119
  Cagliari sustainable urban mobility plan 195–214

mobility (*cont'd*)
car sharing 115–116
Russia 245
Smart Cities Mission, India 225
modelling
distribution network operators 42–43
urban climate 135–136
Modernising Energy Data (MED) coalition 26–27
momentum, energy system change 20–21
Moscow, Russia 270
M-Pesa 55
multi-agent management systems 271
multi-agent systems, urban climate 135–136
multi-service hubs 180–181
municipal governments
climate change 17–18
Climate Emergency Declarations 18
as testbeds 18

n
narratives, smart local energy systems 21–28
national policy
district heating and cooling 319
versus local policy, UK 22–28
National Renewable Energy Policy, Sierra Leone 60–61
nationwide infrastructure planning 133–135
negawatt market 298–299
Neighbourhood Walkability Ratios (NWR) 181–183
Netherlands, Amsterdam 270
network capacity
adequacy 41–42
African contexts 57–58
and connection costs 40–41
investment 41–43
network effects 22
net-zero
Berlin 147–166
carbon dioxide budget 148, 152–154
emissions apportionment 153–154
energy sources 154–155
restrictions-based approaches 155–160
status quo 154–155
strategy recommendations 160–162
framework standards 149–151
Paris Agreement 17–18, 148–154
Russia 245
sub-national actors 17–18
theory and practice 151–152
see also greenhouse gas emissions; renewable energy generation; smart...; sustainability
net-zero energy buildings (NZEB) 339–340
New York City, USA 268, 299
non-state actors, climate change 17–18
nudging, energy poverty 253–255
NWR *see* Neighbourhood Walkability Ratios
NZEB *see* net-zero energy buildings

o
objectives, Cagliari sustainable urban mobility plan **200**
obligatory passage points 36–37
Office for Gas and Electricity Markets (Ofgem) 26–27
Ofgem *see* Office for Gas and Electricity Markets
Ökopark Hartberg 81–84
oligoptic approaches, definition 37
one-minute city 176
onshore wind, socio-technical analysis 44–47
Open Energy 27
opportunity structures, energy poverty 250
organised industry events 277–291
India Smart Grid Week 2016 281–287
research methods 280–281
theoretical background 278–280
Oxford, UK, smart local energy systems 24
ozone, and tree selection 132

p
Paris, 15 minute neighbourhoods 170–171
Paris Agreement
Berlin 148, 152–154
Russia 238–241
sub-national actors 17–18
theory and practice 151–152
urban standards 149–151
parks 132
passive architecture 131–132
passive retrofitting 331–332
path dependency
energy grids 37, 42
replication and upscaling 76–77

patterns, of upscaling 78
PCM *see* phase change materials
PCs *see* production economies
pedestrian infrastructure planning 186
see also walkability
PFER *see* Prospering from the Energy Revolution
phase change materials (PCM) 131
photovoltaics (PV) 330, 336–338
Piani Urbani di Mobilità Sostenibile (PUMS) 195–214
see also sustainable urban mobility plans
pilot projects
Malmö Hyllie 81, 84–88
Ökopark Hartberg 81–84
replication and upscaling 75–91
case studies 81–88
contextual considerations 77–80
fundamental constraints 76–78
socio-technical aspects 78–79
transferability 79–80
transferability 79–80
place-based civic actors, 20-minute idea 187
planning
Cagliari sustainable urban mobility plan 195–214
sub-Saharan Africa 54
planning regulations, wind farms 44–47
policy constraints
long-term alleviation 49
Sierra Leone 60–61
West Africa 58–63
wind farms 44–47
policy innovations
Berlin 147–166
concepts 19
onshore wind 44–45
United Kingdom 26–28
population growth, West Africa 59
Portland, USA, 20-minute city 172–173
poverty
India 227
Sierra Leone 60
sub-Saharan Africa 54, 60
see also energy poverty
PPP *see* public-private partnerships
pre-paid electricity meters
Kenya 57
time of use tariffs 98
principle to policy shift 169
privacy, smart meter data 27, 64, 98
privatisation, and local energy systems 22–23, 25–26
production economies (PCs) 228
Prospering from the Energy Revolution (PFER) innovation challenge, UK 22
prosumers, distributed ledger technology 298–300
public interest initiatives 18
public-private partnerships (PPP), Smart Cities Mission, India 227
public space, autonomous vehicle-induced urbanism 116–117
public transport 133, 245
pulsG mixed-use district, Geretsreid 354–356
PUMS *see Piani Urbani di Mobilità Sostenibile*
PV *see* photovoltaics

r

rational consumers, time of use tariffs 97–98
redevelopment 219–220, 227
refusals, wind farms 45, 47
regional policy frameworks, West Africa 59–60
regional traffic 133
renewable energy generation
African contexts 56–57, 60
asset life extension 46–47
distributed ledger technology 298–300
Europe 21
Manchester 43–47, 49
policy constraints 44–47, 49
repowering 46
Russia 240, 246
United Kingdom 21
verifiability 296–298
see also wind farms
replicability, definition 77
replication 75–91
case studies 81–88
contextual considerations 77–80
difficulties in 22, 25–26
embedding 76
fundamental concepts 77
fundamental constraints 76–78

replication (*cont'd*)
  Malmö Hyllie pilot project 87–88
  Ökopark Hartberg 84
  path dependent constraints 76
  socio-technical aspects 78–79
  and transferability 79–80
repowering 46
resilience, buildings 130–131
resolution, climate models 135–136
restrictions-based approaches, Berlin climate neutrality 155–160
retrofitting 219, 227, 330–332
  active 332
  economics 332
  passive 331–332
right to disconnect 115
robots, definition 112
roll-out, smart grids 79
roof shapes 132
rooftop photovoltaics 330, 336–338
rural areas
  city relationships 37, 42
  West Africa 59
  wind farms 44–47
Russia 237–247
  climate adaptation measures 245
  climate policy 237–241
  digitalisation 245–246
  district heating 246
  e-Government 245–246
  energy sector 240–242
  IQ of Cities index 243–244
  New Moscow 270
  public transport 245
  smart cities 242–246
Rwanda, Kigali 55

S

scalability
  contextual considerations 77–80
  definition 77
  fundamental constraints 76–77
  pilot projects 75–91
  smart local energy systems 22–23, 25–26
  see also upscaling
scale, 20-minute idea 171
Scandinavian countries, district heating 307
scenarios, Berlin climate neutrality 155–160
SCM *see* Smart Cities Mission
SDG *see* sustainable development goals
SDI *see* Smart Data Initiative
sea levels 133
SEC *see* smart energy cities
second generation district heating systems 308
selection process, India's smart cities mission 217–218
sensors
  autonomous cities 113
  bottom-up decision making 349–351
service accessibility 168
  20-minute cities 177–180
service networks
  capacity 41–42
  connection costs 40–41, 48
  contingencies 40–42
  investment 41–43
  Manchester 38–49
  as obligatory passage points 36–37
  path dependency 37, 42
  replication and upscaling 75–91
  socio-technical analysis 35–52
  substations 41, 42
SET *see* European Strategic Energy Technologies
setting, carbon dioxide budgets 149–154
Sherbro Island City Project 63
short-term scenarios, Berlin climate neutrality 157–160
Sierra Leone
  Freetown 54
  policy environment 60–61
  Sherbro Island City Project 63
sinks, climate neutrality 149
SLES *see* smart local energy systems
slums, Smart Cities Mission, India 227
smart buildings
  infrastructure constraints 40–41
  Manchester 39–42
smart charging 101, 103
smart cities, definition 266
Smart Cities Mission (SCM), India 215–235
  city selection 217–218
  economic sustainability 226–228
  environmental sustainability 224–226
  equity 221–224, 227
  financing 218–219, 227–229

implementation and governance framework 218
institutional sustainability 228–231
integrated command and control centres 221
model components 219–221
overview 217–219
social sustainability 221–224
taxation 228
Smart City of Amsterdam 270
Smart City Masterplan, Kigali 55
Smart Data Initiative (SDI), UK 27–28
smart energy cities (SEC)
active buildings 93–110
African contexts 56–58
AI and mobility 111–122
autonomous vehicle-induced urbanism 114–117
behavioural factors 96–98, 102–103
bicycle urbanism 117–119
city brains 112
concepts 19–20
equity 64
ethics 64
Indian mission 215–235
infrastructure 40–41, 48, 102–105, 269–272
Manchester 38–49
rural relationships 43–44
Russia 242–246
scales 19
socio-technical analysis 35–52
street lighting 365–372
UK Industrial Strategy, 2017 21–22
urban climate 125–146
variability of needs 19–20
vision 1–3
West Africa 53–74
smart energy systems
decarbonisation 95–96
demand shifting 97
distributed energy Resources 21
and fairness 22–23, 28, 54, 64
infrastructure 21
replication and upscaling 75–91
time of use tariffs 97–98, 100–102, 328–329
UK Industrial Strategy, 2017 21–22
United Kingdom 21–28
Smarter City Programme, Manchester 39
smart grids
ambiguity of definitions 283–284
consumer cost impacts 284–285
contextual considerations 77–80
definition 277, 283–284
diffusion
patterns of 78
types of 79
distributed ledger technology 293–304
embedding 76
energy security 268–269
internet of things 268–269
interoperability 77
Malmö Hyllie pilot project 81, 84–88
multi-agent management systems 271
Ökopark Hartberg pilot project 81–84
replication and upscaling 75–91
case studies 81–88
contextual considerations 77–80
embedding 76
fundamental constraints 76–77
path dependent constraints 76
Russia 244–245
socially oriented meta-systems 269
storage devices 270–271
technologies 268–269
time of use tariffs 97–98, 100–102, 328–329
transferability 79–80
smart homes 93–110
behavioural factors 96–98, 102–103
communication needs 104–105
and decarbonisation 93, 95–96
demand shifting 97–98, 102–104
infrastructure constraints 102–105
time of use tariffs 97–98, 100–102
see also active buildings
smart local energy systems (SLES)
Bristol 23–24
competing logics 25–26
governance 26–28
London 23
Oxford 24
and privatisation 22–23, 25–26
scalability 22–23, 25–26
United Kingdom 17, 21–28
smart meters
active buildings 97–98, 100–105

smart meters (*cont'd*)
behavioural factors 97–98, 102–103
consumer communication 104–105
data privacy 27, 98
and digital inclusion 104
infrastructure limitations 103–105
smart poles, street lighting 368–369
Smart Spaces 40
smart street lighting 367
socially oriented meta-systems (SoM) 269
social objectives, district heating and cooling 320
social sustainability, Smart Cities Mission, India 221–224
socio-technical analysis 35–52
active buildings 93–110
Manchester 38–49
replication and upscaling 78–79
wind farms 44–47
solar architecture 131
solar panels 330, 336–338
solar thermal energy 338–339
SoM *see* socially oriented meta-systems
sources, climate neutrality 149
Soviet Union, district heating 307
spatially induced cognitive scarcity, energy poverty 253
spatial-structural aspects, transferability 80, 81–82, 84–85
special purpose vehicles (SPV), Smart Cities Mission, India 218–219, 228–229
SPV *see* special purpose vehicles
SRU *see* German Advisory Council on the Environment
stakeholders
20-minute idea 187
inclusivity 54
standards, carbon dioxide budgets 149–151
Stockholm, Sweden, Hammarby Sjostad 270
storage devices 270–271, 329–330
strategic measures
Berlin climate neutrality 160–162
Cagliari sustainable urban mobility plan **201–207**
strategic niche management 78
street lighting 365–372
connected security lighting systems 368
controller systems 367–368
dynamic control systems 369
energy efficiency 366–367
smart 367
smart poles 368–369
sub-national actors, climate change 17–18
sub-Saharan Africa
diaspora-led investment 62–63
equity 54–55, 57–58
external investment 61–63
Ghana 55
integrated approaches 63–65
mobile phone penetration 61
policy environments 58–63
poverty 54
pre-paid electricity meters 57
Rwanda 55
Sierra Leone 54, 60–61, 63
urban planning 54
substations 41, 42
SUMP *see* sustainable urban mobility plans
superintelligence 114
surface activation 334–335
sustainability
20-minute idea 167–193
cities 172–173, 177–180
citizens 175–176, 184–185
Edinburgh 176–185
neighbourhoods 173–175, 180–184
autonomous cities 113–114
bicycle urbanism 117–119
India's smart cities mission 215–235
Manchester 38–49
and mobility 111–122
and network capacity 41–43
and path dependency 37, 42
socio-technical analysis 35–52
urbanity and climate 125–146
wind farms policy 44–47
sustainable development goals (SDG), and energy services 55–56
sustainable urban mobility plans (SUMP)
Cagliari case study 195–214
objectives **200**
strategies and actions **201–207**
principles 195–196
Sweden, Malmö Hyllie pilot project 81, 84–88
system-level innovations, concepts 19

t

taxation, Smart Cities Mission, India 228
TCI *see Terremoto Centro Italia*
technical decision making
  bottom-up digital technologies 347–364
    +EINS project 352–356
    +EQ-NET project 354–356
    challenges 360–361
    sensors 349–351
    user-generated content 349–351, 356–360
technical objectives, district heating and cooling 321
technological aspects, transferability 80, 82, 85–86
technologies
  distributed ledgers 293–304
  district heating systems 308–311
  internet of things 268–269
Terremoto Centro Italia (TCI) project 357–358
thermal mass energy storage 329–330, 332–335
third generation district heating systems 308
timber buildings 129–130
time of use (ToU) tariffs 97–98, 100–102, 328–329
ToU *see* time of use
transferability
  actor constellations 80, 83, 86–87
  institutional factors 80, 83–84, 87
  mission and economic factors 80, 82–83, 86
  pilot projects 79–80
  spatial-structural aspects 80, 81–82, 84–85
  technological aspects 80, 82, 85–86
transport planning
  Cagliari sustainable urban mobility plan 195–214
  Russia 245
  Smart Cities Mission, India 225
  see also mobility
tripe SoM 269
types, of upscaling 79

u

UGC *see* user-generated content
UK *see* United Kingdom
UNFCC *see* United Nations Framework Convention on Climate Change
unit commitment, pricing 328
United Kingdom (UK)
  Bristol 23–24
  competing logics 25–26
  district heating 305–306
  Edinburgh 20-minute idea 167–193
  governance 26–28
  greenhouse gas emissions policy 93–94
  Industrial Strategy, 2017 21–22
  London 23
  Manchester 38–49
  national vs local policy 22–28
  Oxford 24
  privatisation 22–23
  Prospering from the Energy Revolution innovation challenge 22
  renewable energy generation 21
  smart local energy systems 17, 21–28
  urbanisation 20
  vested interests 25–26, 28
United Nations Framework Convention on Climate Change (UNFCC) 150
United States of America (USA)
  New York City 268, 299
  Portland 20-minute city 172–173
upscaling 75–91
  case studies 81–88
  constraints 76–78
  contextual considerations 77–80
  embedding 76
  fundamental concepts 77
  Malmö Hyllie pilot project 87–88
  Ökopark Hartberg pilot project 84
  path dependent constraints 76
  patterns of 78
  socio-technical aspects 78–79
  and transferability 79–80
  types of 79
urban artificial intelligences 112–113
urban blue 132
urban climate 125–146
  atmospheric science studies 127–128
  building design 129–132
  city planning 133
  energy efficiency 128–132
  green and blue spaces 132

urban climate (*cont'd*)
greenhouse gas emissions 128
modelling approaches 135–136
nationwide infrastructure planning 133–135
sea levels 133
urban gardening 132
urban green 132
urbanisation
and climate 125–146
greenhouse gas emissions 128
and infrastructure constraints 20
predictions 266–268
sustainable agglomeration 126
urban planning
20-minute idea 167–193
cities 172–173, 177–180
citizens 175–176, 184–185
Edinburgh 176–185
neighbourhoods 173–175, 180–184
Cagliari sustainable urban mobility plan 195–214
and climate 127
climate considerations 132–133
energy efficiency 128–132, 133
green and blue spaces 132
greenhouse gas emissions 129–130, 131
modelling approaches 135–136
nationwide infrastructure planning 133–135
passive architecture 131
resilience 130–131
sea levels 133
sub-Saharan Africa 54
urban sprawl reduction 133
20-minute idea 167–193
cities 172–173, 177–180
citizens 175–176, 184–185
Edinburgh 176–185
neighbourhoods 173–175, 180–184
urban surfaces 131
USA *see* United States of America
usage, vehicles 115–116
user-generated content (UGC) 349–351, 356–360

v
vehicles
autonomous 114–117
frequency of usage 115–116
sharing 115–116
smart charging 101, 103
see also electric vehicles
verifiability, distributed ledger technology 296–298
vertical scaling 79
vested interests, United Kingdom 25–26, 28
vulnerable road users (VRUs) 117–119

w
walkability, 20 minute neighbourhoods 174, 181, 186
water, in cities 132
West Africa
demographics 54, 59
diaspora-led investment 62–63
electrification 59
equity 54–55, 57–58
external investment 61–63
Ghana 55
integrated approaches 63–65
mobile phone penetration 61
network capacity 57–58
policy environments 58–63
population growth 59
pre-paid electricity meters 57
regional policy frameworks 59–60
Sierra Leone 54, 60–61, 63
smart energy cities 53–74
wind catchers 132
wind farms
asset life extension 46–47
policy constraints 44–47, 49
refusals 45, 47
repowering 46
Royd Moor 45–47
Scout Moor 44–45
socio-technical issues 44–47
wood, as a building material 129–130

y
Yokahama Smart Grid model 270